Environmental Problem-Solving:
Balancing Science and Politics
Using Consensus Building Tools

Environmental Problem-Solving: Balancing Science and Politics Using Consensus Building Tools

Guided Readings and Scenario Assignments from MIT's Training Program for Environmental Professionals

Lawrence Susskind, Bruno Verdini,
Jessica Gordon and Yasmin Zaerpoor

ANTHEM PRESS

Anthem Press
An imprint of Wimbledon Publishing Company
www.anthempress.com

This edition first published in UK and USA 2022
by ANTHEM PRESS
75–76 Blackfriars Road, London SE1 8HA, UK
or PO Box 9779, London SW19 7ZG, UK
and
244 Madison Ave #116, New York, NY 10016, USA

Previously published in the UK and USA by Anthem Press in 2017 and 2020

© 2022 Lawrence Susskind, Bruno Verdini, Jessica Gordon and Yasmin Zaerpoor

British Library Cataloguing-in-Publication Data
A catalogue record for this book is available from the British Library.

ISBN-13: 978-1-83998-612-3 (Pbk)
ISBN-10: 1-83998-612-3 (Pbk)

This title is also available as an e-book.

CONTENTS

ACKNOWLEDGMENTS

This book is meant as a window into a beloved MIT course (11.601) created by Professor Lawrence Susskind. The structure reflects the commitment to nurture environmental professionals through their study of environmental policy in a way that helps them identify their own personal theory of practice. Whether or not they explicitly realized it at the time, generations of students have profoundly influenced the shape of the course over the years, and as such, the book invites readers to explore classics from six different decades as a springboard to further complement and explore new and crucial threads in decades to come.

We, therefore, want to extend our gratitude to all the students in 11.601, with particular mention to those who shared their work as examples in addressing the questions posed by the scenario assignments and exams. We are also grateful for Kathleen Schwind's efforts to secure copyrights for the excerpts, Kelly Heber Dunning's early insights, Takeo Kuwabara's generous planning, Julie Herlihy's careful coordination, as well as Daniel Glenn's impactful editing. This book would not have come to fruition without the good-natured collaboration between the authors in partnership with Anthem and our loved ones.

Thank you!

CREDITS AND PERMISSIONS

Cohen, Steven. 2014. "'Understanding Environmental Policy' and 'A Framework for Understanding the Environmental Policy Issue.'" In *Understanding Environmental Policy*. New York: Columbia University Press. Reprinted by permission of the publisher.

Corburn, Jason. 2005. "Local Knowledge in Environmental Health Policy." In *Street Science: Community Knowledge and Environmental Health Justice*. Cambridge, MA: MIT Press. Reprinted by permission of the publisher.

Costanza, Robert, Rudolf de Groot, Paul Sutton, Sander van der Ploeg, Sharolyn J. Anderson, Ida Kubiszewski, Stephen Farber and R. Kerry Turner. 2014. "Changes in the Global Value of Ecosystem Services." *Global Environmental Change*. Reprinted by permission of the publisher.

DesJardins, Joseph R. 2013. *Environmental Ethics: An Introduction to Environmental Philosophy*. Belmont: Wadsworth. Reprinted by permission of the publisher.

Hardin, Garrett. 1968. "The Tragedy of the Commons." *Science*. Reprinted by permission of the publisher.

Howlett, Michael, M. Ramesh and Anthony Perl. 2009. *Studying Public Policy: Policy Cycles & Policy Subsystems*. Ontario, Canada: Oxford University Press. Reprinted by permission of the publisher.

Kunreuther, Howard, and Paul Slovic. 1996. "Challenges in the Risk Assessment and Risk Management." *Annals of the American Academy of Political and Social Science*. Reprinted by permission of the publisher.

Ludwig, Donald. 2000. "Limitations of Economic Valuation of Ecosystems." *Ecosystems*. Reprinted by permission of the publisher.

Ostrom, Elinor. 1990. "Reflections on the Commons." In *Governing the Commons*. Cambridge: Cambridge University Press. Reprinted by permission of the publisher.

Ostrom, Elinor. 2012. *The Future of the Commons: Beyond Market Failure and Government Regulations*. London: Institute of Economic Affairs. Reprinted by permission of the publisher.

Pearce, David, Giles Atkinson and Susana Mourato. 2006. *Cost Benefit Analysis and the Environment: Recent Developments*. Paris: OECD. Reprinted by permission of the publisher.

Rosa, Eugene A., Ortwin Renn and Aaron McCright. 2014. "Risk Governance: A Synthesis." In *Risk Society Revisited: Social Theory and Governance*. Philadelphia: Temple University Press. Used by permission of the publisher.

Sagoff, Mark. 2008. "At the Shrine or Lady Fatima, or Why All Political Questions Are Not Economic." In *The Economy of the Earth: Philosophy, Law, and the Environment*, 2nd ed. Cambridge: Cambridge University Press. Reprinted by permission of the publisher.

Shapiro, Ian. 2009. "Aggregation, Deliberation, and the Common Good." In *State of Democratic Theory*. Princeton: Princeton University Press. Reprinted by permission of the publisher.

Sterman, J. D. 1991. "A Skeptic's Guide to Computer Models." In *Managing a Nation: The Microcomputer Software Catalog*, edited by Gerald O. Barney, W. Brian Kreutzer and Martha J. Garrett. Boulder, CO: Westview Press. Reprinted by permission of the publisher.

Susskind, Lawrence. 2009. "The Environment and Environmentalism." In *Local Planning: Contemporary Principles and Practice*. Washington, DC: ICCMA Press. Reprinted by permission of the publisher.

Susskind, Lawrence, and Connie Ozawa. 1984. "Mediated Negotiation in the Public Sector: The Planner as Mediator." *Journal of Planning Education and Research*. Reprinted by permission of the publisher.

Susskind, Lawrence, and Jeffrey Cruikshank. 2006. *Breaking Robert's Rules*. Oxford: Oxford University Press. Reprinted by permission of the publisher.

Susskind, Lawrence, Ravi K. Jain and Andrew O. Martyniuk. 2001. "How Environmental Policy Studies Can Be Used Effectively" and "How Policy Studies Should Be Organized." In *Better Environmental Policy Studies*. Washington, DC: Island Press. Reprinted by permission of the publisher.

University of Washington Urban Ecology Research Lab. 2008. "Scenario Planning," "Scenario Methodology" and "Driving Forces." In *Puget Sound Future Scenarios*. University of Washington. Reprinted by permission of the author.

Van Buuren, Arwin, and Sibout Nooteboom. 2009. "Evaluating Strategic Environmental Assessment in the Netherlands: Content, Process and Procedure as Indissoluble Criteria for Effectiveness." *Impact Assessment and Project Appraisal*. Reprinted by permission of the publisher.

Sam Barnard. Permission of the author.
Andrew Binet. Permission of the author.
Nicholas Cohen. Permission of the author.
Johanna Greenspan-Johnston. Permission of the author.
Holly Jacobson. Permission of the author.

Cristina Logg. Permission of the author.
Emily Long. Permission of the author.
Devon Neary. Permission of the author.
Kara Runsten. Permission of the author.
Griffin Smith. Permission of the author.
Shin Bin Tan. Permission of the author.

INTRODUCTION

Environmental problem-solving is at the heart of every decision about allocating natural resources and crafting standards to protect public health and safety. Elected and appointed officials make problem-solving choices every day that affect their constituents' lives, but they cannot ensure that everyone gets what they want and need. Environmental problem-solving involves trade-offs. Sometimes policy makers make decisions that respond to immediate pressures but that sacrifice the interests of future generations. Often, they respond to pressures from some groups but not others. In other words, environmental problem-solving almost always turns out to be harder than it first appears.

For one thing, there are likely to be competing diagnoses of what is causing the environmental problem. In the case of air or water pollution, for example, there may be differences in view about its causes and about how great a threat it poses. In addition, there are sure to be conflicting notions about how best to respond. Is the level of pollution a serious threat to long-term ecosystem survival? Is it a threat to everybody or only to a vulnerable subset of the population? Will it eventually dissipate of its own accord? Or will it continue to get worse, eventually reaching a tipping point where remedial action is no longer possible? Is the effect of the pollution the same everywhere or only in certain locations? Is there a new (pollution control) technology coming along that might render the pollution harmless? Or do we need to ban a particular industrial process until we come up with a less risky way of doing things? What is the likely cost if we do nothing? Who will bear that cost? Do we have to compensate those who have already been adversely affected? Where should responsibility rest for making sure that future pollution levels are safe? How should pollution control measures be enforced? What is a reasonable amount to spend to eliminate or reduce pollution? What funds should be invested in research and development to increase our understanding of pollution risks and the likely effectiveness of alternative cleanup methods? These questions must be answered to attempt to solve even an obvious environmental problem like air or water pollution—and as we all know, many environmental problems are much more complicated! In addition, most environmental problems are beyond what individuals or dedicated groups can handle on their own—only collective action will work.

Even if our policy makers have the best of intentions, environmental problem-solving is difficult. When many stakeholders do not take the problem seriously, it is even harder. For example, some groups may only care about economic growth or some other issue they think must be traded off against environmental protection. Or they may be committed to an ideological point of view that causes them to oppose any and all problem-solving ideas of a certain kind (e.g., "only market mechanisms work" or "only government intervention works"). When stakeholders are opposed to environmental problem-solving in principle, evidence that supports a particular diagnosis will not be convincing to them or their supporters. The same is true of expert advice on the probable effectiveness of alternative solutions. If someone will not admit there is a problem, it is hard to get them to consider the pros and cons of different ways of solving that problem.

This book is for anyone who wants to be involved in environmental problem-solving. Elected and appointed officials only respond if an informed constituency lets them know how they feel. You can play a role in environmental problem-solving regardless of your disciplinary training or whether you are inside or outside of government, industry, or a nongovernmental organization.

In this book, we offer a self-guided tour of environmental problem-solving. Each unit includes reading excerpts that are selected from a highly regarded course for environmental professionals called "Introduction to Environmental Policy and Planning" at the Massachusetts Institute of Technology (MIT) in Cambridge, Massachusetts. These excerpts are preceded by our commentary on some of the most important points that we hope the reader will take from each excerpt. We have included scenario assignments to help you think about how you might apply the ideas in the readings and commentaries to different scenarios. There is a more comprehensive assignment at the end of each unit, designed to assess your mastery of the topics introduced in that unit. We provide two examples of student responses to compare with your own response. The last section includes a Final Exam with written responses to selected questions. The questions on the exam focus on using the material to construct environmental problem-solving responses to actual situations that you may come across in your practice.

By the time you have made your way through all the material in this book, our hope is that you will be more confident and better equipped to address the challenges associated with the fact that (1) there are no easy answers to most environmental problems; (2) the analytical tools our societies rely on to diagnose and prescribe appropriate responses to environmental problems are often limited in what they can do for us; (3) environmental problem-solving is a political activity that needs to be handled in a democratic way in order to be effective and sustainable; and (4) collaborative efforts to solve environmental problems can, at best, only produce provisional agreements that will require continuous monitoring and adaptation.

The contents of the book are divided into four sections. Unit I, "Influencing the Environmental Policy-making Process," focuses on how certain environmental problems can only be solved through efforts to formulate and implement government policies that take both science and politics into account. Using the context of the United States, this unit introduces the steps in the national environmental policy-making process and dissects the ways in which environmental policy is implemented.

Unit II, "Ethical Dilemmas in Environmental Problem-Solving," focuses on the philosophical choices that environmental problem solvers have to make. Every personal theory of environmental problem-solving builds on key ethical assumptions. The unit begins with a general overview of environmental ethics and then reviews classic ethical debates like the tension between utilitarianism and intrinsic value as well as arguments for and against "deep green" approaches to environmental problem-solving.

Unit III, "Developments in Policy and Project Analysis," summarizes the range of analytical tools that environmental practitioners can use to help improve decision-making. These readings and commentaries highlight the strengths and weaknesses of tools such as environmental impact assessment (EIA), cost-benefit analysis (CBA), ecosystem services analysis (ESA), risk assessment (RA), simulation, modeling, and scenario planning. Experts and citizens need to be aware of the underlying assumptions and shortcuts built into each of these methods.

The book concludes with Unit IV, "Collective Action to Solve Environmental Problems." This unit reviews the literature on the difficulties of mobilizing public support for environmental problem-solving. It discusses the basics of democratic decision-making and zooms in on various ways in which the public can be a partner in government-led efforts. We look at methods of collaborative decision-making and more recent ideas about consensus building and environmental dispute resolution. The final writing assignment asks you to advocate for what you perceive to be the best way of engaging the public in a particular environmental problem-solving situation. The underlying questions are whether and how the public interest can be served.

The most important idea embedded in this book is that anyone can teach themselves the basics of environmental problem-solving. This book has been created as a toolkit that can stand alone or can be used in conjunction with courses in many fields and disciplines in diverse educational settings. The most important outcome of studying this material should be greater clarity in your personal theory of environmental problem-solving. You may not realize that you have such a theory, but you do. Everybody does. We want you to examine your ethical assumptions; think hard about the role that science, scientists and the public ought to play in environmental problem-solving; understand the strengths and limitations of various analytical tools; and reconsider the appropriate role of government in helping communities make collective choices.

Scenario Assignments

Included in each unit of the book, the scenarios invite professionals to develop an action plan geared to solve specific environmental problems. In each case, you will need to think carefully about your own personal theory of environmental problem-solving. To respond to the stakes and challenges, you will have to (1) identify the choices that the stakeholders in the scenario need to make; (2) determine which principles the stakeholders should rank most highly; (3) propose a politically plausible strategy; (4) assess potential barriers to implementation; and (5) present a short summary likely to be compelling to the stakeholders who are your clientele.

You may choose to supplement your written response with a video in order to practice your persuasion skills. Your presentation should be for five minutes. When answering the questions, "place" yourself fully in the assigned role. Take a stand and be sure to answer the core themes. Your success will be a function of how well you analyze, substantiate, and deliver your arguments. Rely on the readings as a springboard, but do not try to restate everything that is in them. A good strategy is to demonstrate the ways in which your thinking genuinely ties to one or two of the key ideas in the assigned readings. Then, go complement what others have had to say with your own thoughtful insights. A key part of each assignment is being able to advocate a point of view in the face of substantial uncertainty in ways that engage and inspire the stakeholders to meet the challenge at hand. A lot of the situations described in the scenarios have been faced by environmental professionals, so consider each assignment a "dry run" for a challenge you are likely to face in practice. Good luck!

Unit I

INFLUENCING THE ENVIRONMENTAL POLICY-MAKING PROCESS

Introduction

In this first section, we examine how national governments formulate and implement environmental policies. Fairly frequently, environmental problem-solving, whether at the project, municipal, regional or state level, requires us to work within, against or toward improving a specific policy. When that is the case, it is crucial to know how (and have the tools) to impact the policy-making process. There is a great deal of published work on the subject of public policy-making in general, but our focus is on environmental policy-making in particular. We are especially interested in how practitioners and citizens can analyze environmental problems and affect environmental policy change.

The Steven Cohen excerpt sets the stage by providing an overview of environmental policy-making from both a theoretical and a practical perspective. Cohen explains why every effort to "solve" an environmental problem is fundamentally a values issue (i.e., "right" versus "wrong"); a political issue (i.e., in which winners and losers are selected); a science and technology issue (where uncertainty and innovation are up for discussion); a policy design and economics issue (i.e., in which regulatory strategies and incentives for changing consumer, corporate and citizen behavior must be selected); and a management issue (i.e., where organizational capacity probably needs to be enhanced). In other words, Cohen draws attention to the inherent challenge of environmental policy-making when there are so many factors to consider.

Michael Howlett, M. Ramesh and Anthony Perl take this discussion a step further, exploring numerous theories of policy design that explain how public agendas are set, policy is formed and decisions are made and implemented. They provide one of the most concise and clear descriptions of different models of policy-making. As you read through this excerpt, think about your own opportunity to influence environmental policy-making. If we think of policy-making as a linear process (as is described in the "textbook" model of policy-making), the process starts with agenda setting before going into policy formulation, decision-making, implementation and evaluation. Different actors have more influence at different stages of the policy-making process. For example, those involved in the first stage—agenda setting—define the problem, thereby largely shaping the nature of the policy. Those involved in decision-making, however, allocate resources, thereby empowering a specific set of stakeholders. This is one of the reasons why we believe that broad participation is important in all stages of policy-making. Practitioners recognize that policy-making rarely follows the linear model. In sum, Howlett, Ramesh and Perl highlight various opportunities to affect environmental policy-making at each of these stages and suggest that the process of policy-making is less rational than is often presented.

The Lawrence Susskind reading draws our attention to ways in which environmental professionals can proactively affect policy-making and implicitly raises the issue of participation in the policy-making process. The excerpt introduces the practice of adaptive management and what it takes to sustain it. This piece is meant to facilitate thinking about your personal theory of environmental problem-solving and, more specifically, how you can influence environmental policy.

The final reading, by Elinor Ostrom, provides a framework for analyzing environmental problems. She identifies six key variables that social scientists and policy makers should keep in mind when thinking about environmental problem-solving in different contexts. As you will see in Unit IV, Ostrom emphasizes the importance of matching social action to the particular ecological situation.

In other words, all four readings in this unit emphasize the importance of understanding how policies are made so that you can influence environmental policy in whatever capacity (i.e., as a citizen, scientist, policy maker, academic etc.). You will notice that these readings are based on policy-making in the United States, but their insights are widely transferable to other contexts, let alone because their authors have vast experience and expertise with projects around the world, but due to their core message: our ability to help inform policy will depend on the sociopolitical context we each face at a given moment in time, at a given place. As you read this section, think about your personal theory of practice—where and when do you have the most influence over environmental policy-making in your community? Whom might you want to form coalitions with to help reach your desired policy goal? What strategies will help ensure that the policies are effective and fair?

Once you've explored the readings in detail, we invite you to examine the scenarios. The first one centers on how to evaluate and ensure the credibility of pollution control policies in the eyes of competing scientific, business and environmental advocacy communities. The second scenario focuses on the question of how we can learn from the problem-solving experience of other leaders and stakeholders in different countries.

By the end of the section, it should be apparent that national environmental policy-making is a much less structured and a much more haphazard process than many policy scientists (and politicians) have suggested. The question we urge you to consider is, what is the simplest model of environmental policy-making (for any country) that you can formulate? You will find example student responses to these questions at the end of this section. Before reading those, think about how you would answer them based on your own experience and policy-making context. Use the opportunity to thoughtfully question what you might incorporate into your practice and what you might do differently.

Commentaries and Reading Excerpts

Steven Cohen — "Understanding Environmental Policy" and "A Framework for Understanding the Environmental Policy Issue." In *Understanding Environmental Policy*. New York: Columbia University Press.

Reading Commentary

Environmental issues can be extremely complex. An environmental problem is usually defined as "a set of interconnected issues that determine the sustainability of the planet earth for continued human habitation under conditions that promote our material, social and spiritual well-being." Solving environmental problems at any scale means addressing human needs such as access to clean water, energy, food, clean air and open space. This complex process goes beyond the needs of humans and into taking account of ecosystem survival and biological diversity as intrinsic values (past their immediate importance to the current generation). So, how can we solve environmental problems when there are so many things to consider?

Cohen argues that solving environmental problems begins by acquiring an understanding of environmental policy-making. To achieve this understanding, Cohen believes that an interdisciplinary analytical framework can help. It includes (1) values; (2) politics; (3) science and technology; (4) policy design and economics; and (5) management. His framework starts with an analysis of the causes and effects of environmental problems and how they evolve over time. In particular, he thinks it is useful to focus on how problems are framed, solutions are selected, and ideas are implemented. He also believes that the importance of each of these five factors will shift as the situation changes. For example, climate change is often framed as a scientific problem. This thinking leads to technical or engineering solutions like reducing greenhouse gas emissions from vehicles.

Starting with different values would lead to a different framing of the problem (and, thus, to different solutions). Values are instrumental to the way we think about environmental problems. Ultimately, our values determine how we frame and prioritize them. A decision about whether to use a parcel of land to build a mall or to conserve a wetland is ultimately a question of values. Which is more important? Whatever choice is made leads to the designation of winners and losers. The balance between winners and losers, of course, is shaped by politics, available technologies, enforcement options, the use of financial incentives and who has responsibility for policy implementation.

Cohen argues for keeping multiple perspectives in mind in any environmental problem-solving or policy-making situation. The questions he raises at the end of the reading are meant to help you do that. As you take a step back, how have any of the five factors he identifies played out in an environmental policy-making context that you have studied or experienced?

CHAPTER 1

UNDERSTANDING ENVIRONMENTAL POLICY

Steven Cohen

DIFFERING PERSPECTIVES ON ENVIRONMENTAL POLICY

Environmental and sustainability policy is a complex and multidimensional issue. As Harold Seidman observed in *Politics, Position, and Power: The Dynamics of Federal Organization*, "Where you stand depends on where you sit." That is, one's position in an organization influences one's stance and perspective on the issues encountered. Similarly, one's take on an environmental issue or the overall issue of environmental protection and sustainable economic development varies according to one's place in society and the nature of one's professional training.

For example, to a business manager, the environmental issue is a set of rules one needs to understand in order to stay out of trouble. For the most part, environmental policy is a nuisance, or at least an impediment to profit. It is true that the development of the field and practice of sustainability management is changing corporate understanding of environmental resources; however, many business managers still see environmental stewardship as a set of conditions that impede, rather than facilitate, the accumulation of wealth. For now, most business practitioners see a conflict between environmental protection and economic development, though this view of a trade-off is false. To an engineer, the environmental problem is essentially physical and subject to solution through the application of technology. Engineers tend to focus on pollution control, pollution prevention (through changes in manufacturing processes or end-of-pipeline controls), energy efficiency, closed-system production, and other technological fixes. Lawyers view the environment as an issue of property rights, contracts, and the regulations that are needed to protect them. Economists perceive the environment as a set of market failures resulting from problems of consumption or production. They search for market-driven alternatives to regulation. Some understand the importance of protecting natural resources to maintain wealth, but many do not. Political scientists see environmental policy and sustainability as a political concern. To them, it is a problem generated by conflicting interests. Finally, for philosophers, the environment is an issue of values and differing worldviews.

The environment is subject to explanation and understanding through all of these disciplines and approaches. It is, in fact, a composite of the elements identified by the various disciplines and societal positions, and has dimensions that exist at the intersection points of the disciplines and social perspectives. The difficulty is that each view tends to oversimplify environmental problems, contending with only one facet of the situation. Although such problems are multidimensional, different types of environmental issues are weighted toward different conceptual orientations. One view may explain a greater or lesser share of the problem than another. For example, the problem of electronic waste is not a technical issue, because we know how to safely remove toxics from discarded electronics; the technology need not be developed anew. Neither is it a problem of economics, for many of the parts of discarded electronics can be recycled for continued use. Rather, the fact that e-waste leaks into the environment is primarily a management problem: we have not developed the standard operating procedures needed in order to collect and safely recycle or dispose of this waste.

DEVELOPING A FRAMEWORK TO HELP UNDERSTAND ENVIRONMENTAL ISSUES

The following chapters are intended to contribute to a conversation about the problem of environmental sustainability in general, as well as some specific areas in greater detail. The environmental problem can be defined as the set of interconnected issues that determine the sustainability of the planet Earth for continued human habitation under conditions that promote our material, social, political, and spiritual well-being. In Chapter 2, I develop a framework for understanding the dimensions of the environmental problem and solutions proposed to address the problem. The framework allows us to deconstruct particular environmental issues and programs to increase our understanding of the causes and effects of these issues and programs. The framework examines environmental issues as a multifaceted equation encompassing a variety of factors, including values, politics, technology and science, public policy design, economics, and organizational management. Each aspect of the framework illuminates a specific feature of the environmental issue and at the same time clarifies all the environmental issues examined here. Each separate issue, however, tends to find its main source of explanation in a single factor.

APPLYING THE FRAMEWORK TO A SET OF ENVIRONMENTAL ISSUES

With this rough framework on the table, I'll apply it to a set of environmental policy issues. While any number of issues could have been selected, I tried to choose issues of policy import, which varied by the level of government most involved. I also tried to select issues I had experience in analyzing. In Chapter 3 we will review an environmental issue that is driven by politics: New York City's effort to enact a congestion pricing fee. In examining congestion pricing in New York City, I will be analyzing a policy with proven success elsewhere and comparing the experience in New York City to the successful implementation of congestion pricing in London, noting the differences and the lessons learned. Chapter 4 focuses on the emerging issue of e-waste, or electronic waste: the toxics from discarded computers and cell phones. E-waste is a global issue, yet local actors across many jurisdictions affect the outcome of e-waste. In the United States, no federal regulations exist to recycle e-waste, though a number of states have passed rules regulating it. Electronic waste is a multidimensional problem of management, science and technology, values, and politics. We will look at emerging strategies, including producer responsibility policies and corporate recycling programs. In Chapter 5 we apply the framework to the issue of hydraulic fracturing of natural gas, commonly known as hydrofracking. Under the George W. Bush administration, this practice was exempted from prevailing federal regulation, requiring states to reluctantly and slowly step into the regulatory vacuum that resulted. In Chapter 6, we present the book's final case study as we apply the framework to climate change, an impact that is more difficult to project than many other environmental issues. The complexity in addressing this issue is due to the fact that the causes are global and the impacts are mostly in the future, making it challenging to address the issue politically.

In Chapter 7, I compare the issues and discuss the strengths and limitations of the framework, as well as identify some possible modifications. In chapter 8, I present some suggestions for improving environmental policy and moving toward sustainability.

The issue of electronic waste management is an indicator of the increasing toxicity of the waste stream. In some cases, toxic substances are used in technological devices out of habit, and little or no effort has been made to produce the electronic device without these toxic components. Waste management in the United States is mainly an issue handled by local governments. While the U.S. government does regulate solid waste and hazardous waste management at the federal level, for the most part, municipal solid waste is considered an issue of local politics and policy. In the United States, hydraulic fracturing is an issue that involves all levels of government. However, because the federal government has been hesitant to take on the issue, state and local governments have assumed regulation. Similarly, when dealing with the issue of congestion pricing, New York City cannot regulate its own highways, as they are regulated by New York State. The problem then becomes one of charging people to drive their cars into a part of the state, which creates political issues when trying to get this pricing mechanism passed at the state level.

With the same impulse that drove us to landfill our garbage, we assumed that once we buried old computers and cell phones underground, they were gone forever. Few of us knew how toxic this waste was, and even fewer understood how the toxics in electronic waste materials were transported through the ground, water, and air. Today, engineers have developed a field called industrial ecology, which has the goal of creating products without generating waste. In the early days of the era of mass production of laptops and cell phones, engineers paid almost no attention to the use of toxics when they designed production processes—"You can't make an omelet without breaking some eggs." The rush to production and to new features could not be delayed by concern about the toxicity of the product once it was discarded as waste. In fact, until Deming demonstrated that higher-quality products were made with less waste of time, materials, and labor, most operations engineers and managers spent little time or effort attempting to reduce waste or pay attention to the toxicity of its content (Deming 1986).

When we learned about toxic waste contamination in the late 1970s and early 1980s, we wanted to clean up the places that had been damaged, and prevent new waste sites from being created. When we learned about electronic waste in the past decade, we had to face up to the fact that some of the products that were most important to us contained toxics. We had no idea how much damage had already been done or how expensive and difficult, if not impossible, it would be to detoxify future cell phones and laptops. How did we create such a lethal technology? How did this issue reach the policy agenda? How was it defined? What did the electronic waste issue teach us about environmental problem solving?

In many respects, electronic waste is simply a continuation of the general issue of toxic waste, which three decades ago led us to define environmental protection as a policy area concerned with human health. Environmental policy no longer focused exclusively on preserving mountain streams and protecting wildlife, but was also concerned about keeping poisons out of our land and water. What was the social, political, and economic impact of this change? How did it come about? In chapter 4 we will attempt to deconstruct the electronic waste problem into its component parts.

It is not difficult to understand why "fracking" became a political issue. The hunger for energy in the United States is difficult to satisfy, and much of the natural gas in the Northeast's Marcellus Shale sits beneath the property of people of modest means. Some property owners are eager to lease their land for drilling operations, while some of their neighbors worry about the potential for accidents and damage to the environment and their rural lifestyle. Meanwhile, the Bush administration encouraged unregulated hydraulic fracturing by allowing corporations to keep the chemical composition of their fracking fluid secret. How did this problem emerge as an environmental problem and as a public policy issue? Why does this problem persist? What can be done to address and solve the problem? Can this gas be extracted without damaging the environment?

Problems like hydraulic fracturing remind us of the fragility of some ecosystems, and the ability that humans have to cause inadvertent damage to nature. While some environmental damage is a direct and unavoidable by-product of a production process, leaking gas wells and transportation accidents are caused by human and organizational management errors. Of course, it is possible to probe further and find deeper causes of damage from fracking. These are the value choices involved in our energy-dependent lifestyles.

If we ask why we need so much energy in the first place, then we need to look into the factors that generated suburban sprawl, large living spaces, and energy-intensive home and transportation technologies. These relate to our values and preferences, and are influenced by culture, history, politics, technology, and economics.

The final issue we will examine is the issue of global climate change. In many respects this is the most complex environmental problem ever faced. Earth's biosphere is an extremely complicated system that science does not fully understand. We know that the planet has experienced non-human-induced climate changes throughout time. We do not fully understand those natural cycles, and so in the 1970s and 1980s we were not certain if some of the changes we were noticing were human-made changes or natural ones. By the turn of the twenty-first century, scientific uncertainty was fading, and it was clear that the carbon dioxide emissions from fossil fuels and other gases such as methane from landfills were causing global warming.

We know that pollution in one part of the planet can have an impact on another place far away. Some of the air pollution from power plants in the Midwest impairs the air quality in New York City. Still, there are clear limits to the degree of global impact from air pollution. My home city doesn't appear to get air pollution from Mexico City or Hong Kong, but some of our pollution originates in Ohio, Pennsylvania, New Jersey, and parts of Illinois. Climate change is the first environmental issue that we know about that is truly global in character. Carbon dioxide emitted from an SUV in suburban Houston contributes to raising temperatures planet-wide. This is not to say that greenhouse gas is our only global environmental problem; it is simply the first one that scientists have managed to bring to widespread public attention.

While toxic waste is an issue that can be addressed at the local level, a local approach to climate change can work only if it is part of an effort coordinated throughout the world. The need for action on a global scale presents a challenge to our international system of diplomacy. Upon reflection, it appears that technology has posed at least three threats to the viability of the nation-state. The historic origin of the nation-state derives from the need for security and the ability of this form of governance to provide that security. Threat number one came with the development of the atomic bomb. Nuclear proliferation challenges the nation state's capacity to provide security. Threat number two came from the development of the Internet, containerized and air shipping, bar codes, microcomputers and satellite communication. The technology that has made the global economy possible has had the effect of impairing national economic self-determination. Threat number three comes from the way we generate energy for electricity, climate control, and transport. That technology has resulted in excessive releases of carbon dioxide, may cause other forms of global ecological damage, and has reduced the effectiveness of national environmental policy.

In chapter 6, we will analyze the origin and impact of the climate change issue. We will attempt to characterize the issue and identify its key elements. The impact of climate change is more difficult to project than is the impact of many other environmental issues. The introduction of a chemical pollutant into the environment can be tracked and its effects on human and ecological health can be measured. Climate change will cause a set of changes that are difficult to predict. Some areas may actually benefit from improved agricultural productivity that results from warmer weather and increased rainfall. Other areas could suffer from sea level rise, and still others could be damaged by drought. The impacts will vary in ways that are difficult to predict, and will not resemble the patterns we have seen with other environmental issues.

TOWARD AN INTERDISCIPLINARY UNDERSTANDING OF ENVIRONMENTAL AND SUSTAINABILITY POLICY

The goal of the framework presented here is to engage in a conversation across disciplines. Anyone who seriously seeks to understand environmental policy must learn a modest amount of science, engineering, political science, economics, organizational management, and some things about a variety of other fields as well. Unfortunately, the power and dominance of individual academic disciplines make it difficult for these conversations to take place with the rigor and intensity that we see within disciplines. The explicitly interdisciplinary framework I propose in this book should be seen as an invitation for those with particular disciplinary expertise to critique the framework and improve it. The goal is to develop a more powerful set of tools for understanding this complex issue. This is a theme I will return to in the concluding chapter of the book.

CHAPTER 2
A FRAMEWORK FOR UNDERSTANDING THE ENVIRONMENTAL POLICY ISSUE

Environmental problems cross the boundaries of sovereign states and, in the case of global climate change, affect natural systems that are worldwide in scope. The environmental problem has a great number of dimensions, all linked to the inescapable fact that human beings are biological entities, dependent on a limited number of resources for survival. As Earth's population continues to grow, so too does the stress on finite natural systems and resources. Yet our ability to use information and technology to expand the planet's carrying capacity also continues to grow.

This book is a brief exploration into the fundamental issues of environmental policy. It presents and applies a preliminary or rough framework for a multidimensional analysis of environmental sustainability issues. The cases analyzed range from the issue of hydro-fracking to the complex scientific controversy of global climate change. The cases vary by technical complexity, level of government involvement, and scope of potential impact. They are selected to illustrate the usefulness of examining them from these vantage points. Other cases could easily be selected. In this book's first edition, I presented three other cases: underground tanks, toxic waste cleanup, and New York City's garbage problem. The framework itself is a work in progress. It provides a method for looking at environmental issues from more than one perspective. By applying the framework to specific cases, a practitioner, student, or analyst is able to observe aspects of the issue that might otherwise be easily ignored. For purposes of this analysis, an environmental sustainability problem is conceptualized as:

- *A values issue:* In what type of ecosphere do we wish to live, and how does our lifestyle affect that ecosphere? To what extent do environmental problems and the policy approaches we take reflect the way in which we value ecosystems and the value we place on material consumption?
- *A political issue:* Which political processes can best maintain environmental quality and the economic sustainability of the planet's resources, and what are the political dimensions of this environmental problem? How has the political system defined this problem and set the boundaries for its potential solution?
- *A technology and science issue:* Can science and technology solve environmental problems as quickly as it creates them? Do we have the science in place to truly understand the causes and effects of this environmental problem? Does the technology exist to solve the environmental problem and/or mitigate its impacts?
- *A policy design and economic issue:* What public policies are needed to reduce environmentally damaging behaviors? How can corporate and private behavior be influenced? What mix of incentives and disincentives seems most effective? What economic factors have caused environmental damage and stimulated particular forms of environmental policy? Economic forces are one of the major influences on the development of environmental problems and the shape of environmental policy. In this framework, we view these

economic forces as part of the more general issue of policy design. While most of the causes and effects of policy are economic, some relate to other factors, such as security and political power.

- *A management issue:* Which administrative and organizational arrangements have proven most effective at protecting the environment and promoting sustainable economic production? Do we have the organizational capacity in place to solve the environmental problem and develop a sustainable high throughput economy?

This multifaceted framework is delineated as an explicit corrective to analysts who narrowly focus on one or two dimensions of an environmental problem. Next, there is a discussion of policy and management approaches typically used to "solve" environmental problems. The proposed framework is then applied to a set of environmental problems and solutions that demonstrate specific issues of values, politics, science and technology, policy design, economics, and organizational management.

This approach owes its origin to Graham Allison's classic work *The Essence of Decision* (Allison 1971; Allison and Zelikow 1999). Allison posits three models, or ways, of examining the events of the Cuban missile crisis: the rational actor, organizational process, and governmental politics. He provides different explanations for the events of the crisis depending upon which model he applies to interpret events.[1] He provides an image of an analytic method that I have always found useful, that of "snapping in" an analytic lens in front of our eyes to enable us to interpret events or "facts" through the vantage point offered by that lens. In the case of the missile crisis, the "rational actor" model explains the placement of missiles in Cuba as an act of a rational, goal-seeking decision maker. The "governmental politics" model focuses on the political competition among stakeholders for power, thus explaining the placement of missiles and the U.S. response in terms of the competition for political power. Finally, the "organizational process" model highlights the impact of organizational routine and standard operating procedures in constraining the rationality of decision making.

Similar to the concepts applied by Allison, the framework I propose here also calls for the application of different vantage points when assessing environmental problems, policies, and programs in order to shed light on their different dimensions. The image of snapping a lens into place, like the apparatus used by an optician to test improvements in vision gained by particular lens prescriptions, is what I borrow from Allison's classic work and in a preliminary fashion apply to a set of environmental issues. The power of this approach is that the same facts are reinterpreted from several perspectives and different facts are brought to light by the different dimensions of the framework.

One purpose of this framework is to counter analytic bias deep within the way we understand environmental problems. Economists frequently misunderstand the issues of environmental science, ecology, and technology; engineers often ignore the political factors affecting environmental policy; and just about everyone forgets about issues of ethics and values. Lip service to the notion that environmental problems are inherently interdisciplinary does little to amend the tendency to assume that one's own discipline is the central one. When analyzing an environmental issue, ignoring other fields is an obstacle to improved solutions.

The strength of this proposed framework is that it can be used to understand the causes of environmental problems and the way they are defined on our society's systemic and institutional policy agenda, as well as their evolution over time. Each dimension of the framework illuminates a different aspect of the problem, and as will be demonstrated through the case studies used in later sections, the nature of each problem is weighted more toward certain dimensions than others.

VALUES

Environmental ethics is the most important of the five dimensions we will examine. Ideas about our relationship to the ecological environment derive from our concept of property and a definition of nature as a resource to be used for human material well-being. The domination or taming of the environment has long been a theme in the development of Western politics, economics, society, and religion. In fact, it is central to the definition of what we have termed "civilization." Civilization involves human mastery over the other species and the development of surplus wealth and leisure time needed for thought, reflection, and the transmission of learning. To the extent that we are successful, the natural environment is something that is available for our use: a set of resources to be consumed.

We are more dependent on natural systems than we once thought. We now know we do not have the ability to supplant resources and simultaneously maintain a high-quality existence, as that notion is currently defined. We need ecological systems. Our technology is not sophisticated enough to do without them. The pragmatic argument is compelling, but it is not the only line of reasoning. For instance, according to some environmental philosophers, our very arrogance may be at the heart of the environmental problem. In order to address the root of our environmental problems, they suggest, we must redefine our relationship with the environment and stop looking at other species as resources (Leopold 1949). Although this may be true, it is unlikely that the planet's more than seven billion people will seriously contemplate a return to nature. With more than 50 percent of our population now urban, such a return to nature is no longer feasible. Moreover, other values-based goals that we hope to achieve, such as equity, justice, family, and education, preclude a radical redefinition of our relationship to the biosphere.

Given the current worldwide disparity in wealth, it is difficult to halt economic development and its associated environmental impacts. Instead, some analysts forecast that economic development will result in demographic transitions that reduce population growth and increase the public's stake in protecting the environment (Cohen 1995:47). The idea is that increasing levels of economic development lead to decreased demand and supply of labor, and increased demand and supply of capital. Thus, while in developing nations, children (who represent added labor capacity) are perceived as essential for economic and old age survival, in developed nations children are "decorative" and an economic liability; therefore, there is less economic incentive to have children in developed nations. According to this theory, only economic development can bring population stability to the planet (Ophuls and Boyan 1992:46). The language of economic development in recent years has incorporated the notion of sustainability, which is another way of saying development with sensitivity to environmental impacts. The hope of development advocates is that a fully developed world with low population growth would prove less detrimental to environmental quality than the partially developed world in which we now live.

The desire for economic development is an expression of values. A good life, as we now understand it, includes a high level of resource consumption. It is unrealistic to assume that this concept will change. Though the Western pattern of consumption may disgust some in principle, its seductiveness and appeal are demonstrated facts of modern life. What, then, is the goal of environmental politics and policy? I would argue that it is one that has evolved over time, since the U.S. Environmental Protection Agency (EPA) was established in 1970 to deal with the problems of degradation of the natural environment.

Environmentalism in the United States has roots in late-nineteenth-century anti-urbanism, transcendentalism, and the desire to preserve the productivity of the land for future generations (Rubin 2000:159). In the beginning, concern for the environment was an aesthetic issue and an issue of lifestyle. It included a preference for the virtues of an agrarian and/or rural way of life. Some saw cities as corrupt and evil, in contrast to green open spaces, which could cleanse the soul and stimulate virtuous living. When the EPA was created, it was primarily an anti-air-and-water-pollution agency. Nearly all of the staff in the newly created agency came from the Department of Health, Education, and Welfare's (HEW) air-and-water-pollution-control units. Dirty air and water were regarded as vaguely unhealthy, but decidedly unsightly. As the EPA's mission expanded in the 1970s, it started to work on other issues, such as solid waste, that resulted from urban environmental problems. With the passage of the toxic waste cleanup Superfund program in 1980, the environmental issue began to be defined as a public health issue. Pollution was not just ugly; it could make you sick. This human health orientation continued throughout the 1980s. In the early 1990s, we again saw a shift as the focus turned to international environmental problems, especially global climate change. As holes in the ozone and global warming were discovered, the definition of the environmental problem expanded to include a concern for the viability of the planet itself. In the past decade, the field has continued to evolve, embracing a concern for sustainable economic growth. Environmental protection is no longer concerned exclusively with pollution created at the "end of the pipe"; it now also addresses production using renewable resources and production processes that do not degrade the environment. This "sustainability" perspective is transforming the environmental issue into one that is centrally related to economic development.

In each of these definitions of the agency's mission, the concern has been, and will remain, the protection and advancement of human well-being. We protect the environment in order to make sure we don't kill the goose that lays the golden eggs. Our taste for golden eggs—that is, for economic consumption—continues to grow. The environmental ethic under which we operate requires us to maintain the biosphere for our descendants, not because we care about them, but because environmental deterioration reduces our ability to consume things we desire, such as wholesome, tasty food, fresh air, clean water, and coastal cities that are not submerged because of global warming.

Some have argued that the environmental problem requires a change in the dominant social and political paradigms, and a fundamental change in how we view politics, the environment, and one another. Such dramatic shifts in paradigm are neither necessary nor feasible (Milbrath 1984:81). Instead, environmental policy has focused and will continue to focus on developing less-destructive methods of fulfilling the current consumer ethic. Today's environmentalism results in changed consumption patterns, not a reduction of consumption. For example, compare Internet surfing to cruising around in a gas-guzzling automobile. Ultimately, it does not mean a reduction in economic consumption, and it certainly has not resulted in a reduction in this nation's waste stream. Total production of solid waste in the United States has grown from 2.68 pounds per person per day in 1960 to 4.43 pounds per person per day in 2010 (U.S. EPA 2010b). But during that same period of time, recycling grew from 5.6 million tons per year, or less than 10 percent of total wastes, to approximately 80 million tons per year, or 34 percent of the waste stream (U.S. EPA 2010b). This is additional evidence of the changed nature of consumption and waste management patterns, while consumption continued to increase.

In sum, the environmental ethic that has had the greatest impact in the last four decades, at least in the United States and other developed countries, has been a form of enlightened self-interest. In this value system, environmental protection is not traded off against the value of economic consumption. Though it is actually another form of consumerism and does not signify a break in the culture of consumerism, it has resulted in greater popular awareness of environmental issues in general.

APPLYING THE VALUES DIMENSION OF THE FRAMEWORK

When applying the framework to enhance your understanding of an environmental issue, you investigate the various dimensions in any order you choose. I like to begin with values because they seem fundamental to me. To apply this dimension, you would pose a number of questions. Some may not be answerable, but some will always be of use:

• Does the issue stem from a behavior fundamental to our lifestyle?
• Are issues of right and wrong and/or justice raised by the creation of the problem, or by particular proposed solutions?
• Does the problem or solution require a trade-off between ecological wellbeing and human well-being?
• Does the process that created the problem, and/or the proposed solution to the problem, cause conflicts with ethical and/or religious concepts or precepts?
• Does the problem or its proposed solution raise fundamental issues of conflicting values?
• Can the issue be addressed without facing its fundamental value conflict? Can progress still be made?

Addressing some of these questions will illuminate the value dimension of the issue. It will allow you to see this aspect of the problem and place it in perspective. All policy issues have a values dimension. The analytic task is to determine how fundamental the value is, how important it is, and whether it conflicts with other closely held values. The purpose of applying the framework is to understand the environmental problem as a policy issue. The values lens can provide insight into the potential intensity of the issue's political conflict and its saliency and importance as a political issue.

ENVIRONMENTAL POLITICS

Although the environment as a political issue has not resulted in a major shift in the dominant social and political paradigm, it has added a significant new set of considerations to the policy formulation process. The environmental issue has made significant demands on our political processes and institutions. Americans have called for political processes that develop a consensus about the definition of environmental quality and make decisions about methods for achieving environmental goals. In recent years, we have seen an effort to combine economic development and environmental protection under the broad rubric of sustainability management. In the past forty years, this political process has facilitated a high degree of social learning in the United States. The learning process will continue because new information on human-induced change and ecological conditions is continually becoming available. This information will need to be summarized, disseminated, and understood by decision makers and the broader public in order for policy to adapt to changing conditions. The issue of environmental politics is closely connected to the issue of economic development and worldwide income distribution. Environmental and sustainability policy is about individual and collective patterns of resource deployment, consumption, and degradation. Put simply, we must learn enough about the biosphere to make sure that in our use of it we do not irreversibly degrade or destroy it. Once we know the types of behaviors required to sustain the environment, we must organize ourselves to perform those behaviors.

This learning process creates winners and losers. The assignment and distribution of benefits and costs create political conflict that both impedes and distorts social learning. People and interest groups sometimes present environmental information that is partial or misleading to serve their own particular interests (Sabatier and Jenkins-Smith 1993). Consequently, environmental and sustainability policy never appears as a seamless progression from scientific discovery to implemented public policy. Rather, it looks like a meandering series of disjointed incremental steps, very much like the type of policymaking described by David Braybrooke and Charles Lindblom (1963) in *A Strategy of Decision*. The decision-making strategy they described is "remedial," "serial," and "exploratory." Policymakers move away from problems rather than toward solutions. Braybrooke and Lindblom (1963:104) observed that "analysis and evaluation are socially fragmented, that is they take place at a very large number of points in a society. Analysis of any given problem area, and of possible policies for solving the problem, are often conducted in a large number of centers."

Many environmental scientists and advocates lament the messiness of this type of policy process and believe it is inadequate to the task of addressing long-term, interconnected, large-scale problems such as protecting the environment and making the transition to a renewable economy. In this view, partial answers cannot address the root causes of environmental and sustainability problems. Many environmental advocates criticize incremental environmental policy and the pace of the transition to sustainability management, but are unable to suggest truly viable alternatives. It is not likely that people in developed nations will slow down the input of information, reduce consumption, and get back to the land. It is not realistic to eliminate pluralistic, interest-dominated politics. Certainly even a benign environmental totalitarianism is not a viable alternative form of politics. Mass participatory democracy seems equally unlikely—and might not protect the environment.

Meeting the short-term needs of the mass public could pollute the planet, and as orderly as totalitarianism looks on the surface, resistance to authoritarianism is a virtual certainty. In other words, we are stuck with the messy, partial, and incremental politics that is characteristic of Western democracy. We will need to work within the current political framework if we are going to protect the environment and promote sustainable development.

Even if we were able to achieve a perfect understanding of the environment and the effect of human interaction upon it, our social and political processes cannot absorb and act on the volume and complexity of that information. The exception would be a genuine crisis. Normal politics and incremental policymaking can be suspended for a time during a crisis. In the United States during World War II, for example, the government spent nearly 46 percent of the gross national product (GNP) (U.S. Department of Commerce 1994), and certain civil liberties were suspended. In the weeks after the World Trade Center was destroyed, partisan politics was replaced by an unusual degree of national unity and patriotism. However, crisis management politics and wartime mobilization cannot be sustained indefinitely. Eventually, normal politics resume. Unfortunately, the difficulty with the environmental problem is that if we get to a crisis point, it may be too late to solve the problem.

The answer, to the extent that there is an answer, is to organize politics to accelerate the learning and decision-making process. We need a more rapid incremental political process. Environmental and sustainability management issues must be raised and discussed through the electronic media, public education programs, and active efforts to elicit citizen participation in policymaking. Worldwide environmental education has grown exponentially since 1970 (U.S. EPA 1996:3). In developed countries, young people are raised to understand facts about the biosphere that were unknown when today's Baby Boomers were young. On the other hand, fear of environmental damage has resulted in the reflexive "not in my backyard" (NIMBY) syndrome that sometimes produces greater total environmental impacts in order to avoid lesser impacts to a more powerful or better organized local constituency.

In the United States one result of increased levels of environmental concern and literacy has been a series of successful efforts to protect the environment. As EPA data indicate in figure 2.1, pollution in the United States has decreased dramatically, while population and GDP growth have continued.

The irrational, nonanalytic decision making and political process in the United States has brought about a successful reduction in key pollutants. This does not mean that the environmental problem is solved or has gone away; rather it means that we are "moving away from the problem." Although I cannot state it definitively, I believe that the problems of environmental degradation and solid-waste management in the United States are slightly less severe in 2013 than in 1993. How did we make this progress? In what type of political process did the United States engage? The environment and sustainability management, over time, achieved status on the political agenda. The definition of environmental politics has changed, but it has resisted a number of concerted attacks on its legitimacy, and now appears to be a permanent fixture on the political agenda.

To understand environmental policy one must understand the agenda-setting process. Why do some issues that reach the agenda get acted on while others, through a process that has been termed non-decision making, are ignored (Bachrach and Baratz 1970:44)? Issues can also be denied agenda status altogether, by denying them legitimacy. Typically, powerful interests in a society define an issue as illegitimate not by responding to it substantively but by using their control of the agenda to ensure that it is never heard. For many years, issues such as race and gender bias simply could not be heard. In their book *Participation in American Politics*, Roger Cobb and Charles Elder (1983) took the issue of the policy agenda even further and divided the political agenda into systemic and institutional dimensions. An issue that can be discussed obtains status on the systemic agenda as a legitimate issue. It then must "travel" to the institutional agenda, where it is seen as a legitimate object of government policymaking. Non-decision making is most effective and least visible in keeping items off of the systemic agenda. But it also can be used more overtly to keep issues off of the institutional agenda (Cobb and Elder 1983).

The environment reached the national arena in the United States in several stages. Rachel Carson's *Silent Spring* (1962) and Barry Commoner's *The Closing Circle* (1971) popularized the concept of a global ecosphere threatened by human economic activity such as the application of pesticides and nuclear testing. After the 1968 presidential elections, the environment began to enter the national political agenda in Washington, D.C.

During the 1968 campaign, Maine senator Edmund Muskie carried himself so well as a vice-presidential candidate that he was immediately considered a front-runner for the 1972 presidential nomination of the Democratic Party. Muskie's major issue was protecting the environment, and in 1969 and 1970 he pushed for the passage of an air pollution control act that would set national standards for ambient air quality (Jones 1975:175–182). Though this idea was initially opposed by industry, President Nixon came to support a national air quality

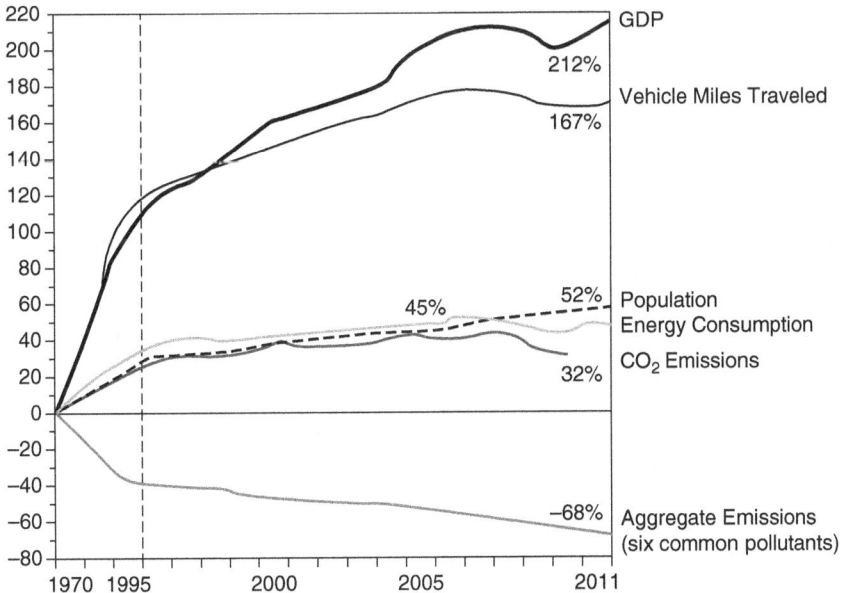

Figure 2.1 Environmental Improvements/Economic and Population Growth
Source: Adapted from U.S. EPA. 2012b. "Air Quality Trends."

bill as a way of countering Muskie's growing political strength. Nixon and others also saw the environment as a safer, less contentious issue than the war in Vietnam. Though some political analysts at the time viewed Nixon's support of the environment as a way to distract people from the Vietnam War, it was an important step on the road to formalizing environmental policy and one that was introduced through the American political machine.

The U.S. EPA was established during the Nixon administration through executive order, not through an act of Congress. This, along with the enactment of the 1969 National Environmental Policy Act (NEPA) and the 1970 Clean Air Act (CAA), provided President Nixon with an environmental record to counter Senator Muskie's in the 1972 campaign. In the end, Muskie failed as a presidential candidate, and with a weak challenge from George McGovern in 1972, Nixon felt confident enough to veto the Federal Water Pollution Control Act, which was subsequently enacted *over* his veto. Despite Nixon's action on the water bill, in the 1970s the environment was typically seen as a foolproof, popular political issue. With the exception of the early years of Ronald Reagan's presidency, environmental protection has continued to be seen as a straightforward issue for politicians. Nowadays, no politico can afford to be perceived as anti-environmental.

In the late 1970s and early 1980s, conservative Republicans from the western part of the United States developed an anti-environmental ideology based on the issue of property rights (Layzer 2002:242). In exchange for their support of Ronald Reagan's presidential campaign, they were given the Department of the Interior (James Watt) and the EPA (Anne Gorsuch, later Burford) to manage. To Reagan's White House team, these were relatively unimportant ministries, so they did not pay a great deal of attention to the selection of the secretary of the interior or the administrator of the EPA. The appointments were used to repay political debts. Moreover, Reagan's senior advisors assumed that popular support for the environment was diminishing. The environment was no longer appearing on the "top ten list" of issues cited by the American public as "important" in public opinion polling. Environmental interest groups recognized the threat to the gains of the 1970s and organized a campaign of opposition that culminated in the resignation of Anne Burford and her replacement by William Ruckelshaus. the first administrator of the EPA (Kraft and Vig 1990:3).

It turned out that the Republican political leadership had misread the polling data. The environment had become a less pressing issue among voters because average Americans thought that reasonable progress was being made to clean up pollution, not because they had lost interest. When it became clear that the leaders of the EPA and the Department of the Interior were attacking some of the programs that had brought those gains, the environment's importance rating in national polls rebounded to levels as high or higher than in the early 1970s (Mitchell 1984:56). Recognizing that it had landed on the wrong side of this issue, and with the 1984 presidential election approaching, the Reagan administration moderated its views, forced out some of the most visible right-wing environmental leaders, and allowed gradual environmental progress to resume.

In the first decade of the twenty-first century, President George W. Bush also struggled with his image on environmental issues. In early 2002, the American public was not sure where President Bush stood on environmental issues. He wasn't yet defined in the public mind as anti-environment. Christine Todd Whitman, Bush's first EPA administrator and former governor of New Jersey, was a moderate with a reasonable environmental record. Her appointment to the EPA was an indication that the lessons of the mid-1980s had not been lost on President Bush and his political team. His second EPA administrator, former Utah governor Michael O. Leavitt, was also an environmental moderate. His third administrator, Stephen Johnson, was an environmental scientist and a twenty-year veteran of the EPA. However, by 2003 Bush began to slip into an anti-environmental position.

Indeed, by the end of 2003, George W. Bush had become the most anti-environmental president since the creation of the EPA. At the start of his second term, this pattern continued, though he made occasional efforts to paint himself as an environmentalist. Although his administration came into office determined to avoid the mistakes that the Reagan administration had made in environmental policy and politics, they were unable to do so. From its Clear Skies legislation, which would slow down the national air cleanup to its Healthy Forests initiative, which would damage the nation's wildlife and forest ecosystems, the Bush administration engaged in a series of subtle attacks on the rules regulating protection of the environment. Congressman Tom Allen of Maine called Clear Skies "a classic case of chutzpah, a triumph of marketing over substance, if I have ever heard of one" (2004). Sierra Club president Carl Pope commented that the Healthy Forests Restoration Act "will certainly succeed in propping up the timber industry" (Jalonick 2003). The public relations terminology used to describe these bills, coupled with revelations in 2003 of White House efforts to tone down the EPA's warnings about air pollution from the destruction of the World Trade Center, were examples of the administration's lack of fervor on environmental issues.

Not only did the first term of President George W. Bush witness an effort to reduce the scope of environmental law, it also reduced enforcement of existing laws. On November 5, 2003, the Environmental Protection Agency dropped investigations into fifty power plants for past violations of the Clean Air Act, a scaling back that at the time was predicted to reduce investment in cleanup equipment by between $10 billion and $20 billion. In response, New York State attorney general Eliot Spitzer, along with a number of his colleagues in other northeastern states, decided to continue legal action (without the federal government) to push these utilities to clean their emissions (Barcott 2004:38). Although most Americans favored the Bush administration's environmental stance in 2001 and 2002, support for its environmental leadership fell to 44 percent in April of 2003 and continued to decline, to 41 percent in 2004 and 39 percent in 2005 (Gallup Organization 2005a, b, c).

The environment's political power tends to lie dormant as long as the public perceives that leaders are serious about promoting pro-environmental policies. However, once a leader is no longer credible on the environment, the issue gains in urgency and importance, and can rise in national polls. The controversy over both Utah governor Michael Leavitt's 2003 and Stephen Johnson's 2005 nominations as EPA administrator was a reflection of the president's reduced credibility on the environment. Even though Governor Leavitt was considered an environmental moderate, and Johnson was a well-regarded environmental professional, Democrats and environmentalists delayed their appointments as a way of focusing attention on the administration's weak environmental record (Stout 2005).

The Obama administration focused less on pure environmental protection regulation and more on the transition to a sustainable, renewable economy through vehicle emissions policy, climate regulation, and funding of renewable energy projects. On January 27, 2009,

President Obama directed the Environmental Protection Agency to allow individual states to set stricter tailpipe emissions regulations than the federal standard. The Obama administration finalized standards in August 2012 that will increase fuel economy to the equivalent of 54.5 miles per gallon for cars and light-duty trucks by model year 2025. Previous standards set by the Obama administration raised average fuel efficiency by 2016 to the equivalent of 35.5 miles per gallon. Obama signed H.R. 1, the American Recovery and Reinvestment Act (ARRA) of 2009, into law on February 17, 2009, which provided $54 billion in funds to encourage domestic renewable energy production, make federal buildings more energy-efficient, improve the electricity grid, repair public housing, and weatherize modest-income homes.

Environmental issues do not typically decide elections because most people assume that just as all elected leaders promote security and safety, they also promote environmental protection. The need to breathe clean air and drink safe water isn't politically controversial. Political analysts often confuse the lack of passion behind the issue with a lack of public interest or concern. That is a mistake. I find that more and more people born since the mid-1980s have internalized aspects of an environmental ethos, and that awareness will soon have a major impact on American politics. While Gallup continues to poll on what I consider the false trade-off between economic growth and environmental protection, even its data report growing environmental awareness, especially among young people.

A recent Gallup poll reported: "For the fifth consecutive year, more Americans are interested in protecting economic growth than in protecting the environment when the two goals are at odds. This year's 48% to 43% split represents a relatively narrow advantage for the economy, similar to last year's reading. But the latest result contrasts with 2011, when a record-high 54% chose the economy as the higher priority."

The item is based on a false premise, asking survey respondents to react to this question: "With which one of the following statements do you most agree? Protection of the environment should be given priority, even at the risk of curbing economic growth; or, economic growth should be given priority, even if the environment suffers to some extent." The connection of environmental quality to economic growth, a central tenet of the concept of sustainability management, is ignored by this question. Environmental protection is not something you must sacrifice for economic growth; it is a key source of economic growth. This is a lesson that China is rapidly learning, and that some Americans learned when we spent hundreds of billions on toxic waste cleanup. When the environment suffers, it costs money when we eventually get around to cleaning up the damage. If we hadn't damaged it in the first place, the money devoted to cleanup could have been used to do something else. Moreover, essential environmental resources such as air and water cannot be used once they are poisoned. If these resources are damaged, they must be filtered before they are used, and the cleansing process requires a lot of energy and money. The Gallup poll question treats the "environment" as if it's a visitor from outer space, instead of the air, water, and soil that human beings require to remain alive. While short-run corporate profits can be made by exploiting natural resources, real economic growth requires high environmental quality to both attract investment and keep our food and water prices affordable.

It is clear from the Gallup data that when the economy is doing well, more people favor breathing healthy air and drinking clean water. When the economy falters, people favor economic growth over just about anything. Until the economic crash of 2008, Americans always favored "environment" when asked this question. The trade-off choice was just as false then, but the response to the poll is certainly statistically valid and is an accurate measure of the public's response to the question.

While Americans are willing to trade off the environment for economic growth, the response to this flawed question varies significantly by age. Young people, between the ages of eighteen and twenty-nine, favor the environment over economic growth by 49 percent to 45 percent. As Americans age, they increasingly select the economy over the environment, with the oldest Americans, those sixty-five and older, favoring economic growth by 53 percent to 37 percent.

Gallup has also polled about government's role in protecting the environment, and has found that "Americans tilt toward the view that the government is doing too little to protect the environment—at 47%—while 16% say it is doing too much. Another 35% say the government's efforts on the environment are about right. These views have not changed much since 2010, although Americans in most years between 1992 and 2006 were more likely than they are today to say the government was doing too little to protect the environment."

While the responses still indicate that average Americans would like to see a stronger government role in protecting the environment, it is not clear if the question is measuring attitudes toward environmental protection or attitudes toward government. The poll reports that 27 percent of all Republicans believe that government is doing too much to protect the environment as compared to 2 percent of all Democrats. I suspect that this poll would find a similar gap between the parties on the role of government in many areas of public policy.

In my view, the questions posed by Gallup are not tapping into the change that is under way. My sense is that many young people have a deep fear that the planet they will inherit from the rest of us may be damaged beyond repair. They do not necessarily see its repair as a function of government, especially our national government. Instead, they are looking for change at the community and municipal level. The fact that young people are moving away from suburbs and back to cities is in part a rejection of a lifestyle that they suspect may not be sustainable. The cars, lawns, and costs of cooling and heating large suburban homes are replaced by biking, walking, mass transit, and smaller apartments where heating and cooling costs tend to be lower.

This is not to say that young Americans are rejecting consumption—far from it—but they are looking to consume in a different way. These emerging consumption patterns reflect their concerns about environmental sustainability. The growing number of people biking to work and shopping at local green markets is an indicator of this change. The number of cities, nonprofits, and corporations engaged in sustainability initiatives is another indicator. People and institutions are thinking about their use of natural resources and energy and the impact of their consumption on the planet. That trend is most pronounced among young people, and that social change will gain momentum as they age.

The generation that grew up in the first part of the twenty-first century will be coming to power in the coming decades. They have grown up with a concern about the sustainability of our economy and the health of our ecosystem. The issue of environment and sustainability has already moved from the fringe of political awareness and our policy agenda to its center. In the next decade, this will only increase as the sustainability generation comes of age.

Public support for environmental protection runs deep in this country, cutting across every demographic category (Gallup Organization 2005a, b, c).

Nevertheless, climate change as an issue was largely absent from the 2012 presidential election until Hurricane Sandy hit the Northeast. President Obama stopped campaigning to head the federal emergency response, and Romney halted his campaign shortly after and encouraged donors to donate to the Red Cross. Though this may have become a last-minute issue during the 2012 election, the focus was more on disaster response than on environmental protection.

Nonetheless, as long as the public sees good-faith progress in environmental protection, the issue does not generate much political heat beyond the environmental community. Issues such as police protection and education act in the same way. Key values such as education, police protection, and environmental protection are nonissues until the public perceives that they are threatened. When that happens, however, the political reaction is rapid and sweeping.

An oddity of President Bush's approach to environmental issues was that it reflected an antiquated understanding of the environment. This is the notion at the heart of the Gallup poll question trading off environmental protection against economic growth. There is a relationship between environmental quality and economic growth, but it is not a trade-off relationship. We are learning that economic growth *depends* on environmental quality: the return on investments in environmental quality pays off. Sound and sustainable economic development requires the maintenance or restoration of environmental quality. A study published by the Office of Management and Budget in 2012 estimated that from 2001 to 2011 we spent $23 billion to $29 billion to clean up our air, resulting in net benefits valued at between $85 billion and $565 billion for that investment (OMB 2012:12). Investments in sewage treatment have increased the value of waterfront property all over the United States. The funds given to upstate property owners to reduce pollution near New York City's reservoirs have saved the city $6–$10 billion that it would have had to spend on a water filtration plant (Helhoski 2009).

Popular support for protecting environmental quality is the basis for the issue's political strength. The importance of environmental protection has been the subject of massive educational and propaganda efforts by scientists, advocates, the media, and professional educators. Though respect for property rights remains strong in the American political culture, especially in the western states, support for environmental protection frequently dominates concern for property rights. Interestingly, one source of political strength for environmental protection is the perception that pollution can diminish the value of private property. In this respect, high environmental quality is a form of property or wealth. This did not happen overnight and is arguably a result of the high level of economic wealth in the United States. Our wealth permits consumption of beach homes, country homes, suburban living and vacations in national parks and rural areas—"goods" that would be diminished by environmental degradation. Maintaining that level of consumption requires the maintenance of environmental quality.

Despite support for curbing the pollution of others, not all consumption patterns in the United States provide behavioral evidence of support for protecting the environment. The increased average weight of the American automobile is an example of behavior that does not promote environmental protection. Nevertheless, the popular consensus for protecting the environment is strong across regional, racial, and socioeconomic categories (Mohai and Bryant 1998:475). Clearly, this type of consensus facilitates social learning about the environment, eventually leading to effective policymaking.

An important dimension of the environmental problem is its status as a political issue. This involves its presence on the political agenda as a problem appropriate to collective societal action.

APPLYING THE POLITICAL DIMENSION OF THE FRAMEWORK

As the preceding discussion indicates, an issue becomes a public policy issue only when it becomes part of a political process. How the issue is defined, how it enters the political agenda, and the views of its key stakeholders determine the issue's definition. Once again, to apply this dimension, there are a number of questions to pose, many of which are unanswerable but essential:

- What is the agenda status of this issue?
- What is the issue's degree of legitimacy?
- What role has the issue played in electoral politics?
- Who are the political stakeholders involved in this issue? What is the nature and style of their participation? How important is the issue to these stakeholders?
- Who are the potential winners and losers in the political competition around this issue? Who have been the winners and losers to date, and how are political victory and loss defined in this issue area?
- To what degree are stakeholders willing to discuss and compromise on this issue?
- Does this issue act independently of other political issues or does it cluster with other key issues?[2]
- In the U.S. context, what level of government is considered primarily responsible for addressing this issue (state, federal, or local)?
- What is the level of controversy and consensus around this issue? What are the areas of agreement and disagreement?
- How does scientific certainty or uncertainty related to the definition of the problem or its potential solution influence the politics of this issue?

The political definition of the issue is critical in shaping how the issue is framed and ultimately addressed. For example, a nuclear power plant could be seen as a source of vital electric energy, as a terrorist target, as a source of potential contamination from a meltdown, or as a source of a difficult-to-discard waste product. It could also be seen as a combination of some or all of those factors. The early part of a political controversy is often a struggle over the definition of the issue. That definition will shape the way the problem is perceived and will set the boundaries for the problem's solution. This definition will persist, often in the face of new scientific information, changed context, and changed behavior. Understanding an environmental policy issue requires a careful and sophisticated analysis of the issue's political dimension.

SCIENCE, TECHNOLOGY, AND THE ENVIRONMENT

Much of the progress we have made in protecting the environment has been the result of the development and implementation of technological fixes to environmental problems. We reduce air pollution by utilizing newly developed environmental controls such as catalytic converters on autos and scrubbers on electrical power plants. We treat sewage before dumping it into waterways. In sum, we use science to fix the mess that science helped make. The question is can we solve problems as quickly as we create them? Thus far, the answer to that question is no. The question then becomes can we fix the most pressing problems fast enough to maintain a habitable environment? Here the answer is more complicated. What do we consider to be a livable environment? Are the shantytowns and slums of developing countries livable? When is an environment so dangerous that we consider it unacceptable? If we create a technology that causes disease in some percentage of people, but we develop a cure for that disease, is our tolerance for a lethal environment expanded?

To some degree, then, the environment is a problem of science and technology. We invent new products and put them to use before we project their effects on human health, the biosphere, and the local ecosystem. Until now, technologically based economic development has raised living standards and increased population around the globe. The benefits are unevenly distributed, but the results are undeniable. Can science and technology keep up? When science cannot develop remedies for new harms caused by new technologies, can we slow down the introduction of new technologies until we can figure out how to use them safely?

Experience over the past several decades provides evidence on both sides of this issue. In the United States, the problem of nuclear waste and reactor safety has limited the use of nuclear power to generate electricity. In addition, a number of toxic substances, such as the insecticide commonly known as DDT, have been banned here. Chlorofluorocarbons, the refrigerants that cause damage to the ozone layer of the atmosphere, are being gradually banned and replaced. Nevertheless, some developed countries, including France, continue to rely on nuclear power for electricity, and DDT is still used as a pesticide in many developing nations where malaria is a problem. My own view is that if the harm is easily proven and a clear technical fix is in place, with strong government intervention, technological solutions can be implemented. If the price of the substitute is too high, or the technology is not fully developed, it is more difficult to eliminate a dangerous technology. For example, if fossil fuels were not relatively plentiful at the moment, the risks of nuclear power would be given less attention than its benefits.

Problems caused by the impact of technological innovation on the environment are not easy to measure. Sometimes the problems take a long time to develop. Other times, a causal relationship between an environmental problem and the introduction of a specific technology is hard to establish. Even more difficult to assess are problems caused by the interactions of one or more technologies in varied ecological settings. Environmental impacts are unavoidable by-products of the strength and power of the scientific method.

The scientific method is based on the concept of the controlled experiment. The researcher first isolates the variables, subjecting only certain specific variables to a particular test. The goal is to identify and understand causal relationships, "all things held equal." However, in the natural ecological environment, nothing is held equal. The interactions and relationships that take place can best be understood at the system level, where controlled experiments are rare. Whereas experiments in the traditional scientific method are reductionist, attempting to reduce the test to a simple causal relationship, ecological systems are holistic and interconnected and cannot be understood through reducing reality to simple relational terms. Such understanding requires the use of models that estimate the interaction effects and account for the multiple feedback loops that characterize living systems. Technologies developed with reductionist methodologies must then be introduced to environmental impact studies that are based on a more appropriate conceptual framework and orientation.

In a controlled experiment, the whole test is designed to demonstrate or rule out an effect. When an effect is discovered and verified, it becomes a fact. When a model is built, we leave the world of scientific certainty and enter one of probability statements and other unknowns. It is difficult for an environmental scientist who suspects a harmful effect to compete with the power and the certainty of a technology's proven benefit. When an environmental modeler thinks he or she may have uncovered a destructive effect, it is initially expressed as a probability statement. In fact, the most persuasive scientific evidence of environmental damage must rely on models to develop hypotheses for relationships that are then tested in reductionist, controlled experiments. Only at that point can the factual basis for environmental damage be conclusively established.

Improvements in environmental measurement technology and in environmental modeling provide some hope that we are still learning and will do a better job of detecting, understanding, and ending practices that damage the environment. However, the information provided about environmental damage is not always factored into decision making. To the degree that the public is educated about environmental threats, they can become a powerful force behind active environmental protection policy. Unfortunately, many of the threats to the environment are long term, difficult to prove, and hard to explain.

APPLYING THE SCIENCE AND TECHNOLOGY DIMENSION OF THE FRAMEWORK

Some environmental problems, such as the Indian Ocean tsunami in late December of 2004, are the result of natural phenomena, but most, like the 2010 BP Deepwater Horizon oil spill, are caused by humans. Some, like Hurricane Sandy in 2012, most likely combine both human (global warming) and natural causes. As we seek to understand an environmental issue, an important dimension to consider is the level of scientific knowledge and certainty associated with the problem and its potential solution. As we seek to understand this dimension of the problem by applying this element of the framework, we should attempt to address some of the following questions:

- Is there scientific certainty about the causes and effects of the problem?
- What are the principal areas of uncertainty and what is the effect of that degree of uncertainty on decision makers?
- Are there cost-effective substitutes for the technologies that are causing harm? What are the prospects for developing such technologies?

- Does the technological cause of the problem need to be halted in order to address the problem?
- Are there "off-the-shelf," proven technologies available to mitigate the impact of the environmental problem? What are the prospects for developing such technologies?
- Are the control or mitigation technologies widely available and do we have experience with their management?
- What is the monetary cost of research to address this issue, and are these funds likely to become available?

Some environmental issues are scientifically complex and some are simple. Some lend themselves to relatively low-cost technological fixes, while others will be expensive to address. The relative scientific certainty, the complexity, and the potential cost of control or remediation technology influence the political definition of the issue.

The issue of climate change, discussed later in this book, provides an excellent illustration of the role played by scientific uncertainty in defining a policy issue. In the first term of the George W. Bush administration, inaction on climate was based on a perceived lack of scientific certainty regarding the causes of global warming. As that uncertainty was reduced and scientific consensus emerged, pressure to address the issue grew. In 2005, the first year of Bush's second term, this pressure continued to increase. The level of scientific uncertainty can therefore influence the type of policy design that is appropriate to address an environmental issue (Layzer 2002:230). Nearly all environmental issues go through an early "problem definition" phase, during which scientific research is funded to increase our understanding of the problem, thus delaying action. Water pollution and air pollution policy in the 1950s and 1960s focused on increasing our understanding of the science of the problem and potential solutions. Often, when environmental programs are established, control technology is fairly primitive. Policymakers hope that the need to comply with new environmental standards will force the development of new technology. Often new environmental rules have had that very effect.

ENVIRONMENTAL POLICY DESIGN AND ECONOMIC FACTORS AS AN INFLUENCE ON DAMAGING CORPORATE AND PRIVATE BEHAVIORS

Economic forces are the primary cause of environmental degradation and are the primary means of environmental preservation, cleanup, and pollution prevention (Schneider and Ingram 1997:99). The environment as a public policy issue should be conceptualized as a form of government regulation of corporate and individual behavior. This section deals primarily with the design of policies that regulate corporate behavior, since that has been the main target of environmental policy and regulation thus far. Toward the end of this section, the regulation of individual behavior and the problem of social learning are discussed.

While other sections of the framework helped explain both environmental problems and solutions, this section of it focuses on understanding solutions, or environmental policies.

The policy approach that we take is, of course, related to the definition of the problem. The policy approach can also influence the evolution of the problem. If progress is made, the problem may come to be seen as routine and less urgent. This section provides a catalog of the variety of policy designs that have been used to solve or address environmental problems.

In the past several decades we have heard a good deal of political, popular, and academic discussion on the concept of regulation. Regulation is criticized as harming the economy by stifling entrepreneurial initiative, discouraging technological advances, and being insufficiently cost-effective. Economists criticize lawyers for being overly formalistic and for not understanding how firms behave. Policymakers criticize economists for proposals that lack political feasibility.

DEFINING REGULATION

Kenneth Meir defined regulation as "any attempt by the government to control the behavior of citizens, corporations, or sub-governments" (1985:1). Regulation is a set of rules or directives intended to induce specific behaviors in target populations. Modifying his definition slightly, substitute the word "influence" for "control." Regulated behaviors represent tendencies and carefully augmented actions rather than goal-seeking, rationally controlled behaviors. "Control" is simply too strong a term. Organizations, for the most part, do not truly control their own actions; instead, these actions are the result of a variety of internal exchange relationships and influences evidenced by explicit and implicit bargains and the deployment of potential and actual incentives. Again, this regulated behavior is merely a tendency toward incremental actions rather than goal-seeking, rationally controlled behavior.

The goal of regulation is to influence individual or organizational behavior. To provide a graphic example, consider the case of automobiles converging on a corner traffic light. Hopefully, the behavior of each driver is influenced by the color of the traffic light. The signal is relatively clear, and when the light turns amber, the drivers are faced with the need to make a rapid decision. Several factors affect each driver's decision to slow down, speed up, or stop:

1. Is the signal working?
2. Does the driver see and understand the signal?
3. Is the driver willing to adhere to the signal?
4. Is the car mechanically capable of stopping and/or accelerating?

Are the regulated parties, in this case the drivers, capable of changing behavior in the desired direction, and are they willing to do so? The goal of regulation is to influence the variables that enter into a regulated party's calculus of the costs and benefits of compliance. What are the incentives and disincentives to stopping at a red light?

1. The possibility of a collision with a fully loaded trailer truck.
2. A traffic ticket from a highway patrol officer for running the light.
3. Belief in the rule of law.

4. Pre-patterned behavior of braking for a red light.
5. A severely ill child sitting in the backseat, and an urgent need to get to the hospital (a disincentive to stop).
6. A second disincentive to stop might arise if one is in a hurry and no traffic is visible.

The goal of traffic regulation is to reinforce the incentives to comply so that they outweigh the potential motivation to pass the red light. Similarly, the goal of regulation is to influence the perceptions and behaviors of regulated parties. Therefore, each regulatory program must be based on a strategy that seeks to understand the motivations of regulated parties and to influence their behavior.

POLICY DESIGN: HOW TO DEVELOP AND IMPLEMENT A REGULATORY STRATEGY

Strategic regulatory planning is an effort by government to develop a comprehensive strategy for influencing behavior. There are two components to this plan. The first is the formal regulation itself; the second is the manner in which the regulatory plan is implemented. Extra-regulatory elements that can be manipulated to encourage compliance include funding, technical assistance, exhortation, and publicity. Since willingness and capacity to comply with regulation can vary widely within a given regulated community, it is critical to have an array of regulatory mechanisms available. It is also important to approach the task of influencing behavior without ideology or preconceptions.

One might argue that it is administratively or legally not feasible to target regulation for maximum influence on specific regulated parties. The administrative argument is easy to counter. First, regulations are now individually tailored through the permit process (Rabe 2000:36–37). Second, it is possible to deal with groups of regulated parties and tailor approaches to classes of regulatory situations rather than to individual organizations. Finally, an approach focused on changing the behavior of regulated parties will tend to he less process-oriented and thereby less administratively complex. It will also utilize strategic alliances between different parties who share a similar interest in the successful implementation of the regulatory program.

The issue of legal feasibility is the argument that the law cannot be adjusted to account for an organization's willingness and capacity to conform to the law's requirements. Regulatory enforcement through the courts, one should note, typically results in bargains that take into account what an organization is capable of and its apparent willingness to move toward compliance. We might as well acknowledge that the application of environmental rules involves these negotiations. The notion that the law is applied without consideration of feasibility is simply untrue. In fact, Cass Sunstein (1990:416) argues that when regulators are compelled to implement rules that do not allow them to consider issues of feasibility, they frequently fail to act. A more typical response than inaction is deal negotiation. Frequently, this involves a compliance schedule and other government concessions.

A strategic approach to regulation would acknowledge the reality of the bargaining process up front and develop compliance strategies with input from the regulated community. Under these circumstances, enforcement and the threat of enforcement are reserved for recalcitrant organizations that willfully violate agreements, engage in deception, or are otherwise unwilling to change their practices.

THE TOOLS OF STRATEGIC REGULATION

"Command-and-control" describes a process in which government commands a regulated party to act in a certain way and then uses the legal system to control behaviors that are not in compliance with the rules. The traditional notion of command-and-control is a simplistic view of regulation. Regulation involves all government policies and programs deployed to influence the behavior of regulated parties. An updated definition of regulation includes command-and-control regulation, plus the use of market mechanisms and a wide variety of other techniques of influence.

There is no need to choose between command-and-control and market mechanisms. Neither is necessarily better than the other. Each target of regulation must be assessed to determine what mix of incentives and disincentives will result in the desired change in behavior (Rosenbaum 2005:167). A variety of techniques of influence are available to government regulators:

1. Market solutions and economic incentives.
2. Insurance programs.
3. Self-regulation.
4. Taxes and fees.
5. Education, information disclosure, and the use of media.
6. Reporting and formal compliance tracking.
7. Licensing.
8. Permitting.
9. Standard setting.
10. Grants, training, and compliance assistance.
11. Assessing penalties.
12. Inspections.
13. Adjudication.

These activities include both coercive and relatively noncoercive actions. In my view, policy design should favor the least-coercive methods that obtain the desired results. The following regulatory actions constitute a partial listing of activities typically available to regulators that influence the behavior of regulated parties.

MARKET SOLUTIONS AND ECONOMIC INCENTIVES
Government sells firms and other private parties permits to pollute, specifying an allowable level of pollution. These permit levels can be traded to other firms, creating a market in pollution allowances. This encourages permit holders to reduce their own level of pollution and maximize the cost-effectiveness of pollution control.

INSURANCE PROGRAMS
Government requires private parties to carry insurance in order to clean up unanticipated releases of pollution and to compensate victims of negative environmental impacts. For example, a gas station owner might be required to carry insurance to pay for the cost of cleaning up gasoline leaks, and to pay third-party liability claims arising from such leaks.

SELF-REGULATION

Government permits an industry to regulate itself. The use of industry codes and professional ethics are examples of such self-regulation.

TAXES AND FEES

Government charges regulated parties for each unit of pollution or waste created. Alternatively, raw materials that eventually cause pollution are taxed, as in the pre-1995 Superfund's tax on oil and chemical feedstocks.

EDUCATION, INFORMATION DISCLOSURE AND THE USE OF THE MEDIA

Government informs the public about regulatory violations or about dangers, causing negative public relations for a company. An example is the warning-label requirement on cigarettes. Government may also use the media to educate about regulatory requirements and their purposes.

REPORTING AND FORMAL COMPLIANCE TRACKING

Government requires regulated parties to report on their compliance with rules. This is less expensive than inspections and can begin the process of creating the institutional capacity in regulated firms to comply with a rule. Whoever fills out the form must at least pay some attention to the regulation.

LICENSING

Government certifies competent professionals who can assist with compliance. The best example of this method is the regulation or licensing of certified public accountants, who facilitate compliance with tax regulations. In the environmental area it might be possible to certify environmental auditors and other professionals who could help a firm reduce and prevent pollution.

PERMITTING

Government requires firms to obtain a permit in order to pollute legally. A permit can call for gradual reductions in pollution. The absence of a permit can result in a judicial order to close a factory.

STANDARD SETTING

This is the traditional command part of command-and-control regulation. There are two basic types of standards. The first type is the performance standard, which requires the accomplishment of specific goals but does not specify how one achieves those goals. A second type of standard specifies a process, technology, or practice that a regulated party must deploy to be in compliance with a rule. This simplifies compliance and oversight of regulatory compliance by requiring a specific, easily measurable activity. However, it also reduces the discretion that a firm has to determine the most cost-effective mode of compliance.

GRANTS, TRAINING, AND COMPLIANCE ASSISTANCE

Many of the targets of regulation are individuals and small businesses that are willing to comply but lack the capability or resources to do so. Sometimes grants, loans, or even loan guarantees can help a small business obtain the capital needed to comply with a regulation. Training and consulting services can also have a large impact, especially in areas where regulation and technologies are new.

ASSESSING PENALTIES
Penalties are typically fines charged against violators. Penalties are particularly complex disincentives that must be used with great care. A penalty that is too low is simply absorbed into the cost of doing business. A penalty that is too high can result in extensive litigation and high transaction costs for the agency. It can also lead to illicit avoidance behavior and/or political opposition to the legitimacy of the regulation and even the regulator. Nevertheless, as the Internal Revenue Service (IRS) has learned by auditing celebrities and Eliot Spitzer has demonstrated by prosecuting large corporations, a well-targeted penalty with sufficient publicity can result in widespread compliance with an agency's rules.

INSPECTIONS
Visits by regulators to regulated parties to determine compliance are an important part of the traditional command-and-control model. Inspections provide evidence that regulated parties are following the rules. A more important use of inspections, especially if they are random and unannounced, is to stimulate compliant behavior due to fear of an impending inspection. How many people keep careful tax records out of fear that one day an IRS tax auditor will inspect them?

ADJUDICATION
Formal adjudication is an administrative or judicial trial to determine if a regulated party has violated a rule. The threat of adjudication can often promote compliant behavior.

A STRATEGIC APPROACH TO REGULATION

The choice between command and control and market-based regulation is a false one. All regulation involves gradual, strategic calculation and bargaining. Command-and-control results in: regulations that adjust the law to reality, permits that interpret regulations in the light of real-world constraints, and judicial and administrative bargains on how permits should actually be implemented. Donald Elliott, former EPA general counsel, notes:

> It is important to recognize that we don't have to have and we don't have an all or nothing system in which we have either an incentive-based system or a health-based system of command and control regulations. Many of our environmental problems, like many of our other legal problems, involve a complex coming together of different goals and different moral norms. The system cannot simply optimize any single value like controlling the total amount of pollution at the least cost but must be responsive to multiple values. Multiple goals for hybrid systems. ... Thus a combination of health-based standards and market-based incentives may be preferable to either standing alone.

(Breger et al. 1991:479)

A broader framework is needed that provides policymakers with a menu of devices depending on what or who is being regulated. Some substances are so toxic that command-and-control is necessary. Some regulated parties are so weak that if they are not paid to comply, they will be driven out of business. In other cases a market can be created and environmental improvement can be accomplished through this mechanism (e.g., recycling, air emissions). The economic causes of environmental problems and the economic impact of proposed solutions vary according to the role the polluting business plays in the nation's economy. The approach to policy design should be as varied as the economic forces the policy is seeking to influence. In some cases, market mechanisms can encourage compliance and avoid the legal and administrative costs of direct regulation. When regulations are necessary, government should provide subsidies, training, and consulting services for organizations that do not have the capacity to comply. On occasion, government may decide that the costs of subsidizing regulation are so high, and the benefits of regulation so important, that a business should be allowed to die in order to protect the environment. These instances should be as infrequent as possible, or the political support for protecting the environment will ultimately erode.

Policy analysts often lament the fact that environmental goals are sold to the public with fear and inadequate risk assessment, and to politicians for their value as "pork." They argue that the goals of legislation and regulation ought to be based on careful scientific consideration of risks (Landy et al. 1990:279–283). Similarly, economists frequently argue that policy designs should reflect a careful assessment of costs and benefits and seek to achieve the maximum possible bang for the minimum possible buck. These ideas seem rational and attractive, but unfortunately they are not always feasible in the messy, pluralistic, federal, divided-power political system in which we operate. Sometimes cost-benefit analysis is difficult to conduct. One problem is that the distribution of costs and benefits can be unpredictable, and distribution effects can be more politically salient than the overall economic effect. Another problem is that some costs and benefits cannot be compared without questionable assumptions about the relative weights assigned to specific cost-and-benefit factors (Layzer 2000:8).

There are no shortcuts. Each regulatory program must be based on a strategy that seeks to understand the motivations of the regulated parties. Whether we decide to employ direct regulation, indirect market mechanisms, or direct subsidies, none of these approaches will work without a profound understanding of the firms being regulated. Developing the administrative capacity in government to make these assessments is far more important than making decisions on which regulatory mechanism is superior. With this knowledge in hand, environmental regulators can develop flexible and dynamic strategies to reduce and prevent pollution in the real world.

POLICY DESIGN THAT REGULATES INDIVIDUAL BEHAVIOR AND STIMULATES SOCIAL LEARNING

To some degree, regulating corporate behavior has the effect of regulating individual behavior, and if the corporation is large enough the impact can be massive. For example, by regulating the pollution produced by all cars manufactured by a single large automaker, government has the ability to change the individual behaviors of all those who drive cars made by this company. The compliance of a small number of parties representing a large number of individuals eases the administrative costs and challenges of convincing millions of people to behave in new ways.

Unfortunately, not all environmental problems can be addressed through the regulation of corporations. Some environmental policies involve reaching individuals, educating them, influencing their values, and changing their behavior. For example, solid-waste reduction and recycling both require the change of individual behavior. To recycle, people need to sort their garbage within the household. Even if the technology of garbage sorting advances, public understanding of the importance of recycling is needed to ensure that government continues to sort waste for reuse.

Most important is the requirement that individuals learn to value the natural environment. Although it is true that the value is expressed as a part of the consumer culture of the West, there is no requirement that people continue to consume the "economic good" of environmental quality. Living without nature may sound like science fiction, but the fact that people go camping, visit the beach, and enjoy nature is an expression of learned values. It is not a form of innate behavior. If we stop valuing environmental quality and passing that value on to our children, the environment will not generate support in the political or economic marketplace. People might decide to experience nature as a virtual rather than a physical reality. Though a world based totally on technology might have some appeal to confirmed urbanites, at this point we do not have the technology or the energy required to totally supplant natural systems with human-made systems. Therefore, our survival depends on the use of natural systems to generate our sustenance. The value of protecting the environment must be learned at the individual level if we are to survive.

The levers to inspire this social erudition include price mechanisms and other economic incentives that guide people to learn about and value the environment (Kolstad 2000). Other tools for social learning include the curricula in our educational system as well as the mass media. All have been used and will need to be continued to reinforce the message of the importance of environmental protection.

APPLYING THE POLICY DESIGN DIMENSION OF THE FRAMEWORK

A key dimension of a public policy issue is the approach taken by the polity to address the issue. The type of policy design that is considered feasible tells you something about the seriousness of the issue, its salience and importance. To understand the policy design dimension of the issue, once again there are a number of questions that can be posed. The aim is to understand the rationale for the approach taken, and how the approach has evolved. As noted above, in the early days of air pollution policy, government convened conferences of scientists and policymakers to discuss the nature of the issue. That very soft approach was suddenly and dramatically modified in the federal 1970 Clean Air Act with the development

38

of the first national ambient air quality standards. This leads to the first policy design issue that needs to be understood: What is the degree of compulsion and coercion included in the policy design?

Other key questions include:

- What is the mix of incentives and disincentives used to influence behavior to reduce damaging the environment?
- What are the economic costs and benefits of the policy design?
- Does the policy design reflect strategic thinking, or is it based on political considerations, stakeholder compromises, or a lucky guess (what Jones referred to as "speculative augmentation")? (Jones 1974:438)
- Does the regulated community understand what they are being asked to do, and are they supportive of the approach taken?
- Are the regulated parties willing to comply with the policy as designed, or will they resist by pursuing noncompliant strategies such as legal challenges and pro-forma compliance?
- Are there other stakeholders who are not regulated parties and do they support the policy design that has been promulgated?
- What resources are available to ensure compliance with the policy design and are these resources likely to be sufficient?
- What is the role of government in general and of specific governmental levels and units in implementing the policy design?
- What type of progress away from the problem or toward a solution is the policy design likely to generate? Why?

The design of the policy helps analysts and practitioners understand the operational definition of the environmental problem. The proposed or adopted solution tells you what part of the problem is considered important enough to be addressed by policymakers. The operational definition of the *problem* is the one that the *policy design* seeks to address, just as the operational definition of the *policy* is the *program* that the management system actually puts into place. Let us turn to that final dimension of the framework, and discuss the issue of management.

ENVIRONMENTAL AND SUSTAINABILITY MANAGEMENT

Once the political dust settles and a policy design is adopted, the environment becomes a management issue. For policy to become meaningful in the real world it must be translated from words to deeds. Policy and management are related. Cumbersome, complex policy designs are typically more difficult to implement than simple designs. Jeffrey Pressman and Aaron Wildavsky demonstrated that point in their classic work *Implementation*. Policy designs that exhort or mandate private action are less certain to be carried out than policy designs that provide concrete incentives or punishments for private actions.

However, even the simplest policy designs can be wrecked through bad management or political interference. For example, take the case of the Federal Water Pollution Control

Act of 1972. Section 208 called for "areawide regional water quality management plans." Section 201 provided a grant program to help local governments build sewage treatment plants. The idea was that "208 plans" would be used to determine the best place to locate sewage treatment plants, and plant siting would be based on environmental rather than political criteria. Unfortunately, President Nixon impounded the funds authorized and allocated by Congress to pay for Section 208 planning grants. Due to the lobbying efforts of local governments and construction companies, he did not succeed in impounding the funds for building sewage treatment plants. Billions of dollars were spent in the 1970s on plants that were sited without assessing their role in regional water quality management. As a result, some plants were located in inappropriate places. Some communities overbuilt capacity and needed to attract development to help pay the cost of plant operation and maintenance. Other areas found their growth and development stalled by inadequate waste treatment capacity.

Despite the lack of planning and the mistakes made, the sewage treatment program was a great success. At the program's peak in 1976, the federal government spent $9 billion annually on grants to local governments to build sewage treatment plants. This amount gradually declined to about $2 billion per year in the 1980s. At that time, the grant program was replaced by the Clean Water State Revolving Fund Program, which provided low-interest loans to cities and other governments for environmental infrastructure needs, including sewage treatment and systems to control nonpoint sources of pollution. The result of the sewage treatment program and regulation of industrial discharges of pollution is obvious and measurable. In 1974, EPA data indicated that only 40 percent of the rivers in the United States were safe for swimming and fishing. By 1998, 60 percent of the rivers were safe for fishing for recreational purposes (U.S. EPA 1998).

How did the U.S. government organize itself to accomplish these results? First, a decentralized federal structure was put in place. The EPA helped stimulate the creation of state-level environmental organizations, and they in turn encouraged the development of local units. Early in the process of regulating industrial polluters, the EPA delegated enforcement implementation to the states. Policy was developed in Washington, but state and local governments carried out the actual monitoring and enforcement. This helped ensure that national rules were interpreted to accommodate local political realities. Although this approach may have slowed down initial efforts at pollution control, over the long run states, with occasional prodding from the EPA's regional and headquarters offices, achieved results.

In the case of municipal sewage treatment, the federal government designed specifications for the plants, recommended contractors, and reviewed their competence, as well as provided significant funding. The strategy of eliciting state and local buy-in through grants worked. Though in the long run the federal share of costs for these sewage facilities was relatively small, initially few governments could resist the "free" federal funding. Ultimately, the operation and maintenance of these facilities were more costly than the initial capital outlay. The growth of the average American homeowner's water bill in the 1980s and 1990s was a direct result of the need to pay these ongoing costs. It is unlikely, or at least less likely, that sufficient political support for sewage treatment would have been generated if people had fully known and understood the total cost of constructing and operating sewage treatment plants (Freeman 1990:97).

From a management perspective, the water pollution control program's strategy had some useful features. First, it focused on the biggest source of pollution—municipal sewage and the pollution by large industrial facilities. Only a few actors needed to change their behavior to implement the program. The decentralized structure and use of private contractors ensured that centralized bureaucratic clearance was not required during sewage treatment plant construction. The public works approach had the advantage of visible, (excuse the pun) concrete accomplishments.

For policy words to become policy deeds, goals must be clear and well understood. The tasks to be carried out must be simple. Joint actions between organizations and even between individuals should be minimized. The technologies required to implement policy should be well developed and available on an off-the-shelf basis. If new technology is needed, not only must it be invented and then debugged for the real world, but it must also be explained to those who are expected to require it, install it, maintain it, and monitor its functioning.

One difficulty with the management of environmental policy is that environmental programs tend to take a piece of a larger problem and subdivide it in order to work on a solution. For example, we use sewage treatment plants to reduce the amount of raw sewage that we dump into the water; however, the treatment plant creates a sludge that must be dumped in a landfill or in the sea or burned. Solving a problem in one environmental medium can create new problems in other media. Our hope is that by gaining a measure of control over the process of releasing the pollutant into the environment we can minimize the damage it might cause. The analysis of management effectiveness and efficiency must move beyond the performance of the organization and its narrow task, and be broadened to consider management of the entire ecological system being maintained. Measures of these environmental outcomes need to be developed and used to influence management decisions.

Finally, there is the question of goals and the definition of success. The 1972 Federal Water Pollution Control Act set the goal of ending discharges of pollution into navigable waters by 1977. A nice thought, perhaps, but a ridiculous, unachievable goal. The 1980 Superfund toxic waste cleanup program had a similar problem in goal setting. After a decade of hard work and billions of dollars of expenditure, the press reported that fewer than a dozen toxic waste dumps had been "cleaned up." What the media did not report was that more than 3,000 threat-removal and emergency-response actions had taken place and millions of people had been moved out of the paths of potential exposure to toxic chemicals (U.S. EPA 1992). Unfortunately, the EPA sold the Superfund program to Congress with the promise that toxic sites could be cleaned up and made usable. The goal of identifying toxic sites and removing people from harm's way was never articulated. At the time, Superfund had no experience in cleaning up a waste site, and learning how to do this was an important accomplishment of its first decade. The Superfund program discovered that the full restoration of a toxic dump site was very expensive, and that often it is more cost-effective to contain the contaminants on-site than to dig them up or flush them out.

The Superfund program accomplished a great deal, though advocates and the media saw it as a failure. Its only actual failure was the lack of a realistic goal. Unfortunately, the political support needed to obtain the resources required to build a program is sometimes won by exaggerating the possibility of success. When that happens, it is important for the program's operating managers to redefine success and try to get buy-in from key players on the new, more realistic set of goals. Although one might argue for stretch targets and the importance of shooting high, in policy areas of great uncertainty, we need to be a bit more modest when defining success. We need to give ourselves time to learn more about the problem, step back from early accomplishments, examine ends and means, and reassess the choices we have made. Politics makes that task difficult, but the needs of program management make it essential.

APPLYING THE MANAGEMENT DIMENSION OF THE FRAMEWORK

An elegant policy design is purely symbolic without organizational capacity to perform the behaviors needed to implement the design. In the final analysis, an environmental problem can be addressed only if managed organizational capacity is put into place to either control or prevent the problem. If we find that policymakers have ignored the issue of organizational capacity, in effect they have ignored the problem altogether. This tells the analyst or practitioner that the policy process had not yet gotten serious about this particular environmental problem.

To apply the management dimension of the framework there are a number of questions that can be posed. While the specific questions that can be addressed will vary by issue, the following provide an indication of issues that should be addressed when applying the management dimension of the framework:

- Does the organizational capacity exist to effectively use technology that *measures* the environmental problem? If so, how much is in place and does capacity exist in the same location as the problem?
- Does the organizational capacity exist to directly utilize or encourage the use of technology or other strategic plan elements needed to *prevent* or *control* the environmental problem? If so, how much is in place and does capacity exist in the same location as the problem?
- How much experience do we have in addressing this issue and/or issues that share its characteristics? Are standard operating procedures in place, and are they well tested and well understood? Do we know how to manage this kind of procedure or is it something we still need to learn how to do?
- What resources are available to develop and maintain needed organizational capacity, and are these resources adequate?
- What is the quality of the leadership in the organizations implementing this program?

The management dimension of the environmental issue is the one that tells us if the issue is considered to be important enough to actually address. While it cannot be examined without an understanding of the other dimensions of the issue, if the management dimension is ignored, it is possible to confuse a symbolic response for a real one.

NEXT STEPS

The preceding discussion illustrates the complexity of the environmental problem, and, I hope, the necessity of viewing it from a variety of perspectives. The next several chapters of this book seek to apply this preliminary or rough framework to a set of environmental policy issues. In Chapter 3, we will examine New York City's attempt to pass congestion pricing. Chapter 4 will examine the national issue of e-waste. Chapter 5 will apply the framework to the issue of hydrofracking, and Chapter 6 will focus on the issue of global climate change. Chapter 7 provides a comparison of these issues and a discussion of the applicability of this framework, and Chapter 8 concludes the book with suggestions for improving environmental policy. One of the more important changes in environmental policy over the past decade has been to integrate environmental protection into the management decisions involved in ensuring sustainable economic development. This sustainability perspective has created a new definition of the environmental policy issue and moved it from the periphery to the center of the public policy agenda.

NOTES

1 Aldo Leopold, "The Land Ethic," in *A Sand County Almanac with Essays on Conservation from Round River,* pp. 237–64 (New York: Oxford University Press, 1949).
2 For an insightful critique of a standard understanding of the role of ethical theory and principles in environmental ethics, see Ken Sayre, "An Alternative View of Environmental Ethics," *Environmental Ethics* 13 (Fall 1991): 195–213. Sayre argues against what he calls the inferential view of ethical theory, a view similar to what we will be calling applied ethics. Like Sayre, this textbook suggests that the inferential/applied view of ethical theory is at best incomplete when addressing environmental issues.

Michael Howlett, M. Ramesh and Anthony Perl — "Agenda-Setting," "Policy Formulation and Policy Design, "Public Policy Decision-Making and Policy Implementation" and "Policy Evaluation: Policy Making as Learning." In *Studying Public Policy: Policy Cycles & Policy Subsystems.* **Ontario, Canada: Oxford University Press.**

Reading Commentary

Howlett, Ramesh and Perl describe the five stages of the public policy-making process: agenda setting, policy formulation, decision-making, implementation and evaluation. Any kind of environmental problem-solving must take place within this context.

Their model of policy-making challenges the idea that public policy-making is a haphazard process by which political choices are made in response to external shocks. By describing each stage in detail, Howlett, Ramesh and Perl hope to draw attention to the fact that people have much more "agency," or control, than they think they do over which problems are added to the public agenda and how decisions are made about what to do. In their description of agenda setting, they rely on John Kingdon's (1984) work on *policy entrepreneurs* and *policy windows*. In sum, policy entrepreneurs (individuals inside and outside the government) can find ways to put a problem on the public agenda by creating or taking advantage of a policy window—an opportunity that arises when "streams" of problems, policies, and politics interact in a certain way. This, again, challenges the notion that policies are nothing more than a response to pressure from powerful groups. Timing, and the astuteness of policy entrepreneurs, may be just as important.

Why is this significant? As you continue to read, we hope you will reflect back on the theoretic ideas introduced in this excerpt. Think about when policies emerge, how they take shape and why implementation is difficult. Any environmental problem-solving effort can be better understood within this framework. The "best" version of a solution may not emerge until a number of conditions align.

CHAPTER 4
AGENDA-SETTING
Michael Howlett, M. Ramesh and Anthony Perl

Why do some issues get addressed by governments while others are ignored? Although sometimes taken for granted, the means and mechanisms by which issues and concerns are recognized as requiring government action are by no means simple. Some demands for government resolution of public problems come from the international and domestic actors discussed in Chapter 3, whereas others are initiated by governments themselves. These issues originate in a variety of ways and must undergo intense scrutiny before they are seriously considered for resolution by government.

Agenda-setting, the first and perhaps the most critical stage of the policy cycle, is concerned with the way problems emerge, or not, as candidates for government's attention. What happens at this early stage of the policy process has a decisive impact on the entire subsequent policy cycle and its outcomes. The manner and form in which problems are recognized, if they are recognized at all, are important determinants of whether, and how, they will ultimately be addressed by policy-makers. As Cobb and Elder (1972: 12) put it:

> Pre-political, or at least pre-decisional processes often play the most critical role in determining what issues and alternatives are to be considered by the polity and the probable choices that will be made. What happens in the decision-making councils of the formal institutions of government may do little more than recognize, document and legalize, if not legitimize, the momentary results of a continuing struggle of forces in the larger social matrix.

John Kingdon, in his path-breaking 1984 inquiry into agenda-setting practices in the United States, provided the following concise definition of this crucial first stage of the policy cycle:

> The agenda, as I conceive of it, is the list of subjects or problems to which governmental officials, and people outside of government closely associated with those officials, are paying some serious attention at any given time.... Out of the set of all conceivable subjects or problems to which officials could be paying attention, they do in fact seriously attend to some rather than others. So the agenda-setting process narrows this set of conceivable subjects to the set that actually becomes the focus of attention. (Kingdon, 1984: 3–4)

At its most basic, agenda-setting is about the recognition of some subject as a problem requiring further government attention (Baumgartner and Jones, 2005). This does not in any way guarantee that the problem will ultimately be addressed, or resolved, by further government activity, but merely that it has been singled out for the government's consideration from among the mass of problems existing in a society at any given time. That is, it has been raised from its status as a subject of concern to that of a private or social problem and finally to that of a *public issue*, potentially amenable to government action. While threats and challenges are more frequently the forces that motivate issue definition in policy agenda-setting, there are also times when policy agendas are set by the attraction of an opportunity—such as the US space program's race to land a man on the moon during the 1960s.

How a subject of concern comes to be interpreted as a public issue susceptible to further government action raises deeper questions about the nature of human knowledge and the social construction of that knowledge (Berger and Luckmann, 1966; Holzner and Marx, 1979), and the policy sciences literature has gone through significant changes in its understanding of what constitutes such problems. Early works in the policy sciences often assumed that problems had an 'objective' existence and were, in a sense, waiting to be 'recognized' by governments who would do so as their understanding and capacity for action progressed. Later works in the post-positivist tradition, however, acknowledged that problem recognition is very much a socially constructed process since it involves the creation of accepted definitions of normalcy and what constitutes an undesirable deviation from that status (McRobbie and Thornton, 1995). Hence, in this view, problem recognition is not a simple mechanical process of recognizing challenges and opportunities, but a sociological one in which the 'frames' or sets of ideas within which governments and non-governmental actors operate and think are of critical significance (Goffman, 1974; Haider-Markel and Joslyn, 2001; Schon and Rein, 1994). Each of these perspectives will be discussed in turn below.

THE OBJECTIVE CONSTRUCTION OF POLICY PROBLEMS: THE ROLE OF SOCIAL CONDITIONS AND STRUCTURES

Most early works on the subject of agenda-setting began with the assumption that socio-economic conditions led to the emergence of particular sets of problems to which governments eventually responded. These include both models that posited that the issues facing all modern governments are converging towards a common set, and those postulating that the interplay of economic and political cycles will affect the nature of issues that make it onto the agenda.

The idea that public policy problems and issues originate in the level of 'development' of a society, and that particular sets of problems are common to states at similar levels of development, was first broached by early observers of comparative public policy-making. By the mid-1960s, Thomas Dye and other observers of differences and similarities in the policies across 50 American states had concluded that cultural, political, and other factors were less significant for explaining the mix of public policies than were factors related to the level of economic development of the society. In his study of US state-level policy development, for example, Ira Sharkansky concluded that 'high levels of economic development—measured by such variables as urban per capita income, median educational level and industrial employment—are generally associated with high levels of expenditure and service outputs in the fields of education, welfare and health.' This conclusion led him to argue that 'political characteristics long thought to affect policy—voter participation, the strength of each major party, the degree of interparty competition, and the equity of legislative apportionment—have little influence which is independent of economic development' (Sharkansky, 1971: 277).

This observation about the nature of public policy formation in the American states was soon expanded to the field of comparative, cross-national studies dealing with the different mixes of policies across countries. Authors such as Harold Wilensky (1975), Philip Cutright (1965), Henry Aaron (1967), and Frederick Pryor (1968) all developed the idea that the structure of a nation's economy determined the types of public policies adopted by the government.

Taken to the extreme, this line of analysis led scholars to develop the *convergence thesis*. The convergence thesis suggests that as countries industrialize, they tend to converge towards the same policy mix (Bennett, 1991; Kerr, 1983; Seeliger, 1996). The emergence of similar welfare states in industrialized countries, its proponents argue, is a direct result of their similar levels of economic wealth and technological development. Although early scholars indicated only a positive correlation between welfare policies and economic development, this relationship assumed causal status in the works of some later scholars. In this 'strong' view, high levels of economic development and wealth created similar problems and opportunities, which were dealt with in broadly the same manner in different countries, regardless of the differences in their social or political structures. Wilensky (1975: 658–9), for example, noted that 'social security effort'—defined as the percentage of a nation's gross national product (GNP) devoted to social security expenditures—was positively related to the level of GNP per capita, a correlation leading him to argue that economic criteria were more significant than political ones in understanding why those public policies had emerged.

In this view, agenda-setting is a virtually automatic process occurring as a result of the stresses and strains placed on governments by industrialization and economic modernization. It matters little, for example, whether issues are actually generated by social actors and placed on government agendas, or whether states and state officials take the lead in policy development. Instead, what is significant is the fact that similar policies emerged in different countries irrespective of the differences in their social and political structures.

The convergence thesis was quickly disputed by critics who argued that it oversimplified the policy development process and inaccurately portrayed the nature of the actual welfare policies found in different jurisdictions, where policies were characterized by significant divergences as well as convergences (Heidenheimer et al., 1975). It was noted, for example, that in comparative studies of policy development in the American states, economic measures explained over one-half of the interstate variations in policies in only 4 per cent of the policy sectors examined. Second, it was intimated that the desire to make a strong economic argument had led investigators to overlook the manner in which economic factors varied in significance over time and by issue area (Sharkansky, 1971; Heichel et al., 2006).

By the mid-1980s, a second, less deterministic explanation of agenda-setting behaviour had emerged, which treated political and economic factors as an integral whole. This *resource-dependency model* argued that industrialization creates a need for programs such as social security (because of the aging of the population and processes of urbanization that usually accompany economic modernization) as well as generating the economic resources (because of increased productivity) to allow states to address this need (through programs such as public pension plans, unemployment insurance, and the like). More importantly, in this view, industrialization also creates a working class with a need for social security and the political resources (because of the number of working-class voters and the ability to disrupt production through work stoppages) to exert pressure on the state to meet those needs. The ideology of the government in power and the political threats it faces are also important factors affecting the extent to which the state is willing to meet the demand for social welfare and the types of programs it is willing to utilize to do so.

While some issues, such as the role of international economic forces in domestic policy formation, were still debated (Cameron, 1984; Katzenstein, 1985), this resource-dependency view offered a reasonable alternative to convergence theory-inspired explanations of public policy. However, it remained at a fairly high level of abstraction and was difficult to apply to specific instances of agenda-setting (see Uusitalo, 1984).

One way that scholars sought to overcome this problem was by reintegrating political and economic variables in a new 'political economy of public policy' (Hancock, 1983). Here it was argued that both political and economic factors are important determinants of agenda-setting and should therefore be studied together, especially insofar as political-economic events can affect the *timing and content* of specific policy initiatives.

One of the most important versions of this line of argument posited the idea of a *political business cycle*. The economy, it was suggested, has its own internal dynamics, which on occasion are altered by political 'interference'. In many countries the timing of this interference could be predicted by looking at key political events such as elections and budgets, which tend to occur with some degree of regularity in democratic states.

The notion of a political business cycle grew out of the literature on business cycles, which found that the economy grew in fits and starts according to periodic flurries of investment and consumption behaviour (see Schneider and Frey, 1988; Frey, 1978; Locksley, 1980). When applied to public policy-making, it was argued that in the modern era governments often intervened in markets to smooth out fluctuations in the business cycle. In democratic states, it followed that the nature of these interventions could be predicted on the basis of the political ideology of the governing party—either pro-state or pro-market—while the actual timing of interventions would depend on the proximity to elections. Policies that caused difficulties for the voting public and hence could affect the electoral prospects of political parties in government, it was argued, were more likely to be developed when an election did not loom on the immediate horizon. As Edward Tufte (1978: 71) put it:

> Although the synchronization of economic fluctuations with the electoral cycle often preoccupies political leaders, the real force of political influence on macroeconomic performance comes in the determination of economic priorities. Here the ideology and platform of the political party in power dominate. Just as the electoral calendar helps set the timing of policy, so the ideology of political leaders shapes the substance of economic policy.

While few disagreed that partisan ideology could have an impact on the nature of the types and extent of government efforts to influence the economy, this approach was criticized for its limited application to democratic countries, and only to a subset of these such as the United States, where electoral cycles were fixed. In many other countries, elections either do not exist, are not competitive, or their timing is indeterminate and depends on events in parliaments or other branches of government, making it difficult if not impossible for governments to make such precise policy timing calculations (Foot, 1979; Johnston, 1986). It was also argued that the concept of the business cycle itself was fundamentally flawed and that the model simply pointed out the interdependence of politics and economics already acknowledged by most analysts (see McCallum, 1978; Nordhaus, 1975; Schneider and Frey, 1988; Boddy and Crotty, 1975).

THE SUBJECTIVE CONSTRUCTION OF POLICY PROBLEMS: THE ROLE OF POLICY ACTORS AND PARADIGMS

In the alternative post-positivist view, the 'problems' that are the subject of agenda-setting are considered to be constructed purely in the realm of public and private ideas, detached from economic conditions or other macro-social processes such as industrialization and unionization (Berger and Luckmann, 1966; Hilgartner and Bosk, 1981; Holzner and Marx, 1979; Rochefort and Cobb, 1993; Spector and Kitsuse, 1987).

It had long been noted in policy studies that the ideas policy actors hold have a significant effect on the decisions they make. Although efforts have been made by economists, psychologists, and others to reduce these sets of ideas to rational calculations of self-interest, it is apparent that, even in this limiting case, traditions, beliefs, and attitudes about the world and society affect how individuals interpret their interests (Flathman, 1966).

As we have discussed in Chapters 2 and 3, sets of ideas or ideologies can have a significant impact on public policy-making, for it is through these ideational prisms that individuals conceive of social or other problems that inspire their demands for government action and through which they design proposed solutions to these problems (Chadwick, 2000; George, 1969). However, different types of ideas will have different effects on policy-making, and especially on agenda-setting. As Goldstein and Keohane (1993b) have noted, at least three types of ideas are relevant to policy: world views, principled beliefs, and causal ideas (see Braun, 1999; Campbell, 1998). These ideas can influence policy-making by serving as 'road maps' for action, defining problems, affecting the strategic interactions between policy actors, and constraining the range of policy options that are proposed.

World views or *ideologies* have long been recognized as helping people make sense of complex realities by identifying general policy problems and the motivations of actors involved in politics and policy. These sets of ideas, however, tend to be very diffuse and do not easily translate into specific views on particular policy problems. While scholars recognized that the general *policy mood* or *policy sentiment* found in a jurisdiction can be an important component of a general macro-policy environment, for example by influencing voting for representatives and yielding a certain political orientation in government (Durr, 1993; Stimson, 1991; Stimson et al., 1995; Lewis-Beck, 1988; Suzuki, 1992; Adams, 1997), the links of these kind of beliefs to agenda-setting remain quite indirect (Stevenson, 2001; Elliott and Ewoh, 2000).

Principled beliefs and causal stories, on the other hand, can exercise a much more direct influence on the recognition of policy problems and on subsequent policy content (see George, 1969). In the policy realm, this notion of ideas creating claims or demands on governments was taken up by Frank Fischer and John Forester (1993) and Paul Sabatier (1987, 1988), among others writing in the 1980s and 1990s. The concept of causal stories, in particular, was applied to agenda-setting by Deborah Stone (1988, 1989). In Stone's view, agenda-setting usually involved constructing a 'story' of what caused the policy problem in question. As she has argued:

> Causal theories, if they are successful, do more than convincingly demonstrate the possibility of human control over bad conditions. First, they can either challenge or protect an existing social order. Second, by identifying causal agents, they can assign responsibility to particular political actors so that someone will have to stop an activity, do it differently, compensate its victims, or possibly face punishment. Third, they can legitimate and empower particular actors as 'fixers' of the problem. And fourth, they can create new political alliances among people who are shown to stand in the same victim relationship to the causal agent. (Stone, 1989: 295)

As Murray Edelman (1988: 12–13) argued, in this view policy issues arise largely on their own within social discourses, often as functions of pre-existing ideological constructs applied to specific day-to-day circumstances:

> Problems come into discourse and therefore into existence as reinforcements of ideologies, not simply because they are there or because they are important for well-being. They signify who are virtuous and useful and who are dangerous and inadequate, which actions will be rewarded and which penalized. They constitute people as subjects with particular kinds of aspirations, self-concepts, and fears, and they create beliefs about the relative importance of events and objects. They are critical in determining who exercise authority and who accept it.

As originally presented by the French social philosopher Michel Foucault (1972), the concept of a political discourse was a tool for understanding the historical evolution of society. Foucault believed that historical analysis should contribute to social theory by explaining the origin, evolution, and influence of discourses over time, and by situating current discourses into this framework. In a policy context this means that policy issues can be seen as arising from pre-existing social and political discourses that establish both what a problem or policy opportunity is and who is capable of articulating it.

In this view, the idea that agenda-setting proceeds in a mechanistic fashion or through rational analysis of objective conditions by social and political actors is considered to be deceptive, if not completely misleading. Rather, policy-makers are a part of the same discourses as the public and, in Edelman's metaphor of the 'political spectacle', are involved in manipulating the signs, sets, and scenes of a political drama. According to the script of these discourses, which is written as the play is underway, different groups of policy actors are involved, and different outcomes prescribed, in agenda-setting (Muntigl, 2002; Schmidt and Radaelli, 2005; Johnson, 2007).

The discursive frames from which actors define policy challenges, of course, are not always widely, or as strongly, held by all policy actors, meaning that the agenda-setting process very often features a clash of frames and a struggle among policy actors over the 'naming' of problems, the 'blaming' of conditions and actors for their existence, and the 'claiming' of specific vantage points or perspectives for their resolution (Felstiner et al., 1980–1; Bleich, 2002). The resolution of this conflict and the elevation of a private or social grievance to the status of a public problem, therefore, are often related more to the abilities and resources of competing actors than to the elegance or purity of their ideas, but these ideas are critical in determining its content (Surel, 2000; Snow and Benford, 1992; Steinberg, 1998, Dostal, 2004).

In this view, then, the policy-making agenda is created out of the history, traditions, attitudes, and beliefs encapsulated and codified in the discourses constructed by social and political actors (Jenson, 1991; Stark, 1992). Symbols and statistics, both real and fabricated, are used to back up one's preferred understanding of the causes and solutions of a problem. Symbols are discovered from the past or created anew to make one's case. When using statistics, policy-makers and analysts know how to find what they are looking for.

Thus, in the post-positivist view, understanding agenda-setting requires understanding how individuals and/or groups make demands for a policy that is responded to by government, and vice versa. In short, policy researchers need to identify the conditions under which these demands emerge and are articulated in prevailing policy discourses (Spector and Kitsuse, 1987: 75–6; McBeth et al., 2005). How the material interests of social and state actors are filtered through, and reflect, their institutional and ideological contexts is a required component of the study of agenda-setting (Thompson, 1990).

COMBINING IDEAS, ACTORS, AND STRUCTURES IN MULTI-VARIABLE MODELS OF AGENDA-SETTING

Alone, neither the pure positivist approach, with its emphasis on structures and institutions, nor the pure post-positivist approach, with its emphasis on ideas, has withstood subsequent testing and examination. Their difficulty in identifying a single source of factors driving public policy agenda-setting led to the development of multivariate models, which attempt to systematically combine some of the central variables identified in these early studies into a more comprehensive, and empirically accurate, theory of agenda-setting.

FUNNEL OF CAUSALITY

One such multivariate model was advanced in parallel efforts by Anthony King (1973) in Great Britain, Richard Hofferbert (1974) in the United States, and Richard Simeon (1976a) in Canada. Each author developed a model of policy formation that sought to capture the relationships among social, institutional, ideational, political, and economic conditions in the agenda-setting process. These models considered *all* these variables to be important but situated them within a *funnel of causality*, in which each factor was 'nested' among other variables.

Rather than considering structural, ideational, and actor-related variables as contradictory or zero-sum relationships, the funnel of causality conception suggested that the substance of government's agenda is shaped by socio-economic and physical environment, the distribution of power in society, the prevailing ideas and ideologies, the institutional frameworks of government, and the process of decision-making within governments (King, 1973). These variables were intertwined in a nested pattern of interaction where policy-making occurs within institutions, institutions exist within prevailing sets of ideas, ideas operate within relations of power in society, and relations of power arise from the overall social and material environment.

This synthetic model pointed to the relations between the material and ideational variables that had been identified by previous positivist and post-positivist studies without bogging down in attempts to specify their exact relationship or causal significance. Such causal diversity is the model's greatest strength because it allows different views on agenda influences to be explored empirically so that specific relationships among the variables can be established. This approach is also a weakness, though, because it does not explain the reasons for these factors influencing agendas in different ways. Why the place of one issue on the public agenda might be influenced by ideas and that of another, for example, by environmental factors is not broached, let alone resolved. The funnel-of-causality model also says very little about how multi-dimensional influences on policy agendas, such as the environmental context, ideas, and economic interests, create any particular effect on policy actors in the agenda-setting process (Mazmanian and Sabatier, 1980; Green-Pedersen, 2004).

ISSUE -ATTENTION CYCLES

Another early example of an agenda-setting model built on the premise that this stage of the policy process involved the interaction of institutions, actors, and ideas was put forward by Anthony Downs in 1972. Downs proposed that agenda-setting often followed what he termed an *'issue-attention cycle'*, much like the media 'news cycle'. In an article focusing on the emergence of environmental policy in the United States in the early 1960s, Downs argued that public policy-making often focused on issues that momentarily capture public attention and trigger demands for government action. However, he also noted that many of these problems soon fade from view as their complexity or intractability becomes apparent. As he put it:

> public attention rarely remains sharply focused upon any one domestic issue for very long—even if it involves a continuing problem of crucial importance to society. Instead, a systematic issue-attention cycle seems strongly to influence public attitudes and behaviour concerning most key domestic problems. Each of these problems suddenly leaps into prominence, remains there for a short time, and then—though still largely unresolved—gradually fades from the center of public attention. (Downs, 1972: 38)

The idea of the existence of a systematic issue-attention cycle in public policy-making gained a great deal of attention in subsequent years and Downs's work is often cited as an improved model for explaining the linkages between political institutions, actors, and beliefs, especially public opinion, and public policy agenda-setting (see, e.g., Dearing and Rogers, 1996). In a democracy, where politicians ignore public demands at their electoral peril, Downs argued, waxing and waning public attention to policy problems results in a characteristic cyclical pattern of agenda-setting, but without necessarily producing policy action on the part of governments.

Despite its frequent citation in the policy literature over the past three decades, however, the idea of Downsian-type issue-attention cycles was rarely subject to empirical evaluation.

In 1985, Peters and Hogwood made an effort to operationalize their own version of Downs's cycle, attempting to assess the relationship between waves of public interest as measured in Gallup polls and periodic waves of organizational change or institution-building in the US federal government. Although they found evidence of major periods of administrative consolidation and change over the course of recent US history, they noted that only seven of 12 instances of administrative reorganization met the expectations of the Downsian model, when dramatic administrative changes occurred during the same decade as the peak of public interest as measured by Gallup survey questions.

On the basis of these results, Peters and Hogwood argued: 'Our evidence supports Downs' contention that problems which have been through the issue-attention cycle will receive a higher level of attention after rather than before the peak' (ibid., 251). However, they were also careful to note that there appeared to be at least two other patterns or cycles at work in the issue-attention process beyond what Downs first identified. In the first type, cycles were initiated by external or exogenous events such as war or an energy crisis and then were mediated by public attention. In this type of 'crisis' cycle, the problem would not 'fade away' as Downs hypothesized. In the second type of 'political' cycle, issue initiation originated in the political leadership and then caught public attention (ibid., 252; see also Hogwood, 1992).

Peters and Hogwood's work emphasized the key role played by state actors in socially constructed agenda-setting processes (Sharp, 1994b; Yishai, 1993). Officially scheduled political events, such as annual budgets, speeches from the throne, or presidential press conferences, can spark media attention, reversing the purely reactive causal linkages attributed to these actors by Downs (Cook et al., 1983; Howlett, 1997; Erbring and Goldenberg, 1980; Flemming et al., 1999). Evidence from other case studies revealed that interest group success and failure in gaining agenda access tended to be linked to state institutional structures and the availability of access points, or policy venues, from which these groups could gain the attention of government officials and decision-makers (Baumgartner and Jones, 1993; Brockmann, 1998; Pross, 1992; Newig, 2004). These insights and others were soon put together in new models that attempted to more accurately reconcile policy theory with agenda-setting reality.

MODES OF AGENDA-SETTING

A major breakthrough in agenda-setting studies occurred in the early 1970s when Cobb, Ross, and Ross identified several typical patterns or 'modes' of agenda-setting. In so doing, they followed the insight of Cobb and Elder, who distinguished between the *systemic* or informal public agenda and the *institutional* or formal state agenda. The systemic agenda 'consists of all issues that are commonly perceived by members of the political community as meriting public attention and as involving matters within the legitimate jurisdiction of existing governmental authority' (Cobb and Elder, 1972: 85). This is essentially a society's agenda for discussion of *individual and social* problems, such as crime and health care. Each society, of course, has literally thousands of issues that some citizens find to be matters of concern and would have the government do something about.

As we have noted, however, only a small proportion of the problems on the systemic or informal agenda are taken up by the government for serious consideration as *public* problems. Only after a government has accepted that something needs to be done about a problem can the issue be said to have entered the institutional agenda. These are issues to which the government has agreed to give serious attention. In other words, the informal agenda is for *discussion* while the institutional agenda is for *action.*

Cobb, Ross, and Ross identified four major phases of agenda-setting that occur as issues move between the informal and institutional agendas. Issues are first *initiated,* their solutions are *specified,* support for the issue is *expanded,* and, if successful, the issue *enters* the institutional agenda (Cobb et al., 1976: 127). Cobb, Ross, and Ross then proposed three basic patterns or modes of agenda-setting based on the previous comparative work of Cobb and Elder (1972). Each of these modes is associated with the different manner and sequence in which issue initiation, specification, expansion, and entrance occurs. In a further step, they identified or linked each mode with a specific type of political regime.

They identified the *outside initiation model* with liberal pluralist societies. In this model, 'issues arise in nongovernmental groups and are then expanded sufficiently to reach, first, the public [systemic] agenda and, finally, the formal [institutional] agenda.' Here, social groups play the key role by articulating a grievance and demanding its resolution by the government. These groups lobby, contest, and join with others in attempting to get the expanded issue onto the formal agenda. If they have the requisite political resources and skills and can out-manoeuvre their opponents or advocates of other issues and actions, they will succeed in having their issue enter the formal agenda (Cobb et al., 1976). Successful entry on the formal agenda does not necessarily mean a favourable government decision will ultimately result. It simply means that the item has been singled out from among many others for more detailed consideration.

The *mobilization model* is quite different and was attributed by Cobb and his colleagues to 'totalitarian' regimes. This model describes 'decision-makers trying to expand an issue from a formal [institutional] to a public [systemic] agenda' (ibid, 134.) In the mobilization model, issues are simply placed on the formal agenda by the government with no necessary preliminary expansion from a publicly recognized grievance. There may be considerable debate within government over the issue, but the public may well be kept in the dark until an official announcement. Gaining support for the new policy is important, since successful implementation will depend on public acceptance. Towards this end, government leaders hold meetings and engage in public relations campaigns. As the authors put it, 'The mobilization model describes the process of agenda building in situations where political leaders initiate a policy but require the support of the mass public for its implementation ... the crucial problem is to move the issue from the formal agenda to the public agenda' (ibid., 135).

Finally, in the *inside initiation model,* influential groups with special access to decision-makers launch a policy and often do not want public attention. This can be due to technical as well as political reasons and is an agenda-setting pattern one would expect to find in corporatist regimes. In this model, initiation and specification occur simultaneously as a group or government agency enunciates a grievance and specifies some potential solution. Deliberation is restricted to specialized groups or agencies with some knowledge or interest in the subject. Entrance on the agenda is virtually automatic due to the privileged place of those desiring a decision (ibid., 136).

This line of analysis identified several typical agenda-setting modes that combined actor behaviour with regime type, and also identified key sources of policy ideas and discourses associated with each mode. In its original formulation, Cobb, Ross, and Ross suggested that the type of agenda-setting process is ultimately determined by the nature of the political system, with outside initiation being typical of liberal democracies, mobilization characteristic of one-party states, and inside initiation reflective of authoritarian bureaucratic regimes. However, it was soon recognized that these different styles of agenda-setting varied not so much by political regime as by policy sector, as examples of each type of agenda-setting behaviour could be found within each regime type (Princen, 2007). Subsequent investigations sought to specify exactly what processes were followed within political regimes, especially in complex democratic polities like the United States with multiple quasi-autonomous policy subsystems. Their results have led to a more nuanced understanding of how agenda-setting modes are linked to actors, structures, and ideas and, ultimately, as set out below, to the actual content of the problems and issues likely to emerge in specific instances of agenda-setting.

LINKING AGENDA-SETTING MODES TO CONTENT: POLICY WINDOWS AND POLICY MONOPOLIES

In the 1980s, John Kingdon (1984) developed an analytical framework for agenda-setting that drew upon his investigations of policy initiation in the US Congress. His model examined state and non-state influences on agenda-setting by exploring the role played by *policy entrepreneurs* both inside and outside of government in constructing and utilizing agenda-setting opportunities—labelled *policy windows*—to bring issues onto government agendas. His model suggested that policy windows open and close based on the dynamic interaction of political institutions, policy actors, and the articulation of ideas in the form of proposed policy solutions. These forces can open, or close, policy windows, thus creating the chance for policy entrepreneurs to construct or leverage these opportunities to shape the policy agenda.

In Kingdon's study of agenda-setting in the United States, three sets of variables—*streams* of problems, policies, and politics—are said to interact. The *problem stream* refers to the perceptions of problems as public issues requiring government action. Problems typically come to the attention of policy-makers either because of sudden events, such as crises, or through feedback from the operation of existing programs (ibid., 20). The *policy stream* consists of experts and analysts examining problems and proposing solutions to them. In this stream, the various possibilities are explored and narrowed down. Finally, the *political stream* 'is composed of such factors as swings of national mood, administrative or legislative turnover, and interest group pressure campaigns' (ibid., 21). In Kingdon's view, these three streams operate on different paths and pursue courses more or less independent of one another until specific points in time, or during *policy windows*, when their paths intersect or are brought together by the activities of entrepreneurs linking problems, solutions, and opportunities.

Thus, in the right circumstances, policy windows can be seized upon by key players in the political process to place issues on the agenda. Policy entrepreneurs play the chief role in this process by linking or 'coupling' policy solutions and policy problems together with political opportunities (ibid., chs 7–8; Roberts and King, 1991; Mintrom 1997; Tepper, 2004).

As Kingdon argues, 'The separate streams of problems, policies, and politics come together at certain critical times. Solutions become joined to problems, and both of them are connected to favourable political forces.' At that point an item enters the official (or institutional) agenda and the public policy process begins.[1]

Kingdon suggests that window openings can result from fortuitous happenings, including seemingly unrelated external 'focusing events', crises, or accidents; scandals; or the presence or absence of policy entrepreneurs both within and outside of governments. At other times, policy windows can be opened by institutionalized events such as periodic elections or budgetary cycles (Birkland, 1997, 1998; Tumber and Waisbord, 2004; Nohrstedt, 2005; Mertha and Lowry, 2006).

Kingdon (1984: 213) characterizes the different types of windows and their dynamics as follows:

> Sometimes, windows open quite predictably. Legislation comes up for renewal on schedule, for instance, creating opportunities to change, expand or abolish certain programs. At other times, windows open quite unpredictably, as when an airliner crashes or a fluky election produces unexpected turnover in key decision-makers. Predictable or unpredictable, open windows are small and scarce. Opportunities come, but they also pass. Windows do not stay open long. If a chance is missed, another must be awaited.

Kingdon differentiates the 'problem' from the 'political' window as follows:

> Basically a window opens because of change in the political stream (e.g. a change of administration, a shift in the partisan or ideological distribution of seats ... or a shift in national mood); or it opens because a new problem captures the attention of governmental officials and those close to them. (Ibid., 176)

He then notes that windows also vary in terms of their predictability. While arguing that random events are occasionally significant, he stresses the manner in which institutionalized windows dominate the US agenda-setting process (Birkland, 2004). As he puts it, 'There remains some degree of unpredictability. Yet it would be a grave mistake to conclude that the processes ... are essentially random. Some degree of pattern is evident' (ibid., 216).

The model established by Kingdon suggests at least four possible window types based on the relationship between the origin of the window—political or problem—and their degree of institutionalization or routinization. Although he does not describe these four window types specifically, the general outline of each type is discernible from examining his work and several of his sources. The four principal window types are:

- *routinized political windows*, in which institutionalized procedural events trigger predictable window openings;
- *discretionary political windows*, in which the behaviour of individual political actors leads to less predictable window openings;
- *spillover problem windows*, in which related issues are drawn into an already open window (May et al., 2007); and
- *random problem windows*, in which random events or crises open unpredictable windows (Cobb and Primo, 2003).

In this model, the level of institutionalization of a window type determines its frequency of appearance and hence its predictability (Boin and Otten, 1996; Howlett, 1997b).

Kingdon's model has been used to assess the nature of US foreign policy agenda-setting (Woods and Peake, 1998); the politics of privatization in Britain, France, and Germany (Zahariadis, 1995; Zahariadis and Allen, 1995); the nature of US domestic anti-drug policy (Sharp, 1994a); the collaborative behaviour of business and environmental groups in certain anti-pollution initiatives in the US and Europe (Lober, 1997; Clark, 2004); and the overall nature of the reform process in Eastern Europe (Keeler, 1993). While a major improvement on earlier models, it has been criticized for presenting a view of the agenda-setting process that is too contingent on unforeseen circumstances, ignoring the fact that in most policy sectors, as Downs had noted, action tends to produce bursts of change that are followed by lengthy periods of inertia (Dodge and Hood, 2002).

One way that policies 'congeal' into lengthy periods of program stability, for example, is that policy windows can be designed to stay closed for extended periods—as occurs, for example, in the multi-year funding authorizations of transportation and military programs which reduce the opportunities available to discuss issues and adjust priorities. Windows can even be locked through fiscal devices such as trust funds and revenue bonding that commit spending and taxing for many years into the future (French and Phillips, 2004).

A key mechanism that provides stability in agenda-setting, through its control over the policy discourse, is the construction of a stable policy subsystem or 'policy monopoly'. Such a subsystem entrenches the basic idea set, many of the actors, and the institutional order in which policy development occurs, 'locking in' a policy discourse or frame in which policy issues are named and claimed. Frank Baumgartner and Bryan Jones (1991, 1993, 1994, 2005), in a landmark series of studies, developed this idea, which modifies Kingdon's work and helps to explain the likely content of the typical patterns or modes of agenda-setting behaviour identified earlier by Cobb and his colleagues.

The key element that differentiates modes of agenda-setting, Baumgartner and Jones argue, revolves around the manner in which specific subsystems gain the ability to control the interpretation of a problem and thus how it is conceived and discussed. For Baumgartner and Jones, the 'image' of a policy problem is significant because:

> When they are portrayed as technical problems rather than as social questions, experts can dominate the decision-making process. When the ethical, social or political implications of such policies assume center stage, a much broader range of participants can suddenly become involved. (Baumgartner and Jones, 1991: 1047)

The primary relationship that affects agenda-setting dynamics in Baumgartner and Jones's view is that between individuals and groups who have power inside existing subsystems and those who are seeking to impact those subsystems by leveraging outside influences. In their model, policy monopolies attempt to construct hegemonic images of policy problems that allow influential actors in a subsystem to practise *agenda denial*—that is, preventing alternate images and ideas to penetrate and thus influence governments (on agenda denial, see Yanow, 1992; Bachrach and Baratz, 1962; Debnam, 1975; Frey, 1971; R.A. Smith, 1979; Cobb and Ross, 1997b). Subsystem members opposed to prevailing conditions and government responses seek to alter policy images through a number of tactics related to altering the venue of policy debate, or other aspects of the prevailing policy discourse, thus undermining the complacency or stability of an existing policy subsystem (Sheingate, 2000).

Baumgartner and Jones posit that actors attempting to alter the official agenda of government will adopt either of two strategies to make subsystems more 'competitive'. In the Downsian strategy, groups can publicize a problem in order to alter its venue through mobilizing public demands for government to resolve it (Baumgartner and Jones, 1993: 88). In a second typical approach, which they term a 'Schattschneider' mobilization (after the early American scholar of pressure group behaviour), groups involved in the policy subsystem dissatisfied with the policies being developed or discussed seek to alter the institutional arrangements within which the subsystem operates in order to expand or contract its membership (ibid., 89; for examples of these strategies in practice, see Maurier and Parkes, 2007; Daugbjerg and Studsgaard, 2005; Hansford, 2004; Pralle, 2003).

The key change that results if either strategy succeeds is the transformation of a policy monopoly into a more competitive subsystem where new actors and new discourses, and thus new issues, can enter into policy debates. A good example of this occurring in recent years can be found in the transformation of smoking from a highly stable issue framed as a personal consumption issue to one that, through venue shifts to courts and other bodies on the part of policy entrepreneurs, was redefined as a social health and welfare issue. This led, ultimately, to the articulation and subsequent implementation of alternative options for tobacco control related to sales, advertising, and workplace bans and other restrictions, which were unthinkable under the previous regime (Studlar, 2002).

The four typical modes of agenda-setting that flow from the analysis of Baumgartner and Jones are shown in Figure 4.1. In this model, the chance for new problems or options to emerge on government agendas depends on whether policy subsystems are monopolistic or competitive and whether new ideas about the nature of a policy problem and its solution can be found in the subsystem. Where a well-established monopoly exists with no new ideas present, agenda denial and a status quo orientation are likely to result: that is, the hegemony of the existing subsystem over the definition and construction of problems and solutions will be maintained. When that same monopoly has some new ideas, these are likely to result in some reframing of issues within the subsystem.

When a more competitive subsystem exists but no new ideas have been developed, contested variations on the status quo are likely to be features of agenda-setting. When the ideas

Figure 4.1 Typical Agenda-Setting Modes

		Subsystem Type	
		Monopolistic	*Competitive*
Ideas	*Old*	**Status Quo** Character: static/hegemonic (agenda denial)	**Contested** Character: contested variations on the status quo
	New	**Redefining** Character: internal discursive reframing	**Innovative** a Character: unpredictable/ chaotic

58

remain old, however, nothing more than proposals for modest changes to the status quo are likely to be raised to the institutional agenda. Only when both situations exist—that is, both a competitive subsystem and the presence of new ideas—are more profound (and potentially paradigmatic) innovative changes in problem definition and identification likely to proceed onto the formal agenda of governments and move forward for consideration in the next stage of the policy cycle: policy formulation.

CONCLUSION: REVISITING AGENDA-SETTING MODES THROUGH A POLICY SUBSYSTEM LENS

This overview of agenda-setting studies has shown how investigations have moved from simple univariate models focused on 'objective' or 'subjective' constructions of policy problems to more sophisticated examinations that link many variables in complex multivariate relationships. It has also shown how contemporary studies have developed a set of agenda-setting patterns or modes that reveal how this stage of the policy process is influenced by key actors in prevailing policy subsystems, the dominant sets of ideas about policy problems they espouse, and the kinds of institutions within which they operate.

The most significant variables influencing modes of agenda-setting turn out to have less to do with automatically responding to changes in the nature of the economy, as Dye, Wilensky, and Sharkansky argued, or with the nature of the political regime involved, as Cobb, Ross, and Ross had claimed. Instead, as Kingdon points out, agenda-setting processes are contingent, but very often are still predictable, involving the complex interrelationships of ideas, actors, and structures. And, as Baumgartner and Jones have suggested, the nature of the actors initiating policy discussions and whether the structures in which they operate allow new ideas to come forward are the most important determinants of the movement of public problems from the informal agenda to the state's institutional agenda (Daugbjerg and Perdersen, 2004).

While the exact timing of the emergence of an issue onto the systemic or formal policy agenda depends, as Kingdon showed, on the existence of a policy window and of the capacity and ability of policy entrepreneurs to take advantage of it, the content of the problems identified in the agenda-setting process depend very much on the nature of the policy subsystem found in the area concerned and the kinds of ideas its members have. Whether subsystem members are capable of creating and retaining an interpretive monopoly on understanding a policy issue, as understood by Baumgartner and Jones, largely determines if the matching of problems with solutions found in the agenda-setting and subsequent policy formulation stages of the policy process will yield consideration of the issue within an existing policy paradigm or in a new ideational framework (Haider-Markel and Joslyn, 2001; Jeon and Haider-Markel, 2001).

NOTES

1 This is a type of threshold model of social behaviour. On these, see Wood and Doan (2003); Granovetter (1978); Schelling (1971).

CHAPTER 5
POLICY FORMULATION: POLICY INSTRUMENTS AND POLICY DESIGN

Policy formulation refers to the process of generating options on what to do about a public problem. In this second stage of the policy process, policy options that might help resolve issues and problems recognized at the agenda-setting stage are identified, refined, and formalized. An initial feasibility assessment of policy options is conducted at this stage of policy development, but these formulation efforts and dynamics are distinct from the next stage, decision-making (discussed in Chapter 6), where some course of action is approved by authoritative decision-makers in government.

WHAT IS POLICY FORMULATION?

Once a government has acknowledged the existence of a public problem and the need to do something about it, that is, once it has entered onto the formal agenda of government, policy-makers are expected to decide on a course of action. Formulating what this course of action will entail is the second major stage in the policy cycle.

As Charles Jones (1984: 7) has observed, the distinguishing characteristic of policy formulation is simply that means are proposed to resolve perceived societal needs. Policy formulation, therefore, involves identifying and assessing possible solutions to policy problems or, to put it another way, exploring the various options or alternative courses of action available for addressing a problem. The proposals may originate in the agenda-setting process itself, as a problem and its possible solution are placed simultaneously on the government agenda (Kingdon, 1984), or options may be developed after an item has moved onto the official agenda. In all cases, the range of available options considered at this stage is always narrowed down to those that policy-makers could accept before these alternatives move on to the formal deliberations of decision-makers. Defining and weighing the merits and risks of various options hence forms the substance of this second stage of the policy cycle, and some degree of formal 'policy analysis' is typically a critical component of policy formulation activity.

Jones (1984: 78) describes other broad characteristics of policy formulation:

- Formulation need not be limited to one set of actors. Thus there may well be two or more formulation groups producing competing (or complementary) proposals.
- Formulation may proceed without clear definition of the problem, or without formulators ever having much contact with the affected groups...
- There is no necessary coincidence between formulation and particular institutions, though it is a frequent activity of bureaucratic agencies.
- Formulation and reformulation may occur over a long period of time without ever building sufficient support for any one proposal.
- There are often several appeal points for those who lose in the formulation process at any one level.

- The process itself never has neutral effects. Somebody wins and somebody loses even in the workings of science.

This picture presents policy formulation as a highly diffuse and disjointed process that varies by case. However, it is possible to say something about the general nature of the formulation process and the activities it involves.

THE PHASES OF POLICY FORMULATION

The formulation stage of policy-making can be subdivided into phases to clarify how various options are considered and to highlight how some options are carried forward while others are set aside. Harold Thomas (2001) identifies four such phases to policy formulation: appraisal, dialogue, formulation, and consolidation.

In the *appraisal* phase, data and evidence are identified and considered. These may take the form of research reports, expert testimony, stakeholder input, or public consultation on the policy problem that has been identified. Here, government both generates and receives input about policy problems and solutions.

The *dialogue* phase seeks to facilitate communication between policy actors with different perspectives on the issue and potential solutions. Sometimes, open meetings are held where presenters can discuss and debate proposed policy options. In other cases, the dialogue is more structured, with experts and societal representatives from business and labour organizations invited to speak for or against potential solutions. Hajer (2005) notes that the structure of engaging input about policy options can make a considerable difference in the effects of that participation, both on the policy process and on the participants themselves. Formal consultations and public hearings tend to privilege expert input and frustrate new participants, while techniques that engage participants from less established organizations and points of view can add energy and enthusiasm to the dialogue over policy options.

At the core of deliberations, the aptly named *formulation* phase sees public officials weighing the evidence on various policy options and drafting some form of proposal that identifies which of these options, if any, will advance to the ratification stage. Such feedback can take the form of draft legislation or regulations, or it could identify the framework for subsequent public and private policy actors to negotiate a more specific plan of action.

Making recommendations about which policy options to pursue will often yield dissent by those who have seen their preferred strategies and instruments set aside during formulation. These objections can be addressed during the *consolidation* phase, when policy actors have an opportunity to provide more or less formal feedback on the recommended option(s). Some actors who advocated alternative options may come around to joining the consensus so that they can stay connected to official policy development efforts. Supporting the policy solutions that are being recommended for further action may provide the opportunity to subsequently influence the ratification and implementation stages from within. Other policy actors will register their continued dissent from specific policy options, hoping to leverage future developments from outside the consensus that has emerged over what is to be done.

Note once again that the limitations that lead policy-makers to reject certain types of options need not be based on facts (Merton, 1948). If significant actors in the policy subsystem believe that something is unworkable or unacceptable, this is sufficient for its exclusion from further consideration in the policy process (Carlsson, 2000). As we have seen with the discussion of agenda-setting in the previous chapter, perception is just as real as reality itself in this second stage of the policy process.

THE GENERAL CONTENT OF POLICY FORMULATION

Like agenda-setting, the nuances of policy formulation in particular instances can be fully understood only through empirical case studies. Nevertheless, most policy formulation processes do share certain characteristics.

Policy formulation involves identifying the technical and political constraints on state action. It involves recognizing limitations, which uncovers what is infeasible and, by implication, what is feasible. This may seem obvious, but it is not reflected in many proposals made about what policy-makers ought to be doing, which often fail to acknowledge the limitations that constrain a proposed course of action. For instance, the public choice theorists' key assumption—that politicians choose policies that best promote their electoral appeal—presumes more room for manoeuvre than is actually available (Majone, 1989: 76). Politicians simply cannot do everything they think might appeal to voters. Other constraints can arise from the state's administrative and financial capacity. For example, governments that have an ownership stake in economic sectors such as energy, finance, and transportation may have more policy options open to them than states where the private sector exclusively delivers these goods and services.

Policy-makers typically face numerous substantive or procedural constraints when considering policy options. *Substantive* constraints are innate to the nature of the problem itself. Thus, policy-makers wishing to eliminate poverty do not have the option of printing money and distributing it to the poor because inflation will offset any gains, and so they must necessarily address the problem in more indirect ways. Similarly, the goal of promoting excellence in arts or sports cannot be accomplished simply by ordering people to be the best artists or sportswomen in the world; the pursuit of these goals requires far more delicate, expensive, and time-consuming measures. The problem of global warming, for instance, cannot be entirely eliminated because there is no known effective solution that can be employed without causing tremendous economic and social dislocations, which leaves policy-makers to tinker with options that barely scratch the surface of the problem. Substantive problems with policy alternatives are thus 'objective' in the sense that redefining them does not make them go away, and their resolution or partial resolution requires the use of state resources and capacities such as money, information, personnel, and/or the exercise of state authority.

Procedural constraints have to do with procedures involved in adopting an option or carrying it out. These constraints may be either institutional or tactical. Institutional constraints, as discussed in Chapter 3, can include constitutional provisions, the nature of the organization of the state and society, and established patterns of ideas and beliefs that can prevent consideration of some options or promote others (Yee, 1996). Efforts to control handguns in the United States, for example, run up against constraints imposed by the constitutional right to bear arms. Federalism imposes similar constraints on German, American, Mexican, and Australian policy-makers in areas of public policy where two levels of government must agree before anything can be done (Montpetit, 2002; Falkner, 2000). How the main social groups are organized internally and are linked with the state also affects what can or cannot be done, especially the nature of political party and electoral systems, which can create 'policy horizons' or limited sets of acceptable choices for specific actors in the policy process (Warwick, 2000; Bradford, 1999). In a similar vein, the predominance of specific sets of philosophical or religious ideas in many societies can lead to difficulties with potential policy solutions that might seem routine in others (DeLeon, 1992). The actual options that will be weighed and measured for potential adoption in policy thus depend a great deal on the people who consider them, their ideas about the world and the issue involved, and the nature of the structures within which they work.

The essence of the search for solutions to a policy problem entails discovering not only which actions are considered to be technically capable of addressing or correcting a problem but also which among these is considered to be politically acceptable and administratively feasible (Majone, 1975, 1989; Huitt, 1968; Meltsner, 1972; Dror, 1969; Webber, 1986). Choosing a solution to a public problem or fulfilling a societal need does not even remotely resemble the orderly process of detached, 'objective' analytical scrutiny of all policy alternatives often proposed by subscribers to rationalist analytical models. As we saw in the preceding chapter, defining and interpreting a problem is often a nebulous process that does not lead to clear or agreed-upon problem definitions, making the identification of solutions equally problematic. Even if policy-makers agree that a problem exists, they may not share an understanding of its causes or ramifications.

Hence, the search for a policy solution will usually be contentious and subject to a wide range of conflicting pressures and alternative perspectives and approaches, frustrating efforts to systematically consider policy options in a rational or maximizing manner. Among other things, certain players in the policy process can be advantaged over others if they are granted some authority in diagnosing a policy ill or in establishing the feasibility of a proposed solution. This is the case, for example, with scientists or government specialists in many policy areas, but such privileged positions can be eroded if these experts are challenged on their neutrality or competence (see Afonso, 2007; Carpenter, 2007; Nathanson, 2000; Heikkila, 1999; Doern and Reed, 2001; Harrison, 2001; Callaghan and Schnell, 2001).

Understanding the ideas and experiences that these actors bring to policy formulation, and the contexts within which they operate, can help explain why some options gain considerable attention while others are ignored. Before delving into the details of the formulation process and the types of policy instrument typically considered at this stage, it is worth highlighting key characteristics of the actors involved in the analysis of alternatives and the kinds of activities they undertake.

THE SUBSTANCE OF POLICY FORMULATION: POLICY INSTRUMENTS

When policy-makers are exploring policy options, they consider not only what to do but also how to do it. Thus, while formulating a policy to tackle traffic congestion, for example, policy-makers must simultaneously consider whether to build more roads, improve public transit, restrict automobile usage, or some combination of these, as well as the tools by which the policy will actually be implemented. These *policy tools*, also known as *policy instruments* and *governing instruments*, are the actual means or devices that governments make use of in implementing policies. Proposals that emanate from the formulation stage, therefore, will specify not only whether or not to act on a policy issue, but also how to best address the problem and implement a solution. For example, in a case such as that of deteriorating water quality, policy options could emphasize public educational campaigns that urge people to refrain from polluting activities, they could embrace regulations that prohibit all activities causing the pollution, they could propose a subsidy to the polluting firms encouraging them to switch to safer technologies, or they could advance some combination of these or other means (Gunningham et al., 1998; Gunningham and Young, 1997).

TAXONOMIES OF POLICY INSTRUMENTS

The variety of instruments available to policy-makers is limited only by their imaginations. However, scholars have made numerous attempts to identify such instruments and classify them into meaningful categories (see Salamon and Lund, 1989: 32–3; Lowi, 1985; Bemelmans et al., 1998).

Most efforts to construct such a typology stem from Lasswell's insight that governments use a variety of policy instruments to achieve a relatively limited number of political ends. Lasswell (1958: 204) argued that governments had developed a limited number of 'strategies' that involved 'the management of value assets in order to influence outcomes'. Understanding these basic strategies and the instrument types that go with them required identifying the resources that governments work with (see also French and Raven, 1959). Cushman (1941) introduced a simple taxonomy of policy instruments based on whether government chose to regulate societal activities or not, and whether those regulations were coercive or not. Dahl and Lindblom (1953) went on to categorize policy tools by their degree of intrusiveness and dependence on state agencies or markets. Theodore Lowi (1966, 1972) blended these insights to create the first encompassing model of how preferences for particular types of policy tools were involved in characterizing epochs or periods of government activity.

Lowi observed in the US case that American governments had tended to favour certain types of instruments for prolonged periods, providing the opportunity to identify major transitions in government activities between these periods. To distinguish the major types and eras of government activity, he proposed a four-cell matrix based on the specificity of the target of coercion and the likelihood of its actual application. The original three policy types he identified included the weakly sanctioned and individually targeted 'distributive' policies; the individually targeted and strongly sanctioned 'regulatory' policy; and the strongly sanctioned and generally targeted 'redistributive' policy. To these three, Lowi later added the weakly sanctioned and generally targeted category of 'constituent' policy.

Although widely read, Lowi's typology was rarely applied because it was not only difficult to operationalize but also somewhat internally inconsistent. Nevertheless, Lowi's central premise of 'policy determining politics' proved alluring and encouraged further efforts to classify and comprehend policy instruments. Anderson's (1971) suggestion that public policy analysis shift from the study of policy problems and inputs to the study of policy implements and outputs was endorsed by scholars such as Bardach (1980) and Salamon (1981), both of whom suggested that policy studies had 'gone wrong' right at the start by defining policy in terms of 'areas' or 'fields' rather than in terms of tools. As Salamon (1981: 256) argued: 'rather than focusing on individual programs, as is now done, or even collections of programs grouped according to major "purpose", as is frequently proposed, the suggestion here is that we should concentrate instead on the generic tools of government action, on the "techniques" of social intervention.' This challenge was taken up in the 1980s and 1990s by the 'policy design' literature (Bobrow and Dryzek, 1987; Dryzek and Ripley, 1988; Linder and Peters, 1984).

Rather than attempt to construct exhaustive lists, which had already produced arcane inventories (such as the scheme for at least 64 general types of instruments in European economic policy produced by Kirschen and his colleagues [1964]), policy design researchers sought ways to group roughly similar types of instruments into a few categories that could then be analyzed to determine the answers to Salamon's questions. Scholars returned to Lasswell's early work on instrument 'strategies' and tried to identify the basic 'governing resources' that different instruments relied on for their effectiveness (Balch, 1980).[1]

A simple and powerful taxonomy known as the 'NATO model' was developed by Christopher Hood (1986a), who proposed that all policy tools used one of four broad categories of governing resources. He argued that governments confront public problems through the use of the information in their possession as a central policy actor ('*nodal*-ity'), their legal powers ('*a*uthority'), their money ('*t*reasure'), or the formal organizations available to them ('*o*rganization') or 'NATO'. Governments can use these resources to manipulate policy actors, for example, by withdrawing or making available information or money, by using their coercive powers to force other actors to undertake activities they desire, or simply by undertaking the activity themselves using their own personnel and expertise. Using Hood's idea of governing resources, a basic taxonomy of instrument categories can be set out.

In the post-Salamon era, studies of instrument choice tended to look at instances of single-instrument selection and, on the basis of such cases, to discern the general reasons why governments would choose one category of instrument over another. These studies, heavily influenced by economists, tended to focus on *substantive* instruments—that is, those tools (such as classical command-and-control regulation, public enterprises, and subsidies) that more or less directly affect the type, quantity, price, or other characteristics of goods and services being produced in society, either by the public or private sector (Salamon, 1989, 2002a; Bemelmans-Videc et al., 1998; Peters and van Nispen, 1998).

Much less attention was paid by analysts of this period to the systematic analysis of their *procedural* counterparts—that is, instruments designed mainly to affect or alter aspects of policy processes rather than social or economic behaviour per se. In these early works, policy instruments had often been defined broadly to include a wider range of tools, or techniques, of governance than in the post-Salamon era. By 2000, however, this neglect had been noted, prompting the emergence of systematic treatments of procedural instruments (Riker, 1983, 1986; Dunsire, 1986, 1993a, 1993b), such that we now have knowledge of both substantive and procedural instruments, their effects, and the reasons they are chosen.

NOTE

1 Bardach (1980), for instance, argued that government ha**d** three 'technologies' at its disposal—enforcement, inducement, and benefaction—and that these required different combinations of four critical governmental resources: money, political support, administrative competency, and

creative leadership. Rondinelli suggested that all policy instruments depended on a limited set of 'methods of influence' that governments had at their disposal: in his case, persuasion, exchange, and authority (Rondinelli, 1983: 125).

CHAPTER 6
PUBLIC POLICY DECISION-MAKING

The decision-making stage of the policy process is where one or more, or none, of the many options that have been debated and examined during the previous two stages of the policy cycle is approved as an official course of action. Policy decisions usually produce some kind of a formal or informal statement of intent on the part of authorized public actors to take, or not to take, some action such as a law or regulation (O'Sullivan and Down, 2001). Acting on this decision is the subject of the next stage of the policy cycle, policy implementation, discussed in the following chapter.

Gary Brewer and Peter DeLeon (1983: 179) characterized the decision-making stage of the public policy process as:

the choice among policy alternatives that have been generated and their likely effects on the problem estimated... It is the most overtly political stage in so far as the many potential solutions to a given problem must somehow be winnowed down and but one or a select few picked and readied for use. Obviously most possible choices will not be realized and deciding not to take particular courses of action is as much a part of selection as finally settling on the best course.

This definition makes several important points about the decision-making stage of the policy cycle. First, decision-making is not a self-contained stage, nor is it synonymous with the entire public policy-making process. Rather, it is a specific stage rooted firmly in the previous stages of the policy cycle. It involves choosing from among a relatively small number of alternative policy options identified in the process of policy formulation in order to resolve a public problem. Second, this definition highlights the fact that different kinds of decisions can result from a decision-making process. That is, decisions can be *'positive'* in the sense that they are intended, once implemented, to alter the status quo in some way, or they can be *'negative'* in the sense that the government declares that it will do nothing new about a public problem but will retain the status quo. Third, this definition underlines the point that public policy decision-making is not a technical exercise but an inherently political process. It recognizes that public policy decisions create 'winners' and 'losers', even if the decision is a negative one.

Brewer and DeLeon's definition, of course, says nothing about the actors involved in this process, or the desirability, likely direction, or scope of public decision-making. To deal with these issues, different theories and models have been developed to describe how decisions are made in government as well as to prescribe how decisions ought to be made. The nature of public policy decision-makers, the different types of decisions that they make, and the development and evolution of decision-making models designed to help understand the relationship between the two are described below.

ACTORS IN THE DECISION-MAKING PROCESS

With the exception of usually infrequent exercises in direct democracy such as referendums (Wagschal, 1997; Butler and Ranney, 1994), the number of relevant policy actors decreases substantially when the public policy process reaches the decision-making stage. Such concentrated engagement is not found at the agenda-setting stage, where virtually any actor in the policy universe could, theoretically at least, become active and involved. The policy formulation stage is also open to numerous actors, though in practice only those who are members of specific policy subsystems tend to participate. But when it comes time to decide on adopting a particular option, the relevant group of policy actors is almost invariably restricted to those with the authority to make binding public decisions. In other words, the public policy decision-making stage normally centres on those occupying formal offices in government. Excluded are virtually all non-state actors, including those from other levels of governments, both domestically and internationally. Only those politicians, judges, and government officials actually empowered to make authoritative decisions in the area in question can participate with both 'voice' and 'vote' at this stage of the policy cycle (Aberbach et al., 1981).

This is not to say that other actors, including non-state ones as well as those belonging to other governments, are not active and even influential during policy decision-making. These actors can and do engage in various kinds of lobbying activities aimed at persuading, encouraging, and sometimes even coercing authoritative office-holders to adopt preferred options and avoid undesirable ones (Woll, 2007). However, unlike office-holders, those other actors have, at best, a 'voice' in the decision-making process, not a 'vote' (see Pal, 1993b; Richardson et al., 1978; Sarpkaya, 1988).

This does not mean that decision-makers, since they occupy positions of formal authority, can adopt whatever policy they wish. As discussed in earlier chapters, the degree of freedom enjoyed by decision-makers is in fact circumscribed by a host of constraining rules and structures governing political and administrative offices, as well as by the sets of ideas or paradigms and the social, economic, and political circumstances within which they work. As we have seen, key rules and structures that affect configurations of political power and resources in both state and non-state actors range from the country's constitution to the specific mandates conferred on individual decision-makers. Specific decision-makers, such as judges and civil servants, must act within specific sets of laws and regulations governing their behaviour and fields of competence (Markoff, 1975; Page, 1985a; Atkinson and Coleman, 1989a), while the different actors with which they must contend in so doing were discussed in Chapter 3.

Countries have different constitutional arrangements and distinct sets of rules governing the structure of public agencies and the conduct of government officials. Some political systems concentrate decision-making authority in the elected executive and the bureaucracy, while others permit the legislature and judiciary to play a greater role. Parliamentary systems tend to fall in the former category and presidential systems in the latter. Thus, in Australia, Britain, and Canada and other parliamentary democracies, the cabinet and bureaucracy are often solely responsible for making many policy decisions. They may at times have decisions imposed on them by the legislature in situations when the government does not enjoy a parliamentary majority, or by the judiciary in its role as interpreter of the constitution, but these are not routine occurrences. In the United States and other presidential systems, although the authority to make most policy decisions rests with the executive (and the cabinet and bureaucracy acting on the president's or governor's behalf), those decisions requiring legislative approval often involve intense negotiation with the legislators, while some are modified or overturned on a regular basis by the judiciary on constitutional or other grounds (Weaver and Rockman, 1993b).

At the micro level, various rules usually set out not only which decisions can be made by which government agency or official, but also the procedures that must be followed. As Allison and Halperin (1972) have noted, over time such rules and operating procedures often provide decision-makers with 'action channels'—a regularized set of *standard operating procedures* for producing certain types of decisions. These rules and standard operating procedures help explain why so much of the decision-making in government is of a routine and repetitive nature. Nevertheless, while rules and normal procedures circumscribe the freedom available to some decision-makers (especially those in administrative or judicial positions), others (especially political decision-makers) retain considerable discretion to judge the 'best' course of action to follow in specific circumstances. Since decision-makers themselves vary greatly in terms of background, knowledge, and the beliefs that affect how they interpret a problem and its potential solutions (Huitt, 1968), different decision-makers operating in similar institutional environments can respond differently even when dealing with the same or similar problems. Hence, even with standard operating procedures in place, exactly what process is followed and which decision is considered 'best' will vary according to the structural and institutional context of a decision-making situation.

CHOICES: NEGATIVE, POSITIVE, AND NON-DECISIONS

Regardless of which actor or actors make a policy decision, whether a relatively large group of legislators in a partisan political arena or a single civil servant in a more insulated bureaucratic setting, the results of this process will fit into just a few categories. That is, although the substance of decisions can be infinitely varied, their fundamental effect will be to either perpetuate the policy status quo or alter it.

Traditional *'positive' decisions* that alter the status quo receive considerable attention in the decision-making literature and are therefore accorded most attention in this chapter. However, it is important to note that other kinds of decisions uphold the status quo. Here we can distinguish between *'negative'* decisions, in which a deliberate choice is made to preserve the status quo, and what are sometimes termed *'non-decisions'* (discussed earlier in the book) in which options to deviate from the status quo are not considered at the policy formulation or agenda-setting stages (see Zelditch et al., 1983; R.A. Smith, 1979).

Non-decisions have been the subject of many inquiries and studies by scholars interested in tracing the effects of ideologies, religions, and other similar factors that blind decision-makers to the need to act on a public problem; similarly, power allows decision-makers to ignore certain issues despite public clamour for change (see Bachrach and Baratz, 1962, 1970: ch. 3; Debnam, 1975; Bachrach and Baratz, 1975; Zelditch and Ford, 1994; Spranca et al., 1991; Oliviera et al., 2005).

Very little research into negative decisions, however, exists. This is partly due to the difficulties associated with identifying instances in which policy options to alter the status quo are explicitly rejected in favour of its maintenance (see Howlett, 1986). Nevertheless, these decisions can be examined in terms of their affect on the functioning of the policy cycle. That is, negative decisions are instances of *arrested* policy cycles. Unlike non-decisions, in which certain options are filtered out at earlier stages of the policy process and may thus never be identified in policy deliberations, negative decision-making does go through agenda-setting and policy formulation efforts that place alternative courses of action before those with the authority to decide. However, when a negative decision is made, the policy process does not move onto the implementation stage but simply confirms that the status quo is appropriate and halts at that point (van der Eijk and Kok, 1975).

The outcome likely to emerge from the decision-making stage thus depends both on the operation of earlier stages in the cycle, which serve to filter out some policy options while allowing others to proceed, and on the exact configuration of decision-making actors, their beliefs, and the context in which they work. The nature of the relevant subsystem and the kinds of constraints under which decision-makers operate will have a significant effect on the type of decision that will emerge in different situations and, especially, will affect the particular weight accorded to the option of simply retaining the status quo (Bardach, 2006; Kay, 2006; Genschel, 1997).

CHAPTER 7
POLICY IMPLEMENTATION

After a public problem has reached the policy agenda, various options have been proposed to address it, and the government has set policy goals and decided on a course of action to attain them, it must put the decision into practice. The effort, knowledge, and resources devoted to translating policy decisions into action comprise the policy cycle's implementation stage. While most policy decisions identify the means to pursue their goals, subsequent choices are inevitably required to attain results. Funding must be allocated, personnel assigned, and rules of procedure developed to make a policy work.

Policy implementation often relies on civil servants and administrative officials to establish and manage the necessary actions. However, non-governmental actors who are part of the policy subsystem can also be involved in implementation activities. In some countries, like Sweden, there may be a tradition of non-governmental actors directly implementing some important social programs (Ginsburg, 1992; Johansson and Borell, 1999). In other countries, like the US, which have only recently attempted to implement some programs through community and religious ('faith-based') groups (Kuo, 2006), non-governmental actors are typically involved in the design and evaluation of policies but not their actual administration and management.

ACTORS AND ACTIVITIES IN POLICY IMPLEMENTATION

Once the direction and goals of a policy are officially decided, the number and type of actors involved begin to expand beyond the small subset of actors making policy decisions to encompass the policy universe of interested actors. Policy subsystems then become important contributors to implementation as their participants apply knowledge and values to shaping the launch and evolution of programs implementing policy decisions. Usually, however, only a narrow range of subsystem actors become involved in the implementation process. Bureaucrats are the most significant actors in most policy implementation, bringing the endemic intra – and inter-organizational conflicts of public agencies to the fore of this stage in the policy cycle (Dye, 2001).

Different bureaucratic agencies at various levels of government (national, state or provincial, and local) are usually involved in implementing policy, each carrying particular interests, ambitions, and traditions that affect the implementation process and shape its outcomes, in a process of 'multi-level' government or governance (see Bardach, 1977; Elmore, 1978; Bache and Flinders, 2004). Implementation by public agencies is often an expensive, multi-year effort, meaning that continued funding for programs and projects is usually neither permanent nor guaranteed but rather requires continual negotiation and discussions within and between the political and administrative arms of the state. If their preferred solution to a problem was not selected, this creates opportunities for politicians, agencies, and other members of policy subsystems to use the implementation process as simply another opportunity for continuing the conflicts they may have lost at earlier stages of the policy process. Such processes, of course, greatly complicate implementation and move it further away from being simply a 'technical' issue of decision-processing (Nicholson-Crotty, 2005).

While politicians are significant actors in the decisions that lead into the implementation process and can play an active role in subsequent oversight and evaluation efforts, most of the day-to-day activities of policy administration typically fall within the purview of salaried public servants. This is because of the key role played by laws codifying the results of decision-making and empowering specific state agencies to put those decisions into practice (Keyes, 1996; Ziller, 2005).

In most countries, traditional or *civil or common laws* form a 'default' or basic set of principles governing how individuals interact with each other and with the state in their day-to-day lives. These laws are often codified in writing—as is the case in many continental European countries—but they may also be found in less systematic form in the overall record of precedents set by judicial bodies, as is the case in Britain and its former colonies. Even in these so-called 'common-law' countries, *statutory laws* are passed by parliaments to replace or supplement the civil or common law (Gall, 1983; Bogart, 2002). These statutes take the form of Acts, which, among other things, usually also create a series of rules to be followed in implementing particular policies, as well as a range of offences and penalties for non-compliance with the law.

Statute law usually also designates a specific administrative agency or ministry as empowered to make whatever *'regulations'* or administrative rules are required to ensure the successful implementation of the principles and aims of the enabling legislation. Regulations giving effect in specific circumstances to the general principles codified in laws are then prepared by civil servants employed by administrative agencies, often in conjunction with target or 'clientele' groups (Kagan, 1994). Regulations cover such items as the standards of behaviour or performance that must be met by target groups and the criteria to be used to administer policy. These serve as the basis for licensing or approval and, although unlegislated, provide the de facto source of direction for most implementation processes. As was discussed in Chapter 5, this general form of implementation is sometimes referred to as 'command-and-control' regulation whereby a command is given by an authorized body and the administration is charged with controlling the target group to ensure compliance (Sinclair, 1997; Kerwin, 1994, 1999; Baldwin and Cave, 1999).

Although many efforts have been made in recent years to supplement or replace this mode of governance with others in which implementation is based more on collaboration or incentives (Freeman, 1997; Armstrong and Lenihan, 1999; Kernaghan, 1993), in the modern era, such legal processes continue to form the basis for implementation in all but the worst instances of dictatorship or personal rule. Instructions may be issued through compliant legislatures but also directly from the executive to the administration. These types of legal processes are a necessary part of adapting general statements of intent, which usually result from the decision-making stage, to the specific circumstances and situations that administrators face on the ground in attempting to alter societal behaviour in the direction desired by decision-makers. Even in the case of efforts to develop more collaborative relationships with target groups, administrative actions must still be based on legal authority provided by legislatures and executives (Grimshaw, 2001; Klijn, 2002; Phillips, 2004).

The usual form of such administrative venues is the *ministry* or *department,* and the actual practice of administering policy and delivering services is performed overwhelmingly by civil servants in such agencies. However, other forms of quasi-governmental organizations (quangos) (Hood, 1986; Koppell, 2003) ranging from state-owned enterprises (Stanton, 2002; Chandler, 1983; Laux and Molot, 1988) to non-profit corporations and bodies (McMullen and Schellenberg, 2002; Advani and Borins, 2001) and public–private partnerships (English and Skelern, 1995; Hodge and Greve, 2007), as discussed in Chapter 5, can also be vehicles for service delivery.

This does not, however, exhaust the types of state agencies involved in implementation, which also include organizations designed to perform specific tasks related to service delivery without being directly or indirectly involved in its management. Among these are various kinds of tribunals, such as *independent regulatory commissions,* that exist at arm's-length from the government and develop the rules and regulations required for administration (Cushman, 1941; Braithwaite et al., 1987; Christensen and Laegreid, 2007). Another form of implementing agency is the administrative appeal *board* and other forms of *commissions and tribunals* created by statute or regulation to perform many quasi-judicial functions, including appeals concerning licensing, certification of personnel or programs, and the issuance of permits. Appointed by government, administrative tribunals and boards usually represent, or purport to represent, some diversity of interests and expertise and are expected to moderate the public–private interface in goods and service delivery without displacing non-state actors in the production and distribution of various kinds of goods and services. *Public hearings* may be statutorily defined as a component of such administrative process and operate to secure regulatory compliance. In most cases, however, such hearings are held at the discretion of a decision-making authority and are often after-the-fact public information sessions rather than true consultative devices (Talbert et al., 1995; Grima, 1985). *Specialized advisory boards and commissions* (Brown, 1955, 1972; Smith, 1977) often supplant public consultations, yielding more expert views on specific regulatory activities than open public hearings would typically provide, but also allowing some subsystem members to exercise inordinate influence on policy implementation (Dion, 1973).

Thus, while state officials remain an important force in the implementation stage of the policy process, advisory and quasi-governmental agencies allow them to be joined by members of the relevant policy subsystems, as the number and type of policy actors return to resembling those found at the formulation stage (Bennett and McPhail, 1992). Just as at that stage, *target groups,* that is, groups whose behaviour is intended or expected to be altered by government action, play a major role in the implementation process (Donovan, 2001; Kiviniemi, 1986; Schneider and Ingram, 1993). The political and economic resources of target groups certainly have an impact on the implementation of policies (Montgomery, 2000). Powerful groups affected by a policy can influence the character of implementation by supporting or opposing it. Thus, regulators will commonly strike compromises with groups, or attempt to use the groups' own resources in some cases, to make the task of implementation simpler or less expensive (Giuliani, 1999). Although this is typically done informally in some jurisdictions, such as the United States, more formal efforts have been made in many sectors to incorporate regulator–regulated negotiations in the development of administrative standards and other aspects of the regulatory process (Coglianese, 1997). Changing levels of public support for a policy can also affect implementation. Many policies witness a decline in support after a policy decision has been made, enabling administrators to vary the original intent of a decision (see Hood, 1983, 1986a). Despite the rule of law, bureaucrats thus possess considerable influence, whether they seek it or not, in realizing the policy initiatives they are called upon to implement.

CHAPTER 8
POLICY EVALUATION: POLICY-MAKING AS LEARNING

Once the need to address a public problem has been acknowledged, various possible solutions have been considered, and some among them have been selected and put into practice, a government often assesses how the policy is working. At the same time, various other interested members of policy subsystems and of the general public are engaged in their own assessment of the workings and effects of the policy in order to express support for or opposition to the policy, or to demand changes to it. The concept of *policy evaluation* thus refers broadly to the stage of the policy process at which it is determined how a public policy has actually fared in action. It involves the evaluation of the means being employed and the objectives being served. As Larry Gerston (1997: 120) has defined it, 'policy evaluation assesses the effectiveness of a public policy in terms of its perceived intentions and results.'

How deep or thorough the evaluation is depends on those initiating and/or undertaking it, and what they intend to do with the findings.

After a policy has been evaluated, the problem and solutions it involves may be completely reconceptualized, in which case the cycle may swing back to agenda-setting or some other stage of the cycle, or the status quo may be maintained. Reconceptualization may consist of minor changes or fundamental reformulation of the problem, including terminating the policy altogether (DeLeon, 1983). How evaluation is conducted, the problems the exercise entails, and the range of results to which it typically leads are the concerns of this chapter. It then outlines the patterns of policy change that typically result from different types of policy evaluation.

POSITIVIST AND POST-POSITIVIST POLICY EVALUATION

For the most part, policy evaluation has been the analytical domain of those who view such assessment as a neutral, technical exercise in determining the success (or failure) of government efforts to deal with policy problems. David Nachmias (1979: 4), an influential figure in the field's early development, captured this positivist spirit in defining policy evaluation as 'the objective systematic, empirical examination of the effects ongoing policies and public programs have on their targets in terms of the goals they are meant to achieve'. Discerning readers will have no difficulty detecting the rationalist premise underlying this definition. It specifies explicitly that examining a policy's effects on the achievement of its goals should be objective, systematic, and empirical, the hallmarks of the positivist approach to policy analysis. However, as we have mentioned before, public policy goals are often neither clear nor explicit, necessitating subjective interpretation to determine what exactly was achieved. Objective analysis is further limited by the difficulties encountered in developing neutral standards by which to evaluate government success in dealing with societal demands and socially constructed problems in a highly politicized environment.

After much work in the 1960s and 1970s to develop quantitative systems of policy evaluation, it became clear (Anderson, 1979a; Kerr, 1976; Manzer, 1984) that developing adequate and acceptable measures for evaluating policy was more contentious and problematic than was previously believed. Astute observers also noted that it was naive to believe that policy evaluation was always intended to reveal the effects of a policy. In fact, it is at times employed to disguise or conceal certain facts that a government fears will show it in a poor light. It is also possible for governments to design the terms of evaluation in such a way as to lead to conclusions that would show it in a better light. Or, if it wants to change or scrap a policy, it can adjust the terms of the evaluation accordingly. Similarly, evaluation by those outside the government is not always designed to improve a policy, but often to criticize it in order to gain partisan political advantage or to reinforce ideological postulates (Chelimsky, 1995; Bovens and t'Hart, 1995).

As a result, more recent thinking tends to view policy evaluation, like other stages of the policy process, as an inherently political activity, albeit, like the other stages, with a technical component. In its extreme, post-positivist form, it has been argued that since the same condition can be interpreted quite differently by different evaluators, there is no definitive way of determining the correct evaluation mode. Which interpretation prevails, in this view, is ultimately determined by political conflicts and compromises among the various actors (Ingram and Mann, 1980b: 852).

This is not to suggest that policy evaluation is an irrational or a purely political process, devoid of genuine intentions to assess the functioning of a policy and its effects. Rather, it serves as a warning that we must be aware that relying solely on formal evaluation for drawing conclusions about a policy's relative success or failure will yield unduly limited insights about policy outcomes and their assessment. To get the most out of policy evaluation, the limits of rationality and the political forces that shape it must also be taken into account, without going so far as to believe that the subjective nature of policy assessments allows no meaningful evaluation to take place.

POLICY EVALUATION AS POLICY LEARNING

One way of looking at policy evaluation, which combines elements of both the positivist and post-positivist perspectives, is to regard it as a very significant stage in an overall process of *policy learning* (Grin and Loeber, 2007; Lehtonen, 2005). That is, perhaps the greatest benefits of policy evaluation are not the direct results it generates in terms of definitive assessments of the success and failure of particular policies per se, but rather the educational dynamic that it can stimulate among policy-makers as well as others less directly involved in policy issues (Pressman and Wildavsky, 1984). Whether they realize it or not, actors engaged in policy evaluation are often participating in a larger process of policy learning, in which improvements or enhancements to policy-making and policy outcomes can be brought about through careful and deliberate assessment of how past stages of the policy cycle affected both the original goals adopted by governments and the means implemented to address them (see Etheredge and Short, 1983; Sabatier, 1988; Lehtonen, 2006).

The concept of 'learning' is generally associated with intentional, progressive, cognitive consequences of the education that results from policy evaluation. However, policy learning also has a broader meaning that includes understanding both the intended and unintended (see Merton, 1936) consequences of policy-making activities, as well as both the 'positive' and 'negative' implications of existing policies and their alternatives on the status quo and efforts to alter it. From a learning perspective, public policy evaluation is conceived as an iterative process of active learning about the nature of policy problems and the potential of various solutions to address them (Rist, 1994; Levitt and March, 1988). This view shares some similarities with the idea of policy-making as a 'trial-and-error' process of policy experimentation, but with the added idea that successive 'rounds' of policy-making, if carefully evaluated after each 'round', can avoid repeating mistakes and move policy implementation ever closer towards the achievement of desired goals.

Like other concepts in policy science, there are differing interpretations of what is meant by 'policy learning' and whether its source and motivation are within or outside existing policy processes. Peter Hall makes the case for *'endogenous'* learning, defining the activity as a 'deliberate attempt to adjust the goals or techniques of policy in the light of the consequences of past policy and new information so as to better attain the ultimate objects of governance' (Hall, 1993: 278). Hugh Heclo, on the other hand, suggests that learning is a less conscious, *'exogenous'* activity, often occurring as a government's response to some kind of external or exogenous change in a policy environment. According to Heclo, this often takes the form of an almost automatic process, as 'learning can be taken to mean a relatively enduring altera- tion in behaviour that results from experience; usually this alteration is conceptualized as a change in response made in reaction to some perceived stimulus' (Heclo, 1974: 306). The two definitions describe the same relationship between policy learning and policy change, but differ substantially in their approach to the issue. For Hall, learning is a part of the normal public policy process in which policy-makers attempt to understand why certain initiatives may have succeeded while others failed. If policies change as a result of learning, the impetus for change originates within the normal policy process of the government. For Heclo, on the other hand, policy learning is seen as an activity undertaken by policy-makers largely in reaction to changes in external policy 'environments'. As the environment changes, policy- makers must adapt if their policies are to succeed.

Regardless of its external or internal causes, however, most scholars agree that several types of learning can result from different kinds of evaluations. Assessments can be car- ried out by both governmental and non-governmental actors at this stage of the policy cycle, since the number of actors involved in evaluation expands towards the size of the policy universe existing during agenda-setting (Bennett and Howlett, 1991; May, 1992; Sabatier, 1988; Hall, 1993; Etheredge, 1981; see also Argyris, 1992; Argyris and Schon, 1978). Some lessons are likely to concern practical suggestions about specific aspects of the policy cycle, based on the actual experience with the policy on the part of policy implementers and target groups. These include, for example, their perceptions of the lessons they have learned about which policy instruments have 'succeeded' in which circumstances and which have 'failed' to accomplish expected tasks or goals, or which issues have enjoyed public support in the agenda-setting process and which have not, and therefore which are likely to do so in future. Richard Rose (1988, 1991) defined one such relatively specific and limited type of learning as *lesson-drawing*. This type of learning originates within the formal policy process and is aimed primarily at the choice of means or techniques employed by policy-makers in their efforts to achieve their goals; in Rose's formulation this often involves the analysis of, and derivation of lessons from, experiences in other sectors, issue areas, or jurisdictions.

Other lessons probe broader policy goals and their underlying ideas or paradigms, or the 'frames' in which lesson-drawing takes place. This is a more fundamental type of learning, which is accompanied by changes in the thinking underlying a policy that might result in a policy being terminated or drastically revised in light of new conceptions and ideas developed through the evaluation process. Following Hall (1993), this type of learning is often referred to as *social learning*. It tends to originate outside the formal policy process and affects the policy-makers' capacity to change society.

EVIDENCE-BASED POLICY-MAKING AS POLICY LEARNING

'Evidence-based policy-making' is a term that has come into use in recent years as policy practitioners have struggled to enhance the rationality of policy deliberations and promote improved policy learning on the part of governments. It represents an effort to reform or restructure policy processes by prioritizing data-based evidentiary decision-making criteria over less formal or more 'intuitive' or experiential policy assessments in order to avoid or minimize policy failures caused by a mismatch between government expectations and actual, on-the-ground conditions. The evidence-based policy movement (Pawson, 2006) is thus the latest in a series of efforts undertaken by reformers in governments over the past half-century to enhance the efficiency and effectiveness of public policy-making through the application of a systematic evaluative rationality to policy problems (Sanderson, 2006; Mintrom, 2007).

Exactly what constitutes 'evidence-based policy-making' and whether analytical efforts in this regard actually result in better or improved policies are contentious (Packwood, 2002; Pawson, 2002; Tenbensel, 2004; Jackson, 2007). Through a process of theoretically informed empirical analysis consciously directed towards promoting policy learning, however, proponents of this approach believe that governments can better learn from experience, avoid repeating the errors of the past, and better apply new techniques to the resolution of old and new problems (Sanderson, 2002a, 2002b).

ASSESSING POLICY SUCCESS OR FAILURE

Policy evaluation is made challenging by the difficulties that arise in assessing the success or failure of policy initiatives. As Bovens and t'Hart (1996: 4) have argued, 'the absence of fixed criteria for success and failure, which apply regardless of time and place, is a serious problem' for anyone who wants to understand policy evaluation. Policies can succeed or fail in numerous ways. Sometimes an entire policy regime can fail, while more often specific programs within a policy field may be designated as successful or unsuccessful (Mucciaroni, 1990; Moran, 2001; Gundel, 2005). And both policies and programs can succeed or fail either in substantive terms—that is, delivering or failing to deliver the goods—or in procedural terms—as being legitimate or illegitimate, fair or unfair, just or unjust (Bovens and t'Hart, 1995; Weaver, 1986; McGraw, 1990; Hood, 2002).

'Success' is always hard to define. In some instances of an unequivocal disaster, like an airplane crash or nuclear reactor meltdown, analyses can pinpoint obvious causes such as technical failures, managerial incompetence, or corruption (Bovens and t'Hart, 1996; Gray and t'Hart, 1998). Evaluation can also uncover lesser-known causes of breakdown such as 'practical drift', in which increasingly large deviations from expected norms are allowed to occur until, finally, significant system failure occurs (Vaughan, 1996). Although some of the lessons drawn from these spectacular accidents—such as the significant potential for failure of complex organizational systems when elements are either too loosely or too tightly coupled (Perrow, 1984)—can be translated into policy studies, the causes behind more typical policy failures, such as overspending on project development or the unintended consequences of a policy initiative, are harder to pin down.

Failures can occur at any stage of the policy cycle and do not necessarily have their source in the same stage (Michael, 2006). Thus, an overly ambitious government may agree to address intractable ('wicked') problems (Pressman and Wildavsky, 1973; Churchman, 1967) at the agenda-setting stage, a decision that can lead to failure at any succeeding stage of the policy cycle. Failure can also arise from a mismatch between goals and means at the formulation stage (Busenberg, 2000, 2001, 2004a, 2004b), or it can result from the consequences of lapses or mis-judgements at the decision-making stage (Bovens and t'Hart, 1995, 1996; Perrow, 1984; Roots, 2004; Merton, 1936). Another set of pitfalls arises through various 'implementation failures' in which the aims of decision-makers fail to be properly or accurately translated into practice (Kerr, 1976; Ingram and Mann, 1980). Policy failure can also arise from a lack of effective oversight by decision-makers over those who implement policy (McCubbins and Schwartz, 1984; McCubbins and Lupia, 1994; Ellig and Lavoie, 1995). Finally, failure can stem from governments and policy-makers not effectively evaluating policy processes and learning useful lessons from past experiences (May, 1992; Scharpf, 1986; Busenberg, 2000, 2001, 2004a, 2004b).

In many circumstances, the operation of a policy system is too idiosyncratic, the actors too numerous, and the number of outcomes too small to permit clear and unambiguous post-mortems of *policy outcomes*. Nevertheless, such efforts are made by many actors with varying degrees of formality and the results of these investigations, whether accurate or not, are fed back into the policy process, influencing the direction and content of further policy cycle iterations.

The role of actors at this stage is crucial. Different types of evaluations can be undertaken by different sets of actors and can have very different impacts on subsequent policy deliberations and activities (Fischer and Forester, 1987). As Bovens and t'Hart (ibid., 21) note, ultimately 'judgements about the failure or success of public policies or programs are highly malleable. Failure is not inherent in policy events themselves. "Failure" is a judgement about events.' These judgements about policy success and failure often depend partly on imputing notions of intentionality to government actors, assuming that there was a 'method to the madness' and that policy actors meant to achieve what their actions produced. Intentionality makes it possible to assess policy-making results against expectations. However, even with this rational assumption, assessment is not a simple task (see Sieber, 1981: ch. 2). First, as we have seen, government intentions may be vague and ambiguous, or even potentially contradictory or mutually exclusive. Second, labels such as 'success' and 'failure' are inherently relative and will be interpreted differently by different policy actors and observers. Moreover, such designations are also semantic tools used in public debates to seek political advantage. That is, policy evaluations affect considerations and consequences related to assessing blame and taking credit for government activities at all stages of the policy process, all of which can have electoral, administrative, and other consequences for policy actors (Bovens and t'Hart, 1996: 9; Brandstrom and Kuipers, 2003; Twight, 1991; Hood, 2002; Hood and Rothstein, 2001).

Such judgements, by nature, are at least partially linked to factors such as the nature of the causal theories used to frame policy problems at the agenda-setting and policy formulation stages and the conceptual solutions developed at the formulation stage. The expectations of decision-makers about likely program or policy results and the extent of time allowed for those results to materialize before evaluators make their assessments are other important factors (Bovens and t'Hart, 1996: 37). Policy evaluation processes, recognizing these built-in biases, often simply aim to provide enough information to make reasonably intelligent and defensible claims about policy outcomes, rather than offering definitive explanations that build airtight cases concerning their absolute level of success or failure.

Lawrence Susskind — "The Environment and Environmentalism." In *Local Planning: Contemporary Principles and Practice.* Washington, DC: ICCMA Press.

Reading Commentary

In a world in which roles and relationships are constantly changing, how should we structure environmental problem-solving efforts? Given that boundaries are fluid and uncertainty is widespread, what kind of individual and collective actions are likely to be most helpful? Can traditional political and legal hierarchies be supplemented by ad hoc power-sharing arrangements? Will ad hoc power-sharing arrangements produce better results? And if so, which analytic tools should we rely on in these ad hoc arrangements?

Susskind outlines a set of responsibilities, tools and partnerships for environmental problem-solving. He allocates different tasks to concerned citizens, whom he calls pioneers, public-venture capitalists, superintendents, mediators, and stewards of the common good. His aim is to foster effective collaboration and achieve sustainable solutions. The efforts he has in mind avoid the search for the mythical right answers and focus instead on enhancing flexibility and resilience in response to ever-evolving (and often unexpected) man-made and natural impacts.

Though many environmental problems can be addressed by modifying individual behavior (i.e., recycling), most require the adoption of new collective standards (i.e., the siting of noxious facilities). In either case, environmental challenges require us to account for different attitudes toward risk. The best way to cope with these differences is to treat proposed environmental solutions as experiments in which we commit to continuously monitoring the effects of our interventions and plan for frequent adjustments.

THE ENVIRONMENT AND ENVIRONMENTALISM

Lawrence Susskind

Environmental concerns have received increasing attention in the United States since the 1970s. This attention derives, in large part, from three sources: (1) educational efforts, symbolic appeals, and lobbying undertaken by a range of non-governmental organizations; (2) corporate efforts to increase market share by appealing to environmentally conscious consumers; and (3) the explosion in scholarly research documenting environmental threats to public health and human survival.

THE COSTS OF MISMANAGING NATURAL RESOURCES

Decisions about the use of natural resources, and regulatory or spending decisions designed to protect health and safety, allocate costs and benefits in ways that are rarely distributed evenly across the population or across geographic areas. While it is difficult to calculate these costs and benefits in a precise (and noncontroversial) way, there is no longer any debate about the damage that can result from failure to manage natural resources effectively, particularly when the impacts are concentrated in specific locations or fall on vulnerable segments of the population. In fact, the field of environmental planning emerged, in large part, in reaction to the extreme costs created by mismanagement of natural resources and development pressures. Recently, ecological economists have begun to calculate the benefits of "nature's services"; such calculations assign a monetary value to basic ecological functions, such as the cleansing performed by wetlands. This should make it easier to make effective resource allocation decisions.

INDIVIDUAL AND COLLECTIVE RESPONSIBILITY FOR THE ENVIRONMENT

Many environmental concerns can be addressed only by encouraging individuals to change their behavior: to set their thermostats lower; drive less; reuse or recycle as much as possible; and use their purchasing power to demand more ecologically sustainable production, shipping, and waste disposal. But many decisions about using resources and managing waste cannot be addressed by individuals; they must be addressed collectively. If a community needs a sewage-treatment plant, for example, no individual, no matter how highly motivated, will single-handedly construct and operate it. Instead, the plant will be constructed as a public good. Collective action usually requires the imposition of enforceable standards, the adoption of powerful incentives, or both. In part, this is a response to the "free rider" phenomenon, which assumes that every person will pursue his or her own self-interest – even if that produces environmentally reckless decisions – because each assumes that his or her individual action will not be significant enough to be noticed.

Merely encouraging consumers and businesses to be guided by an ethic of sustainability is not enough: collective action at the neighborhood, municipal, state, federal, and international levels is essential to develop sustainable resource management practices. Nevertheless, it is often unclear how best to collectively pursue environmental protection or sustainable development. Given this uncertainty, it may be wise, in the short run, to treat each environmental policy decision as an experiment: that is, when the risks of doing nothing are worrisome, but the likely costs and benefits of taking action are hard to estimate because of the complexity of the systems involved, we should probably take small steps in what we believe to be the right direction. If we commit to monitoring what happens as a result of each move and assume that continuous adjustments will be needed, we can move in the right direction even if we don't know exactly where we are trying to go. This approach is known as adaptive environmental management.

ENVIRONMENTAL PLANNING AT THE LOCAL LEVEL

A distinction is often made between government-led environmental protection and a market-driven approach. Under the government-led model, government agencies set specific resource management and public health objectives, specify the means that will be used to achieve them (including the choice of acceptable technologies), and mandate reporting time-tables and testing procedures. This approach also presumes that government agencies will allocate the funds and personnel needed to ensure enforcement. Under the market-oriented approach, consumers and investors, rather than government agencies, decide whether, how, and when to invest in environmental protection.

In practice, however, there is really no choice between the two approaches. The public expects the government to set and enforce standards to protect health and safety. At the same time, the ingenuity of private entrepreneurs is essential to the invention of increasingly effective ways to meet such standards. Business wants the predictability and level playing field provided by environmental protection standards and even-handed enforcement. "First movers" who are committed to innovative green technologies need government subsidies – that is, incentives – to support their entrepreneurial efforts. In short, a balance between regulatory and market mechanisms is necessary to ensure both fairness and efficiency. Thus, government must formulate environmental protection objectives while simultaneously unleashing the power of the market, stimulating investment in research and development, offering incentives for innovation, and ensuring that all information is shared.

A VALUES-BASED APPROACH

Many conflicts over the use of natural resources, the siting of necessary but noxious facilities, and the pattern and style of development can be traced to differences in values. For example, some members of the public take a utilitarian view of natural resources: they believe that such resources should be drawn on as needed in order to foster economic growth. Others advocate an ethic of environmental stewardship, in which resources are carefully husbanded to ensure their availability for future generations. Similarly, whereas some members of a community may be comfortable taking environmental risks in order to proceed with development, others prefer taking precautions to minimize environmental risk. Finally, some local constituencies assign the highest priority to the preservation of individual property rights, whereas others are willing to sacrifice those rights (with or without appropriate compensation) to achieve community-wide objectives. Municipal governments need to find ways of reconciling these competing views to win broad-gauged political support for environmental policies or resource management decisions.

COLLABORATIVE ENVIRONMENTAL DECISION MAKING

The era is long since past when formal hearings were held merely to ratify decisions that had already been made. Since the 1980s, in the face of growing public demand for direct involvement in environmental decision making – and the willingness of the courts to grant standing to an increasingly broad set of plaintiffs – stakeholder groups of all kinds have been invited to engage in a range of environmental planning activities. Public participation or civic engagement, dispute resolution, consensus building, and other collaborative approaches to environmental decision making are now par for the course. Sharp differences remain, however, between collaborative processes that are purely advisory and those that guarantee genuine joint decision making. Engagement that is limited to figurehead advisory groups, selected by agencies that intend to make all final decisions on their own, is quite different from collaborative processes that invite self-identified stakeholders to participate in consensus building.

TOOLS FOR ENVIRONMENTAL PLANNING

Planners must be familiar with a growing array of environmental tools. Table 2.1 offers a partial list of the techniques that are regularly used in local environmental planning, three of which – impact assessment, sustainability analysis, and joint fact finding – are discussed in this section.

Impact assessment – the analysis of the environmental or socioeconomic effects of various project designs or policy options – has been part of environmental planning since the enactment of the National Environmental Policy Act (NEPA) of 1969 and dozens of related state environmental review procedures. Other regulations require similar but more limited assessments in advance of decision making. Whatever the source of the requirements for an assessment, planners need to be able to forecast multiple types of impacts; to scale, weight, and integrate them; to suggest appropriate mitigation strategies; and to interact with an array of stakeholders who want to participate in such investigations and to help make decisions that are based on their findings.

The public participation requirements embedded in NEPA have forced federal (and many state) agencies to consider a richer set of project and policy options and more extensive mitigation strategies than they would otherwise have pursued. Higher public expectations – for participation and more effective mitigation of adverse environmental impacts – have spilled over to the local level as well. On the other hand, even though tens of thousands of impact

Table 2.1 Tools for environmental planning

	Tools and their purposes		
How the tool is used	**Modeling: What is the current situation? How do the relevant systems work?**	**Forecasting: What might happen?**	**Decision analysis: What should we do?**
By the agency on its own	Case studies, cost-benefit analyses, system dynamics models	Scenario casting, Delphi exercises, risk assessment, sustainability (ecological footprint) analysis	Expert brainstorming
In collaboration with stakeholders	Multiagent interactive models	Joint fact finding, impact assessment	Charrettes, policy dialogues, consensus building

assessments have been prepared, there has been little improvement in their accuracy; nor has there been an increased commitment to choosing the least environmentally harmful course of action when decisions have to be made. Much of the forecasting has relied on very simple (and not very well-grounded) models. The public continues to demand more reliable assessments, based on in-depth studies of human-ecosystem interactions, but even after hundreds of lawsuits, there is no clear mandate for meeting technical standards of excellence.

Sustainability analysis, sometimes called ecological footprint (EF) analysis, calculates the "draw" on ecologically productive land and marine areas required to sustain a population, manufacture a product, or undertake a particular activity.[1] Such assessments use an accounting procedure, similar to that used in life-cycle analysis, in which the projected consumption of energy, biomass, building materials, water, and other resources is converted to a normalized measure of the land area required to produce or sustain whatever is being proposed. A per capita calculation of the land area required (per capita EF) is then used to portray relative consumption levels. EF analysis offers the potential to evaluate long-term carrying capacity and to address the policy implications of above-average consumption levels.

Environmentalism and sustainability

The start of the twenty-first century marked a seminal shift in public policy from a focus on environmental protection to a focus on sustainability. Minimizing the adverse environmental impacts of proposed development, infrastructure investment, or public policy is no longer enough if such initiatives undermine the long-term sustainability of key ecological systems. We can no longer take the narrow view, limiting our focus to arbitrarily defined project areas, thinking primarily of the current moment, and tending mostly to short-term budgetary effects. Instead, decisions about resource use and waste management must take into account ecosystem-wide effects, cross-media impacts (i.e., the ways in which efforts to protect one resource inadvertently undermine efforts to protect another), and the implications of our actions for future generations.

At the heart of this transformation is the concept of resilience. When we plan human settlements and try to manage natural resources, our goal should be to increase the capacity of ecological and built systems to respond effectively to surprises (including disasters), both man-made and natural.

Joint fact finding (JFF) draws together contending interest groups to ask questions and to engage in joint modeling and collaborative data assessment in advance of decision making. In the first step of JFF, known as conflict assessment, planners engage in extensive interviewing of prospective stakeholders to generate a credible list of experts and nonexperts who will be involved in each environmental planning process and will scope out the issues of greatest concern. Next, experts work with stakeholders and government agency staff to jointly frame questions, design data-gathering procedures, review preliminary findings, explore the policy implications of these findings, and evaluate how sensitive the findings are to slight changes in key analytic assumptions and data gaps. JFF stands in stark contrast to more traditional approaches, in which technical experts determine what analyses are required, conduct the analyses, and submit the results to decision makers. In today's more participatory environment, all planners should have the skills to conduct JFF efforts.

NEW KINDS OF PARTNERSHIPS

Intergovernmental arrangements continue to move away from the stratified or "layer cake" model, in which federal agencies have one set of responsibilities, states another, and local governments still another. Similarly, the boundaries between public and private, and between government and civil society, continue to blur. New partnerships are emerging in which historical roles and responsibilities are reassigned.

In the intergovernmental arena, for example, the hierarchy may be clear in theory, but practice is closer to a "marble cake" than to a layer cake. In large part, federal agencies rely on the states to implement national programs or to adopt regulations that are sometimes more demanding than federal law – and states, in turn, often cede responsibility to local governments to make key decisions about resource allocation.

The traditional approach, in which the boundaries of the "problem shed" were determined by political and legal authority, is giving way to temporary, negotiated agreements that transcend geopolitical boundaries. Long-standing governmental entities and their boundaries won't be erased any time soon, but that doesn't mean that environmental planners can't bring together institutional actors and other stakeholders to work out ad hoc power-sharing arrangements for environmental management. Such arrangements can take a number of forms, including intergovernmental agreements and memorandums of understanding.

The fact is, it doesn't matter which level of government or segment of society is in the lead or handles particular tasks, as long as they are all working together. A team of scholars at the Massachusetts Institute of Technology has dubbed this the PENs – Public Entrepreneurship Networks – model. Effective partnerships need to include

- *Pioneers* who recognize opportunity, seize initiative, and catalyze action by making commitments
- *Public-venture capitalists*, public officials who understand and embrace risk and can pull together the necessary financial, social, and human capital to meet project-driven needs
- *Superintendents* who provide a setting in which innovation can flourish by fostering formal and informal relationships
- *Mediators* who build consensus on goals and facilitate the resolution of conflicts that threaten to stall new ventures
- *Stewards of the common* good who maintain high standards of responsible behavior and bring together disparate groups in support of agreed-upon actions.

In the environmental planning field, as long as each of these roles is handled effectively, environmental planning objectives can be achieved.

NO RIGHT ANSWERS, ONLY INFORMED AGREEMENTS

Environmental planners must get used to preparing plans, policies, and programs in conjunction with others: experts and non-experts, participants with conflicting values, and agencies and organizations with conflicting mandates. Such endeavors can be time-consuming and frustrating, but any attempt by a single agency or policy maker – no matter how knowledgeable or confident – to act unilaterally will lead to political opposition, legal challenges, and deadlock. Although communication can be difficult in the midst of conflict, the only way to succeed is to facilitate face-to-face conversation.

As this article has made clear, multiparty problem solving is at the heart of environmental planning. Professional planners must balance science and politics, reconcile conflicting values, help a variety of stakeholders take account of constantly changing roles and relationships, and take responsibility for championing sustainability.

NOTE

1 Mathis Wackernagel and William Rees. *Our Ecological Footprint: Reducing Human Impact on the Earth* (Gabriola Island. B.C.: Canada New Society Publishers July 1995).

Elinor Ostrom — "The Future of the Commons: Beyond Market Failure and Government Regulations." In *The Future of the Commons: Beyond Market Failure and Government Regulations.* **London: Institute of Economic Affairs.**

Reading Commentary

In this excerpt from *The Future of the Commons*, Ostrom develops an overview of the way in which societies and ecologies fit together (i.e., in "social – ecological systems," or "SES"). She explains how groups of stakeholders can manage common-pool resources. She also suggests that the only way to do this is through "a common framework of language."

Ostrom delineates six key variables that should be included in any effort to analyze efforts to manage common-pool resources: the (1) resource system (e.g., forest); (2) resource units (e.g., trees within the forest); (3) actors that affect the long-term sustainability of the system; (4) governance system; (5) "action situation," or interaction among individuals or groups that generate outcomes; and (6) resulting outcomes.

The SES framework enables researchers to identify and evaluate instances of common-pool resource management in a way that will yield generalizable lessons without sacrificing the importance of context-specific characteristics. This is an extension of Ostrom's earlier work on the Institutional Analysis and Development (IAD) framework (2005) and part of her broader effort to develop systematic methods for conducting policy analysis. We include this reading in Part 1 of the book because it relates to the Howlett, Ramesh and Perl (2009) reading. Both offer a framework and a lexicon for policy analyses. Do you believe that these frameworks can overcome the "panacea trap" described by Ostrom?

CHAPTER 3
THE FUTURE OF THE COMMONS:
BEYOND MARKET FAILURE AND
GOVERNMENT REGULATION

Elinor Ostrom

INTRODUCTION

Earlier studies of the commons focused on small-to-medium-sized common-pool resources, such as irrigation systems, fisheries and forests, into which we and many others undertook a great deal of research. But many of the studies of particular common-pool resources were by people working in a particular discipline without comparison with other studies and without any theoretical foundation.

But as we looked at those studies in our own research and as we did some empirical work, we were able to get a good sense of how small – and medium-sized common-pool resources were managed by common-property institutions. Now it turns out, especially after 2009, that there is considerable interest in our research on small, medium, large and global environmental systems. And researchers, citizens and officials are asking for some kind of a general framework that puts people and societies together and explains the ways in which they are able to manage common-pool resources.

When we put people and ecologies together, we can think of the results as a 'social-ecological system' (SES). Academic teams do not tend to share resources across disciplines, so as you are trying to study these things you have to learn about the terminology from other disciplines. For example, I am now working with a group of ecologists studying forest resources and I had to learn what DBH meant (it is diameter breast height) and how to measure the size of a tree. I also had to learn a variety of other technical things because we are measuring the condition of the forests in a scientifically very careful way besides looking at the social systems and how they are organised.

Without understanding both the social systems and the technical aspects of the management of a resource, we cannot conduct work that enables us to understand the conditions that help produce sustainable management. We need to have a common framework of language that will enable us to help develop sustainable systems and achieve the sustainability of diverse commons.

CHALLENGES IN ACHIEVING SUSTAINABILITY

So what are some of the challenges that we face in achieving sustainability? The first one that I will talk about shortly is overcoming what I call the 'panacea trap'. A second one is developing a multidisciplinary, multi-tier framework for analysing sustainable social-ecological systems that people across disciplines can use. We need to build better theories for explaining and predicting behaviour. We need to find ways of collecting data over time, but we have got to learn which variables we should be studying in a consistent way to have good studies over time. And we need to understand design principles and why they work.

This is a very big agenda. They all point us to the importance of institutional diversity. In this lecture I will only be able to provide an overview of these, but I will be very glad to pursue one or another of them in questions if people want to do so.

Challenge one, as I mentioned, is the panacea problem. A very large number of policy-makers and policy articles talk about 'the best' way of doing something. For many purposes, if the market was not the best way people used to think that it meant that the government was the best way. We need to get away from thinking about very broad terms that do not give us the specific detail that is needed to really know what we are talking about.

We need to recognise that the governance systems that *actually have worked in practice* fit the diversity of ecological conditions that exist in a fishery, irrigation system or pasture, as well as the social systems. There is a huge diversity out there, and the range of governance systems that work reflects that diversity. We have found that government, private and community-based mechanisms all work in some settings. People want to make me argue that community systems of governance are always the best: I will not walk into that trap.

There are certainly very important situations where people can self-organise to manage environmental resources, but we cannot simply say that the community is, or is not, the best; that the government is, or is not, the best; or that the market is, or is not, the best. It all depends on the nature of the problem that we are trying to solve.

Challenge two that we also need to be working on is the development of a multidisci-plinary, multiple-tier framework for analysing social-ecological systems. And what we have done here is identify and analyse four very large encompassing variables that are at what we call a 'focal level'. These generate together an action situation in which individuals and groups interact and produce outcomes. When we talk about market relationships between buyers and sellers, we are looking at an action situation. And that focal level is affected by and affects larger and smaller ecosystems as well as larger and smaller social, economic and political systems.

So let us look at the first tier of that framework. We can think broadly of a resource system and a governance system – these are the sub-parts. A resource system sets conditions for an action situation, but we can think of resource units as part of that. So when we talk about a forest, part of the resource units are trees. If we talk about a fishery, the resource units are fish. They differ dramatically in their characteristics, but both are the resource unit that is being harvested.

We can also think of a wide diversity of actors who are participating in one or another of the action situations that affect the long-term sustainability of that system. They act within the governance system that sets the rules. This is a very broad framework, I am going to unpack it in a few minutes, but it is now being used by a number of people for current studies. So how does the framework help us build and test better theories?

That is the third challenge that we are facing. And the important thing is that the frame-work helps identify multiple variables that potentially affect the structure of action situations; the resulting interactions between the governance systems; the actions of the resource users and the resource system; and the outcomes in terms of the sustainable management of the resource.

And so this framework is one way that we can study similar systems that share some variables in common but that do not share all variables in common. It helps us to look at quite different systems. The framework then avoids the problem of people overgeneralising as they do in the literature, suggesting, for example, that all resources should be privately owned or that all resources should be government owned. If you read the original work on the tragedy of the commons, that was Garrett Hardin's conclusion. And in many contemporary textbooks, the Hardin argument is repeated.

There is also a problem of over-specification. Researchers can fall into the trap of pretending that their own cases are completely different from other cases. They refuse to accept that that there are lessons that one can learn from studying multiple cases. In reality, to diagnose why some social-ecological systems do self – organise in the first place and are robust, we need to study similar systems over time. We need to examine which variables are the same, which differ and which are the important variables so that we can understand why some systems of natural resource management are robust and succeed and others fail.

THE IMPORTANCE OF SECOND-TIER VARIABLES

Thus, part of our need is to look beyond the first tier of variables and to begin to develop the language more thoroughly by going on to examine a second tier of variables. Many of the second tiers have third and fourth tiers – but I am not going to get down to that level tonight: we are working on that diagnostic framework further. You can see a version of this in my 2009 *Science* article, and Mike McGinnis and I are currently working on a paper that is looking at all this.

Figure 2 shows the second-tier variables that are important under each first-tier variable for a social-ecological system. I am going to warn you that when people see this for the very first time, there is a kind of worried reaction at its complexity. This looks very complex. When you start thinking about what is involved in a resource system, you need to know what sector you are talking about (for example, forest, pasture, fish). You need to know where the clear boundaries for the resource are (for example, how are the boundaries defined if the resource is mobile?). You need to know how big the resource is, what kind of human-constructed facilities there are, and so on. Similarly, if you are going to talk about a governance system, we need to know whether we are talking about government organisations and about the kind of non-governmental organisations that could be involved in the governance system. We also need to think about the various kinds of property rights systems, the monitoring and sanctioning rules, and so on – all these things are very important. Then there are very important problems relating to the attributes of the resource units. For example, there is a difference between fish that move independently and fish that move in channels, and a difference between both of these types of resources and, for example, trees, which do not move at all.

And then there are the attributes of the kinds of actors involved. How many are there? What kind of socio-economic attributes do they have? What is their history of use of the resource? Where are the actors – are they in a similar location to the location of the resource, or in places far away? What kind of leadership is there? And so on.

Figure 2 Second-tier variables of an SES

Social, Economic, and Political Settings (S)
SI – Economic development. S2 – Demographic trends.
S3 – Political stability.
S4 – Government resource policies. S5 – Market incentives.
S6 – Media organisation.

Resource Systems (RS)

RSI – Sector (e.g., water, forests,
pasture, fish)
RS2 – Clarity of system boundaries
RS3 – Size of resource system
RS4 – Human-constructed facilities
RS5 – Productivity of system
RS6 – Equilibrium properties
RS7 – Predictability of system dynamics
RS8 – Storage characteristics
RS9 – Location

Governance Systems (GS)

GS1 – Government organisations
GS2 – Nongovernment organisations
GS3 – Network structure
GS4 – Property-rights systems
GS5 – Operational rules
GS6 – Collective-choice rules
GS7 – Constitutional rules
GS8 – Monitoring and sanctioning rules

Resource Units (RU)

RU1 – Resource unit mobility
RU2 – Growth or replacement rate
RU3 – Interaction among resource units
RU4 – Economic value
RU5 – Number of units
RU6 – Distinctive characteristics
RU7 – Spatial and temporal distribution

Actors (A)

A1 – Number of actors
A2 – Socioeconomic attributes of actors
A3 – History of use
A4 – Location
A5 – Leadership/entrepreneurship
A6 – Norms (trust-reciprocity)/social capital
A7 – Knowledge of SES/mental models
A8 – Importance of resource (dependence)
A9 – Technology used

Action Situations: Interactions (I)

I1 – Harvesting
I2 – Information sharing
I3 – Deliberation processes
I4 – Conflicts
I5 – Investment activities
I6 – Lobbying activities
I7 – Self-organising activities
I8 – Networking activities
I9 – Monitoring activities
I10 – Evaluative activities

Outcomes (O)

O1 – Social performance measures
(e.g., efficiency, equity, accountability,
sustainability)
O2 – Ecological performance measures
(e.g., overharvested, resilience, biodiversity,
sustainability)
O3 – Externalities to other SESs

Related Ecosystems (ECO)
ECOI – Climate patterns. EC02 – Pollution patterns.
EC03 – Flows into and out of focal SES.

This framework for the management of the commons is a broad framework, just like when you learn economic theory more generally. You do not need to look at all the variables in an economic theory for all the questions that you are going to look at. You need to learn how to pick out the variables that are important for the analysis of particular questions.

This approach does give us a sense of some of the variables that have been identified repeatedly as being important when determining whether people are able to govern a resource and do so sustainably. They are useful for that purpose.

QUESTIONS THAT CAN BE ADDRESSED IN OUR RESEARCH FRAMEWORK

We can address three broad sets of questions within this research framework. The first set examines the patterns of interactions and outcomes that you might expect from a given set of rules for the governance and use of a particular resource system. This includes the question of how much overuse there will be; what kind of conflict there is likely to be among those governing the system; and whether a system with a particular set of attributes is likely to collapse or not. In other words, we are looking to find which rules generate sustainable outcomes for particular kinds of resources and looking at how to distinguish different kinds of resources that require different rules for their management. What we have learned is that the rules that are often used with regard to grasslands and pastoral institutions often generate overuse and collapse. We need to understand which ones do that and why. We need to understand which rules generate adaptation. And we need a framework of that sort to develop good research and good theories as we move along.

The second type of question is for a particular resource in a particular setting. What is the likely endogenous development of different governance arrangements, use patterns and outcomes with and without externally imposed rules or financing? This helps us to answer the important question of whether we need to impose institutions from outside.

We have been studying irrigation systems and forestry resources around the world. I have just finished a paper with an Indian colleague looking at the lakes in Bangalore in an urban area and comparing their sustainability. We need to know when we can expect the local people to be able to develop their own rules so that we know when to worry about whether a particular situation is one in which we are going to need to impose rules from outside. In what situations are the local people going to develop well-tailored rules of their own and how do we predict that they are going to do so? This depends on the autonomy of people living in a particular setting and using a resource, and their history.

The third type of question is how robust and sustainable is a particular configuration of users, resource systems, resource units and governance systems to external and internal disturbances? In other words, we need to look at the long-term sustainability both of resource units and governance, what kind of disturbances – such as climate change or population change – we are potentially going to see and whether we need to worry about them.

So all three of these are part of a long-term and big research programme, but all of them are enhanced by having a common framework for understanding social-ecological systems and the management of natural resources.

In researching these problems, a major challenge is to find comparable data over time for testing theories. This is another situation where challenges in terms of doing research are exacerbated by very tall walls between disciplines in terms of approaches and language. People do studies coming from one academic discipline and those studies can be very hard for someone in a different discipline to understand. As such, we need a common taxonomy of core variables in the social-ecological framework that will help us build more empirical research that we can all study.

Individual researchers have written a large number of individual case studies, but there has not been as much accumulation of scientific knowledge as we need. We need a large number of case studies because we see such variation between situations where natural resources are managed. If the variation is across only one or two variables, you do not need a large number of studies. But when you find more than a hundred different combinations of variables, as we have, you need large, large studies.

One of the initial things that we have been doing over time has been to study these cases and these combinations of variables. We developed a database early on in which we coded a lot of information about irrigation systems and fisheries. I thought that I was going to be able to analyse a series of cases using statistical analysis, but found out that I had to move up in my level of generality and look at a broader way of thinking about the problem. Instead of the details of a boundary rule, we had to look at whether they had a boundary rule at all. Instead of the details of collective-choice mechanisms that they might use, we had to ask whether they had the right to make their own rules and so on.

DESIGN PRINCIPLES FOR THE MANAGEMENT OF NATURAL RESOURCE SYSTEMS

Back in the 1980s, my co-researchers and I were struggling to try to find statistical relationships between features of social-ecological systems and outcomes. I developed a series of rules that were more general than specific, having failed to find the specific rules that were always successful in terms of producing sustainable outcomes for the management of a natural resource. I called these general rules 'design principles'. At times, I think that I should have called them something else because people confused that term with the idea that we are trying to design something from the beginning. However, I was really undertaking a study of robustness of systems that already existed. I presented the principles, which are discussed in great detail, in my 1990 book *Governing the Commons*.

I am very pleased to report that Cox, Arnold and Tomas have finished a very interesting article, published in *Ecology and Society* in 2010, where they searched the literature for people who had overtly studied whether or not the design principles that we identified actually characterised the case studies that they were looking at.

People had indeed done studies, and the authors looked at whether or not the management of a resource was successful and whether the design principles were helpful in bringing about that success. Cox, Arnold and Tomas looked at more than ninety studies and they did find very strong empirical support for the original design principles. The authors then suggested a better way of framing the design principles than I had done originally. For example, when I talked about boundary rules, I did not make a distinction between a clear set of boundaries of the resource and a clear set of boundaries for the users. Sometimes systems have clear boundaries for the resources but not for the users or vice versa and, in some of the case studies that were reported, that was a problem. So Cox, Arnold and Tomas crafted and clarified three of the design principles. They distinguished between clear boundaries of the resource users (that is the membership) and clear boundaries of the resource itself. So hopefully we will use that in our future work.

A second design principle is congruence with local and environmental conditions. Here, I am talking about the distribution of benefits and costs to the social structure, and I did not distinguish between the social part and the ecological part as Cox, Arnold and Tomas have.

In terms of monitoring, they distinguished monitoring of the resource conditions as well as monitoring of users' actions. Besides the boundaries and congruence and monitoring, my original design conditions also talked about graduated sanctions, conflict-resolution mechanisms and a recognition of the right of users to make their own rules, and, if it was a larger system, whether it was nested. They found very strong support for all of these and no need to distinguish them over time.

One important question is, why do the design principles work? Why do they enhance institutional robustness? One thing that we find is that the participants in a system that is characterised by the design principles know that the rules are being followed by others because they are monitored. A second reason is that those who are most knowledgeable about the effects of what is going to happen are the ones who are making the rules. A third is that they lead to a system where it is possible to resolve conflicts before they escalate.

We also find that a diversity of governance units trying to solve a fishery or irrigation or other resource problem stimulates learning and increases performance over time. And, as you study these things over time, you see people passing information about how they are doing and about why what they are doing is working. We find that both large and small units back each other up. So that is one of the important sets of findings from our research.

WHAT HAVE WE LEARNED?

In general, then, what have we learned? The attributes of the users that are conducive to their self-organising and managing a resource sustainably include that the users ask questions and that they view the resource as highly salient. They then usually have a relatively low discount rate in terms of the benefits obtained from the resource so that they are not over-exploiting the resource in the current time period. Over time, the users have developed high levels of trust and reciprocity and have the autonomy to determine at least some of their own rules. They are nested in complementary, multiple-tier systems. Usually in these kinds of settings, those organising the system have prior organisational experience; they have well-developed social capital and they have local leaders who are able to take on that very tough job. They also share some common understanding about the resource. These are the attributes that we are finding in systems that are sustainable.

At the same time, we are finding that the rules devised by self-organising communities differ in important ways from a lot of our traditional textbook remedies. For example, many of the textbook recommendations for regulating fisheries, if they are not for government to regulate them, are for individual transferable quotas. The key thing, it is argued, is regulating the quantity of the quota allocated. Yet, what we find in practice in many self – managed fisheries is that the fishers regulate the time when users of the resource can go and fish and they regulate the space where it is appropriate to harvest and the technology that should be used. The sustainable remedies in practice differ from the traditional textbook solutions, so those managing resources in practice are actually using different attributes from those suggested in the literature. Many of the rules that people develop or their methods of interrelationship are designed to encourage growth of trust and reciprocity. They tend to rely on unique aspects of a local resource and the local culture when developing their approach to managing the resource.

THE RELATIONSHIP BETWEEN LARGER AND SMALLER UNITS OF GOVERNANCE

We also find that larger regimes can facilitate local self-organisation so that we are not thinking about little tiny units self-organising without any relationship to larger units. And very large units can be important in providing accurate scientific information to help the smaller units interact. For example, in the groundwater basin that I studied in southern California, the national United States Geological Survey has done some important research that helps local people figure out the boundaries of their resource.

Larger jurisdictions can also provide important conflict resolution. For example, court systems provided by larger jurisdictions are very important for helping resolve basic conflicts. Larger jurisdictions can provide technical assistance, which is effective if they view the local users as partners. It is important that they do not just assume that the locals do not know very much and tell them what to do! If there is some respect for the local user, the technical information provided by larger units can be very helpful. And the larger units can provide mechanisms for backing up monitoring and sanctioning efforts.

We have also looked at larger units that are donor-assisted units that are supported through the US government via USAID and by development agencies of one kind or another. We conducted a major study of them and produced a book entitled *The Samaritan's Dilemma*. What we found is that, tragically, they do not have a good foundation based on either theoretical or empirical knowledge. They frequently encourage a national government to give resources back to local people. But the resources have been taken away, degraded and then given back in a one – or two-hour meeting. I have been to some of those meetings and it is rather incredible. They bring the local people into a hall. They say now you own x'; they give them a little bit of background of what they must do now; tell the people that they are responsible; and then walk away.

Frequently in these kinds of situations, the governments retain formal ownership so that they are not passing on the ownership but only the management. Furthermore, they expect the users to perform rapidly what government agencies have not been able to do for years. So there is a very grim history out there in terms of donor-assisted handover projects of natural resource systems to local people.

One of the things that we have repeatedly found is the importance of what we call polycentric systems. This is where systems exist at multiple levels, with some autonomy at each level. So, we can think about a region where there is a government agency responsible for the large region, but there is a lot of local autonomy in the management of local resources in that region. If we create a polycentric system, then it retains many of the benefits of local-level systems because there are people at a local level making decisions about many of the rules. But it also adds overlapping units to help monitor performance, obtain reliable information and cope with large-scale resources. Indeed, I argue very strongly for the need for polycentric institutions to cope with climate change.

CONCLUSION

I have given you a very rapid overview of a vast amount of research. The final question is, so what? One of the things that we have found in our large-scale studies, much to the surprise of many people, is that local monitoring is one of the most important factors affecting resource conditions and the success of resource management systems in fisheries, pastures, forestry, water and so on. We now have studies published in *Science* on our forestry work, looking at situations where nobody was thinking that local users could be important monitors. But we examine this because we have found it so important in many studies. The local people pay attention to what is happening in the forest if they have some rights to collect. Local users are in the forest from time to time, and monitoring is not very expensive when it is done in this way. If you have to hire government officials to be the monitors, that is very expensive. And frequently, when you hire government officials, they cannot be paid very much so you have problems of corruption.

We are now working with colleagues on the social-ecological systems framework. We continue to fine-tune the framework, and will have reports on about ten updated studies in a special issue that is forthcoming later this year or the beginning of next year in the journal *Ecology and Society*.

We are also working on getting definitions of key terms done, and how this affects the development of theories. We are studying forestry, water resources and fisheries over time. We are trying to study which propositions hold with regard to diverse resources on diverse scales. So, that has given you a very fast overview.

Scenario Assignment:
Policy Evaluation

You are a senior policy analyst on the staff of the U.S. Secretary of the Interior's office. Your academic background includes a master's degree in planning (with an environmental policy specialization) and a doctoral degree in applied microeconomics. You have been asked to review a series of studies produced by the conservative Heritage Society that emphatically argue that the command-and-control policies used to protect air quality, water quality and endangered species for several decades are no longer needed. These studies indicate that now that we understand how these systems work and have internalized the need for pollution control and the protection of biodiversity, it is no longer necessary for the federal or state government to do more than emphasize these objectives through "soft" policy statements and educational programs. According to the Heritage Society, we certainly do not need the national government setting and enforcing detailed standards (including technological specifications of various kinds). It argues that this just locks us into outmoded technologies, inhibits innovation and applies a one-size-fits-all mentality when what we need is to "let a thousand flowers bloom." The Secretary is not convinced. He wants you to sketch a major national study that would rebut what the Heritage Society is saying.

1. How should such a study be organized? Who should be asked to do it? (Assume that the agency has upward of US$2 million to spend.)
2. What are the most important criteria and methods to use in preparing such a policy evaluation?
3. Given that the Secretary has already indicated to you that he does not agree with what the Heritage Society is saying, how should that factor into the design and implementation of the study?
4. Please list the steps you would take to ensure the credibility of the study in the eyes of the scientific community, the environmental advocacy community, the business community and Congress.
5. Is it appropriate to begin a policy evaluation with an outcome in mind?

Scenario Assignment:
Comparative Policy Analysis

The Europeans, Japanese and Australian governments have different approaches than the United States in their efforts to protect agricultural land and to ensure the quality of food production. Obviously, the history and culture of each country play a key role in shaping national policy. Assume you are the director of a not-for-profit institute that is focused on the preservation and expansion of agricultural production in the United States. Your own background includes growing up in a farming community, a graduate degree in agricultural science, ten years of experience managing a large agribusiness and a five-year stint living and working in Germany with their Ministry of Agriculture. Your board of directors includes a wide range of well-known industry, academic and nongovernmental organization (NGO) leaders. They are distressed that the United States is not doing nearly enough to protect farmland, particularly in rapidly developing areas of the Northwest, the West and the Southeast. They want you to take a close look and see what the United States ought to learn from the experiences of the European Union, Japan and Australia. They will fund you to make separate two-week trips to each of those three parts of the world. You can bring along one of your senior staff members. You can collect any information you need before you go.

1. Who do you want to see in each country and what do you want to try to find out?
2. What problems do you anticipate with regard to making sense of what you are told and what you observe?
3. What problems do you anticipate with regard to generalizing across your findings from other regions of the world? Your Board expects you to enumerate the opportunities for and the obstacles to cross contextual learning.
4. To what extent do you think that your findings will reveal ideas and strategies that will be helpful in the United States? What makes you think so?
5. How ought you deal with the skeptics on your board who feel that the laws, customs, regulations, ecosystems and political systems in each country are too different for anything useful to come of a comparative analysis?

**End of Unit I Written Assignment:
National Environmental Policy-Making**

It should now be clear that national environmental policy-making is a much less structured and a much more haphazard process than many policy scientists have suggested. Given what you have read in Unit I, provide the simplest model you can of national environmental policy-making for any country you choose.

In developing your model, include a one-page diagram of your model and address the following questions: (1) What are the key variables and forces at work in your model and why have you selected them? (2) Which variables or forces are unique to "environmental" policy-making, if any (as compared to public policy-making in general)? (3) To the extent that future environmental policy is largely a product of previous policy and practice, what is your sense of how major shifts in national environmental policy might occur?

First Example Response to Assignment:
National Environmental Policy-Making in the United Kingdom

In this paper, I present a model to describe the environmental policy-making process in the United Kingdom. I identify three factors that are characteristic of the UK system in particular. These are (1) a notable culture of rural conservatism and countryside heritage, (2) a strong-handed and influential tabloid media and (3) the dominance of an elite, club-based political structure. I argue that these factors generally act to limit the array of plausible policy options open to policy makers, contributing to a general trend of political stasis or incrementalism within British environmental policy that is only overcome on rare occasions when a number of ephemeral factors combine to cause a substantive reframing of the issue in question.

My model is based on the stages of policy-making described by Howlett, Ramesh and Perl (2009): agenda setting, policy formulation and policy decision-making. I present these stages as an increasingly tight funnel through which policy ideas must pass if they are to reach implementation. The funnel represents the policy subsystem surrounding a particular environmental issue. At each stage of the process, policy ideas mingle, are compared and compete. Some ideas succeed in passing through to the next stage, while others are discarded. Ideas may join with others or be cycled back into previous stages of the process. The early stages of the process are critical because they largely determine the options that will be on the table further down the line and frame how they will be discussed. The movement and development of ideas is primarily dependent on the efforts of "policy entrepreneurs" (Kingdon 1984) actively engaged in a particular policy subsystem. At each stage competing policy entrepreneurs work either to open up or to narrow the breadth of policy ideas that are considered worthy of serious consideration, represented in my model by the funnel's width.

I include three factors that are distinctive of the British policy-making process. These factors all result in a limited scope of options that policy makers can plausibly consider, indicated in my model by a narrower policy funnel than would otherwise be the case. First, and most relevant to environmental policy-making specifically, Britain has a strong and long-standing largely middle-class cultural emphasis on the protection of a bucolic image of the countryside and associated way of life. Despite the fact that a dwindling number of people now live in rural areas, this emphasis has been powerful in contributing both to the establishment of strong environmental regulations, such as "greenbelt" planning rules restricting urban encroachment into rural areas as well as a ubiquitous pattern of "NIMBYism" that has severely constrained efforts to, for instance, expand onshore wind and solar farms. Second, these conservative, heritage-focused values are propped up by a very influential tabloid media. British politicians are terrified of provoking the wrath of the *Daily Mail* or *The Sun*, which are owned by science-skeptical individuals and are so dominant in steering the opinions of a large swath of "Middle England." Inaccurate coverage by the tabloids of climate change science (Boykoff and Mansfield 2008), for instance, has played an important role in limiting the British government's ability to take more ambitious policy action on this front. Finally, and in comparison to more open policy-making cultures such as the United States, UK politics in general tends to operate based on "the collegiate traditions and club-like instincts of the British policy-making elite" (John and Jennings 2010, 565), which act to diminish the range of actors with access to the policy-making process, and therefore the range of policy options that can be legitimately considered.

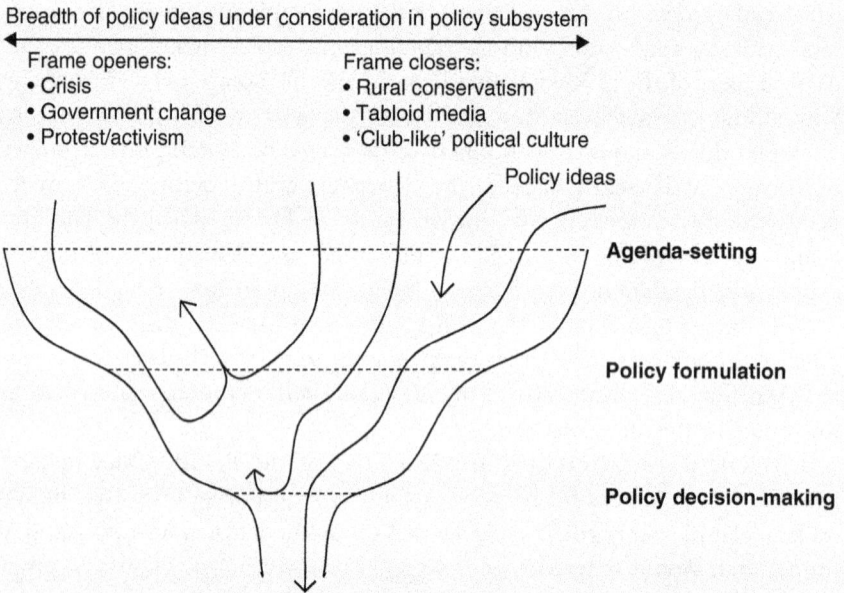

Breadth of policy ideas under consideration in policy subsystem

Frame openers:
• Crisis
• Government change
• Protest/activism

Frame closers:
• Rural conservatism
• Tabloid media
• 'Club-like' political culture

Policy ideas

Agenda-setting

Policy formulation

Policy decision-making

I believe that the most critical factor in determining policy stasis or change is the level of control that authorities exert over a particular issue's framing. The dynamics of issue framing are central to the tendency for policies to stay relatively stable most of the time, interspersed with rare instances of significant change. With their influential punctuated equilibrium (PE) theory, Frank Baumgartner and Bryan Jones (1993) present convincing evidence to support this assertion for the United States. According to PE theory, policy makers seek to exert maximum authority over the policy subsystem under their management, restricting the number of plausible policy options under consideration to the existing arrangement or alternatives that they favor. For example, this might be achieved by reinforcing the framing of an issue as a purely technical problem that should be limited to the realm of "experts" to manage. Conversely, proponents of change seek to reduce this control, redefining the problem in different terms and opening up the policy funnel to a broader range of options. Most of the time, dominant negative feedback processes act to keep policy positions stable. However, on a relatively rare basis external "focusing events," structural political changes or significant technological developments can act to overcome these negative tendencies and produce substantial change (ibid.).

More recent work by other scholars indicates that PE theory also holds largely true for the United Kingdom. P. John and W. Jennings (2010) note that most British policy communities spend most of their time shielded from the glare of media attention and public opinion. At this time a subsystem's policy funnel could be considered closed, or slightly ajar, if the subsystem is undertaking some internal considerations over small policy tweaks. However, if a swing in the wider political agenda brings that policy area abruptly into the spotlight, then its policy funnel might be suddenly wrenched open. In instances like this the framing of the issue in question is critical in determining how the policy-making process will play out. If the government is unable to maintain control over an issue's framing, then a definitional vacuum will open up in which other groups will seek to reframe the issue in their own way, undermining the government's authority to effectively handle its management (Kingdon 1984).

Recent debates over nuclear energy in the United Kingdom are an instructive example. The Fukushima nuclear crisis in March 2011 elicited a global focus on nuclear safety. In the United Kingdom, the government, which has nuclear as a key pillar of its long-term energy strategy, was successful in controlling the framing of issue to one of technical safety, and was able to limit the breadth of legitimate voices to the small number of experts tasked with carrying out a technical report on the implications (if any) of the accident for British energy policy. Public protest based on alternative ethical or moral framings of the threats of nuclear power were kept to a minimum, and the government was able to continue with its current policy while avoiding significant political controversy. The ongoing debate over the question of whether to construct the first new nuclear reactor in a generation at Hinkley Point in Somerset provides a contrast. The government has justified this proposal largely on the basis of economic and national sovereignty framings, which are far more contestable than the technical framing employed post-Fukushima. The result has been an intense public controversy, with a wide range of critics disputing the government's arguments and depicting it as both inept and deluded. Interestingly, despite this outcry, the level of central control that the government holds over energy policy means that it looks likely to be able to continue with its plans nonetheless.

The ongoing controversy over shale-gas extraction, or "fracking," in the United Kingdom provides a further illustration of the importance of framing in determining policy outcomes in the United Kingdom. The viability of fracking at scale has only been demonstrated fairly recently in the United States, and now drilling companies are seeking to exploit opportunities in the Britain. Local protest movements have sprung up across the country in response to applications for well-drilling permits, but the decision of these activists to frame their arguments against fracking primarily in terms of its likely exacerbation of global climate change has limited their success (Hilson 2015). This is because a global framing of the problem does not fall within the boundaries that the local regulatory and planning agencies adjudicating on the permitting decisions can consider, and so has been mostly unable to influence their decisions. If the process for local planning decisions were revised to consider a wider range of factors then, in my model, the policy funnel would become more open. However, the authorities are unlikely to accept the diminishment of control that such a change would represent.

I have presented a model for environmental policy-making in the United Kingdom that envisions multiple ideas competing to pass through a multistage policy funnel. In this model, those in control seek to limit the breadth of ideas that can enter and pass through the funnel, usually by attempting to control an issue's framing. Dramatic changes in policy are only possible when policy entrepreneurs are able to shift the popular framing of the issue in question to one that implies a substantively different policy response. My model includes three wider societal factors particular to the United Kingdom that serve to limit the likelihood of such considerable reframings occurring in British environmental policy-making processes.

References

Baumgardner, F., and B. Jones. 1993. *Agendas and Instability in American Politics.* Chicago: University of Chicago Press.

Boykoff, M., and M. Mansfield. 2008. "'Ye Olde Hot Air': Reporting on Human Contributions to Climate Change in the UK Tabloid Press." *Environmental Research Letters* 3: 1–8.

Hilson, C. 2015. "Framing Fracking: Which Frames Are Heard in English Planning and Environmental Policy and Practice?" *Journal of Environmental Law* 27, no. 2: 177–202.

Howlett, M., M. Ramesh and A. Perl. 2009. *Studying Public Policy: Policy Cycles and Policy Subsystems*. 3rd ed. Oxford: Oxford University Press.

John, P., and W. Jennings 2010. "Punctuations and Turning Points in British Politics: The Policy Agenda of the Queen's Speech, 1940–2005." *British Journal of Political Science* 40: 561–86.

Kingdon, J. 1984. *Agendas, Alternatives and Public Policies*. Boston: Little, Brown.

Second Example Response to Assignment:
A Model of Environmental Policy-Making in the United States

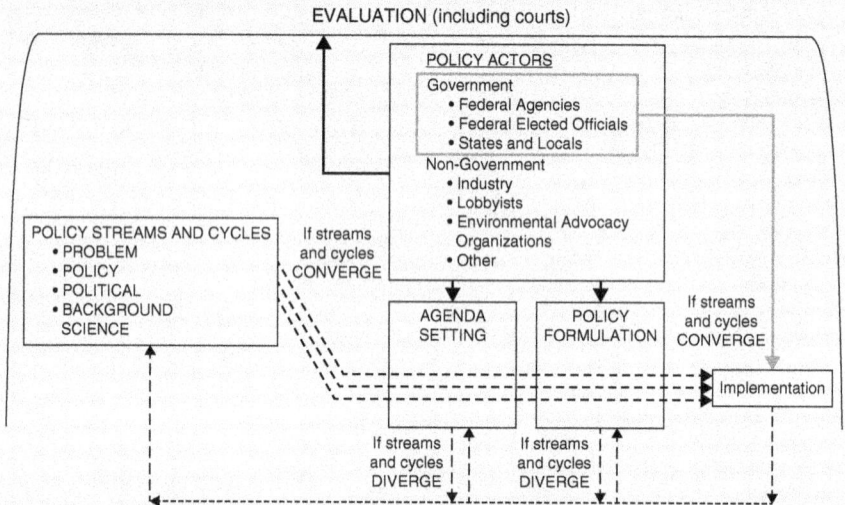

EVALUATION (including courts)

POLICY ACTORS

Government
• Federal Agencies
• Federal Elected Officials
• States and Locals

Non-Government
• Industry
• Lobbyists
• Environmental Advocacy Organizations
• Other

POLICY STREAMS AND CYCLES
• PROBLEM
• POLICY
• POLITICAL
• BACKGROUND SCIENCE

If streams and cycles CONVERGE

AGENDA SETTING

POLICY FORMULATION

If streams and cycles CONVERGE

Implementation

If streams and cycles DIVERGE

If streams and cycles DIVERGE

This paper covers a model of environmental policy-making in the United States. I have copied the basic framework of policy agenda setting, policy formulation and policy decision-making from *Studying Public Policy: Policy Cycles & Policy Subsystems* by Michael Howlett, M. Ramesh and Anthony Perl (2009). Onto that framework, I have added the policy actors and other variables that shape each step.

The diagram has been drawn linearly to show a progression in the process, but importantly, there are also cyclical and nonlinear elements. Above all, this format demonstrates that there is a range in predictability among the elements of environmental policy-making, some of which involve the influence of policy actors and some of which do not.

For overall structure, I have included a modified version of John Kingdon's metaphor of a policy window (Howlett, Ramesh and Perl 2009, 103–5). In my diagram, if the policy streams *and* cycles and policy actors' influences converge, a policy moves forward. If not, it returns to influence other steps in the process.

The first section (left) of the diagram includes Kingdon's (1984) "streams" and cycles. In a slightly differently way than Kingdon frames the process, in order to capture features of the process that influence policy-making but occur outside the influence of policy actors and before a given policy gets made, in this diagram I have included the streams as separate from but leading to agenda setting. Like Kingdon, I have included problem, policy and political streams (ibid., 103). In the case of environmental policies in the United States, problems—especially environmental crises—often drive the policy process. After Superstorm Sandy in 2012, there was increased interest in storm resilience policies at the federal level. The storm dictated the environmental policy, beginning with agenda setting (increased interest in resilience), but the storm itself obviously occurred before and outside the federal policy-making process.

Similarly, there are streams of policy, such as studies, reports, analyses and even previous policies, that influence policy-making but may be introduced before or unrelated to a specific policy being made. In my diagram, political streams would only include those elements before and outside the control of the political actors, such as public perception of an issue before the policy process begins (ibid., 103). I have added cycles to streams because many of these elements are affected by features with repeating schedules, such as the federal budget (ibid., 95–6). Additionally, I have included background science as a stream, reflecting scientific understanding before a policy gets made. Scientific analyses done for a particular policy would be influenced in a later step by a particular policy actor. Therefore, these streams cover the various elements that influence policy-making, but are not influenced by the policy actors and occur outside the process for making a particular policy.

In my diagram, agenda setting is where the process for making a certain policy begins. Howlett, Ramesh and Perl (2009) describe this step as the purview of government, but in my diagram, I included the influence of all policy actors because these actors set their own policy agendas, which can directly influence the government's policy agendas. On the nongovernmental side, advocacy organizations shape their agendas around the issues they think will gain the most traction with politicians and donors. Industry and lobbyists engage in similar agenda setting in establishing priorities based on needs, threats, opportunities and strategies. For example, with increasing momentum around federal climate change legislation in 2009, many environmental advocacy organizations increased their attention on that issue, while several groups in the fossil fuel industry and associated lobbyists ramped-up their efforts to discount the science around climate change (*Climate of Doubt*). Such programs represent agenda setting, as both sets of groups determined that among all of the issues that influence their work, climate change legislation was an important factor. In doing so, the groups sought to shape the federal policy agenda. Further, I have also included "Other" in this section to represent nongovernmental actors, such as the United Nations, that may be involved in influencing certain policy agendas but are not always active (Kraft 2007, 64).

As in agenda setting, I included these same nongovernmental actors in policy formulation because many times they bring specific policies, or at least ideas for policies, to the federal government that shape governmental policy formulation. Advocacy organizations often draft specific policies that they submit to government officials to be introduced, or at least to inspire future policies. Similarly, in some cases, industry representatives and lobbyists actually formulate policy. In 2001, an investigation by the *New York Times* uncovered that Vice President Dick Cheney met with 100 energy industry officials to write the administration's energy report, along with the president's energy policy suggestions (Van Natta Jr. and Banerjee 2002).

I have also included state and local governments as policy actors to capture the influence they can have on federal agenda setting and policy formulation, in terms of learning from experiences and/or applying best practices. As Justice Louis Brandeis argued, states can serve as "laboratories of democracy," with opportunities to experiment with policy and then scale the policy to the federal level (John 1994, 16–17). California frequently establishes more stringent and innovative environmental policies than the federal government. Following successes, Congress frequently mimics California state and local environmental laws in drafting federal versions (ibid., 70). Thus, state and local actors influence agenda setting and policy formulation through experiences and test cases.

The final concrete step in the diagram is decision-making and implementation. In the case of implementation, I have included only government as influencing the process because government is the only entity that actually puts the policy in place, even if other actors are affected by it. States and federal partners often implement the policy together, as in many features of the Clean Water Act. Following implementation, a policy influences future policymaking by providing lessons learned for all previous steps.

I have also included evaluation, not as a step but as an umbrella over the whole process, because evaluation influences future policy-making and can be undertaken by any actor at any time. Policy actors might evaluate previous streams to inform future policy, or they might evaluate a recently passed policy to determine its effectiveness. Additionally, courts can serve as evaluators and therefore as policy makers in mediating disputes and determining the legality of policies (Sussman, Daynes and West 2002, 236). These evaluations can influence any step in the process, and thus evaluation serves as an overarching feature of policy-making rather than a specific step.

Overall, this diagram highlights the different roles and steps in federal environmental policy-making in the United States. With linear and nonlinear elements, as well as consistent and nonconsistent parts, it reflects a combination of predictability and uncertainty in environmental policy-making. Due to the common nature of some aspects of all policy-making in the United States, this diagram may reflect other policy-making sectors, not just environmental policy. However, because of the complex nature of environmental policy, with overlapping agencies' jurisdictions, unique political arrangements and industry regulation, a more precise diagram for other policy-making is needed. Additionally, because of the model's cyclical nature and the entrenched American political processes, this model of environmental policy is unlikely to change in structure, and yet the balance among the steps and actors within the structure do change.

References

Climate of Doubt. 2012. Directed by Catherine Upin. PBS WGBH/Boston Frontline. Accessed September 29, 2016. http:// www.pbs.org/wgbh/frontline/film/climate-of-doubt/.

Cohen, Steven. 2006. *Understanding Environmental Policy*. New York: Columbia University Press.

Howlett, Michael, M. Ramesh and Anthony Perl. 2009. *Studying Public Policy: Policy Cycles and Policy Subsystems*. 3rd ed. New York: Oxford University Press.

John, DeWitt. 1994. *Civic Environmentalism: Alternatives to Regulation in States and Communities*. Washington, DC: CQ Press.

Klyza, Christopher McGrory, and David Sousa. 2008. *American Environmental Policy, 1990–2006: Beyond Gridlock*. Cambridge, MA: MIT Press.

Kraft, Michael E. 2007. *Environmental Policy and Politics*. 4th ed. New York: Pearson Longman.

Layzer, Judith A. 2006. *The Environmental Case: Translating Values into Policy*. 2nd ed. Washington, DC: CQ Press.

Sussman, Glen, Byron W. Daynes and Jonathan P. West. 2002. *American Politics and the Environment*. New York: Longman.

Van Natta, Don Jr., and Neela Banerjee. 2002. "Top G.O.P. Donors in Energy Industry Met Cheney Panel." *New York Times*, March 1, sec. US. https:// www.nytimes.com/2002/03/01/us/top-gop-donors-in-energy-industry-met-cheney-panel.html.

Unit II

ETHICAL DILEMMAS IN ENVIRONMENTAL PROBLEM-SOLVING

Introduction

All environmental problem solvers, whether public officials or not, have to make ethical assumptions. These are usually informed by basic philosophical beliefs about what is right and what is wrong. Discussions of environmental ethics usually lead to a cascade of follow-up inquiries that often dominate environmental policy-making and problem-solving: What is the proper relationship between humans and the environment? Do humans have a moral obligation to protect the environment? How can such an obligation be reconciled with our need to survive and our desire to prosper?

The inherent conflict posed by these questions constitutes what philosophers call an ethical dilemma. Multiple answers are available depending on the lens we rely on. For example, on one side are those who think in purely utilitarian terms (i.e., the instrumental value of the environment). They believe that humans can and should do whatever will enable them to derive the most benefit or satisfaction from the environment. An alternative view is that humans have a responsibility to serve as stewards of the natural environment (i.e., the intrinsic value of environment). This group of thinkers believes we should do whatever we can to protect the environment. We do not presume to know which choices are correct, but we are convinced that all environmental problem-solving needs to confront them.

From our standpoint, three other ethical dilemmas are equally compelling: (1) Should knowledge derived through the application of formal scientific methods outweigh local or indigenous knowledge when it comes to decisions about natural resource management? (2) Should economic concerns outweigh long-term sustainability of depletable natural resources and ecosystem services? And, finally, (3) Should we err on the side of precaution when there is uncertainty about the possible effects of proposed human actions, or should we have confidence that human ingenuity (and technological innovation) will enable us to repair any damage our actions might cause?

This unit illuminates a range of arguments on all sides of these dilemmas, from advocates of "deep ecology," who believe that "reverence for life determines who we are," to "pragmatists," who believe that "the satisfaction of human preferences should be the overriding goal of public policy."

We believe a solid understanding of environmental ethics is necessary both in terms of recognizing how your own ethics influence your decision-making process and of communicating with others. That is why we begin with an excerpt by Joseph DesJardins, one of the best environmental philosophers we know. We recommend reading the entire book when you have a chance, but for the purposes of this book we have included passages that question the utilitarian ethics that typically drive policy-making.

The second reading is an excerpt about the use of "street science," or locally collected or driven knowledge, to help better inform policy-making. Science plays an important role in policy-making, but the uncertainty inherent in science—in relation both to how the boundaries of a given problem (e.g., determining risk) are defined and to the complexity inherent in each challenge—is rarely acknowledged or communicated. Jason Corburn illustrates the important role that the local community can play by being involved in assessing environmental health risk. We believe that a joint approach—where technical experts work with the local community on fact-finding—can significantly improve the environmental decision-making process, rendering it more flexible and resilient.

The first scenario asks whether it makes sense to rely on "the precautionary principle" given scientific uncertainty regarding a dwindling natural resource. The second asks you to define sustainability in the context of a specific effort to promote economic development. The third asks you to describe a process for tapping local knowledge in deciding on the desirability of permitting mining on indigenous land (where the community has suffered the adverse health impacts of such mining in the past).

The unit concludes with a writing assignment intended to test your grasp of the key philosophical and ethical concepts presented in the readings and scenarios. You can assess your answers by reviewing some of the MIT student responses.

Commentaries and Reading Excerpts

Joseph R. DesJardins — *Environmental Ethics: An Introduction to Environmental Philosophy*. Belmont: Wadsworth.

Reading Commentary

Environmental problem-solving is often portrayed as a series of rational and objective choices that will lead to a "desired" outcome. But how should we determine what is desirable?

There are a number of ethical assumptions or philosophical positions that environmental problem solvers can use to justify their preferred actions. They can draw on utilitarian ideals, which emphasize maximizing the gains that humans can derive from the use of nature or natural resources, or they can base their arguments on deontological ideals, which give greater weight to the duties or obligations that humans have to the natural environment or to each other. Thinking about alternative environmental philosophies forces us to reflect on the first principles that guide our actions. For example, are humans inherently worth more than the rest of the natural world? Do humans have an obligation to maintain or sustain nature's endowment for future generations? Do animals and other nonhuman species have a right to exist?

DesJardins, a philosopher, challenges the conventional approach to environmental decision-making—one based on an anthropocentric or human-oriented ethic aimed at achieving the "greatest good for the greatest number." This conventional view is known as utilitarianism. DesJardins offers deep ecology as an alternative to utilitarianism. The eight-tier deep ecology platform, as developed by Norwegian philosopher Arne Naess and American environmentalist George Sessions in 1984, maintains that all life has inherent worth and calls for a balance between human and nonhuman life. This "balance" probably requires scaling back a whole range of human activities—both in terms of consumption levels and the size of the human population. Although DesJardins does not come out in favor of, or against, deep ecology, his point is well taken—there is nothing inherently ethical about our utilitarian approach to environmental decision-making or the tools we use to ensure "efficient" allocation of natural resources. The clash between utilitarianism and deep ecology—two ways of approaching these and related questions—illuminates the choices that must be made whenever we engage in environmental problem-solving or policy-making.

The third unit of this book highlights the implicit value judgments inherent in the selection of tools like environmental assessment or cost-benefit analysis. For example, costs and benefits can only be calculated after decisions have been made regarding the time frame that will be used for the analysis. This hinges on non-objective judgments because there is no scientifically or technically correct choice. If you think that the long-term implications of natural resource management decisions are very important, you would structure your analysis accordingly. If you think that maximizing immediate "return" to a subset of the population is most important, you would arrange your analysis differently.

We hope that this excerpt encourages you to consider your underlying philosophical assumptions. How might they affect your choice of environmental decision-making tools? After reading this section, take a moment to think about whether or not there is a difference between the philosophy that you would advocate in theory and what you would do in practice. Finally, is there a difference between your personal philosophy and the philosophy you would use to garner others' support for a particular policy decision?

CHAPTER 2

PART 2.2 PHILOSOPHICAL ETHICS: GETTING COMFORTABLE WITH THE TOPIC

Joseph R. DesJardins

The word *ethics* is derived from the Greek word *ethos*, meaning something like "customary" or "habitual." In this sense, ethics consists *of the general beliefs, attitudes, or standards that guide customary behavior.* Thus any society will have its own ethics in the sense that it will have certain typical beliefs, attitudes, and standards that determine what is customary. Whether consciously or not, the behavior of every individual is also guided by certain beliefs, attitudes, or standards. But from the earliest days of Greek philosophy when Socrates challenged Athenian authorities, *philosophical ethics* has not been satisfied simply to accept as right that which is customary. Ethics as a branch of philosophy seeks a reasoned examination of what custom tells us about how we ought to live. Indeed, it is fair to say that Western philosophy was born in Socrates's lifelong critical examination of the customary norms of Greek society.

This critical examination involves stepping back or abstracting ourselves from ordinary experience. (Perhaps this explains the common perception of philosophy as "too abstract.") We all hold certain beliefs, attitudes, and values about our ordinary, customary experience. Philosophy asks us to step back from this experience to reflect critically on it: Why do we believe the things we believe? Should we change our attitudes? Are our values justified? At this first level of abstraction, customary behavior is examined by appeal to some norm or standard of what *ought to* or *should* be done. The difference between ordinary experience and the first level of philosophical abstraction is the difference between what *is* done (or valued or believed) and what *ought to* be done (or valued or believed).

One of the first challenges in any study of ethics involves identifying an issue as an ethical issue. We all need to practice this stepping back in order to recognize ethical issues in our everyday experience. For example, in the classic environmental essay *The Land Ethic*, Aldo Leopold retells the story of Odysseus's return from the Trojan War.[1] Odysseus hanged a dozen women slaves whom he suspected of misbehavior. Because Greeks saw slaves as property, they apparently saw nothing ethically wrong with this action. Leopold uses this example to call for an "extension of ethics" to include human relations to the land. Just as Odysseus was ethically insensitive to the evil of killing slaves, we fail to notice the wanton destruction of the land. Leopold's point is that sometimes we need to work intellectually in order to be capable of noticing an ethical issue. The creation of the Endangered Species Act might be seen as another example of a previously unrecognized ethical issue coming to public attention. A major contribution of such notable environmental writers as Aldo Leopold and Rachel Carson was precisely this ability to get us to notice ethical and philosophical issues where previously we were limited by our customary beliefs, attitudes, and values.

Uncovering the limitations of our ethical and environmental consciousness is a regular theme in the chapters that follow. This may also turn out to be the cause of some of the frustration that characterizes ethical discussions and prevents us from following through on the careful thinking required in ethics. Many environmental controversies rest on different attitudes and values concerning our world. Few things are as frustrating as having our fundamental perspectives challenged. But we need to be open to the possibility that, like Odysseus, we might suffer from an ethical blindness. A primary goal of philosophical ethics is to stretch our understandings, shift our perspective and consciousness, and help us escape the limitations implicit in customary ways of thinking.

To make ethical judgments, give advice, and offer evaluations of what ought to or should be is to engage in *normative ethics*. This first level of abstraction is the type of ethical reasoning that most people associate with ethics. Normative judgments prescribe behavior. "We should reduce the level of carbon dioxide emissions." "Pesticide use should be reduced." "Endangered species ought to be protected." "Nuclear power plants should not be located in flood zones." Normative judgments implicitly or explicitly appeal to some norm or standard of ethical behavior. Many environmental controversies involve disputes of normative ethics. One side believes that the spotted owl ought to be protected even it this costs jobs in the timber industry, and the other side believes that jobs for human beings are more important than the life of some obscure bird. Each side cites evidence and appeals to certain norms to support its judgments. Normative disputes can be frustrating when ethical discussions are left at this level, with disagreements and controversies abounding.

However, philosophy insists that we not remain at the level of normative ethics. Resolving controversy requires us once again to step outside of, or abstract from, specific disagreements in order to examine the values in conflict and the competing factors that underlie the conflict. Moving to this more abstract level of thinking is moving from normative to philosophical ethics.

Philosophical ethics is a next level of generality and abstraction, at which we analyze and evaluate normative judgments and their supporting reasons. This is the level of the general concepts, principles, and theories to which we appeal in defending and explaining normative claims. This is the level at which philosophers are most comfortable and have the most to offer. The essence of philosophical ethics involves evaluating reasons that support normative judgments or seeking to clarify the concepts involved in such judgments. In this sense, environmental ethics is a branch of philosophy engaged in the systematic study and evaluation of the normative judgments that are so much a part of environmentalism.

As used here, the term *ethical theory* refers to any attempt to provide systematic answers to the philosophical questions raised by descriptive and normative approaches to ethics. These questions are raised from both an individual moral point of view and the point of view of society or public policy. Individual moral questions include: What should I do? What kind of person should I be? What should I value? How should I live? Questions of social philosophy or public policy include: What type of society is best? What policies should we follow as a group? What social arrangements and practices will best protect and promote individual well-being? What should be done when individuals disagree? From the earliest days of Western philosophy, ethics has included questions of both individual and social morality. Thus, in this broad sense, ethical theory includes philosophical analyses of moral, political, economic, legal, and social questions.[2]

Four general considerations make theory relevant to the study of environmental ethics. First, ethical theories provide a common language for discussing and understanding ethical issues. Environmental ethics is characterized by deep and numerous controversies. Clearly, a necessary first step in examining and resolving controversy is to understand these disputes fully and accurately. The basic concepts and categories of ethics—rights, responsibilities, utility, the common good, and the relationships among these concepts—can provide a basis for mutual understanding and dialogue. Ethical theories make explicit and systematize the common beliefs and shared values that are often implicit in specific controversies. By learning the language of philosophical ethics, we become better able to understand, evaluate, and communicate. This, in turn, can empower us to become full participants in environmental debates. Philosophical ethics can contribute to the common language that is essential to reasoned dialogue.

Second, because various ethical theories have played a major role in our traditions, they tend to be reflected in the ways in which many of us think. By learning about ethical theories, we become more aware of the patterns in our ways of thinking and the assumptions reflected therein. Thus we become better able to articulate our views and better able to defend them. Of equal importance, we gain a philosophical perspective that makes possible a critical examination of our ways of thinking. Having made these patterns explicit puts us in a better position to see and understand issues as ethical issues.

Third, one traditional function of an ethical theory is to offer guidance and evaluation. We can apply theories to specific situations and use them to generate specific recommendations. The long history of ethics gives us a reasonable and strong basis from which to analyze and offer advice. As we work our way through environmental controversies, it will be helpful if we do not have to reinvent the wheel at every step. People reason about ethics in standard ways, many of which match standard ethical theories. Because philosophers have spent considerable time pondering these theories and uncovering their strengths and weaknesses, knowledge of theory is an important resource for the debates that follow.

Finally, familiarity with ethical theories is important because some commentators claim that these theories, embedded as they are in common ways of thinking, have actually been responsible for some of the environmental problems we face. That is, the practice of environmental ethics occasionally involves challenging the very theories of ethics that philosophers have been busy defending. Some argue that these theories are part of the problem and have misled us. Thus an important part of environmental ethics is examining philosophical theories about ethics. In this way, environmental ethics not only benefits from traditional ethical theory but also contributes to the development of this branch of philosophy.

CHAPTER 3

PART 3.7 ETHICAL ANALYSIS AND ENVIRONMENTAL ECONOMICS

To the degree that contemporary economic analyses of environmental problems reflect a utilitarian ethics, philosophers have much to offer in the evaluation of this ethical theory. Several standard criticisms, mentioned briefly in Chapter 2, are a useful starting point for this evaluation.

Utilitarians face several problems when they attempt to quantify and measure consequences. These problems arise again in the use of cost-benefit analysis. One aspect involves the attempt to quantify qualitative goods. We have seen, for example, the challenge posed by trying to determine standards for clean or safe water and air quality. These qualitative goods find no place in the economic approach, because they cannot be easily quantified. A second problem is the resulting tendency to translate qualitative goods into categories that can be measured. Thus we find Baxter translating discussions of cleanness and safety into a discussion of risks, the probabilities of which can be quantified and calculated. We find O'Toole, in a manner consistent with typical applications of the cost-benefit method, translating qualitative goods into economic terms. The value of wilderness or recreation areas is understood as measurable by the willingness of users to pay for them. A final measurement problem involves the tendency to artificially restrict the range of relevant subjects. As presented in the chapters that follow, some critics claim that this tendency systematically ignores the well-being of animals, future generations, trees, the biosphere, and the like. We examine the charge that the economic approach is overly anthropocentric, or human centered, in greater detail elsewhere. The general point of these measurement problems is to raise the possibility that economic analysis seriously distorts or ignores important environmental issues.

In the book *The Economy of the Earth*, Mark Sagoff develops an insightful and convincing case against the use of economic analysis as the dominant tool of environmental policymakers.[1] In the remainder of this chapter, we will use Sagoff's evaluation as an example of the best that applied ethics has offered. Although his book offers a variety of subtle and powerful arguments, we concentrate on three major challenges to the use of economic analysis.

Sagoff argues that much economic analysis rests on a serious confusion between wants or preferences, on the one hand, and beliefs and values on the other. Economics deals only with wants and preferences because these are expressed in an economic market. The market can measure the intensity of our wants by our willingness to pay (by price), measure, and compare individual wants (through cost-benefit analysis), and determine efficient means for optimally fulfilling wants. But markets cannot measure or quantify our beliefs or values. Because many environmental issues involve our beliefs and our values, economic analysis is beside the point. When economics is involved in environmental policy, it treats our beliefs as though they were mere wants and, thereby, seriously distorts the issue. In an early article, Sagoff makes the following claims:

> Economic methods cannot supply the information necessary to justify public policy. Economics can measure the intensity with which we hold our beliefs; it cannot evaluate those beliefs on their merits. Yet such evaluation is essential to political decision making. This is my greatest single criticism of cost-benefit analysis.[2]

What exactly is the distinction between wants and beliefs, and why is it important?

When individuals express a want or personal preference, they are stating something that is purely personal and subjective. Another person has no grounds to challenge, rebut, or support my wants. Wants are neither true nor false. If I express my preference for chocolate ice cream, someone cannot challenge that and claim, "No, you don't." I have a certain privileged status with regard to my wants. In the public sphere, they are taken as a given. This is the way economists treat human interests. Willingness to pay measures the intensity with which I hold my wants (I will not pay more than a few dollars for a dish of chocolate ice cream), but willingness to pay says nothing about the legitimacy or validity of that want.

Beliefs, on the other hand, are subject to rational evaluation. They are objective in the sense that reasons are summoned to support them. Beliefs can be true or false. It would be a serious mistake (a "category mistake" in Sagoff's terms) to judge the validity of a belief by a person's willingness to pay for it. To put a price on beliefs is to profoundly misunderstand the nature of belief.

Sagoff reminds us that when environmentalists argue that we ought to preserve a wilderness area or an Alaskan fishing ground for its aesthetic or symbolic meaning, they are not merely expressing a personal want. They are stating a *conviction* about a public good that should be accepted or rejected by others on the basis of reasons, not on the basis of who is most willing to pay for that public good. Because economics has no way to factor them into its analysis, beliefs and convictions are either ignored or treated as though they were mere wants.

Essentially, O'Toole's marketization solution to environmental problems does exactly this. Remember that O'Toole's goal is to provide all the wilderness and the like that the American people "want." But he equates this goal with what the American people (in their roles as timber users, hikers, hunters, and so forth) are willing to pay. If recreational users are unwilling to pay a user fee that is economically competitive with the fees paid by the timber industry, then by definition they must not want recreation as much as timber users want the lumber. Returning to the discussions at the beginning of this chapter, if preservationists in the Gulf of Mexico area are not willing to pay as much for the land as are the oil companies, then they must not want the wilderness as much as consumers want oil. Likewise, if a community is unwilling to spend any more tax money to reduce air and water pollution, its residents must not want cleaner air and water as much as they want lower taxes or other public projects.

This tendency to reduce all beliefs and values to wants and preferences also seriously distorts the nature of the human being. That distortion treats people at all times as *consumers*.

People, at least insofar as the economist or policy maker is concerned, are simply the locations of a given collection of wants. People care only about satisfying their personal wants, and the role of the economist is to determine how to maximally attain this end.

The alternative that is ignored by economic analysis treats humans as thinking and reasoning beings. The market leaves no room for debate, discussion, or dialogue in which we can defend our beliefs with reasons. It ignores the fact that people are active thinkers, not merely passive "wanters." Most important, by ignoring the distinction between wants and beliefs, economic analysis reduces the most meaningful elements of human life—our beliefs and values—to matters of mere personal taste or opinion. To the degree that they are held with equal intensity, all desires equally deserve to be satisfied, no matter what the desire is.

This leads to a second major challenge to economic analysis. By ignoring the distinction between wants and beliefs, market analysis threatens our democratic political process. By treating us as always and only *consumers*, market analysis ignores our lives as *citizens*. As consumers, we may seek to satisfy personal wants. As citizens, we may have goals and aspirations that give meaning to our lives, determine our nature as a people and a culture, and define what we stand for as a people. Ours is a liberal democratic society—liberal in the sense that we value personal liberty to pursue our individual goals, but democratic in the sense that collectively we seek agreement about public goods and shared goals. Thus our political system leaves room for both personal *and* public interests. We are all, at one and the same time, both private individuals and public citizens. Market analysis ignores this public realm and thereby undermines our democratic political institutions.

According to Sagoff:

> Our environmental goals—cleaner air and water, the preservation of the wilderness and wildlife, and the like—are not to be construed, then, simply as personal wants or preferences; they are not interests to be "priced" by markets or by cost-benefit analysis, but are views or beliefs that may find their way, as public values, into legislation. These goals stem from our character as a people, which is not something we choose, as we might choose a necktie or a cigarette, but something we recognize, something we are. These goals presuppose the reality of public or shared values that we recognize together, values that are discussed and criticized on their merits and are not to be confused with preferences that are appropriately priced in markets. Our democratic political processes allow us to argue our beliefs on their merits.[3]

Economic analysis seems to assume a particular view of democracy wherein representatives passively follow the demands of the electorate, seeking to balance competing demands in a manner that satisfies the majority. The role of the politician in this model is to read the public opinion polls and act accordingly. But this neglects the more participatory nature of democracy in which citizens exchange views, debate their merits, learn from each other, and reach agreement.[4] The participatory model encourages a view of elected officials as active leaders rather than passive followers. We are committed not only to the personal freedom that Baxter's analysis assumes, but also to a system in which we mutually define and pursue a vision of the good life. A healthy, beautiful, undeveloped, and inspiring environment may not benefit me as a consumer, but it may be quite valuable to me as a citizen. This participatory model of democracy would reject the views of the new resource economists that O'Toole approvingly quotes: "It is a common misconception that every citizen benefits from his share of the public lands and resources found thereon."[5]

Many economists reject the notion of a public welfare or public good, because they view people solely as consumers. Not every citizen "consumes" the Alaskan wilderness, for example. But this fails to recognize that we are citizens as well as consumers and that we can benefit from the environment as citizens. The Alaskan wilderness can be valuable to us as citizens because of what it means to us, because of what it says about our self-image and self-respect. These benefits are not and cannot be priced in the market, so they are ignored by the type of economic analysis offered by O'Toole, the new resource economists, and Baxter.

A final challenge denies that economic analysis has any ethical basis at all. Despite the appearance that markets are committed to utilitarian ends, in actuality the goal of efficiency lacks any coherent and substantive ethical basis. Remember the role that economic analysis plays in many contemporary environmental issues. Unquestionably, economic and cost-benefit analyses are the major public policy methodologies used in reaching environmental decisions. Economics tells us as individuals, as a society, and as a government what we should do. Why should we follow this advice? Presumably because doing so will lead us to a better state of affairs. At first glance, this better state of affairs—economic efficiency—appears to be the utilitarian goal of providing the greatest good for the greatest number. But does economic efficiency provide the greatest good for the greatest number? Again, Sagoff is persuasive in claiming that it does not.

What is the goal of economic efficiency? As suggested earlier, efficiency implies optimal satisfaction of consumer preferences. An efficient market is one in which more people get more of that for which they are most willing to pay. But why should we, as a society and especially when we are concerned with environmental issues, take the satisfaction of individual preferences as our overriding goal? Why should this be the goal of public policy, when we recognize the obvious and acknowledge that many individual preferences are silly, foolish, vulgar, dangerous, immoral, criminal, and the like? Why should we think that satisfying the preferences of a racist, criminal, fool, or sadist is a good thing?

What is so good about satisfying preferences? The only options seem to be that satisfying preferences is good in itself or that it is a means to something that is good. In terms that we used in describing utilitarianism in Chapter 2, preference satisfaction is either intrinsically good or instrumentally good. Given the wide variety of harmful, decadent, and trivial preferences that exist, surely no one could claim that satisfying preferences is good in itself. Surely it is not good in itself that child molesters or rapists have their preferences satisfied. If not good in itself, what other good is brought about instrumentally by satisfying preferences?

Typically, this economic approach uses such terms as *utility, welfare, well-being*, or *happiness* to explain the goal of satisfying preferences. However, to explain the value of preference satisfaction by simply defining it in these ways is to beg the question by offering a trivially true explanation. On the other hand, if utility, welfare, happiness, and well-being are more thoroughly defined, the claim that preference satisfaction always leads to these goods is false. Satisfying my preference for a cigarette does not always make me happy in a nontrivial sense. Sometimes having my preferences frustrated can be in my best interest by teaching me patience, diligence, or modesty. Sometimes satisfying preferences is disappointing. Sometimes I might have all that the market can supply, but I might still lack what is important

("What would it profit a man to gain the whole world if he loses his soul?"). The economic methodology assumes that all other things being equal, for people to get what they want is a good thing. A more realistic and honest assumption would seem to be that whether what I want is a good thing depends on what it is that I want.

Thus, even if (and it is a big *if*) economic analysis could overcome the measurement problems and all the other problems associated with applying market analyses to the real world, and if the market succeeded in attaining its goal, we still would have no reason for accepting preference satisfaction as an ethical goal. An efficient allocation of resources is not itself an ethical goal at all.

CHAPTER 5

PART 5.5 DO TREES HAVE STANDING?

Before turning to more systematic attempts to extend ethical consideration to animals, we should examine another early and influential attempt at extending rights to nonhuman natural objects. Law professor Christopher Stone argues to extend legal, if not moral, rights to "forests, oceans, rivers and other so-called 'natural objects' in the environment—indeed, to the natural environment as a whole."[1] Unlike many defenders of animal rights, Stone bases his claim for standing less on the characteristics of humans and more on the nature of legal rights.

The occasion for Stone's defense of the rights of natural objects was the legal dispute concerning Mineral King Valley. The Sierra Club had filed suit to prevent Walt Disney Enterprises from building a large ski resort in the Sierras. This suit was rejected in California courts because the Sierra Club lacked standing. That is, members of the Sierra Club could not show that they would suffer any legally recognized harm by the development of Mineral King Valley. As this case made its way on appeal to the U.S. Supreme Court, Stone wrote an essay titled "Should Trees Have Standing?" Stone hoped to support the Sierra Club's case by arguing that the natural objects, such as trees and mountainsides, that would be destroyed in this development should be given legal standing. The Sierra Club could then be seen as a legal guardian of these rights.[2]

Stone's analysis begins with an examination of the nature of legal rights. Implicitly rejecting the view that rights are somehow there in nature to be discovered, Stone emphasizes the evolutionary development of rights. Rights exist when they are recognized by "some public authoritative body [that] is prepared to give some amount of review" to violations of that right. Citing Darwin's observation that "the history of man's moral development has been a continual extension in the objects of his social instincts and sympathies," Stone shows how the recognition of legal rights is witness to a parallel development.[3] Rights function to protect rights-holders from injury, and the list of rights-holders has been continually expanded. He reminds us that at one time only landowning white adult males enjoyed full legal rights. Legal standing now includes people who do not own land, women, blacks. Native Americans, and such things as corporations, trusts, cities, and nations. It is time to extend this protection to natural objects.

Stone argues that recognition by some authoritative body alone is not enough to establish the existence of rights.

As I shall use the term, "holder of legal rights," each of three additional criteria must be satisfied. All three, one will observe, go towards making a thing *count* jurally—to have a legally recognized worth and dignity of its own right, and not merely to serve as a means to benefit "us." They are, first, that the thing can institute legal actions *at its behest;* second, that in determining the granting of legal relief, the court must take *injury to it* into account; and, third, that relief must run to the *benefit of it*.[4]

The proposal to give legal rights to trees and other natural objects satisfies all three criteria. How can natural objects "institute legal actions" on their own behalf? Noting that corporations and mentally incompetent humans have legal standing, Stone argues that a guardian or conservator or trustee could be appointed to represent the interests of natural objects. Just as a comatose person has a legal guardian, for example, or a corporation, a board of trustees, forests, streams, and mountains could be legally represented by humans who are charged with representing their interests.

But do natural objects have interests that (1) we can agree on and (2) can be harmed in a legally recognizable way? Stone thinks that they do. Again noting the parallel with corporations, Stone believes that we can "know" the interests of and acknowledge the injuries to natural objects with at least as much certainty as we do in corporate cases.

The guardian-attorney for a smog-endangered stand of pine could venture with more confidence that his client wants the smog stopped, than the directors of a corporation can assert that "the corporation" wants dividends declared.[5]

Similarly, Stone believes that we can give meaning to the concept of a legal remedy that can provide relief to the injured natural object. As a guiding principle, we could adopt a common legal standard and aim to make the environment whole. Just as when a person is injured in an automobile accident and is compensated for medical costs to return that person to health, so we could require the responsible party to compensate the natural object by returning it to health. In this sense, "environmental health" would be the state in which the environment existed before the injury.

Consider how this proposal might work. During the summer of 2010, the British Petroleum (BP) deepwater oil drilling platform, the Deepwater Horizon, exploded sending millions of gallons of oil into the Gulf of Mexico. The three month-long oil spill affected coastlines from Florida to Texas, causing extensive damage to fisheries, wetlands, beaches, and wildlife habitats.

Under current legal guidelines, the door is open for injured humans to file for damages against BP, which was responsible for the oil drilling. Landowners along the coast or businesses that depend on tourism and fishing, for example, might argue that they deserve to be compensated for certain losses. Under Stone's proposal, representatives of the shoreline and of the fish and wildlife killed by the oil could also sue for damages. Thus, not only would humans be compensated for their injuries, but also the coast itself should be "made whole"—that is, returned to its pre-spill state.

There are, of course, challenges for this proposal to overcome. First, despite Stone's suggestions, it is not at all clear that we can agree on the interests of natural objects. For example, some believe that the Gulf of Mexico should immediately be restocked with fish from hatcheries. Others argue that the Gulf should be allowed to restock itself. Good reasons can be given to support both options. Which is in the best interest of the Gulf?

A second challenge follows from this. Perhaps Stone's response would allow the shoreline's guardians to make that decision in the same way that a legal guardian might decide what is best for an orphaned child. But who should this guardian be? The Wilderness Society would have one view of the shoreline's interests, and a local fishing industry might have another view. Choosing the guardian would also be to choose the theory of interests that is ascribed to natural objects. Should the Sierra Club represent the interests of Mineral King Valley or the Pebble Mine region? Should a lumber company or mining company?

None of this suggests that Stone's approach cannot work. But it does suggest that more work needs to be done to articulate and defend a view of nature's interests. Stone's proposal essentially relies on society's reaching a consensus about the extension of legal standing to natural objects. Legal standing is. after all, something that needs to be "recognized" by a public body. But it would seem that this consensus can be reached only after the public has already reached a consensus about the nature and value of natural objects. This consensus, regrettably, is still to be achieved.

CHAPTER 6

PART 6.2 INSTRUMENTAL VALUE AND INTRINSIC VALUE

One way to understand the philosophical shift that is occurring among environmental philosophers is to contrast questions of morality with more general questions of value. Morality, narrowly understood, has always taken human well-being and the relationship between humans as its focus. Morality seeks to understand the rights and responsibilities of humans, human well-being, and the good life for human beings. Therefore, it is perhaps not surprising that philosophers have difficulty granting environmental concerns moral consideration. Environmental concerns simply do not fit within the traditional domain of morality.

But understood more broadly, philosophical ethics asks more general questions about the good life and about human flourishing. These questions involve wider concerns of *value*. From this perspective, environmental concerns are more legitimately ethical concerns, because they raise a wide variety of value questions that establish norms for how we ought to live. Not all value questions concern moral value (narrowly understood). We also recognize aesthetic, spiritual, scientific, and cultural values as worthy and deserving of respect.[1]

Thus, central to a comprehensive environmental philosophy is a consideration of the nature and scope of value. A full account of value determines the ethical domain by helping to define what objects have moral relevance or what objects deserve consideration. Ethics is concerned with how we should live, how we should act, and the kind of persons we should be. Defining the full scope of these "shoulds" is to give an account of all that has value or worth. Consider the example described in the preceding discussion case.

Uncounted species of insects are becoming extinct as a result of destruction of the rain forest. Many people find the wanton destruction of diverse life-forms offensive. But what exactly is wrong with causing the extinction of millions of insects? Insects do not feel pain, are not conscious, and are not subjects-of-a-life. They are not, in any obvious way, moral beings. What seems to be wrong is that something of value is lost—indeed perhaps wantonly destroyed—by human activity. Too often these values are lost for the sake of greed or out of sheer ignorance. A similar explanation might be given for the destruction of the rain forest itself, as well as for the loss of wilderness areas, wetlands, trees, lakes, oceans, fish, and plants. Why value insects? Why protect a wilderness area? Why care about plants? As we noted in Chapter 2, some religious traditions are exploring ways to answer these questions from theological starting points, but how are they to be answered philosophically? The shift from an ethics of animal welfare to a more holistic environmental philosophy can perhaps best be understood as a shift from a narrow conception of morality and moral value to a broader concern with value itself.

Philosophers often have discussed moral value in terms of interests. Among the philosophers considered in the last chapter, Joel Feinberg, Christopher Stone, Peter Singer, Tom Regan, and Kenneth Goodpaster all use the concept of interests to decide what sorts of things deserve moral consideration. To say that an object has interests is to say that it has a "sake of its own" (Feinberg), a "worth" in its own right (Stone), a "welfare" of its own (Bentham and Singer), "inherent value" (Regan), or its own "well-being" (Goodpaster). All this is to say that these objects have a value or worth that is independent of the value and worth ascribed to them by human beings. This implies that we do something wrong when we treat an object that has a value in itself and of its own as though it has value only in relation to us. This difference is typically expressed in the important distinction between instrumental and intrinsic value. *Instrumental value* is a function of usefulness. An object with instrumental value possesses that value, because it can be used to attain something else of value. A pencil is valuable, because I can write with it. A dollar bill is valuable, because I can use it to buy something. The instrumental value of an object lies not in the object itself but in the uses to which that object can be put. When such an object no longer has use, or when it can be replaced by something of more effective or greater use, it has lost its value and can be ignored or discarded.

Thinking of natural objects in terms of "resources" is to treat them as having instrumental value. For example, Gifford Pinchot's conservation movement emphasized the instrumental value of forests and wilderness areas. We should protect and conserve the wilderness, because it is the repository of vast resources that humans can use. Pinchot and other progressives argued that the value of national resources was too often unfairly distributed or wasted, which is to say improperly used. Many other environmental concerns rest on the instrumental value of the environment. Clean air and water are valued, because without them human health and well-being are jeopardized. The preservation of plant and animal species is valued by many because of the vast potential therein for medical and agricultural uses. Virtually any utilitarian or economic proposal is based on the instrumental value of nature. The stewardship tradition in religious ethics also has a strong instrumental predisposition. Likewise, synthetic biologists view the life forms that they are creating as valuable in this instrumental sense.

Appealing to the instrumental value of the environment can be an effective political strat-egy. Public opinion is often most responsive to claims of lost opportunities, wasted resources, and the like. Yet an environmental ethics that is based solely on the instrumental value of the environment may prove unstable. As human interests and needs change, so too will human uses for the environment. The instrumental value of the Colorado River as a water and hydroelectric power source for southern California will quickly override its instrumental value as a scenic wilderness or recreation area. Emphasizing only the instrumental value of nature means, in effect, that the environment is held hostage by the interests and needs of humans, and it immediately evokes the necessity to make trade-offs among competing human interests.

An object has *inherent or intrinsic value,* on the other hand, when it is valuable in itself and is not valued simply for its uses.[2] The value of such objects is intrinsic to them. To say that an object is intrinsically valuable is to say that it has a good of its own and that what is good for it does not depend on outside factors. Thus its value would be a value found or recog-nized rather than given. Not all things that we value are valued instrumentally. Some things we value, because we recognize in them a moral, spiritual, symbolic, aesthetic, or cultural importance. We value them for themselves, for what they mean, for what they stand for, and for what they are, not for how they are used.

Some examples can help explain this distinction. Think of friendships. If you value a friend only for her usefulness, you have seriously misunderstood friendship, and you would not be a very good friend. Consider also historical monuments or cultural and aesthetic objects. The Liberty Bell, the Taj Mahal, and Michelangelo's *David* possess value far beyond their usefulness. Clearly, many of our environmental concerns rest on the intrinsic value that we recognize in nature. Life itself, in the view of many, is intrinsically valuable, no matter what form it takes. Wilderness areas, scenic landscapes, and national parks are valued by many people, because, like the Liberty Bell, they are a part of our national heritage and his-tory. (This is essentially the argument Mark Sagoff made, which we examined in Chapter 3.) Grizzly bears may have little instrumental value, but many people value knowing that the bears still exist in Yellowstone National Park. The symbolic value of the bald eagle tran-scends any instrumental value it might have. Undeveloped and unexplored wilderness areas are highly valued, even by people who will never visit, explore, or use these areas. John Muir's disagreement with Gifford Pinchot was a disagreement between one who saw an intrinsic value in wilderness (Muir) and one who did not (Pinchot). Muir spoke of the great sequoia groves as a cathedral, suggesting that they possess a spiritual and religious value far above their economic usefulness.

When we say that human activity degrades the environment, we are often referring to the loss of or disrespect for intrinsic value. When sections of the Grand Canyon are eroded by flooding caused by water released from hydroelectric dams upriver, when acid rain eats away at ancient architecture in Greece and Rome, or when shorelines are replaced by boardwalks and casinos, human activity destroys some of the intrinsic goods that we find in nature.

For a number of philosophers working in environmental ethics, the greatest challenge is to develop an account of intrinsic value that can counter arguments based on instrumental values. As we saw in Chapter 5, John Passmore calls for an emphasis on the sensuous to offset the materialism and greed dominant in modern culture. After criticizing the dominant economic model (which recognizes only instrumental value), Mark Sagoff summons philosophers to articulate the cultural, aesthetic, historical, and ethical values that underlie our environmental commitments. These values, Sagoff tells us, determine not just what we *want* as a people but what we *are*.

The development of a more systematic environmental philosophy, then, often involves a shift from a narrow focus on moral standing or moral rights and responsibilities to a more general discussion of value, especially intrinsic value. Unfortunately, appeals to intrinsic value often meet with skepticism. We seem to lack the language for expressing intrinsic value. Many people think that such value is merely subjective, a matter of personal opinion: "Beauty is in the eye of the beholder." Thus, when a measurable instrumental value (such as profit) conflicts with intangible and elusive intrinsic value (such as the beauty of a wilderness), the instrumental value too often wins by default. The remainder of this chapter considers various views that reflect the conviction that life itself possesses intrinsic value and that, accordingly, humans have some responsibilities to it.

CHAPTER 9

PART 9.2 DEEP ECOLOGY

Unlike the land ethic, deep ecology has not developed out of one primary source, nor does it refer to one systematic philosophy. Deep ecology has been used to describe a variety of environmental philosophies, ranging from a general description of all nonanthropocentric theories to the highly technical philosophy developed by the Norwegian philosopher Arne Naess.[1] For many people involved in radical environmentalism as a political movement (members of Earth First! and the Earth Liberation Front, for example), deep ecology also provides the philosophy that legitimizes their form of activism. In recent years, the phrase *deep ecology* has come to refer primarily to the approach to environmental issues developed in the writings of academics Naess, Bill Devall, and George Sessions, which is how it is used in this chapter.[2]

Deep ecologists trace their philosophical roots to many of the people and positions that this textbook has already examined. The debate between Gifford Pinchot and John Muir examined in Chapter 3 was an early version of the tension between shallow (Pinchot) and deep (Muir) approaches. Rachel Carson's critique of anthropocentrism in *Silent Spring* and Lynn White's critique of western Christianity, along with the nineteenth-century romanticism of Thoreau, were precursors of deep ecology.[3]

Arne Naess first introduced a distinction between deep and shallow environmental perspectives in 1973.[4] Naess characterized the shallow ecology movement as committed to the "fight against pollution and resource depletion." He maintained that it is an anthropocentric approach with the primary objective of protecting the "health and affluence of the people in developed countries." Deep ecology looks to more fundamental issues, at what it calls the "dominant worldview," which underlies such issues as pollution and resource depletion. Their critique is based on two positions we have examined: ecocentrism and nonanthropocentrism.

Deep ecologists attempt to work out an alternative philosophical worldview that is holistic and not human-centered.

But any call for a radical change in people's philosophical worldview immediately faces a major challenge. How do we even begin to explain the alternative if, by definition, it is radically different from the starting point? How do we step outside our personal and cultural worldview or ideology to compare it with something radically different?

Deep ecologists use a variety of strategies to meet these challenges, including reliance on poetry, Buddhism, spiritualism, and political activism via civil disobedience and ecosabotage. Perhaps the best way to begin exploring this movement is to consider the practical principles that Naess and Sessions drew up to articulate the ideas on which all its adherents agree. This platform serves as a core around which the diverse deep ecology movement can be unified.

PART 9.3 THE DEEP ECOLOGY PLATFORM

Deep ecologists are committed to the view that solutions to the current grave environmental crisis require more than mere reform of our personal and social practices. They believe that a radical transformation in our worldview is necessary. Naess and Sessions developed the deep ecology platform as a statement of shared principles. The platform is intended to be general enough to allow for a diversity of philosophical interpretations and specific enough to distinguish the deep from the shallow approach to practical matters.[5] As developed by Naess and Sessions, the platform includes these principles:

(1) The flourishing of human and nonhuman life on earth has intrinsic value. The value of nonhuman life-forms is independent of the usefulness they may have for narrow human purposes.

(2) The richness and diversity of life-forms are values in themselves and contribute to the flourishing of human and nonhuman life on earth.

(3) Humans have no right to reduce this richness and diversity except to satisfy vital needs.

(4) Present human interference with the nonhuman world is excessive, and the situation is rapidly worsening.

(5) The flourishing of human life and cultures is compatible with a substantial decrease of the human population. The flourishing of nonhuman life requires such a decrease.

(6) Significant change of life conditions for the better requires changes in policies. These affect basic economic, technological, and ideological structures.

(7) The ideological change is mainly that of appreciating *life quality* (dwelling in situations of intrinsic value) rather than adhering to a high standard of living. There will be a profound awareness of the difference between *big* and *great*.

(8) Those who subscribe to the foregoing points have an obligation directly or indirectly to participate in the attempt to implement the necessary changes.[6]

We can see how these principles could serve to explain and support a wide range of specific positions on practical environmental controversies. In working against the continued destruction of rain forests, for example, we could appeal to the first three principles.

Principles 5 and 7 would be important in developing an energy policy that would address such issues as resource conservation, population growth, consumer demand, and nuclear energy.

An important point is that the platform also reflects the ways in which the science of ecology influences deep ecology. In some sense, ecological science would provide direct support for principles 4 and 5. Ecology would also be relevant in explaining and defending principles 1 and 2. But ecology is also important for deep ecology in that it provides a model for a non-reductionist, holistic worldview.

More specifically, the conclusions reached in ecology and conservation biology are often "statements of ignorance." "Only rarely can scientists predict with any certainty the effect of a new chemical on even a single small ecosystem," Naess writes. Given this pervasive scientific ignorance, the burden of proof should rest with anyone who proposes a policy that intervenes in the natural environment.

> Why does the burden of proof rest with the encroachers? The ecosystems in which we intervene are generally in a particular state of balance which there are grounds to assume to be of more service to mankind than states of disturbance and their resultant unpredictable and far-reaching changes. In general, it is not possible to regain the original state after an intervention has wrought serious, undesired consequences.[7]

Accordingly, ecology contributes to deep ecology in the same ways in which scientific understanding has often contributed to ethical analysis. We gain a better understanding of the world, and on the basis of this understanding, we are in a better position to offer ethical evaluations and prescriptions. Because ecological understanding offers new insights, an ethics that relies on ecology can be expected to offer new evaluations and prescriptions.

PART 9.4 METAPHYSICAL ECOLOGY

Like the land ethic, deep ecology relies on the science of ecology in a variety of ways. Ecology provides a good deal of information about how natural ecosystems function. Ecology helps us to diagnose environmental disorders and to prescribe policies that can resolve these disorders. Ecology provides us with an understanding of natural ecosystems, and this understanding in turn is the basis from which we can make evaluations and recommendations. Ecology also cautions against any quick-fix technological solution to environmental problems. Echoing a theme found in Aldo Leopold's work, Naess argues for a humble and constrained approach to environmental change.

But Naess also is aware of the limits of science and warns against too great a reliance on ecology. There are dangers in what Naess calls "ecologism," the view that takes ecology as the ultimate science. The danger arises when we rely too heavily on ecology for solutions to specific problems. To treat ecology as just another science that can offer scientific answers to specific problems is to be tempted by the standard shallow hope for a technological quick fix. As is consistent with his commitment to deep ecology, Naess believes that environmental issues such as wilderness destruction and species extinction point to fundamental questions about how we ought to live. The fear is that the recent development of an "ecological conscience," will be used simply to substitute one shallow quick fix for another. In this view, ecology would simply be a new means for treating only the symptoms. Thus it would subvert attempts to probe more deeply into the underlying causes of the environmental crisis. Ecology might then become a diversion from these more fundamental issues. The risk is that ecology will be used as part of a political strategy to derail movements that question the fundamental assumptions of our culture. In Naess's words, we need to "fight against depoliticization" to stay focused on the political nature of the deep ecology movement.[8]

Scientific ecology provides a model for thinking about the deep fundamental issues that underlie the environmental crisis. Inspired by ecology, deep ecologists seek to develop alternative worldviews that echo ecological insights into such issues as diversity, holism, interdependencies, and relations. Deep ecology traces the roots of our environmental crisis to fundamental philosophical causes. Solutions can come only from a transformation of our worldview and practices and from our answers to fundamental questions. What is human nature? What is the relation of humans to the rest of nature? What is the nature of reality? These questions are traditionally identified as *metaphysical* questions. Deep ecology, therefore, is as concerned with questions of metaphysics and ontology (the study of what is) as it is with questions of ethics. Deep ecologists trace the cause of many of our problems to the metaphysics presupposed by the dominant philosophy of modern industrial society. Deep ecology is concerned with a *metaphysical ecology* rather than a scientific one.

The dominant metaphysics that underlies modern industrial society is *individualistic* and *reductionistic*.[9] This view holds that only individuals are real and that we approach a more fundamental level of reality by reducing objects to their more basic elements. These most basic elements, whatever they turn out to be, are related according to strict physical laws. But this dominant worldview also sees humans as essentially different from the rest of nature. Individual human beings possess a "mind" or "free will" or "soul" that exempts them from the strict mechanical determinism characteristic of the rest of nature. Thus the dominant worldview rejects the position identified in Chapter 7 as metaphysical holism.

Rejection of these dominant beliefs is central to the metaphysics of deep ecology. Taking its cue from ecology, the metaphysics of deep ecology denies that individual humans are separate from nature. Humans are fundamentally a part of their surroundings, not distinct from them. Humans are constituted by their relationships to other elements in the environment. In an important sense, the environment—by which the deep ecologists mean both the biotic and the abiotic constituents—determines what human beings are. Without the relationships that exist among humans and between humans and nature, human beings would literally become different sorts of beings. A philosophy that "reduces" humans to "individuals" that are somehow distinct from their social and natural environment is radically misguided.

This point has been expressed by Warwick Fox, an Australian philosopher and deep ecologist.

> It is the idea that we can make no firm ontological divide in the field of existence: that there is no bifurcation in reality between the human and the non-human realms … to the extent that we perceive boundaries, we fall short of Deep Ecological consciousness.[10]

Thus, echoing the metaphysical holism of Callicott described in Chapter 8, deep ecologists deny the reality of individuals—at least as they are typically understood in Western philosophy. There are no individuals apart from or distinct from relationships within a system.

Human "nature" is inseparable from nature. Viewing human beings as individuals is how the dominant worldview has understood humans and has broken up reality, but it is a dangerous and misleading metaphysics.

No doubt, thinking like this tempts many people to relegate deep ecologists to the fringe of environmental philosophy. These views certainly represent a radical shift from mainstream Western thinking. But any call for a radical shift in perspective faces difficulties in being understood. We can, perhaps, approach deep ecology in a variety of ways in the hope that we can begin to understand this alternative outlook.

In the spirit of deep ecology, we might begin by taking a hint from scientific ecology. If we think of ecosystems as energy circuits through which solar and chemical energy flow, we might begin to think of individual organisms as less permanent and less real than the chemical and biological processes themselves. Individual organisms come and go, but the process goes on as long as environmental conditions permit. Individual organisms can be thought of as the location at which these chemical processes occur.

Another way of approaching this conclusion is to consider what it means to say that an individual organism is alive. Minimally, an individual organism is alive only if certain chemical and biological processes are occurring. When these processes cease to occur, the organism ceases to live. Thus the processes are necessary for the existence of the organism. On the other hand, when the processes are occurring, life exists. Thus the processes are sufficient for life. Because chemical and biological processes are both necessary and sufficient for the existence of life, we have some reasons for saying that the processes are at least as real as, if not more real than, individual living organisms.

Biophysicist Harold Morowitz makes a similar point.

> Viewed from the point of view of modern [ecology], each living thing is a dissipative structure, that is, it does not endure in and of itself but only as a result of the continual flow of energy in the system... From this point of view, the reality of individuals is problematic because they do not exist per se but only as local perturbations in this universal energy flow... An example might be instructive. Consider a vortex in a stream of flowing water. The vortex is a structure made of an ever-changing group of water molecules. It does not exist as an entity in the classic Western sense; it exists only because of the flow of water through the stream. If the flow ceases the vortex disappears. In the same sense the structures out of which the biological entities are made are transient, unstable entities with constantly changing molecules dependent on a constant flow of energy to maintain form and structure.[11]

Finally, we might better appreciate metaphysical ecology by considering the language of "individualism." Ordinarily, we seem confident that we know what we refer to when we speak of individuals. But although we often use *individual* as a noun *(an* individual, *the* individual), the word is perhaps more precisely used as an adjective (an individual person, an individual tree, and the like).

Imagine being asked to go out and count the individuals that you see. We might assume that the assignment involves individual humans, or we would be well advised to ask, "Individual what?" This suggests that when we speak of individuals, we have already adopted a worldview or metaphysics that has divided our experiences up in one way rather than in another way. Our ordinary language seems to presuppose a metaphysics in which separate and isolated organisms are most real. But note that we might just as well refer to individual communities, individual ecosystems, individual species, and individual chemical cycles. We might just as well refer to individual body parts, individual organs, individual cells, individual molecules, individual atoms, and so forth. The individual human person can be seen either as a part of some larger individual (such as a species or an ecosystem) or as a collection of other individuals (such as organ systems or cells).

The implication of this is that the world does not come already broken down into categories such as individuals and wholes. Rather, particular ways of understanding the world and the particular needs served by understanding it in those ways determine what is to count as an individual and what is to count as a whole. Deep ecology argues that the dominant worldview assumes an artificial distinction between individuals and their surroundings. The ecological and environmental devastation that has followed from this particular metaphysics has proved it to be dangerous. An alternative metaphysics, one inspired by scientific ecology, can offer an opportunity for reversing this devastation.

PART 9.7 CRITICISMS OF DEEP ECOLOGY

It sometimes seems that deep ecology acts as a lightning rod for environmental criticism and backlash. Because deep ecology does critique the dominant worldview, we should not be surprised to find significant critical reaction. As mentioned at the start of this chapter, the term *deep ecology* does not refer to one specific and systematic philosophy. It refers to an assortment of philosophical and activist approaches to ecological issues that share some fundamental ecocentric and nonanthropocentric assumptions. It is perhaps best thought of as a movement that encompasses both philosophical and activist sides.

Given this diversity, it is difficult to offer any precise criticisms of deep ecology. A critique of, for example, the tactics of Earth First! could be rebuffed by deep ecologists as beside the point, because not all deep ecologists agree with these tactics. Likewise, a critique that accuses deep ecology of being too abstract and vague on issues such as "Self-realization" might be rejected by deep ecologists who are more inclined toward political activism.

Of course, this ambiguity itself can be grounds for criticism. In some ways, the claims of deep ecology are so sweeping and general as to become empty. A "movement" that can claim inspiration from such diverse sources as Taoism. Heraclitus, Spinoza, Whitehead, Gandhi, Buddhism, Native American cultures, Thomas Jefferson. Thoreau, and Woody Guthrie is certainly eclectic at best. At worst, it becomes unintelligible.

This ambiguity can be frustrating for critics as they try to focus on specific claims, only to find their target shifting. It can also lead to an end of dialogue because critics can be dismissed as missing the point, as irrelevant, or as misrepresenting deep ecology.[12]

Another criticism echoes the fascism charge raised against holistic and nonanthropocentric ethics in earlier chapters. Biocentric equality would seem to suggest treating human interests as equal to the interests of other living things, as well as of the more general biotic community. However, when this equality is combined with the metaphysical claim that individuals are not real and with the charge that humans alone are responsible for significant environmental destruction, deep ecology can seem misanthropic (as hating humanity).

Humans are no better than other living things and, in fact, are guilty of great environmental malevolence. Thus human well-being is not a moral priority. Some of the better-known examples of misanthropic remarks include Edward Abbey's claim in *Desert Solitaire* that he would rather shoot a human than a snake[13] and Dave Foreman's suggestion that we should not aid starving Ethiopians and should allow them to die.[14]

Deep ecologists disavow such claims. Fox, for example, points out that deep ecologists criticize "not humans per se (i.e., a general class of social actors) but rather human-centeredness (a legitimating ideology)."[15] This claim amounts to the view that deep ecologists do not deny intrinsic value to humans. They simply deny that only humans have intrinsic value.

But the same challenge can be issued here that was raised in Chapter 6. What is to be done when human interests conflict with the interests of elements of the nonhuman natural world, as so often is the case with environmental issues? In such cases, if we favor humans, we seem to abandon nonanthropocentric holism. If we favor the nonhuman world, we approach the misanthropic position that deep ecologists want to deny. Again, this requires deep ecologists to work out a clear hierarchy of vital needs.

Another challenge is raised in various forms by a diverse group of critics. The problem with deep ecology, in their view, is that it has overgeneralized in its critique of human-centeredness, anthropocentrism. and the dominant worldview. From this point of view, not all humans and not all human perspectives are equally at fault for environmental problems. When deep ecologists critique "the" dominant worldview, they fail to acknowledge that many humans are not part of that dominance. Thus deep ecologists are too broad in their critique and, consequently, too broad in their positive program.

One version of this critique is raised by the Indian ecologist Ramachandra Guha.[16] Guha argues that despite its claims to universality, deep ecology is uniquely an American ideology, essentially a radical branch of the wilderness preservation movement. In Guha's view, if it were put into practice, deep ecology would have disastrous consequences, especially for the poor and agrarian populations in underdeveloped countries. Describing India as a "long settled and densely populated country in which agrarian populations have a finely balanced relationship with nature," Guha reasons that a policy of biocentric equality and wilderness preservation would effectively result in a direct transfer of wealth from poor to rich and a major displacement of poor people.

Applying the deep ecology platform to societies in underdeveloped countries smacks of Western imperialism. "We (environmentalists in the West) know what is best for you. Let us generalize from our experiences and our culture and tell you why you should live in ways that we suggest. Stop treating nature as resources, even if you are living at a mere subsistence level. Preserve and respect nature for its own sake."

Guha also faults deep ecology for its appropriation of Eastern philosophies and traditions. Citing Hinduism, Taoism, and Buddhism as though they were a single consistent Eastern worldview and one that is more in tune with environmentalism "does considerable violence to the historical record." Eastern cultures, as well as Western cultures, have manipulated nature and caused significant ecological destruction.

Accordingly, deep ecology is not very helpful to the environmental concerns of peoples of underdeveloped countries. At best, it is irrelevant. At worst, it can be harmful to the very people who already are victimized by social and political dominance.

Similar critiques of deep ecology have been offered by thinkers associated with *ecofeminism*. This perspective agrees that in the search for the "deep" underlying causes of the environmental crisis, deep ecologists have focused their attention at too abstract a level. The more significant causes can be located at a much more localized level: the social, economic, and patriarchal structures of contemporary societies. In faulting anthropocentrism, deep ecologists fail to recognize important distinctions between people. If there is " 'a" dominant worldview, deep ecology must recognize that many humans are also oppressed by it. Not all humans are equally at fault for environmental destruction, and not all humans were included in the "human-centered" dominant worldview. Instead of looking at some abstract dominant worldview, these critics seek to specify the particular practices and institutions that dominate both human and nonhuman alike. Guha calls our attention to the perspective of people, especially poor people, in underdeveloped countries. Ecofeminists suggest that the causes of environmental domination and the oppression of women are connected. We turn now to an examination of ecofeminist environmental philosophies.

NOTES

Chapter 3

1 Mark Sagoff, *The Economy of the Earth* (New York: Cambridge University Press, 1990).
2 Mark Sagoff, "Economic Theory and Environmental Law," *Michigan Law Review* 79 (1981): 1393–1419.
3 Sagoff, *The Economy of the Earth*, pp. 28–29.
4 For an examination of representative and participatory democracy, see Jane Mansbridge. *Beyond Adversarial Democracy* (New York: Basic Books, 1980).
5 O'Toole, *Reforming the Forest Service*, p. 189.

Chapter 5

1 Christopher Stone, *Should Trees Have Standing? Toward Legal Rights for Natural Objects* (Los Altos, Calif.: William Kaufmann, 1974), p. 9.
2 For a brief account of this history, see Roderick Nash, *The Rights of Nature* (Madison: University of Wisconsin Press, 1989), pp. 128–31.
3 Stone, *Should Trees Have Standing?* p. 11.
4 Ibid.; the emphasis is Stone's.
5 Ibid., 24.

Chapter 6

1 Perhaps the best philosophical account of the values that support a comprehensive environmental ethics can be found in the writings of Holmes Rolston. See especially *Environmental Ethics* (Philadelphia: Temple University Press, 1988).
2 The philosophical debate over the meaning and legitimacy of instrumental, intrinsic, and inherent value continues. The understanding of instrumental value is relatively trouble-free. In what follows, I adopt Susan Armstrong and Richard Botzler's suggestion that an "emerging consensus" is settling on the meaning of intrinsic and inherent value. In this view, intrinsic value is independent of the presence of a valuer. An object has intrinsic value when it has

129

value both in and for itself. Inherent value, on the other hand, requires the presence of a valuer who confers the value on the object. Thus, although it may be valued for its own sake (its value does not come from its usefulness), it does not have value in itself (if there were no one around to value it, it would be without value). For example, children have intrinsic value (they are valued in and for themselves), whereas a family heirloom has inherent value (it is valued for itself and not for its economic worth) but would be valueless, if there were no family around to value it. If we accept this distinction as meaningful, the major question becomes whether natural objects and nonhuman living beings have intrinsic value or inherent value. For the emerging consensus claim, see Susan Armstrong and Richard Botzler, *Environmental Ethics: Divergence and Convergence* (New York: McGraw-Hill, 1993), p. 53. For a defense of intrinsic value, see Holmes Rolston, *Environmental Ethics* (Philadelphia: Temple University Press, 1988). For a defense of inherent value, identified as intrinsic value in a "truncated sense," see J. Baird Callicott, "The Intrinsic Value of Nonhuman Species," in *The Preservation of Species*, ed. Bryan Norton (Princeton, N.J.: Princeton University Press, 1986), Ch. 6.

Chapter 9

1 See, for example, the distinction between "deep" and "shallow" offered by Donald VanDeVeer and Christine Pierce in *People, Penguins, and Plastic Trees* (Belmont. Calif: Wadsworth. 1986), pp. 69–70. Here "shallow ecology" is taken as the view that "nature has no value apart from the needs, interests, and good of human beings," and "Deep Ecology holds that nature has value in its own right independent of the interests of humans."

2 See Arne Naess, *Ecology, Community, and Lifestyle*, trans. and rev. David Rothenberg (Cambridge, England: Cambridge University Press, 1989), and Bill Devall and George Sessions, *Deep Ecology: Living As If Nature Mattered* (Salt Lake City, Utah: Peregrine Smith Books. 1985).

3 For these citations and others, see *Deep Ecology for the Twenty-First Century*, ed. George Sessions (Boston: Shambala Publications, 1995), p. ix.

4 Arne Naess, "The Shallow and the Deep, Long-Range Ecology Movement," *Inquiry* 16 (1973): 95–100.

5 This description relies heavily on David Rothenberg's introduction to Naess's *Ecology, Community, and Lifestyle.*

6 This platform is presented in Devall and Sessions, *Deep Ecology: Living As If Nature Mattered*, Ch. 5, as the "basic principles" of Deep Ecology. These principles are also presented as a "platform" in Naess, *Ecology, Community, and Lifestyle*, Ch. 1.

7 Naess, *Ecology, Community, and Lifestyle*, pp. 26–27.

8 Ibid., 130–33.

9 For one sustained account of the "dominant modern worldview," see Devall and Sessions, *Deep Ecology: Living As If Nature Mattered*, Ch. 3.

10 Warwick Fox. "Deep Ecology: A New Philosophy for Our Time?" *Ecologist* 14 (November–December 1984): 194–200, as quoted in Devall and Sessions, *Deep Ecology: Living As If Nature Mattered*, p. 66.

11 Harold Morowitz, "Biology as a Cosmological Science," *Main Currents in Modern Thought* 28 (1972): 156. This material is quoted and discussed in J. Baird Callicott, "Metaphysical Implications of Ecology," *Environmental Ethics* 9 (Winter 1986): 300–15. Callicott's essay suggests that a "consolidated metaphysical consensus" might be emerging from ecological science that would supplant the mechanical model that emerged from seventeenth-century physics.

12 For example, George Sessions dismisses a range of critics in less than one page in his preface to *Deep Ecology for the Twenty-First Century*, pp. xii–xiii. Criticism of Deep Ecology is "based on misinterpretation and misunderstanding." It seems to me that this is a theme in many deep ecologists' responses to critics.

13   The Edward Abbey quote is "I'm a humanist; I'd rather kill a man than a snake." And is taken from *Desert Solitaire: A Season in the Wilderness*. (New York, McGraw-Hill, first ed.) 1968.

14   The Foreman quote is taken from *Confessions of an Eco-Warrior* (New York Crown Publishers), 1991. "An individual human life has no more intrinsic value than does an individual Grizzly Bear life. Human suffering resulting from drought and famine in Ethiopia is tragic, yes, but the destruction there of other creatures and habitat is even more tragic."

15   Warwick Fox, "The Deep Ecology—Ecofeminism Debate," in Sessions, ed., *Deep Ecology for the Twenty-First Century*, p. 279. This is a version of an essay of the same title previously published in *Environmental Ethics* 11 (Spring 1989): 5–26.

16   Ramachandra Guha, "Radical American Environmentalism and Wilderness Preservation: A Third World Critique," *Environmental Ethics* 11 (Spring 1989): 71–84.

Jason Corburn — "Local Knowledge in Environmental Health Policy." In *Street Science: Community Knowledge and Environmental Health Justice*. Cambridge, MA: MIT Press.

Reading Commentary

What if the search for expert knowledge is not taken as an end in itself but rather as a means for raising crucial questions that make explicit the distinct and plural positions of professionals and the lay public? Corburn (2005) argues that expert knowledge should not be unquestioningly accepted as fact and truth. Rather, we should strive for an interaction between traditional forms of evidence and local insights, or local knowledge.

Corburn carefully lays out the concept of street science to draw our attention to the distinction between traditional policy-making, focused on formal decision-making by professional experts and frequently seen in the news, and deliberative policy-making, which incorporates the ideas and practices of people "on the ground." Street science emphasizes what people have learned from their lived experience. It also suggests that the search for knowledge is a political inquiry and involves political action. The parties involved in environmental problem-solving need to come to terms with the fact that changes in prevailing policy should reflect continuous feedback and adjustment.

Corburn's argument is that both professional experts and laypeople have essential contributions to make in decisions about public health. He asks us to consider what happens when expertise is coproduced in both the scientific and the political realm. He highlights the advantages of contingent decisions in the public arena: inherently open to renegotiation, responsive to new feedback and preparing for unexpected change. As such, street science invites us to consider that public officials should rethink their traditional approach to guaranteeing environmental health and safety, recognizing the value of new voices and ideas. The potential benefit would be to ensure that the disenfranchised and least politically powerful have more of a say.

As an example, Corburn illustrates that risk has long been the dominant lens through which environmental health is viewed in the United States (and elsewhere). Risk assessment implies that problems can be defined, quantified and managed by weighing the costs associated with hazards of various kinds against the benefits of alternative control strategies. This framing of risk tends to favor certain types of evidence and ignores other types of expertise, preferring formal and quantitative information gathered by professional experts and bypassing the informally gathered and value-laden information collected by residents. The risk frame, thus, puts the lay public at a disadvantage, limiting its ability to participate in the decisions that affect them directly. As a result, the public interest tends to be defined by organized professional interests rather than by those who actually bear the consequences of public decisions.

In sum, the concept of street science invites us to conceive of our world as provisional and improvisational, where flexibility and adaptability are crucial, and people are encouraged to learn, test and share their insights through public discourse.

CHAPTER 1
LOCAL KNOWLEDGE IN ENVIRONMENTAL HEALTH POLICY

Jason Corburn

THE TENSIONS BETWEEN COMMUNITIES AND PROFESSIONALS

How do environmental-health professionals typically deal with a situation like the controversy over air quality and public health after the World Trade Center collapse described in the introduction? Typically, environmental health seeks to identify the specific pollutants in the medium of concern. In this example, scientists attempt to delineate the individual toxins in the air. Once the pollutants are identified, they are assessed for their toxicity, or their potential danger and deleterious effects on humans generally. Next, each individual pollutant identified is assessed for its potential impact on humans exposed to the air pollution from the World Trade Center. Determinations of human-health impacts in a specific place generally include assumptions about the routes of exposure (e.g., inhalation in the case of air pollution), how much pollution certain groups are inhaling (e.g., children versus construction workers), and how long certain groups are exposed. The toxicity information and the exposure assumptions are combined to estimate the human-health risk from each individual pollutant contained in the World Trade Center air. This process of identifying each hazard and its toxicity to humans, estimating an individual's exposure to the hazard in a particular place, and extrapolating from this information an estimate of potential harm, is called *risk assessment*.

Risk has been the dominant frame through which environmental health is analyzed in the United States for at least the last thirty years (Fiorino 1989).[1] Risk, and its correlate risk assessment, implies that a problem can be clearly defined, quantified, and therefore managed. Once some version of health risk is generated, the "benefits" from the source of the pollution or hazard are weighed against the pollution's "costs" to human health. At this stage, policy analysts and planners are charged with the often-inevitable task of "risk management," or deciding how to weigh "costs and benefits" and inform policymaking.[2]

In the best risk-management processes, the analyst consults with the public that is being asked to bear a "risk" from the beginning of the hazard assessment (Krimsky and Plough

1988). However, more often analysts—perhaps feeling that professional training gives them ultimate discretion to carry out and implement decisions—omit the public from the decision-making process. Additionally, the analysts may find it difficult to divine what the scientists really found in their study, how the legislature, governor, or mayor wants the "costs and benefits" to be interpreted and administered, and what course is consistent with the "public interest." The analysts may feel that their agency is "captured" by private interest groups that are seeking to influence the analysis and any resulting regulation (Lowi 1969). The "captured agency" then might substitute private goals for those of the public at large because the constituency opposing the private sector may not be organized, the agency may rely on the private sector for resources necessary to implement particular programs, or because of the powerful influence industry has in local, state, and national politics.

In the midst of these potentially conflicting interests, the analysts or planners often decide that the tacit operating rule is that the best public is a quiescent one. The analysts might desire to faithfully represent the values and interests of citizens but be unsure what "representation" actually entails. They may ask whether political representation requires that an agency allow local people to participate in analyses and decision making. Recognizing that the success of environmental-health policy is often contingent on the willingness of ordinary citizens to accept the validity of official policy framings, the analysts might hold a public hearing. Hearings tend to open up to unlimited critical scrutiny expert findings that were generated in closed worlds of formal inquiry. These processes are often recipes for unending debate and spiraling distrust, leaving most participants unsatisfied and frustrated that, for instance, technical uncertainties were left unresolved. Thus, the planners may be torn between holding a public hearing that might merely act as a forum to placate the demands of competing special interests groups or organizing some other public process that they have no experience in managing. Public officials, unsure of how to deal with these tensions and competing commitments, often try to work quietly, get the job done without disturbing the public "peace," and then often reassure everyone "out there" that there is no reason to be concerned or involved (Reich 1988, 124).

This description might oversimplify the risk-management process, but it highlights some of the tensions environmental-health professionals face when determining how best to use scientific analyses while simultaneously committing to democratic decision making.[3] One way to resolve this tension is to return to and challenge the "risk framework" that tends to dominate environmental health. In the risk frame, certain types of evidence and expertise are valued and other evidence and expertise is ignored. The risk frame tends to prefer formal and quantitative information and the participation of a select group of professionals trained in certain disciplines. For example, Jasanoff (1990) has noted how expert advisors in policymaking are chosen based on their technical competence, ability to construct "objective science," and political independence and neutrality. Experts protect their authority to deal with the uncertain science of risk though a sociological mechanism known as "boundary work."

Boundary work is a process where experts assign the array of issues and controversies lying between the two ideal typical poles of "pure science" and "pure policy" to one or the other side of the policy-science boundary (Gieryn 1995, 405). As Jasanoff observes:

When an area of intellectual activity is tagged with the label "science," people who are not scientists are *de facto* barred from having any say about its substance; correspondingly, to label something "not science" [e.g., mere politics] is to denude it of cognitive authority. (Jasanoff 1990, 14)

As a result, risk-based problem framing and decision-making processes largely ignore evidence that is more informal, experiential, tacit, and explicitly value laden (Wynne 1996; Irwin 1995). Lay publics, even when granted "entry" into policymaking through formalized public hearings, are required to offer evidence in a "voice" or language that mirrors that of experts. As a result, the quantitative risk frame in environmental health puts lay publics at a disadvantage from the outset and limits their ability to participate in and influence decisions when compared to scientists and other professionals.

ANTECEDENTS TO STREET SCIENCE

Attempts to bring local or lay knowledge into environmental health decision making are not new. From nineteenth-century Progressive Era reformers to 1960s and 1970s anti-toxics activism, today's street scientists are building on ideas and community-based practices that emerged over a century ago (Gottlieb 1993). While taking slightly different approaches and being labeled everything from "shoe-leather epidemiology" to "people's science," community-based science has played a role in shaping environmental-health research and political action. Yet, even before Progressive Era reformers enrolled local knowledge to address the health problems afflicting the urban poor, public-health work in Europe highlighted the importance of considering the social and community aspects of health.

A series of studies in the mid-nineteenth century gave rise to modern movements for community-based environmental health. For example, one of the first modern epidemiological studies of neighborhood health was performed by Louis René Villermé, who used statistics to study Paris neighborhoods in 1840 and demonstrated a clear connection between ill health and neighborhood poverty. In 1848, Rudolf Virchow documented the social causes of a typhus epidemic in Germany. He is credited for linking the biologic, social, and economic underpinnings of health and emphasizing that medicine and public health fail when they ignore the plight of the poor and working class (Rosen 1993).

Perhaps most influential on American reformers was the 1842 publication of Edwin Chadwick's *Report on the Sanitary Conditions of the Laboring Population in Great Britain,* and similar reports that soon followed documenting conditions in New York and Massachusetts (Duffy 1990). These reports stimulated the Sanitary movement in public health and highlighted how inferior living and working environments for the poor and immigrant populations were a key factor in their poor health (Melosi 2000). The Sanitary movement was part of a host of Progressive Era reforms that focused public-health interventions on cleaning up urban neighborhoods and workplaces (Duffy 1990). One of the most well-known reform movements of this time was the Settlement House movement, best exemplified by Hull House in Chicago, where reformers such as Jane Addams, Alice Hamilton, and Florence Kelley founded the modern epidemiologic methods of occupational and community health.

At the time Jane Addams founded Hull House in 1889, pollution in cities and the workplace was seen as a sign of progress and opportunity, not potential harm. In this context, the public-health work by the women at Hull House was revolutionary because it not only challenged this idea, but also because these reformers used research methods that included the lived experiences and knowledge of those experiencing the greatest suffering. The methods of reformers at Hull House applied the information gleaned from workers and community residents to more detailed investigations (Deegan 1990). An important aspect of their public-health philosophy was encouraging community residents to record and share their experiences with others in the community, the general public, and decision makers. As Jane Addams stated in her introduction to the classic 1895 work *Hull House Maps and Papers*:

> The residents of Hull-House offer these maps and papers to the public, not as exhaustive treatises, but as recorded observations which may plausibly be of value, because they are immediate, and the result of long acquaintance. (*Hull House Maps and Papers* 1895, vii)

For Addams and others at Hull House, the knowledge community residents provided was a vital resource for both understanding and changing the unhealthy conditions of the urban environment.

Alice Hamilton, one of the first American specialists in the field of occupational disease and a long-term Hull House resident, pioneered the use of local knowledge to inform her work toward ameliorating common workplace hazards of the day, such as mercury poisoning of felt-hat workers and lead poisoning (Hamilton 1943). Refusing to see workers as appropriate guinea pigs for the discovery of the health effects of industrial chemicals, Hamilton listened to workers' accounts of the workplace experience to help her hypothesize why certain occupations and industrial processes were hazardous (Hamilton 1943). While workers often were reluctant to talk out of fear of losing their jobs, Hamilton met them on their own time, visiting homes to conduct informal interviews and to listen to their stories of workplace horrors (Sicherman 1984). Hamilton's style of fieldwork, which came to be known as "shoe-leather epidemiology," helped her piece together dangers in the workplace that were routinely underreported by factory owners and physicians (Sicherman 1984).

Florence Kelley, another Hull House resident, also pioneered the use of local knowledge in environmental health investigations. Kelley, like Hamilton, took her investigations into the street and canvassed the neighborhood around Hull House to document hazardous living conditions. One of her major achievements was documenting the "sweating system," or the dangerous garment-work women and children who lived in tenement houses performed (*Hull House Maps and Papers* 1895, 31).

The work of Addams, Hamilton, Kelley, and other reformers at Hull House aimed to understand how, in an unjust world, health is driven by social and economic inequalities. They understood that in order to change inequitable social conditions, one must first learn from the vulnerable groups how they described their suffering, because these stories hold clues about causes and effective interventions. These pioneers of local knowledge also encouraged the use of lay practitioners, such as midwifes and sanitation inspectors, to supplement the work of physicians and engineers (Deegan 1990). Importantly, women were at the forefront of early community-based social reforms and, as chapters 3–6 show, continue to lead most street science investigations.

While the Progressive Era reforms continued through the early years of the twentieth century, the public support for this work waned as germ theory came to dominate public health.

Germ theory held that specific agents of infectious disease exist, in particular microbes, and that these agents correspond one-to-one with specific diseases (Tesh 1988). Research and interventions driven by laboratory investigations of microbes quickly replaced the sanitary, social, and political reforms advocated by Progressive Era reformers. Public-health interventions focused on specific immunization plans, with physicians emerging as the new class of public-health professionals, leaving community organizers and lay people with little room to participate in this expert-centered discourse.

One important exception to this dynamic, where local knowledge was integrated into community health, was the neighborhood-health-center movement that emerged around 1910 but declined rapidly after World War I. Seeking in part to replicate the success of settlement workers, city governments began "demonstration projects" where health – and welfare-agency work was bought together and relocated "from city hall to the neighborhood" to better serve the neediest populations (Rosen 1985). The neighborhood health center aimed to replicate the values of "acquaintance" with "active participation" of the local population in delivering services that had proved so successful for the Settlement House movement (Bamberger 1966). Health centers were started in immigrant neighborhoods of Milwaukee and Philadelphia, the Mohawk-Brighton district of Cincinnati, New York's Lower East Side, and the West End of Boston. A key component of all the health centers was the creation of block committees, which allowed residents to raise neighborhood-specific problems to the nurses, physicians, and other professionals staffing the center (Burnham 1920). According to Rosen (1985), in a radical step for the time, the health officer for the Lower East Side center was a Jewish physician who understood the people, their language, and culture.

The cessation of large-scale immigration during the war years, and accusations that the self-governing aspects of the health centers were a "Red plot" and "socialized medicine," eliminated municipal support for neighborhood-based health programs (Rosen 1985). In addition, antagonism toward lay involvement in delivering health services by the American Medical Association helped eliminate funding for community-based prenatal and child health services provided for under the Sheppard-Towner Act of 1921 (Meckel 1990). By the 1930s lay participation in community-health issues was almost nonexistent because most epidemiologic investigations ignored social factors or treated them as nuisance variables in statistical models that focused on isolating germs. In the classic epidemiology framework of host-agent-environment, interventions focused on immunizing the "host" (e.g., individuals) because the "environment" (e.g., the world outside of microorganisms) was seen as harder to control.

While professionals increasingly adopted the biomedical model of disease—which attributed morbidity and mortality to individual behaviors, biology, and genetics—impoverished communities organized to address health issues with the help of organizations such as The Highlander Folk School, later renamed the Highlander Research and Education Center, in Tennessee founded by Myles Horton (Horton 1971). Horton and the Highlander Institute brought local people together from impoverished communities in the Appalachian region to investigate and take action to change their conditions. Describing one meeting at Highlander, Horton recalled the power of local knowledge:

> I remember they wanted to know about farm problems. They wanted to know about getting jobs in textile mills. They wanted to know about testing wells for typhoid. We discussed

these things. To my amazement my inability to answer questions didn't bother them...
That was probably the biggest discovery I ever made. You don't have to know the answers.
You raise the questions, sharpen the questions, get people discussing them. And we found
that in that group of mountain people a lot of the answers were available if they pooled
their knowledge. (Horton 1971, 16–17)

Highlander used a method called "popular education" to empower thousands of community members to collectively tap their own experiences and expertise to change social conditions. Many who attended Highlander, such as Rosa Parks, Ralph Abernathy, and Martin
Luther King Jr., would return home to organize for civil, labor, economic, and human rights
(Horton 1998).

As McCarthyism lost its sting by the late 1950s and 1960s, academic and social movements questioned previously unchallenged assumptions about science, namely its positivist
claims of neutral fact-finding disassociated from social values. In academia, social medicine
emerged as a legitimate field of inquiry, reintegrating social science ideas and notions of lay
participation into medical research and practice (Porter 1997). The social movements of the
1960s also reengaged local people into the public-health discourse primarily by highlighting that despite rising prosperity and increased access to medical care, inequalities in health
persisted for some, particularly for the rural and urban poor.

One example of a civil rights group reconnecting local and professional knowledge for
community environmental health is the work of the Young Lords, a group of New York City
Puerto Rican activists in El Bario, or East Harlem. The Young Lords organized street clean-
ups after the sanitation department refused to collect neighborhood garbage for weeks. They
convinced local professionals to train them to perform door-to-door lead-poisoning screening and tuberculosis testing (Abramson et al. 1971). The group started day-care programs
in local churches, provided breakfast in neighborhood schools, organized tenants to demand
housing improvements, and occupied a neighborhood hospital to highlight its inadequate
service to the local population. Merging the social, political, and environmental aspects of
health, the Young Lords combined local knowledge with professional techniques to address
health disparities in their neighborhood (Melendez 2003).

Community mobilizations to address health disparities in the 1960s helped reinvigorate
the movement for neighborhood health-centers that had begun fifty years earlier (Schorr
and English 1974). Spurred on by the passage of Medicaid and Medicare in 1965 and the
Office of Economic Opportunity's Community Action Program (CAP), the neighborhood-
health-center movement promoted the health and well-being of impoverished and medically
underserved communities by building clinics, developing preventative programs based on
team medical practices that involved local people, investigating the environmental causes
of poor health, and not limiting their work to categorical disease programs (Hollister et al.
1974). While municipal and state health and welfare departments focused on treating
individuals at several locations and departments, neighborhood health centers established
"one-stop" locations for clinical and social services, establishing neighborhood institutions
run by local people capable of linking existing community resources with newly decentralized governmental programs (Kotler 1969). Neighborhood health centers during this time
included the Columbia Point Health Center in a public-housing development in Boston,

the Tufts-Delta Health Center in the rural Mound Bayou in the Mississippi Delta, and the North East Neighborhoods Association Health Center in New York City's Lower East Side (Geiger 1967).

During the same time period, a more general public interest in environmental health emerged after a series of highly publicized environmental disasters, such as the contamination of Boston Harbor and the burning Cuyahoga River. These events, combined with the 1962 publication of Rachel Carson's *Silent Spring*, repopularized the nineteenth-century themes of linking industrial pollution and environmental health. The public trust that science was working in the public interest, so dominant in the first half of the twentieth century, had given way to skepticism, citizen action, and calls for new governmental regulations. As Gottlieb notes:

> while an earlier critic of the chemical industry, Alice Hamilton, laid the groundwork for discussing environmental themes in an urban-industrial context, Rachel Carson, with the evocative cry in *Silent Spring* … brought to the fore questions about the urban and industrial order that a new environmentalism prepared to face. (Gottlieb 1993, 86)

This new environmental activism included community members engaging with and confronting expert views of environmental health hazards, particularly when the hazards were in one's own backyard.

Perhaps the best-known precursor to street science is the grassroots environmental-health activism by residents of Love Canal and of Woburn, Massachusetts. The infamous case at Love Canal, New York, where a concerned mother named Lois Gibbs triggered nationwide interest in the link between landfill contamination and children's health, is the now-classic story of residents organizing to perform and influence science. With the help of Dr. Beverly Paigen, a cancer researcher from Buffalo, Gibbs and other "citizen scientists" were trained to perform telephone and door-to-door health and environmental surveys. This community-driven research found elevated rates of disease but was dismissed by state health officials. Despite the professional rejection of their work, residents pursued, and through the Love Canal Homeowners Association they successfully convinced public officials and scientists to reexamine the environmental health issues in their community. By the summer of 1980 the state and federal government concluded that the neighborhood was unsafe and residents should be relocated.

The Love Canal controversy is an important example of a community struggling to grapple with unexpected health problems because it highlights the challenges local people, public officials, and scientists face when trying to understand the relationships between environmental exposures and health outcomes. Perhaps ironically, the intense scrutiny given to studies of Love Canal residents lead to more rigorous agency peer review, supposedly to ensure the integrity of studies. While at first glance appearing to open up science to public scrutiny, peer review affirmed the proposition that only scientists were qualified to judge the validity of work done by their professional peers. As Jasanoff has noted, self policing not only has enhanced the autonomy and social prestige of science, but it also has encouraged scientists to be accountable to standards considered acceptable by other professionals, not necessarily the general public (1985, 22).

On the heels of the Love Canal controversy another community concerned with sick and dying children, this time in Woburn, Massachusetts, organized residents to investigate the link between local pollution and illness. The story of Woburn citizens engaging in epidemio-logic studies, and enrolling scientists from Harvard to help them, also is well documented (Brown and Mikkelsen 1990; Harr 1996). What this case revealed was that residents with no prior scientific training not only could competently engage in complex science, but that they had unique information about exposures and health outcomes that, when combined with traditional epidemiologic methods, could improve scientific inquiry. When a community organizes to enlist the methods and resources from professional epidemiologists and com-bines these with insights from residents, they are engaging in a process Brown and Mikkelsen have called "popular epidemiology" (1990, 2). When communities engage in science, inject their own knowledge, and reorient investigations, outcomes, and actions, they often are in the process of seeking environmental health justice.

ENVIRONMENTAL-HEALTH JUSTICE AND STREET SCIENCE

The environmental-health-justice movement combines citizen activism and environmental-health problem solving with demands for civil and human rights (Bullard 1990; Di Chiro 1998; Cole and Foster 2000). While this book focuses on one community seeking environmental-health justice, similar communities around the world are engaging in *street science*, often forging research and action partnerships with outsiders, to address the prob-lems they face. A brief review of some of this work suggests that my study of one neighbor-hood in Brooklyn is part of the larger movement for environmental-health justice across the United States.

In Los Angeles, Communities for a Better Environment (CBE) has organized poor Latinos to monitor air toxics and address children's health. Partnering with researchers from the University of California, CBE activists formed a "bucket brigade" to take street-level air samples, to analyze these data according to local conditions, and to use these data to address respiratory-health issues facing local Latino children. These bucket brigades are groups of local activists that use a low-tech method for taking air samples "on the street," or where one breathes. CBE has used young people and other community members to take samples of toxic emissions from oil refineries in Contra Costa County. The brigades rely on local knowl-edge, such as reports of foul odors, seeing or hearing a release from the plant, and reports of nausea, eye and throat irritation, or other health symptoms, in order to determine when and where to take samples.

In Boston another environmental justice organization, Alternatives for Community and Environment (ACE), is collaborating with professional scientists, including some from the Harvard School of Public Health, to address asthma and air pollution in the Roxbury section of Boston (Loh and Sugerman-Brozan 2002). ACE organized students to map neighborhood land uses and found 15 diesel bus and truck garages within one-half mile of an elementary school. The organization then tapped the knowledge of high-school students to count truck traffic at a neighborhood intersection and identified over 150 diesel vehicles passing through neighborhood streets every hour. Combing the knowledge of young people, their maps, and traffic surveys, ACE partnered with Harvard and the Northeast States for Coordinated Air Use Management to take particulate samples of their own, further documenting the air-pollution problem in their neighborhood. The street science of ACE activists has lead to a state-funded but locally operated comprehensive air-monitoring system, which provides hour-to-hour data on particulate matter pollution over the Web and via telephone.

In San Francisco, the People Organizing to Demand Environmental and Economic Rights or PODER, have organized low-income residents within the Mission District of San Francisco to address environmental, public health, and redevelopment concerns and to help build a land-use agenda within the larger environmental justice movement. As part of their involvement in the Mission Anti-Displacement Coalition, PODER and its members helped develop a grassroots, comprehensive plan for the Mission that was presented to the San Francisco Planning Commission, Planning Department, and Board of Supervisors in July 2003. PODER also has developed a model for EJ groups to partner with one another, and they helped coordinate a report entitled "Building Healthy Communities from the Ground Up: Environmental Justice in California" in coalition with Communities for a Better Environment and the Environmental Health Coalition, another EJ group located in San Diego.

In Albuquerque, New Mexico, the SouthWest Organizing Project (SWOP) and the Southwest Network for Environmental and Economic Justice (SNEEJ), have collaborated with one another to organize residents in Veguita, New Mexico, to address water contamination issues. The organizations trained residents to test their drinking-water wells and perform a community survey of water and illegal-dumping concerns in the South Valley of Albuquerque. This work eventually convinced the U.S. Environmental Protection Agency (EPA) to issue a half-million-dollar grant to the local community and water district to plan, build, and maintain a water-distribution and sanitary-sewer system. SWOP also organized residents to perform air monitoring around the Intel Corporation's Rio Rancho facility as a way to pressure the company to address environmental-health issues for workers and communities along the U.S.-Mexico border. SWOP is a unique EJ group because their partnerships span multiple issues (water and air quality, workers rights, globalization) and multiple constituencies (low-income, Latino/as, youth and elderly, immigrants).

The work of all these groups aims to combine environmental-justice organizing with issues of population health. Each group has forged a collaborative research partnership with one or a host of outside professionals to help them combine community knowledge and experience with professional methods of researching and documenting inequitable environmental-health burdens. When community organizations such as these, and the ones in Brooklyn described in this book, engage in the science of environmental health, they grapple not only with understanding complex environment–human health interactions, but also with how to create more democratic partnerships with scientific and political elites that have traditionally ignored their concerns.

DEMOCRACY AND LOCAL KNOWLEDGE

A fundamental aspect of environmental-health justice is the creation of more democratic partnerships between professionals and the public. This ongoing challenge was perhaps best articulated by John Dewey, in his 1954 work *The Public and Its Problems*, where he highlighted the struggle or "problem" of engaging a citizenry in political processes increasingly dominated by technically elite professionals. Dewey's response was a division of labor; experts would analytically identify problems and citizens would set a democratic agenda for addressing them. The central challenge for Dewey was to devise methods and conditions of public debate, discussion, and persuasion where experts and citizens could integrate their knowledge and understandings. He called for participatory processes to increase the democratic character of decisions, where experts were not asked to judge the efficacy of particular policies, but to act as "interpreters and teachers" to help citizens debate in a way that would reflect the "public interest" (Dewey 1954).

While Dewey's analysis remains important for understanding the democratic challenge presented by street science, his analysis did not fully anticipate the influence of the specialized analyst, operating largely removed from any public discourse, on public policy. Nor did Dewey find the information and knowledge that experts (or lay people for that matter) have problematic; science and expertise for Dewey offered a body of facts and methods that only entered the rhythms and influences of politics at a later stage. Finally, Dewey focused on the optimal procedural conditions for reciprocal dialogue among scientists and lay people, but he did not fully anticipate that the content of the scientist-lay conversation might be problematic; scientists may be unable to translate their information into the ordinary language of everyday practice and publics may be unable to translate their knowledge into the specialized language of science. Thus, the rise of the professional analysts, or technocrat, and an uncritical faith in science as facts and truths, are key components for understanding why professionals tend to ignore community knowledge in environmental-health decision making.

TECHNOCRACTS, SCIENCE, AND LOCAL KNOWLEDGE

Theda Skocpol, in her book *Civic Engagement in American Democracy*, notes that "today's professionals see themselves as experts who can best contribute to national well being by working with other specialists to tackle complex technical and social problems" (1999, 495). Skocpol continues that these privileged professionals no longer see their role as "working closely with and for non-professional fellow citizens" or helping to lead "locally rooted" associations for problem solving. The view that public problems ought to be analyzed by a group of autonomous, highly trained and specialized professionals, who offer their dispassionate findings to decision makers, is partially rooted in the belief that facts and values can be separated easily. The positivist view of neutral fact-finding as informing value-laden politics remains a powerful decision-making model in environmental politics (Fischer 2000; Habermas 1970). Perhaps most influential in this view is that one form of rationality has come to dominate environmental politics—where science is the only legitimate form of expertise. Technocrats argue that experience in a given area and training in the specialized collection and systematic analysis of information allow them as professionals to tackle issues with neutrality and dispassionate objectivity (Benveniste 1972).

Yet, political scientists have regularly challenged the technocratic model. For example, Charles Lindblom and David Cohen, in their polemic 1979 book *Usable Knowledge: Social Science and Social Problem Solving*, argue not only that has social policymaking relied too heavily on professionals, but that professional knowledge has not contributed any more than ordinary knowledge to social problem solving. In their strong claim, Lindblom and Cohen (1979) argue for *useable knowledge*, as opposed to the professional knowledge that dominates modern policy-making. The problem with professional knowledge is that it has not delivered on its promise of making better, more efficient, cheaper, more fair or more just social decisions. Nor have the policy sciences contributed a great deal, they argue, to solving some of our most pressing social problems. Lindblom and Cohen (1979) argue for a reintegration of "ordinary knowledge" into policymaking in order to make it more responsive to the needs of the public and to remove the barriers between professional policy makers and citizens.

According to policy analysts like Linblom and Cohen, professionals should not be entrusted to speak for lay publics, especially concerning complex environmental-health controversies. Richard Sclove echoes these concerns in his 1995 book *Democracy and Technology*. Sclove claims that professionals are ill-suited to ensure that science and technology serve democracy because experts normally are more preoccupied with the mechanisms of science and not its structural bearing on society. Sclove also notes that since "experts enjoy a privileged position within today's inegalitarian political and economic structures, they tend to share with other elites an unstated, and usually quite unconscious, interest in suppressing general awareness of technologies' public, structural face" (1995, 50–51). Additionally, since scientists often have similar backgrounds, professionally socialize, and tend to acquire specialized competence at the expense of integrative knowledge and experience, they are unrepresentative of the "public" and should not be expected to understand or communicate the everyday knowledge of lay people.

Clearly, scientific and technical professionals hold important contributions for environmental-health problem solving, but they alone cannot be expected to ensure science and its results serve the larger society, particularly the least well-off. Lay people often are in a better position than professionals to make judgments over the democratic character of science because they experience how science impacts their everyday lives, from the repetitive mechanical tasks on the factory floor, to navigating inadequate mass-transit systems, to substandard housing and inferior medical care. Thus, to be scientifically and technologically "literate" is to have knowledge and experience not only about a technology's internal principles of operation, but also about how it influences democracy and social justice within the context where it is deployed (Nelkin 1984). Lay people are not only well-situated for this task, they are often more knowledgeable than professionals and therefore ought to be considered "local experts" in their own right.

THE CO-PRODUCTION OF EXPERTISE

Since both professionals and lay people have "expert" contributions to make to environmental health decisions, we might think about expertise as being "co-produced." Jasanoff and Wynne (1998) refer to "coproduction" to describe the interdependence of scientific knowledge and political order. As mentioned above, in the co-production model, scientific knowledge and social order evolve jointly; science is understood as dependent on the natural world, as well as on historical events, social practices, material resources, and institutions that contribute to the construction, dissemination, and use of scientific knowledge. Political decision making, in the co-production framework, does not take "scientific knowledge" as a given, but seeks to reveal how science is conducted, communicated, and used. The co-production model problematizes knowledge and notions of expertise, challenging hard distinctions between expert and lay ways of knowing. Finally, the co-production model emphasizes that when science is highly uncertain, as in many environmental-health controversies, decisions are inherently "trans-science"—involving questions raised by science but unanswerable by science alone (Weinberg 1972; Jasanoff 1990).

Decision making in the co-production model requires a negotiation among the always partial and plural positions of professionals and lay people (Haraway 1991; Harding 1991). The co-production model also destabilizes the dominant view in science policymaking that science can be uncritically accepted as "fact" and "truth." The destabilizing stories and emphasis on the need for "negotiating expertise" suggest that a deliberative politics is necessary for the co-production of expertise.

In an attempt to articulate how science might be co-produced, Funtowicz and Ravetz call for an "extended peer community" where professionals and publics collaboratively review evidence aimed at improving scientific knowledge:

> When problems lack neat solutions, when environmental and ethical aspects of the issues are prominent, when the phenomena themselves are ambiguous, and when all research techniques are open to methodological criticism, then the debates on quality are not enhanced by the exclusion of all but the specialist researchers and official experts. The extension of the peer community is then not merely an ethical or political act; it can possibly *enrich the process of scientific investigation*. (Funtowicz and Ravets 1993, 752–753; emphasis added)

The explicit recognition of both professional information and local knowledge—and that neither ultimately can put to rest the uncertainty of environmental-health problems—can encourage decision makers to acknowledge the necessity of renewal, flexibility, and adjustment as key elements of decision-making success. Instead of portraying themselves as the "source of certainty," professional decision makers can highlight the necessity for contingent decisions that must be open to renegotiation as new information becomes available. This means that the professional's role must be reconceptualized from "guarantor of safety" to "guarantor of recognition"—of new knowledge, new voices, new ideas, new possibilities, and new directions for interventions.

Robert Reich gives an eloquent account of how this practice of public deliberation can spur civic discovery. He suggests that professionals seize the opportunity for the public to deliberate over what it wants by:

> convening of various forums ... where citizens are to discuss whether there is a problem and, if so, what it is and what should be done about it. The public manager does not specifically define the problem or set an objective at the start... Nor does he take formal control of the discussions or determine who should speak for whom... In short, he wants the community to use this as an occasion to debate its future.
>
> Several different kinds of civic discovery may ensue... The problem and its solutions may be redefined... Voluntary action may be generated... Preferences may be legitimized... Individual preferences may be influenced by considerations of what is good for society... Deeper conflicts may be discovered... Deliberation does not automatically generate these public ideas, of course, it simply allows them to arise. Policy making based on interest group intermediation or net benefit maximization, by contrast, offers no such opportunity. (Reich 1988, 144–146)

Both Reich's vision and the process articulated by Funtowicz and Ravetz help frame what the co-production process might look in practice.

However, if co-production requires a negotiation between experts and local people, communities should be weary and enter with caution. As Arnstein's (1969) classic essay on the "ladder of citizen participation" highlighted, public participation can often backfire when the professionals controlling such processes do little to understand the residents of disenfranchised, low-income communities and do even less to meaningfully listen to and include them in decisions. Arnstein wrote that "there is a critical difference between going through the empty ritual of participation and having the real power needed to affect the outcome of the process" (1969, 216).

According to Judith Innes, a professor of urban planning at the University of California, Berkeley, urban planners are attentive to the power dynamics that occur in public dialogues and increasingly "depict planners as embedded in the fabric of community, politics, and public decision-making" (1995, 183). Drawing from critical theory and communicative ethics, this view of planning attempts to ensure, much like Dewey's original problem, that public processes are structured to allow the least powerful, politically disenfranchised to meaningfully participate. In order to accomplish this, a distribution of extra resources, assistance, and guidance to disenfranchised groups by planners may be necessary in order for meaningful and fair public deliberations (Habermas 1984; Forester 1989). The communicative view of planning is employed most often when finding an acceptable policy solution depends on appealing to and mobilizing citizens' knowledge of local or regional conditions, when policy issues have a strong ethical component, and when experts are strongly divided over an issue (Yearley 1999). As planning practitioners are increasingly asked to mediate between professionals and disenfranchised communities in local environmental-health decision making, understanding the benefits and limits of communicative practice becomes a necessary component of the co-production process.

Yet, deliberative forums, especially those involving environmental decisions, rarely have found a way to avoid granting science and technical expertise a privileged position in the discourse (Ozawa and Susskind 1985; Amy 1987). Even some of the most collaborative processes advanced by advocates of consensus building, such as joint fact-finding, have been unable to place science and technical expertise on par with lay knowledge, and these advocates instead recommend not pursuing joint fact-finding when "significant power imbalances among the parties" in a policy dispute exist (Ehrmann and Stinson 1999). Technical language remains a prerequisite for most deliberative forums, often creating an intimidating and "disciplining" barrier for lay citizens seeking to express their disagreements in the language of everyday life (Foucault 1977). Speaking the language of science, as well as the jargon of a particular policy community, remains an essential, but often tacit, credential for participation in environmental health decision making—even in the new deliberative forums. The process of *street science* offers a model for interconnecting and coordinating the different but inherently interdependent discourses of citizens and professionals through the co-production process.

STREET SCIENCE AS A PRACTICE

While traditional policymaking focuses on "problems" and "decisions," deliberative policy science has emphasized *practices* as its unit of analysis (Fischer and Forester 1993). Practice is admittedly a difficult concept. The concept of practice is an attempt to develop a unified account of knowing and doing (Dewey 1944). Practice emphasizes that knowledge, knowledge application, and knowledge creation cannot be separated from action; knowing and doing are intimately related (Putnam 1995).

This book argues that *street science* is a practice; a practice of science, political inquiry, and action. Street science is not merely a synonym for action. Street science integrates the actor, her resources, and her external environment in one "activity system," in which social, individual, and material aspects are interdependent (Callon 1986; Latour 1993). The focus in such activity systems is on the way the different elements *relate* to each other rather than just on the elements themselves. As Keller and Keller put it:

> An individual's knowledge is simultaneously to be regarded as representational and emergent, prepatterned and aimed at coming to terms with actions and products that go beyond the already known. Action has an emergent quality, which results from the continual feedback from external events to internal representations and from the internal representations back to enactment. (Keller and Keller 1993, 127)

Street science in this view acknowledges that the world in which we operate is always to a large extent provisional and improvisational. Action never is controlled completely by the actor, but is influenced by the contingencies of the physical and social world (Putnam 1995).

An important aspect of street science is its social character. Street science originates and evolves in a community—whether community is defined geographically, culturally, or socially. Street science also distances itself from mentalistic and subjectivistic views of judging, assessing, and knowing (Putnam 1995). Street science is a public process that originates and has meaning within a particular community. People learn about the world in shared public processes in which they test what they have learned, often through public discourse.

Central to the communicative dimension of street science are stories. Stories are central to the generative, emergent quality of action in context. Actors negotiate reality by telling *stories* about their own and other people's actions within the various elements of their community. Stories, however, are not merely representations of actions and consequences; stories are generative. As a form of discourse, by telling stories actors simultaneously shape, grasp, and legitimate both their actions and the situation that gave rise to their actions (Throgmorton 1996).

While the co-production model and deliberative practice offer frameworks for how street science might happen, they hardly help with understanding its content. How does local knowledge extend science and improve democracy? The next chapter answers this question by detailing what *local knowledge* means and by showing how it acts as the foundation of the *street science* method of inquiry.

NOTES

Many of the quotations in this book are from the author's personal communications. conversations, and meeting notes. Many of the individuals wished to remain anonymous.

1 Jasanoff (1999) and others note that in Europe, the precautionary principle is another way to frame environmental problems. Tesh (2000) also notes that the discourse of risk in

environmental-health policy shifted the focus of regulation from eliminating harms to managing risks.

2 A series of additional ideological and methodological challenges to risk assessment are described in detail in chapter 3.

3 In general, the book aims to avoid suggesting that all professionals act a certain way in all situations, just as it avoids essentializing what lay people know and do in all situations. The book highlights dominant tendencies and seeks to uncover why these trajectories persist rather than assume that all members of particular groups act similarly.

Scenario Assignment:
The Precautionary Principle

Assume you have been hired as a staff member at a new nonpartisan Center for Environmental Ethics and Public Policy that is well funded with a private endowment from a not-very-well-known philanthropist. You have been hired because you have a background in political philosophy and environmental ethics and because you have some experience in the environmental planning and policy-making world. The leadership of the center, including the founder (who made her money as the manager of a giant hedge fund), wants you to put together a report making the case for greater reliance on the precautionary principle as the basis for environmental policy-making. They want you to argue against the many experts and activists who have raised doubts about the viability of the precautionary principle. The reason the leadership of the center likes the precautionary principle, however, is because it appears to offer a pragmatic approach to environmental protection. It basically argues that it makes sense to act as if the worst case might occur and that we should not wait until scientific certainty has been established, by which time it might be too late to save a dwindling or irreplaceable resource.

What they want you to do is outline a philosophical argument for the precautionary principle that draws on both utilitarian ideals as well as the core beliefs of deep ecology.

Based on what you have read by DesJardins, do your best to lay out the case they are looking for.

Scenario Assignment:
Sustainability versus Economic Development

Imagine you work for the new office of Sustainable Development at the World Bank. The Bank has been criticized in the past for failing to pay sufficient attention to environmental considerations in its grant-making and investment decisions in the developing world. The governing board of the Bank has made it clear that the Bank still remains committed to the goal of economic development. Some external critics argue that environmental protection and economic development are incompatible and that either one objective or the other must be compromised. There are other critics, however, who do not agree. They point out that there is an important distinction between economic growth and economic development, arguing that sustainable development and economic development are not incompatible. The new office needs to justify, in philosophical terms, its commitment to both sustainability and economic development.

1. Prepare a brief presentation explaining what the office means by sustainable development as it relates to global investment in a wide range of projects in the poorest of the poor countries.
2. Discuss the distinction between economic growth and economic development.
3. In broad philosophical brushstrokes, lay out the reasons that the World Bank ought to put a premium on investments that promote sustainable development.

Indicate the reasons that economic growth is less important than economic development and explain why the Bank ought to take environmental protection seriously.

Scenario Assignment:
Local Knowledge versus Expert Knowledge

With the proposed revitalization of the nuclear power industry under discussion in the United States, there may well be pressure for renewed uranium production. It seems like the climate change debate has emboldened nuclear power advocates.

One of the sites at which uranium was mined years ago is in Navajo Nation in Arizona. There are hints that various corporations want to renew that activity. Anyone who has spent time in Navajo Nation has heard stories about the disastrous impacts of uranium mining on workers who helped dig uranium out of open pits (sometimes with their bare hands). Because appropriate safeguards were not enforced and workers tracked yellowcake back into their homes every night, cancer rates among these workers were extraordinarily high. A great many families lost loved ones. What these families would consider adequate compensation has never been paid. The federal government has not accepted responsibility for the illnesses and deaths experienced by tribal members involved in the uranium industry. Even after the mining sites were shut down, the federal government and the corporations involved did not cap the sites properly. Polluted water supplies, blowing airborne dust, and animals and children wandering through these unprotected sites continue to pose unacceptable risks.

Assume you work for the Navajo Environmental Agency. You want to be sure that "local knowledge" about the adverse effects of uranium mining in Navajo Nation is given serious consideration when and if the US government or its industry partners try to revive uranium mining.

1. How would you go about capturing this knowledge and ensuring that it is taken seriously?
2. Undoubtedly, you will be up against public health and other experts who will claim that the adverse impacts of uranium mining were minimal, or that if they did occur it was only because the proper rules and regulations were ignored. What is your "theory" about the best ways to incorporate indigenous knowledge into public policy-making?

End of Unit II Written Assignment: Environmental Ethics

Traditionally, economists have argued that humans are utility maximizers, although some behavioral economists and social psychologists have recently raised questions about this. With this in mind, please write a paper addressing the following questions: (1) When environmentalists and environmental planners argue that each generation (and we as individuals) has a stewardship obligation toward the natural environment, is that just one more utilitarian argument? Or, is there a different ethical principle at stake? (2) Please explain, with reference to the literature you read in Unit II, where you stand on the utilitarian versus deep ecology debate. (3) Also, do you think the distinction between economic growth and economic development resolves the apparent tension between arguments on behalf of environmental protection and arguments on behalf of economic well-being? (4) What is the strongest ethical argument you can make on behalf of sustainable development?

Try to draft a response to these questions on your own before reading the two student responses to this assignment that follow.

First Example Response to Assignment:
Environmental Ethics and Sustainable Development

Sustainable development has been described as a necessary alternative to the current mode of development, which is seen as undesirable because of its adverse impact on the natural environment and human well-being. Unsurprisingly, therefore, both sustainable development and environmental protection often share the same advocates, who use similar ethical arguments to articulate their beliefs. This essay examines these ethical arguments.

1. Stewardship Obligation toward Nature: Beyond Utilitarianism.
When environmentalists and environmental planners argue that humans have a stewardship obligation toward the natural environment, they sometimes adopt a utilitarian framework. However, beyond utilitarianism, many other ethical principles also support similar conclusions about human stewardship over the environment.

For utilitarians, what is deemed "good" is defined by its consequences, which should maximize overall welfare, whether this be pleasure, satisfaction of preferences or the reduction of suffering. Utilitarians advocating for environmental stewardship have thus argued that nature should be protected so that it can continue to improve human welfare through the provision of aesthetic enjoyment, natural resources, ecosystem services and other quantifiable benefits (DesJardins 2006, 30–1).

A "rights" perspective, where humans are seen as having a fundamental right to a liveable environment, also supports the idea that humans have an obligation to protect the environment. Insofar as humans have duties to all other fellow humans who would be adversely affected by environmental degradation, they should also be guardians of nature (ibid., 101–2).

Religion also offers another nonutilitarian argument in support of human obligation to the natural environment. Many religions view the world as a gift entrusted to humans by a creator. Thus, humans should treat nature with reverence and respect not necessarily only because of the benefits that can be reaped from nature but also out of respect for nature's origins from God (ibid., 39).

Diverging from these anthropocentric perspectives is the "biocentric" ethical principle, which argues that instead of treating nature as having "instrumental value," nature should be respected as having "intrinsic value" (ibid., 130) and that all living beings are "teleological centers of life" (Taylor 1986, in ibid., 136). Humans as moral beings have a duty toward things of inherent worth, which living beings and nature have. Biocentric ethics thus provide an argument for humans' stewardship of the environment.

Deep ecology takes typical biocentric ethical arguments one step further, to argue for "biocentric equality." According to deep ecology, humans and organisms alike are part of a community, and humans are not individuals but rather part of a greater "self." Thus, humans should respect and care for nature because it is part of who twwnatural life forms (ibid., 206–18).

2. Utilitarianism or Deep Ecology? While both utilitarianism arguments and deep ecology principles have been used to explain why safeguarding the environment is necessary, they stand at opposite ends of the spectrum. Utilitarian viewpoints are individualistic, rational and largely anthropocentric, whereas deep ecology argues for the dissolution of the distinction between the individual and nature as well as between the "subjective" and the "objective" (ibid., 212–16).

I find the utilitarian view a useful one because it adopts a familiar framework of logic that is easily understood. It provides a language that allows the virtues of environmental protection to be sold to those who may not necessarily have sympathies toward nature.

However, I also find the utilitarian arguments to be limiting. While utilitarian arguments can be stretched to cover the protection of an individual animal's welfare, it is difficult to extend it more holistically to cover the protection of nonsentient beings or species as a whole (ibid., 114)—something that I instinctively believe is important.

Similarly, while I do not fully agree with every aspect of deep ecology, particularly its emphasis on a biocentric equality that lacks clear guidance on how to prioritize the competing needs of humans versus the rest of nature (ibid., 217), it nevertheless offers an interesting perspective that helps me provide a better articulation and explanation of my instincts toward nature, which utilitarian arguments cannot.

Thus, I am most inclined to take a more pluralistic view of both ideologies. Instead of picking either one or the other as more or less "correct," I would prefer to acknowledge the inherent difficulty in reconciling both of these viewpoints, while recognizing the value and usefulness of both (ibid., 263).

3. Environment versus Economic Well-being: Achieving Both with Development. Oftentimes, environmentalists bump up against opponents who argue that environmental protection curtails economic growth. As economic growth is typically seen as the solution to poverty, economic well-being and environmental protection are thus often framed as incompatible objectives (Daly 1996).

Ecological economist Herman Daly offers a different perspective by first making a clear distinction between economic growth and development. The

153

former he defines as the *quantitative expansion* of throughput, or of total value of goods and services, which is responsible for environmental degradation. For the latter, Daly defines development as *qualitative improvement*, such as improvements in resource efficiency, and posits that it is more "sparing of the environment" because it allows a reduction in throughput (ibid.).

Daly (1996) argues that economic growth is necessarily limited and thus cannot be relied on to continue delivering economic well-being because the environment and its resources are finite. In contrast, economic development is potentially inexhaustible, as there are no obvious limits to its continuation. Daly thus argues that economic development without economic growth both provides economic benefits and protects the environment. Through this formulation, he neatly resolves the tension between economic well-being and environmental protection.

However, phasing out economic growth in favor of economic development, whether through a combination of technical improvements, reduction in world population or wealth and income distribution, involves major change from the current status quo and likely significant dislocation that would cause significant drops in the economic well-being of affected groups. For instance, workers in traditional resource extraction industries (e.g., oil, mining) may lose their jobs should the world move away from heavy resource use, and may not be able to easily find employment elsewhere, given their skill sets, geographic location and other constraints. Thus, the transition toward Daly's ideal state of greater environmental protection, where there is no economic growth but still economic development, will not be without a negative impact on economic well-being in the shorter term.

4. Strongest Ethical Argument for Sustainable Development. In its broadest definition, sustainable development is development that occurs in a way that can be sustained into the future. While there is consensus on this fairly vague definition, more specific formulations of the term differ. Some, like the United Nations World Commission on Environment and Development, which is responsible for the Brundtland Report, see continued economic growth as sustainable (Bartlett 1998), whereas others, like Daly and Albert Bartlett, adopt a more stringent definition of sustainable development, which requires both economic growth and population growth to stop, in order to allow humans now and in the future to prosper while staying within earth's carrying capacity.

If one defines the "strongest ethical argument" as one that can change the minds of the greatest number of people, a pluralist argument that combines different perspectives arguing for sustainable development, whether utilitarian, teleological or deep ecology, could have greatest traction. An example of such an argument comes from Pope Francis, when he addressed the United Nations General Assembly in September 2015 about the need for action on climate change. He made a powerful appeal for action that highlighted the unfair suffering of the poor as a result of environmental degradation, and the intrinsic "right of the environment" that stems from human existence as part of the environment as well as the intrinsic value of living creatures.

Conversely, if the "strongest ethical argument" for this more stringent definition of sustainable development is defined as one that is airtight and the hardest to refute, then a biocentric ethos, such as in deep ecology, would fit the bill. As highlighted earlier, the shift from current development patterns—continuing economic growth, expanding resource use and population increase—to one with a "steady-state" economy and no further population increase would likely cause short-term human dislocations and disutility. Furthermore, implementing population controls as recommended by Daly (1996) and Bartlett (1998) places the "burden" of impact on developing nations with growing populations, rather than on developed nations, which have mostly reached a population plateau.

Thus, one could conceivably argue against an unqualified push for sustainable development as defined above, if one adopts a purely utilitarian or human rights point of view.

However, if one moves away from an anthropocentric point of view, toward a biocentric view that we need to protect nature as an end in itself, then stopping current practices that degrade the environment and adopting sustainable development becomes absolutely imperative even if doing so may have some negative impact on people. If the sanctity of nature is placed at the center of considerations, then the need for sustainable development becomes very hard to refute.

References

Bartlett, A. 1998. "Reflections on Sustainability, Population Growth and the Environment—Revisited." *Renewable Resources Journal* 15, no. 4: 6–23.

Daly, H. 1996. *Beyond Growth*. Boston: Beacon Press.

DesJardins, J. 2006. *Environmental Ethics; An Introduction to Environmental Philosophy*. 4th ed. Belmont, CA: Thomson Higher Education.

Second Example Response to Assignment:
The Ethics of Sustainable Development

Discussions on environmental protection frequently touch on the idea of a stewardship obligation toward the natural environment. The premise of this argument is that the earth provides benefits to humans and that we should ensure the continued viability of those benefits. Thus the focus is on balancing anthropocentric harm and benefits. The "Global North" commonly takes this utilitarian view in analyzing environmental problems, but DesJardins (2006) offers several different philosophies, including one known as deep ecology, with which to understand humans' role in the larger ecosystem. The tension between utilitarianism and deep ecology results in many of the disagreements in environmental policy-making.

According to utilitarianism, an action is ethically good if it tends to maximize the overall good or to produce the greatest good for the greatest number. Public policy in Western civilization, especially environmental policy, tends to rely on utilitarian reasoning to weigh harm and benefits. It is the foundation of technical analyses like risk assessments that precede policy decisions. The problem with this argument is that it permits harm to a small number of people or the environment if it tends to maximize good for *a lot more people*. For example, a utilitarian argument would allow for a nuclear power plant, as it can power hundreds of thousands of homes and the potential to harm the 2,000 residents who live in the neighboring community is *very, very small* (assuming risks can be entirely localized). Thus, there is a fundamental flaw with a set of ethics if it allows for an action to be deemed "ethically good" despite the harm it produces.

When environmentalists argue we have a stewardship obligation, it is not a utilitarian argument. I would argue instead that deontological and religious principles are involved. As a species, we have a long tradition of self-sacrifice that cannot be explained in utilitarian terms. Instead, it is rooted in our inherent sense of duty, fundamental rights and justice. Utilitarian thinking cannot capture these values. In this frame, the stewardship obligation is an obligation toward future generations. We believe everyone has a fundamental right to a life free of suffering, and we extend that fundamental right to strangers across the globe as well as to unborn generations (just as we hope others do the same for us). These values are also common to many religions, which extend our duty to include the natural world. In the Judeo-Christian religion, God entrusts humans with the natural world with which to sustain them, but in turn, they must nurture and protect it. In the Buddhist and Shinto traditions, the natural world is revered and protected as it provides a place to seek inspiration and connect to the divine. For these reasons, temples and shrines are often in remote locations. And so according to deontological and religious principles, "good" is not defined by a calculation of the total good stewardship produces but rather stewardship is inherently *ethically good*, and not caring for the natural world is *ethically bad*.

Unlike utilitarianism, deep ecology takes a non-anthropocentric view, emphasizing that all life is valuable in itself, regardless of its value and use to humans. Thus, all life forms are intrinsically valuable, and it is ethically wrong if we encroach on their ability to thrive. According to DesJardins, the ethics of deep ecology are exemplified by two ultimate norms: self-realization and biocentric equality. The former, self-realization, is the process in which people come to understand that there is no firm ontological divide between humans and nonhumans (2006, 217), whereas the latter is the understanding that all life has equal intrinsic worth. Just as there is something instinctively wrong with sacrificing a few humans for the sake of more humans, deep ecology would argue that we should have that same gut feeling when we are talking about a species or an ecosystem.

Despite the unprecedented levels of wealth and consumption in the Global North, numerous indicators suggest that we are not physically or mentally healthier (see research by Tim Kasser, David Myers and Edward Diener). Our material wealth has also come at a significant cost to the environment, and challenges like climate change threaten dangerous and irreversible impacts, disturbing the ecological stability humans and nonhumans need to thrive. For the natural world—and thus humans—to flourish, we need to integrate ethics from deep ecology into our culture and radically shift from our obsession with economic growth to an emphasis on sustainable development.

The problem underlying capitalist society is the faith in economic growth to address the ills of the world, including the issue of finite resources. The common argument for economic growth is that we need growth to help increase wealth in impoverished nations. According to Daly (1996), the solution to poverty is not growth but development. Daly's distinction between economic growth and sustainable development is that growth is a focus on quantitative increases (e.g., gross domestic product (GDP)), whereas development is a focus on qualitative improvements (e.g., standard of living). By shifting the focus from growth to development, Daly presents an alternate society wherein the flourishing of humanity is not at odds with the flourishing of nature; in fact, they are co-dependent.

Daly offers several prescriptions to facilitate the transition from a focus on growth to a focus on development. To achieve sustainable development, we need population control, redistribution of wealth and income, and increases in resource productivity. All three solutions must consider ecological limits as the constraining factor. In the first solution, the ultimate goal is reducing consumption, which is the product of population and per capita consumption, and thus he envisions nations individually deciding whether to have higher populations or higher per capita consumption. The important point in his second solution, redistribution of wealth, is that wealth must include the valuable assets of natural resources, including the "right to deplete or pollute up to the scale limit" (Daly 1996, 14). And his final solution relies on technical improvements to use resources more efficiently. Ultimately, however, all of Daly's prescriptions still view natural resources in terms of their utility to humans and depict the environmental degradation caused by humans as a technological problem. It would seem that we can degrade portions of the environment so long as it does not impinge on the availability of that resource for human use. This also exposes some inherent flaws of a utilitarian argument—it denies that nonhuman life forms have intrinsic value and the impossibility of calculating an objective value of benefits and harm.

While Daly's recommendations are certainly an improvement over the dominant view of economic growth and are more prescriptive than deep ecology itself, they are still anthropocentric and privilege humans over environmental protection. In addition, it is clear that two of Daly's prescriptions would be politically impossible in the Global North. A stronger argument for sustainable development therefore would be one that integrates Daly's concepts with ethics from deep ecology.

Data on ecological limits and carrying capacity will not convince many countries of the urgency of achieving sustainability. While it may be rational, we require an ethical argument to produce the sweeping cultural shifts needed in Daly's model of sustainable development. At the core of this shift is for humanity to recalibrate its moral compass and achieve self-realization as called for in deep ecology. One of the principles developed by Naess and Sessions (1985) states "[t]he ideological change is mainly that of appreciating *life quality* (dwelling in situations of intrinsic value) rather than adhering to a high standard of living. There will be a profound awareness of the difference between *big* and *great*" (italics in original; quoted in DesJardins 2006, p. 207). The challenge is that our society currently equates *big* with *great* and is under the false impression that the next material purchase will bring happiness. We need to encourage people to drop out of the consumer rat race and reconnect with the values that are important to them. Some current trends suggest this is slowly happening (e.g., local food movement, urban gardening, demand for do-it-yourself goods), but we need to ensure this is not just a trend. We need to help people achieve self-realization, and once people recognize the interconnectedness of humans and the natural world, and the delicate balance required to sustain all forms of life, sustainable development will naturally follow as we redefine what is morally good.

Signs of environmental degradation are everywhere: the unprecedented background extinction rate, coral bleaching, global warming, and the list goes on. Utilitarian arguments have failed to bring about the policies needed to de-escalate economic growth and slow environmental degradation. Thus, we need a radical philosophical shift to principles that reimagine humans living in harmony and as equals with the natural world and not as the ruler of the natural world. Only with a radical realignment of values will we see truly sustainable development.

References

Daly, H. 1996. *Beyond Growth*. Boston: Beacon Press.
DesJardins, J. 2006. *Environmental Ethics; An Introduction to Environmental Philosophy*. 4th ed. Belmont, CA: Thomson Higher Education.
Devall, B., and G. Sessions. 1985. *Deep Ecology: Living As If Nature Mattered*. Layton, UT: Gibbs Smith, chap. 5.

Unit III

DEVELOPMENTS IN POLICY AND PROJECT ANALYSIS

Introduction

This unit introduces some of the best-known tools and methods for analyzing the prospective and actual effects of alternative policies or proposed solutions to environmental problems. These include environmental impact assessment, cost-benefit analysis, ecosystem services analysis, risk assessment, computer-based simulation and modeling, and scenario planning. Tools like these are aimed at reducing uncertainty and ensuring that decision makers and stakeholders have the information they need to participate effectively in resource allocation and environmental problem-solving.

We know that the way risks are framed can have a decisive impact on the outcome of public policy-making and environmental problem-solving. The more serious people think the risks are, the more inclined they are to support whatever actions are proposed. In the same way, we know that efforts to spell out the economic costs associated with policy alternatives and solutions will quickly shape the trajectory of debates and discussion in the public eye. Fairly frequently, findings associated with sophisticated analyses tend to be appropriated by different parties to support their side, often in tandem with other political arguments. There are other occasions, however, where a balanced result prevails. The unit begins with an excerpt by Lawrence Susskind, Ravi K. Jain and Andrew O. Martyniuk, in which they highlight successful examples of sound environmental studies and demonstrate how they have been used to inform better environmental policies for a wider set of stakeholders.

The readings that follow highlight some of the judgments that can influence the findings of an environmental study. For example, if the goal of a solution to an environmental problem is to maximize public benefits and minimize public costs (in the way we discussed in Unit II), then how such benefits and costs are measured is crucial. Environmental impact assessments (EIAs), required by law in many countries prior to designs or policies being enacted, are meant to ensure that the advantages and disadvantages of certain projects or policy options are considered.

Arwin van Buuren and Sibout Nooteboom similarly emphasize the importance of including stakeholders in the EIA process from the beginning and of conducting the EIA before action is taken. The goal is to allow for effective adjustments to be made to mitigate any negative impacts of proposed projects or policies. The authors point to a case in which the Dutch government, as a result of a well-conducted EIA process, abandoned plans to develop a proposed high-speed train that would have had significant detrimental environmental and social impacts. The case illustrates how an EIA can be used successfully for the wider benefit of a community in high-stakes decision-making.

EIAs often include cost-benefit analyses (CBAs), which convert proposed actions and impacts into monetary units in an effort to facilitate and legitimize collective judgments. David Pearce, Giles Atkinson and Susana Mourato recognize some of the inherent flaws of the valuation process, despite being proponents of CBA as an aid for decision-making.

For example, Robert Costanza and his collaborators point to the potential of ecosystem services analysis (ESA) as more accurately capturing the (often overlooked) value of the vast array of hidden services that ecosystems provide to humans. In other words, ESAs, they argue, provides greater weight to the environment than cost benefit analyses. Donald Ludwig, in contrast, argues that we do not know enough to assess the economic value of ecosystems. We include both perspectives because we recognize there is no easy answer to the question of how best to include environmental sustainability alongside other development goals. Communities need to grapple with these questions in depth and weigh the different trade-offs in light of their goals and values.

A fourth tool, risk assessment (and risk management), seeks to attach societal costs to the best, worst and most likely outcomes when a decision has to be made. Though helpful, the hurdle is that such forecasts still hinge on tentative estimates of the likelihood of long chains of events occurring or not occurring. In this vein, Howard Kunreuther and Paul Slovic argue that, although risk assessment is a useful tool in driving conversations about handling uncertainty, its inherent "inexactness" (as is the case in many of these environmental decision-making tools) makes it vulnerable to politicization.

An increasingly prevalent tool, simulation and modeling, seeks to provide decision makers the opportunity to simplify the complexity of real-life systems for purposes of "testing" alternative policies and solutions. Oversimplification, though, can undercut the reliability of the forecasts that are produced. John Sterman points out that models are only as good as the assumptions they are based on. Making models as transparent and accessible as possible (even to those without a technical background) enables stakeholders to adjust these assumptions more quickly and more frequently, in order to more accurately reflect our evolving understanding of the conditions on the ground.

Finally, scenario planning offers a method of comparing alternative futures when there is substantial uncertainty involved. We favor this method because we think it helps nonexperts quickly grasp the significance of the most important choices that need to be made.

All six methods—EIA, CBA, ecosystems services analysis, risk assessment, simulations and modeling, and scenario planning—require users to make what we call "nonobjective" judgments. For instance, quite frequently, there is not a correct time frame, or even geographic boundaries, that all parties agree should be the basis for analysis. Yet, judgments like these have to be made to set the initial scope, even if these can be modified on subsequent rounds. Experts are likely to disagree with some items, and nonexperts will want some say too, since there are no perfect answers. None of this negates the usefulness of such tools and their associated studies, but the fact that nonobjective judgments have to be made, does raise two important questions. Who should make them? And, how sensitive are the final results to the judgments that are made along the way? We believe that these studies can be very useful as an aid for decision-making, as long as the processes and assumptions built into these tools are made transparent.

This Unit also includes a scenario assignment as well as the general and a sample of the confidential instructions of a role-play simulation. These materials are meant to encourage you to think through how these different tools might be used to determine how to clean up and redevelop a polluted site. You will have to decide who should be offered a role in making the nonobjective judgments that will guide the analysis. You will also want to consider how sensitive the final analysis is to these nonobjective judgments, in the midst of a negotiated agreement. The written assignment offers you a final opportunity to address the two questions posed above.

In all, there are eight excerpts we ask you to read in this section. Most are short. Each discusses an instance in which one of the analytic techniques we present was used in practice. These excerpts will also give you a sense of the products that each method tends to produce. While these readings will not make you an expert on each method, they should enable you to make well-informed choices about which analytical method to use under different sets of circumstances and to recognize the potential shortcomings of each method. Overall, an increased sensitivity to what these tools are like and how they can be used should enhance your environmental problem-solving capabilities.

Commentaries and Reading Excerpts

Lawrence Susskind, Ravi K. Jain and Andrew O. Martyniuk — "How Environmental Policy Studies Can Be Used Effectively" and "How Policy Studies Should Be Organized." In *Better Environmental Policy Studies.* **Washington, DC: Island Press.**

Reading Commentary

If you turn on a cable news network, you are likely to see a congressional commission calling for a study (or discussing the validity of a study) of the latest environmental problem. Have you ever wondered whether these studies are of any help? And if so, how? Susskind, Jain and Martyniuk analyze the evidence provided by an array of landmark studies on a wide range of topics, including ecosystem management of late-succession forests, regulation of pesticide levels in food, the reduction of lead in gasoline and the clean-up of nuclear weapons production. After interviewing the key decision makers and policy insiders at the national, regional and state levels involved in each study, the authors lay out a number of important dynamics through which environmental studies seem to achieve disproportionate influence on new legislation and serve as sound bases for implementing new policy.

First, environmental studies can define a policy problem in an especially helpful way, describing its nature and extent, evaluating why it persists, analyzing who gains and who loses from its existence and reviewing relevant forecasts. Second, the most influential studies describe the full range of potential policy responses to the problem, informing the public debate and helping clarify executive and legislative priorities. Third, they can help overcome agency resistance to change by providing insights into possible institutional, financial and technical barriers. Fourth, they can foster collaborative inquiry, develop trust and enhance government credibility through partnerships (on advisory committees and technical review panels). Fifth, especially influential environmental studies enhance the legitimacy of key policy choices by jointly outlining costs and benefits, acknowledging trades and potential compensation. Sixth, these studies can work out solutions to political disagreements and address overlapping jurisdictional concerns by sharpening the choices that need to be made.

The authors also suggest the best ways of organizing environmental policy studies. They argue that conveners should pay special attention to selecting and using a range of experts; shaping the relationship between sponsors and experts; choosing the right institutional auspices; peer-reviewing their results; fostering learning; and setting a research agenda for the future. While the authors acknowledge that many studies are primarily aimed at bolstering the political arguments of those in power, some of the most influential environmental studies have underscored common interests and helped shape informed public opinion.

CHAPTER 3
HOW ENVIRONMENTAL POLICY
STUDIES CAN BE USED EFFECTIVELY

Lawrence Susskind, Ravi K. Jain
and Andrew O. Martyniuk

SIX EFFECTIVE POLICY STUDIES

Many environmental policy studies are prepared each year, but, as discussed in the previous chapters, the vast majority are not used in a meaningful way to make or change policy. What makes one study more effective than another? What studies have truly changed the course of environmental policy?

In order to answer these questions, we interviewed key environmental policy makers and political insiders at the national, regional, and state level. Two dozen strategically placed individuals were selected based on their past involvement in and knowledge about environmental issues and national policy making. An additional fifteen policy makers and analysts were contacted based on recommendations received from the first group. Individuals were interviewed by mail, detailed telephone interviews, and office visits (Schimek & Merrigan, 1994). We asked our interviewees to nominate environmental policy studies they thought were particularly effective. Specifically, we were interested in how a study changed the way they thought about a problem, framed an issue for public debate, or influenced subsequent laws, regulations, or agency behavior.

Six studies were most often mentioned as having the greatest impact. While there is no statistical significance to our selection, we were impressed by the overlap this open-ended method of inquiry achieved. Indeed, each of the studies described here was mentioned numerous times. And, as we set out to learn all we could about the studies we had selected, it soon became clear that the environmental policy "community" shared our sense that these were, indeed, exemplary. In this chapter, we analyze and evaluate the following six studies to learn what we can about them, particularly what they have in common:

- *Regulating Pesticides in Food: The Delaney Paradox*—by the Board on Agriculture of the National Research Council of the National Academy of Sciences
- *Costs and Benefits of Reducing Lead in Gasoline*—by the U.S. Environmental Protection Agency, Office of Policy Analysis
- *Complex Cleanup: The Environmental Legacy of Nuclear Weapons Production*—by the Office of Technology Assessment
- *Reducing Risk: Setting Priorities and Strategies for Environmental Protection*—by the Science Advisory Board of the Environmental Protection Agency
- *New Farm and Forest Products: Responses to the Challenges and Opportunities Facing American Agriculture (AARC)*—by the Task Force on New Farm and Forest Products.
- *Alternatives for Management of Late-Successional Forests of the Pacific Northwest (Spotted Owl)*—by the Scientific Panel on Late Forest Ecosystems

For each, we will provide a description of the study's history, approach, findings, and eventual impact on policy. Our discussion of study impact includes longitudinal analysis; we show how, over time, these studies led to significant new legislation and how they were used as a basis for implementing environmental policy. We then summarize the "uses" of effective policy studies, drawing upon examples from each case. Finally, we close the chapter with a discussion from an opposing frame, by addressing arguments that challenge our findings drawn from these studies. This analysis leads to our discussion in Chapters 4 and 5 of techniques for organizing studies to enhance their effectiveness.

THE "USES" OF EFFECTIVE POLICY STUDIES

Knott and Wildavsky argue that the true effectiveness of policy studies can be gauged not necessarily from the ultimate outcome of a particular policy decision, but from the various roles or uses the study served (1980) (see also Chapter 2). In the past, the roles and behaviors of individual decision makers have been emphasized while the various processes that aid in the dissemination and diffusion of policy knowledge, particularly when national level policies are concerned, have not. The seven standards of utilization proposed by Knott and Wildavsky take a broader view of the impact of policy studies, allowing judgments concerning effectiveness to be ascertained by looking at uses at differing levels in the policy making process (Webber, 1992).

Our six case studies cover a broad spectrum of environmental issues, from ecosystem management, to pesticide levels in food, to nuclear cleanup. At the time of their release, these studies set the agenda for debate on key issues. Since their publication, several have led to significant new legislation. Most importantly, each was used by environmental managers as a basis for implementing policy. Thus, these six policy studies catalyzed and guided the efforts of federal agencies to handle their environmental responsibilities. A review of these cases reveals six important "uses" of policy studies.

First, these studies have played a critical role in *defining a policy problem in a helpful way*. To define the problem, the authors needed to understand its nature and extent. Important questions that the authors asked were: Why does the problem persist? Who gains and who loses from the problem's existence? Who favors or opposes which solutions or policy options and why?

Just the act of initiating a study can change the way that policy makers and the public view the seriousness and boundaries of a problem. Moreover, a study team legitimizes some and deflates other conceptions of a problem when the team articulates its research objectives, identifies information gaps, frames questions for analysis, and constructs empirical models to generate forecasts. This process can have a major impact on the range of recommendations that agencies, interest groups, or the public view as legitimate. Thus policy studies, when properly organized, can shape congressional debate and public discourse about the direction that governmental action should take. In many instances, effective policy studies can widen or alter commonly held understandings, thereby shifting the agenda for policy reform.

Second, policy studies are invaluable for *describing the full range of potential policy responses to a problem*. A study that thoroughly analyzes the potential impacts of various policy options can generate a new set of legislative or executive priorities, inform public debate, and enhance the understanding of policy makers about which course of action they prefer.

Third, policy studies can help to *overcome agency resistance to change*. In some cases, a study may provide insights into the barriers that retard movement in new directions, and at the same time generate a set of arguments and approaches to overcome these barriers. In other instances, policy studies offer integrated sets of recommendations that respond to anticipated institutional, technical, or financial arguments, thus easing implementation.

Fourth, policy studies can provide important opportunities for *engaging stakeholders in collaborative inquiry*. When stakeholders, included as members of advisory committees or technical review panels, are asked to contribute data or advice regarding the design of new programs, it not only helps to educate them, but also adds credibility to the study process, thereby cultivating support for subsequent action.

Fifth, a thorough policy study can *enhance the legitimacy of a particular action*, and thereby create an unassailable justification for certain recommendations. For example, the use of cost–benefit analysis—together with a discussion of long-term environmental preservation—can elucidate some of the benefits of a new regulation in a way that counterarguments are quickly deflated. Thus research can be used to anticipate the challenges that opposing groups will raise.

Sixth, policy studies can help to resolve conflicting needs and *set resource priorities* in the face of budget constraints. Agencies faced with limited budgets and broad mandates can use policy studies as a vehicle for setting priorities and guiding the allocation of resources and personnel. Policy studies can also clarify the responsibilities of different agencies with respect to competing or overlapping policy jurisdiction.

Lead in Gasoline and the Reduction of Lead Levels in Ambient Air

Issue: Toxic pollution
Sponsor: U.S. Environmental Protection Agency
Responsible Agency: EPA Office of Policy Analysis

The elimination of leaded fuel in motor vehicles is one of the biggest environmental victories since the founding of the EPA. Ambient air lead levels have been reduced by over 90 percent since 1970. The use of leaded fuel was the largest source of these emissions. Regulation of leaded gasoline began with the Clean Air Act of 1970. Over a decade later, the phase-out of leaded gasoline occurred only after the publication of a policy study entitled *Costs and Benefits of Reducing Lead in Gasoline*.

HISTORY

The Clean Air Act Amendments of 1970 provided the administrator of the EPA broad authority to "control or prohibit the manufacture ... or sale of any fuel additive" if its emissions (1) cause or contribute to "air pollution which may be reasonably anticipated to endanger the public health or welfare," or (2) "will impair to a significant degree the performance of any emission control device or system ... in general use." In its effort to comply with this mandate, EPA began examining the use of lead in gasoline. In 1971, EPA initiated regulatory proceedings to phase out the lead in gasoline. Two reasons were given for EPA's action. First, human health concerns were raised. Several studies suggested a strong relationship between the use of leaded gasoline and the level of lead in human blood. The studies found that lead could be absorbed into the body from ambient air and that lead was toxic to the human body. Second, leaded fuel was found to be incompatible with the catalytic converter, a device required on all post-1975 automobiles to reduce hydrocarbon emissions.

As expected, there was significant opposition to any lead phase-out. Powerful refiners, who used lead as a gasoline additive to increase octane, joined together with manufacturers of the lead additives in opposition to all lead restrictions on gasoline. On the other side, several environmental organizations, most prominently the NRDC, pressured EPA to move forward with the phase-out.

EPA issued regulations calling for incremental reductions in average lead content of all gasoline sold over a five-year period. However, several factors conspired to slow EPA's lead phase-out activities. Due to the oil crisis in the late 1970s, President Carter directed EPA to postpone the phase-out until October 1980. The EPA ruling was also challenged in the courts. Consequently, the standard was not put into effect until October 1980.

However, EPA was quick to make the standard more stringent. Four factors contributed to EPA speeding up the phase-out. First, a 1982 EPA survey found that 12 percent of consumers who had catalytic converters designed for unleaded gasoline were nevertheless fueling their automobiles with the cheaper leaded gasoline. This resulted in misfiring that caused significant emissions problems. Second, new studies were generated that further linked leaded gasoline with health problems. Third, the U.S. Court of Appeals for the D.C. Circuit, in a case challenging the provisions of the 1982 lead standards, ruled that available evidence "would justify EPA in banning lead from gasoline entirely." Fourth, alcohol fuel advocates were pushing for restrictions on lead in order to make alcohol more attractive as an octane-boosting fuel additive.

Most important, however, was that the EPA needed a strong policy stand to reestablish the agency's credibility. During the early years of the Reagan administration, the EPA's reputation was shattered due to unethical activities by several political appointees. Professionals within the agency were frustrated and were searching for an issue to restore EPA's reputation as protector of the environment. In 1983, taking a tough environmental stand, EPA issued a final rule tightening the lead limit in gasoline. The rule came as a surprise to the environmental community, which hailed the move as a major victory.

Months later, EPA deputy administrator Alvin Alm noted that the studies on lead in gasoline hinted that an even faster phase-out would further improve air quality and health. Although no one expected EPA to mandate further lead reductions, in September 1983, Alm asked the EPA Office of Policy, Planning, and Evaluation to undertake a cost–benefit study of reducing or eliminating lead in gasoline. Joel Schwartz, an analyst who had worked on the 1982 lead regulation, was asked to provide a "back of the envelope" calculation on the benefits of a total ban on lead additives. When he completed his calculations, Schwartz told AlmAlm that early indications suggested that the benefits of such an action would be more than twice the cost (Schwartz, 1993).

APPROACH

Alm directed the Office of Policy Analysis to undertake a comprehensive lead study. A small team of analysts carried out his orders quickly and quietly. In total, only six staff members contributed to the study, with no more than three people working full time at any particular point. From its inception to the issuance of the final EPA rule on lead, the study lasted a total of eighteen months. The key question for the study team was whether the benefits of an additional lead phase-out would exceed the cost.

The *Lead in Gasoline* study was the first real cost–benefit analysis undertaken by the EPA. In 1981, Executive Order 12991 required all regulatory agencies to prepare a regulatory impact analysis for any regulation likely to affect the economy by $100 million or more per year, but EPA had yet to implement the ruling. Although there were no model cost–benefit analyses upon which to base the lead study, the staff took less than four months to develop a preliminary draft to send to outside experts for peer review. Such a peer review process was unusual at the time. The outside experts included in the peer review were automotive engineers, economists, biostatisticians, toxicologists, clinical researchers, transportation experts, and a psychologist. The staff asked that the review be completed in secrecy. In January 1984, the team refined its analysis and incorporated peer review comments into the final report.

FINDINGS

The resulting study was lengthy and highly technical. It contained the underlying model, a description of the team's methodology, and many charts and graphs. EPA estimated the benefits to be gained from reducing lead in gasoline and categorized the benefits into four major categories: (1) children's health and cognitive effects associated with lead; (2) blood pressure-related effects in adult males due to lead exposure; (3) damages caused by excess emissions of hydrocarbons (HC), nitrogen oxides (NOx), and carbon monoxide (CO) from misfueled vehicles; and (4) impacts on vehicle maintenance and fuel economy. The analysis was conducted using existing databases. First, EPA modified and enhanced a model of the petroleum refining industry originally developed for the Department of Energy (DOE). Second, EPA used one of its existing health databases that had been constructed for another purpose.

The lead study was released to the public in three stages. First, a draft report was released in March 1984, after six months of staff work. The Senate Committee on Environment and Public Works held a hearing to discuss the implications of the study. In August 1984, EPA formally proposed tightening the lead limit as recommended in the cost–benefit analysis. EPA also indicated that it was considering a complete ban on lead to be enacted by the mid-1990s.

In January 1985, a paper by Joel Schwartz on the health effects of lead on hypertension in middle-aged white men was released. After incorporating these additional impacts, the final lead study was released in February 1985. The following month, EPA issued a final rule that established a new standard for lead in gasoline. It was considered by many analysts to be a radical phase-out. In August 1986, the House Committee on Energy and Commerce released a committee print that supported EPA's controversial lead reduction plan.

IMPACT

As one might expect, industry was not too happy with the proposed lead rule and applied pressure to members of Congress to oppose it. However, the study analyzed so many different scenarios that policy makers quickly came to consider the new lead rule as the only reasonable course of action. The cost–benefit analysis was so clear-cut that industry arguments were quickly dismissed. A document produced by the House Committee on Energy and Commerce cites the study as the key reason behind its members' support for the new lead rule.

The lead study helped establish the credibility of two innovative policy approaches. First, the study represented the first real EPA effort to do a cost–benefit analysis. Despite the environmental community's suspicion that cost–benefit analysis would always work against them, the study actually supported a tightening of the lead standard. This eliminated some of the prejudice against cost–benefit analysis and encouraged EPA to make greater use of it in future policy efforts. Second, the new lead rule made use of a new regulatory approach analyzed in the study's model: It allowed for the trading of pollution rights. EPA was willing to test this approach because the study team had provided the analytical backing to demonstrate that it could be effective. Since its inclusion in the lead rule, emissions trading has been used many times by EPA.

The lead study was used by the agency to justify its lead phase-out rule. It was also used by Congress to evaluate policy options, and over time, by policy makers who sought successful examples of cost–benefit analysis and tradable emissions allowances. Analysts throughout the federal government consistently point to the lead study as the clearest example of a policy study leading to agency action. Joel Schwartz, the study director, says "there was no pressure on EPA whatsoever to do anything about lead" (Schwartz, 1993). The EPA rule on lead is directly linked to the study. The release of the first draft study in March 1984 was followed closely by a proposed rule in August 1984. Release of the final study in February 1985 was accompanied by a final rule in March 1985. By all accounts, the study is the single explanatory factor for EPA's sudden and dramatic lead phase-out.

USE I: DEFINING THE PROBLEM IN A HELPFUL FASHION

The lead study is a case where analysis clearly preceded agency decision making, and determined the course of the dramatic EPA lead phase-out. The cost–benefit analysis defined the key policy issues for decision makers and, as Weimer and Vining (1989) write, moved the debate "beyond disputes over predictions to explicit considerations of values."

There was no external pressure on EPA to initiate a study on lead or to further phase down its use. This lack of external pressure allowed the cost–benefit analysis alone to define the problem and guide agency action. Joel Schwartz contrasts this with other policy efforts generated by "political heat." In those cases, Schwartz argues that policy analysis is not as important as it was in the lead decision because external pressure "sort of gives you a hint as to what… to do and how much" (Schwartz, 1993).

Energy consultant Albert Nichols argues that more cost–benefit analysis should be undertaken before key agency decisions are made. He writes that when cost–benefit analysis is carried out during or subsequent to the development of regulations, "the results usually are not available until most of the fundamental regulatory decisions have been made. At that point, bureaucratic momentum, often coupled with strong external pressures or deadlines, makes it very difficult for the analysis to have much influence" (Nichols, 1985).

USE 2: DESCRIBING THE FULL RANGE OF POLICY OPTIONS

The lead study evaluated a wide range of potential lead standards. This analysis used different assumptions about the impacts various rules would have on misfieling, or using leaded gas with catalytic converters. The possible impact scenarios ranged from the complete elimination of misfueling to no change in misfueling rates. In addition, the study computed net benefits both with and without the estimates of the blood-pressure-related benefits of reducing lead levels in gasoline. The final report presented all of the data generated using different assumptions. Regardless of the particular assumptions about misfueling, and whether or not blood pressure-related benefits were included, the study found that net benefits were maximized with the most stringent of the alternative standards under consideration.

Albert Nichols (1985) notes that the full presentation of data had an important impact on the policy process. He says, "At every major meeting, decision makers were presented with quantitative estimates of the effects of alternative options." Weimer and Vining (1989) also note that the EPA analysts were constantly refining their findings, a strategy they believe is a hallmark of successful quantitative analysis.

USE 3: HELP OVERCOME AGENCY RESISTANCE TO CHANGE

Two factors are cited as important in having eased the way for the implementation of the lead study recommendations.

First, it was important that the study was done internally so that the EPA policy staff was fully cognizant of every decision and assumption made in the model. The staff was then able to answer all levels of questions throughout the implementation stage. Joel Schwartz contrasts this with the frequent agency practice of hiring outside consultants to do most of the policy analysis: "We knew what we were doing as opposed to contracting out for all the policy analysis—the standard thing. When you do that you get a report and you really don't understand the details of what went into things" (Schwartz, 1993). Study team member Jane Leggett agrees: "What the policy office does now is spend lots of money on consulting—we don't have people in-house and we spend most of our time pushing paper and managing the contracts. Spending time on details is really critical if it's going to lead to real action because you have to be able to respond when people raise questions and be able to look at the data and do the analysis in a different way. Consultants aren't privy to those kinds of discussions" (Leggett, 1993).

Second, the study analysts carefully scrutinized the potential arguments that would be raised by opponents to the lead phase-out. Numerous calculations were undertaken to provide analysts with the answers necessary to defend the EPA's approach over others. Weimer & Vining cite the "repeated re-analysis to rule out alternative explanations" offered by opponents as critical. They note that the analysts "drew relevant evidence from a wide variety of sources to supplement their primary data analysis." They also point out that the analysts gave serious attention to possible confounding factors, considering both internal tests (such as subsample analyses and model re-specifications) and external evidence to see if they could be ruled out. As a consequence, opponents of the proposed policy were left with few openings to attack its empirical underpinnings (Weimer & Vining, 1989).

USE 4: PROVIDING OPPORTUNITIES TO ENGAGE STAKEHOLDERS

Due to its particular circumstances, the lead study stands as counter-evidence to public involvement in the study process. In this case, EPA decided that its analysis should be conducted in isolation from stakeholders. Stakeholders were viewed as a threat to the integrity and success of the policy effort, and the study was conducted quickly and in secret. Weimer & Vining (1989) write that Administrator Alm "urged speed in order to reduce the chances that word would get out to refiners and manufacturers of lead additives, the primary opponents of a ban, before the EPA had an opportunity to review all the evidence."

Joel Schwartz and Jane Leggett were both briefed on the political sensitivity of their efforts and the importance of keeping the study behind closed doors. Schwartz notes, "The reason we needed to do it fast was that [Alm] was worried that if word got out that we were working on this, the White House would tell us to stop." Leggett adds, "It was highly political. And this was the first Reagan term and there was still a bias against regulation in general. It was conveyed to me that we wouldn't be able to do the work if it were known ... and that before we went public with anything we better be very careful about our results." Although the lead study was conducted in secrecy, a draft of the study was subjected to peer review and comment.

USE 5: ENHANCING THE LEGITIMACY OF PARTICULAR ACTIONS

The lead study showed environmentalists that cost–benefit analysis would not automatically discount environmental and human health concerns. Instead, it proved to be a powerful method that supported their goals. Use of this tool allowed the EPA to show that eliminating lead in gasoline was the only reasonable approach.

The study clearly justified the EPA's radical lead phase-out. Robert Bamberger, energy policy specialist at the Congressional Research Service, recalls how no one expected EPA to further reduce lead levels in gasoline and that there was almost no pressure put on the agency in this area. Bamberger recalls, "I remember how surprised I was at the time. The [EPA lead] action was counterintuitive to my expectations" (Bamberger, 1993). Joel Schwartz noted that, although no one was pushing the agency to do something on lead, the study was undertaken. Therefore, the "analysis turned out to completely determine the total form of the regulation in all of its details."

Joel Schwartz argues that the cost–benefit analysis cleared the way for agency action. He said, "It's a lot easier to use cost–benefit analysis than it is to tell people they shouldn't do what they want to do." Not only did it move the EPA administrator to act, but the study was crucial in winning the support of other federal agencies. Leggett (1993) reports, "Even the OMB (Office of Management and Budget) said it was a great study."

Weimer & Vining (1989) write of the importance of the lead study in establishing cost–benefit analysis in policy analysis: "As a society we might very well be willing to sacrifice considerable efficiency to achieve redistributional or other goals, but we should do so knowingly. Introducing efficiency as a goal in the evaluation of policies is one way that policy analysts can contribute to the public good."

USE 6: SETTING RESOURCE PRIORITIES

The ability to share data within EPA and between the EPA and the DOE was critical to the study's success. Agencies often jealously guard their data. This not only makes policy analysis more difficult, it also prevents analysts from developing key insights into existing databases. For example, the EPA policy analysts discovered a strong correlation between high blood pressure in middle-aged white men and very low levels of lead. The agency personnel responsible for the database had not seen the connection between these issues because, as Jane Leggett points out, "If you hadn't asked the question, you would have never found that out."

Much of the success of the lead study can be attributed to a multidisciplinary team approach. But EPA analysts advise that unless current reward structures are altered, there will continue to be infighting within agencies and other disincentives for team research. Leggett laments, "The whole incentive basis of the federal government does not reward team research. There is no incentive for someone to help someone else with a study; if the other branch succeeds, then they are rewarded with more research. It is not in anyone's interest to acknowledge help. You need a manager who is committed to team research despite the disincentives."

The success of the lead study's team approach points to the need to increase efficiency in resource utilization within and between agencies, especially in view of diminishing budgets and personnel. Interdisciplinary cooperation is a more effective way to conduct policy studies that will minimize environmental and health risks to the public.

THE SPOTTED OWL AND ECOSYSTEM MANAGEMENT IN THE PACIFIC NORTHWEST

Issue: Endangered species and forest management
Sponsor: Agricultural Committee and Merchant Marine and Fisheries
Committee of the U.S. House of Representatives
Responsible Agency: Scientific Panel on Late Forest Ecosystems

In June 1990, the Fish and Wildlife Service (FWS) placed the northern spotted owl, a native of Pacific Northwest forests, on the "threatened" species list under the guidelines of the Endangered Species Act (ESA) of 1973. Environmental groups and the timber industry began to take vocal and opposing positions on the issue. Several of the agencies involved in species protection and forest management—the Forest Service, the Bureau of Land Management (BLM), the FWS, and the National Park Service (NPS)—jointly created a panel called the Interagency Scientific Committee to Address the Conservation of the Northern Spotted Owl (ISC), which was empowered to develop an owl conservation strategy. Subsequently, the Department of the Interior (DOI) was also required to develop an owl conservation plan to meet the statutory requirements of ESA.

HISTORY

By May 1991, a series of different plans were "on the table." Working jointly on the problem, the House Agriculture and the House Merchant Marine and Fisheries Committees found that certain basic information was lacking. First, no one knew with any accuracy the range and location of old-growth forests. Second, there was little hard data on the potential environmental and economic impacts of any particular owl preservation plan.

To address these information gaps, the committees turned to four government and academic scientists. The Gang of Four, as they came to be called, were well respected among members of Congress and environmental groups. One of the four, Jack Ward Thomas, had already been the team leader of the Interagency Scientific Committee to Address the Conservation of the Northern Spotted Owl (ISC), the body that had produced one of the competing owl plans. The four scientists presented their initial findings to Congress in July 1991. Their written report, *Alternatives for Management of Late-Successional Forests of the Pacific Northwest (Spotted Owl)*, was submitted in October of the same year.

APPROACH
The Gang of Four decided that the problem was one of ecosystem management rather than only owl preservation. Because species are interdependent, the four scientists presumed that preservation efforts targeting a single species would not be successful in the long run. Furthermore, the scientists observed that as ecosystems deteriorate, many species are simultaneously threatened. Since individual species protection plans were time consuming and generated significant political attention, the Gang of Four redefined the objective of the study as how to determine the best forest management policies given certain overarching public policy objectives.

FINDINGS
In the *Spotted Owl* report, the scientists presented fourteen different policy options, ranging in their objectives from high timber yield to high forest protection. These alternatives covered all of the forest management options already "on the table" and analyzed several new ones as well. Scenarios were also considered for each of three different management options for lands located outside old-growth reserves. For each different scenario, the report presented the associated impacts on old-growth forest maintenance and species preservation. The report demonstrated the changes in timber harvest and regional income associated with each scenario.

IMPACT
At the same time that the Gang of Four's *Spotted Owl* report was published, the Endangered Species Committee, sometimes called the "God Squad," was created to consider exempting some timber sales on public lands despite the spotted owl's mandated protection under ESA. The Endangered Species Committee decided in May 1992 to allow sales in a few areas, but the decision was subsequently overturned by a federal court. During the spring of 1992, forest protection bills based on the Gang of Four's analysis were passed by both the House Agriculture and House Interior Committees. Neither of these versions proceeded any further in that Congress.

The new Clinton administration convened a "timber summit" in April 1993. Representatives of the stakeholders met in Portland, Oregon, to discuss the issue. The Spotted Owl report became "the standard reference point, the starting point for any discussion," and has provided a menu of options for further action on forest management (Owens personal communication, March 11, 1993). The report shifted the focus of environmental protection from the traditional approach—drawing boundaries on a map or counting trees—to designing integrated strategies to preserve ecosystems (Lyons personal communication, February 1993).

In 1995, Congress passed the Emergency Supplemental Appropriations and Rescissions Act (P.L. 104–19). Section 2001 of this law exempted the U.S. Fish and Wildlife Service from preparing an environmental impact statement under the National Environmental Policy Act (NEPA) to ease prohibitions against the incidental taking of spotted owls on nonfederal lands. Although utilization of Section 2001 has been slowed by litigation regarding its implementation, the so called Timber Rider or Salvage Rider has resulted in additional environmental alternatives analyses and draft rules (FWS, 1996). Against this backdrop, hearings held on March 19, 1998, before the House Subcommittee on Forest and Forest Health indicated that since 1990, the year the spotted owl was designated a threatened species, the area of known habitat has increased each year (U.S. Congress, 1998). Although land has been set aside for the spotted owl in national forests where logging is restricted, litigation continues regarding habitat on nonfederal lands.

USE 1: DEFINING THE PROBLEM IN A HELPFUL FASHION

The Gang of Four argued strongly that ecosystem management would accomplish the overly limited goal of owl preservation. This argument was based largely on ecological principles. As species are interdependent, preservation efforts targeting a single species would not be successful in the long run. In their report, the Gang of Four stated, "We have described the beginnings of a practical 'ecosystem approach' to conserving biological diversity. Nature does things in twos and threes rather than singly. So should we in seeking to preserve or mimic nature" (Johnson et al., 1991).

The argument for a broader problem definition also had a practical justification. As ecosystems deteriorate, many species are threatened simultaneously. It is time consuming for agencies to prepare multiple protection plans for many different species in the same habitat. This problem is magnified when the consequences of forest management for human activities are large enough that each conservation plan generates significant political attention. For Congress, this reality points to the value of avoiding the "endangered species of the month" syndrome (Lyons & Gordon, 1993).

The Gang of Four was able to define the problem broadly in part because they were working for two committees with different species under their respective policy jurisdictions. This gave the scientists the impetus to study the impacts of forest management on a range of species besides owls. For example, they discovered that certain fish species were particularly threatened. In addition, the congressional committees gave the scientists a broad mandate to pursue a "best-science" approach to the problem, wherever it would lead them (Lyons, 1993; Gordon, 1993). In response, the scientists identified a range of forest resource management needs that lay beyond the scope of owl preservation, but are strongly related nonetheless. These needs included such items as information gathering and analysis, watershed restoration, and prescribed burning.

USE 2: DESCRIBING THE FULL RANGE OF POLICY OPTIONS

The authors of the *Spotted Owl* report felt strongly that scientists should not make policy decisions. Instead, they wanted to present the effects of particular policy options on people and the environment. The authors write, "We have provided a sound basis for decisions, given the time and information limits within which we operated. Science (at least as exemplified by the four of us and those who assisted us) has done what it can. The process of democracy must go forward from here" (Johnson et al., 1991). The study analyzed fourteen alternatives and included the impact of each level of ecosystem protection on timber harvest and regional income in its calculations. The wide variety of options in the report made it an effective and credible starting point for further discussion.

USE 3: PROVIDING OPPORTUNITIES TO ENGAGE STAKEHOLDERS

Although the Gang of Four consulted with a wide range of experts, including industry representatives and environmental groups, the final recommendations and results are their own opinions. Stakeholders were not involved in drafting or reviewing the report. In some ways, the lack of constituent involvement was advantageous for this kind of policy study. The scientists' well-known professional reputations helped to give the report credibility in the eyes of stakeholders and members of Congress.

USE 4: ENHANCING THE LEGITIMACY OF PARTICULAR ACTIONS

Before the study, Congress was at an impasse due to divided public opinion and a lack of hard data on the potential environmental and economic effects of different strategies. The Gang of Four's report has been called "a watershed event in terms of quantifying the issues and the choices that would have to be made" by Representative George Miller (D-California), chair of the Interior Committee. The study's thorough analysis prompted action by congressional committees and gave them a credible and scientifically sound analysis to justify their policy choices. The two congressional committees that commissioned the study used it as the basis for a new legislative initiative to protect the northern spotted owl.

The scientists felt that it was important to move the debate away from the traditional arguments that pitted owls against jobs or square miles of owl habitat against board feet of lumber. Environmental groups were initially skeptical of this undertaking, in part because some of the scientists had government positions. However, they were generally pleased with the report's conclusions (Owens, 1993). The lumber industry, on the other hand, was critical of the study's results. Despite the Gang of Four's desire to minimize the polarization of different stakeholder groups, much of the continuing debate on the subject pits industry against environmentalists.

USE 5: SETTING RESOURCE PRIORITIES

The report shifted the focus of policy analysis and resource allocation from individual species protection to integrated ecosystem management. Even though the debate surrounding reauthorization of the Endangered Species Act is heating up again, the focus on integrated ecological management is very much in place. The policy study addressed the prior lack of coordination among agencies responsible for protecting endangered species and managing forest resources. The scientists broadened the scope of their initial mandate from devising a plan for protecting a single species to developing a range of strategies for managing forest ecosystems as a whole. They argued that it simply did not make sense for preservation efforts to focus exclusively on particular species. Such an approach multiplied the work of agencies developing species conservation plans while ignoring the fundamental interdependence of ecosystems. To correct this inefficiency, the report provided information and analysis useful to all of the agencies with overlapping responsibilities for forest management.

THE CHALLENGES

The six policy studies discussed in this chapter were identified by decision makers and policy insiders at the national, regional, and state level as being particularly effective. Once we identified these studies, we looked longitudinally at their impact. We can understand why some might challenge our assertion that these were particularly effective. After all, without tangible evidence to show that these six policy studies produced desirable outcomes (i.e., substantive environmental improvements), how can we say they were effective? This is a difficult challenge since we do not have the necessary evidence to demonstrate that the outcomes in each instance were desirable, or that the public policies in question (and not other factors operating independently) produced the intended results. The fact is, there is no evidence to show that the results of these policy-making efforts were consistent with what was envisioned, or that what was envisioned was, in retrospect, desirable. All we can say is that these studies helped to provoke subsequent changes in policy. We cannot prove that these changes were ideal. Our definition of effective focuses more on the process of policy making than on the impacts of the policies themselves.

We can, however, confirm what we had previously theorized about the uses of environmental policy studies; namely, that they can help to shape public debate, bring stakeholders into the conversation, suggest new policy options, enhance the legitimacy of whatever actions are subsequently taken, or clarify resource allocation policies (to help ensure implementation). We are not asserting, however, that all effective environmental policy studies always lead to the results envisioned by their advocates, or that the outcomes of effective policy-making efforts are always cost-effective or even desirable.

Most policy studies probably bolster the technical basis for the political arguments that those in power were inclined to make anyway. Effective environmental policy studies rarely force decision makers to change their views. They may, though, allow certain decision makers to hold their views more comfortably or, in rare instances, force those with opposing predispositions to hold their political fire.

The media play a role in focusing public attention or shaping public opinion about proposed environmental policy changes. Media reaction may well be an important factor in explaining changes in environmental policy. There may be other factors that come into play as well, but even if we have not isolated all the factors that do play a role in shaping environmental policy-making efforts, the important point here is that the six studies we present were indeed effective in influencing public policy making.

Why do the uses or effectiveness of environmental policy studies matter when ultimately public policy decisions are implemented by "fixers"—political figures with the clout to reshape environmental policy during implementation (Bardach, 1977)? We do not believe the mere existence of a "fixer" invalidates the utility of effective environmental policy studies. Such studies undoubtedly make a fixer's job easier or harder, depending on the overlap between the fixer's intentions and the conclusions of the policy study. Moreover, there is no reason to impute a malevolent motive to a fixer. The reality is that the future holds uncertainties that impact the implementation and ultimate effectiveness of environmental policy. The fixer may have a vital role to play in the implementation of many environmental policies, but effective environmental policy studies are still an important factor that the fixer will have to deal with during implementation.

Why conduct rational analyses of subjective issues that cannot be answered by positive analysis? The argument against rational analysis (and policy studies in general) hinges on the belief that rational analysis cannot alter policy or beliefs. However, rational analysis can influence beliefs if done well. Our review is not comparative; thus our ability to compare the impact of effective versus ineffective policy studies on environmental policy making is limited. We can, however, look at the six studies presented here and conclude that effective environmental policy studies tend to influence subsequent national policy-making efforts when they are used in certain ways.

CONCLUDING REMARKS

We have examined six environmental policy studies in detail. The six case studies we have selected and analyzed are particularly noteworthy because of the impact they have had on actual decision making. Of course, the degree of impact varies considerably. As we have seen, some studies produced recommendations that were not adopted for several years, although they did focus debate by sharpening policy choices at a key moment. We have also discussed the six ways in which effective environmental policy studies tend to be used. As our analysis suggests, effective studies can be used to

- define the problem in a helpful fashion,
- describe the full range of policy options,
- help to overcome agency resistance to change,
- provide opportunities to engage stakeholders,
- enhance the legitimacy of particular actions, and
- help set resource priorities.

Arguably, the more of these functions that a policy study fulfills, the more effective it will be and the greater impact it will have. In Chapter 4, the discussion shifts to how such studies should be organized to maximize their effectiveness.

CHAPTER 4
HOW POLICY STUDIES
SHOULD BE ORGANIZED

In Chapter 3 we looked primarily at the way effective environmental policy studies influenced policy. In this chapter we address some of the organizational issues involved in mounting effective environmental policy studies.

SIX ORGANIZATIONAL TASKS

Each time an agency or a legislative committee decides to initiate a study, it must decide what to study and how the research and analysis should be conducted. These decisions cover every facet of the study design and involve six organizational tasks: (1) selecting and using experts (researchers, analysts, and "doers"); (2) shaping the relationship between sponsors and experts; (3) choosing the right institutional auspices; (4) reviewing policy study results using a peer-review process to build technical credibility (incidentally, the study authors must carefully specify the form that recommendations should take in order to ensure that technological advances and improved understandings of hazards and risks will be absorbed; these choices shape the content of the study, its reception by policy makers and the public, and its ultimate impact on policy); (5) learning from policy studies—individuals and organizations should use the study process itself as a learning tool and possibly as an aid; and (6) setting the policy research agenda.

SELECTING AND USING EXPERTS

Expertise can reside within the organization sponsoring or conducting a study, or it can be brought in from outside. In selecting experts, the study sponsor must first identify the most crucial scientific, technical, economic, social, and political issues involved. Experts should have credibility in these areas. In particular, when policy studies involve significant technical complexity, the experts identified should have in – depth knowledge of the relevant fields. In the cases presented earlier, scientists, engineers, and policy analysts with recognized expertise were engaged from the National Academy of Sciences, Office of Technology Assessment, and the Science Advisory Board of the Environmental Protection Agency (EPA).

In selecting among technical experts, preference should be given to those who are genuinely disinterested (i.e., experts who do not stand to gain or lose personally from a study). This is necessary for the panel to have credibility with the many competing interests involved. Technical and analytical skills alone are insufficient; for an expert to be fully effective, highly developed communication skills, group facilitation skills, and political sensitivity are also essential. Conducting policy studies is both an art and a science, and integrating technical knowledge with social, economic, and political considerations is a unique skill that not all experts possess.

Once experts are identified, they can participate in a number of ways. They can give testimony at hearings, orchestrate data collection and analysis, construct empirical models, oversee the conduct of a study team, advise on implementation, and participate in peer review. Their participation is critical for several reasons. First, the use of experts often enhances the credibility of policy studies in the eyes of policy makers and the public, easing the way for the adoption of legislative reforms, regulations, or other measures. Second, experts from a range of backgrounds can help to ensure that a problem is viewed from numerous perspectives. Third, experts can expedite a study by drawing on past experience and the lessons of parallel research efforts. Finally, experts can help create a persuasive justification for moving forward with new policies, by presenting independent (as opposed to self-serving) arguments for reform. This may be especially important when recommendations are highly political, or likely to be viewed with suspicion by opponents.

In sum, experts in environmental policy studies can

- enhance credibility,
- bring a range of backgrounds to the policy problem,

- draw on parallel research efforts,
- justify moving forward with proposed actions, and
- provide peer review and advice.

CREDIBILITY

The *Delaney Paradox* called upon the expertise of the Board on Agriculture of the National Research Council (NRC) of the National Academy of Sciences (NAS). The Board set up a committee that included experts in pest control, public health, food science, law, and regulatory policy. Because organizations such as the NAS have high standards, their participation in the Delaney study lent significant authority and respect.

However, there are some disadvantages to having such a board undertake a policy study. Jack Moore notes the downside to the NAS approach: "It is very expensive and the NAS delivers on their own time schedule, not yours" (Moore interview, January 3, 1993). Richard Wiles also points out that the *Delaney Paradox* was a "big stretch" for the NAS because, unlike other studies, it is short on science and long on policy. It was not clear whether the NAS would undertake additional policy efforts (Wiles interview, January 26, 1993).

Overall, however, most observers agree that EPA made the right decision in commissioning the Board to undertake the study. As a group of nonpartisan scientists, the Board had the credibility to present controversial findings and withstand the wrath of affected industry groups. Representative Pat Roberts (R-Kansas) reflects on the importance of this choice: "If an 'ag' group does the research, it is just viewed as the fox in the chicken coop." He believes the Board is viewed as "quasi-governmental and independent" and explains, "If you're going to come down from the mountain with a tablet, it's nice to have the NAS name on it" (Foreman & Harsch, 1989).

RANGE OF BACKGROUNDS

Experts from a range of backgrounds can ensure that the problem is viewed from numerous perspectives. Expertise is by nature specific and indepth. Different fields of study employ different methodologies and emphasize different relations and concepts. Bringing experts together from different fields allows for greater questioning of assumptions and for increased creativity in the study's approach.

For the *Spotted Owl* study, Congress selected four government and academic scientists (the "Gang of Four") who had an intimate knowledge of forest ecosystems. They, in turn, consulted with one hundred experts from a range of disciplines. Expertise was needed both to establish credibility and to provide a body of accurate information within a very tight time frame. John Gordon, one of the four scientists, argues, "Different constituencies with different value systems need a fact base from external sources. Scientific analysis should come in before an issue is too pointed—in this case, it didn't happen until almost too late" (Gordon, personal communication, March 23, 1993).

PARALLEL RESEARCH EFFORTS

Experts can expedite a study by drawing on past experience and the lessons of parallel research. Organizations and bureaucracies are very resistant to change. Often, the lessons of experience within an organization can disappear as personnel change and knowledge shifts.

Experts in a particular field keep better track of what has and has not worked than do agency personnel. Competition between experts in a field requires them to keep up with the current state of research. Experts have often conducted research on questions that are similar (if not identical) to the question addressed by the policy study.

In the *Spotted Owl* study, the Gang of Four's task was not to "do science" (i.e., to undertake new experiments), but rather to review existing scientific evidence and make a series of judgments about the likely results of different policies. Gordon describes this process as "science-based assessment" and defines it as "answering a question from outside science based on a synthesis of scientific information" (Gordon, 1993).

MOVING FORWARD WITH PROPOSED ACTIONS

Experts can create a persuasive justification for moving forward. This persuasiveness stems from the continuity of specific, detailed knowledge among professionals in a field. In the *Reducing Risk* study, outside experts were used to strengthen a consensus already emerging in the EPA. The study was conducted by EPA's Science Advisory Board (SAB), a standing panel of academic and professional experts in a variety of scientific and technical disciplines relating to environmental problems (Alm, 1991). The Board also formed a special committee, which consisted of thirty-nine scientists, engineers, and managers with expertise in environment and health problems. The committee was divided into three subcommittees: Ecology and Welfare, Human Health, and Strategic Options. Twelve public meetings were held. Each subcommittee issued a report, which was then included in the main report as an appendix.

None of the ideas in the final report was entirely new. The role of experts was more to synthesize and articulate an already emerging consensus within the agency on the need to redirect EPA's environmental policy to reflect new analytical and policy processes. The breadth and depth of the expertise that contributed to the study lent considerable weight to EPA's call for reshaping priorities. One commentator on the report writes: "The SAB report does not reveal any blinding new insights or divine revelations ... It is nevertheless a very influential document. Never before has such a distinguished group of scientists reached such a strong consensus on the need for new directions" (Alm, 1991).

PEER REVIEW AND ADVICE

Sometimes experts are not consulted during the production of an environmental policy study. In these cases, experts can be consulted to review the results after the policy study is written. Outside experts were not used in the *Lead in Gasoline* study, but when it was finished, it was secretly subjected to intensive peer review. The peer review allowed the analysts to broaden their understanding of their own research, refine their work, and address knowledge gaps prior to public evaluation. Leggett reports, "We went through a peer review process which was probably not all that common in that period. We looked for people who stood on all sides of the issue and asked them to provide peer review in secrecy. We got comments back and we incorporated them" (Leggett interview, January 28, 1993).

Expertise plays a large role even when agencies conduct policy studies in-house. Experts are used in these cases either in the review process for the study or in an advisory capacity to guide the work within the agency. In the case of the *Complex Cleanup* study. Congress turned to the Office of Technology Assessment (OTA) for expert advice. Until 1996 the OTA acted as an analytical resource for members of Congress. The OTA's creation in 1972 came out of the growth of the environmental movement, as citizens began to question the technical basis of various federal policies. While OTA analysts accomplished most of the actual work for each study, an advisory committee made up of experts and stakeholders guided the formation of study questions and reviewed analytical methods.

SHAPING THE RELATIONSHIP BETWEEN SPONSORS AND EXPERTS

The relationship between the sponsors of an environmental policy study and the experts (researchers, analysts, and "doers") of a policy study should depend on the nature of the study. Conscious choices about how involved the sponsor should be in the actual production of the policy study must be made. Issues that should guide this choice include the level of controversy already existing on the topic, the ultimate purpose of the policy study, the resources available within the sponsoring organization, and the credibility of the sponsoring agency.

There are many different ways to structure a study process. In some instances, the agency or legislative committee sponsoring a study may elect to use an established research organization with a proven track record. Independent bodies typically use time-tested procedures and carry out unbiased analyses. Sponsors may also use a smaller team of behind-the-scenes analysts, including those from within the sponsoring agency itself. The benefits of this approach include the flexibility and the speed that informality permits. Agencies opt for this tactic when they want their staff to be completely familiar with all the technical ins-and-outs of the analyses so they can handle subsequent challenges to the legitimacy of the research. Finally, an advisory board of experts identified by different stakeholder groups can review the work of informal agency teams. This can create an opportunity to involve individuals whose support will be essential to future implementation.

The six cases demonstrate a gradation of possible relationships from sponsor-prepared studies to the use of independent research bodies. The relevant options include:

- Engaging an independent research organization
- Involving stakeholder groups
- Appointing a commission or task force
- Hiring experts directly
- Doing the study in-house

Which institution arrangement will lend the most credibility to the results? There are clearly advantages to sponsoring a formal report prepared by an organization that does not have an immediate stake in the study's outcome. On the other hand, agencies may opt to produce a report internally. Either of these two approaches may include soliciting the advice of stakeholders. Yet a third approach is to invite stakeholders to join in a truly collaborative study design effort. All three approaches can generate influential reports with broad public support.

ENGAGING A MAJOR RESEARCH ORGANIZATION

Probably the most typical way of performing an effective policy study is to get one of the major research organizations such as the OTA, the Congressional Research Service, or the

NAS to conduct the study. This approach was used in two of the case studies: *Complex Cleanup* and the *Delaney Paradox*. In both cases, the reputation and independence of these organizations was thought to be necessary for any progress to be made on these issues.

EPA sought out the NAS to perform the *Delaney Paradox* study. Stakeholders were already quite polarized on this issue. Therefore, EPA sought out the most authoritative, independent body it could find. According to Jack Moore, assistant administrator of the EPA for Prevention, Pesticides, and Toxic Substances at the time of the study, "we needed the imprimatur of an NAS committee in order to do something radical in the eyes of some observers" (Moore, 1993). The *Delaney Paradox* was carried out by the Board on Agriculture. The Board selected a seventeen-member committee. Funding was provided by EPA along with some assistance from the Collage Foundation.

The OTA prepared the *Complex Cleanup* study, which used the following methodology:

- Approval of the study methodology and design
- Selection of an advisory committee
- Workshops with experts and stakeholders
- Commissioning consultant research when necessary

The use of OTA to conduct the study was particularly important, especially in light of its most important recommendation for institutional reform of DOE. The Senate Armed Forces Committee, by using OTA to conduct the study, recognized that it is often difficult for any agency to consider and then implement fundamental changes in its management priorities without some external prodding. Moreover, the recognized need for congressional oversight in addition to internal agency reform spurred the Senate Committee's use of OTA.

INVOLVING STAKEHOLDER GROUPS

The *Alternative Agricultural Research and Commercialization (AARC)* study was funded by the USDA. The Task Force on New Farm and Forest Products included a wide circle of outside representatives from the groups most likely to benefit from a national program to promote new products. USDA succeeded in lowering the costs by appointing numerous university and private sector members to the Task Force who were financially supported by their respective institutions. The opportunity thus resulted in a diverse task force representing a broad range of interests.

The Federal Facilities Environmental Restoration Dialogue Committee (FFERDC), which conducted a participatory dialogue over the cleanup of nuclear weapons production facilities, is yet another model for prodding agencies to change. Instead of marshaling state-of-the-art expert opinion from OTA staff, the FFERDC model attempts to push for reform by involving all relevant stakeholders. These stakeholders would collaborate with the Department of Energy (DOE) in conducting the study and would ultimately serve as the watchdogs holding DOE responsible for conducting a "fair" study and ensuring "proper" execution of any resulting recommendations selected for implementation. Whereas the Senate Committee's selection of OTA to conduct the study constituted a "top-down" approach, ultimately requiring DOE compliance subject to congressional oversight, the FFERDC method would have constituted a "bottom-up" approach, holding the DOE accountable to affected stakeholders.

Stakeholder participation may be advisory, as with site specific advisory boards, or it may constitute a true partnership in the ultimate selection of a policy option. Stakeholders can help an agency see shortcomings within its structure and mindset, and thereby may encourage institutional change.

APPOINTING A COMMISSION OR TASK FORCE
Another approach to the relationship between sponsors and experts is for the sponsor to appoint a commission to do the study. Such a commission can be made up of experts with or without stakeholder representatives. This approach is usually used when one of the aims of a policy study is to build a constituency for a new policy direction. This relationship can be seen in both the *Reducing Risk* case study, where EPA turned to its in-house commission of experts, and the *AARC* case, where the USDA put experts and stakeholders together in one body.

The report, *Reducing Risk*, was successful to a much greater degree than the earlier EPA report, *Unfinished Business*, in part because it was not an in-house EPA report. EPA's SAB assembled a study team of outside experts from diverse academic, technical, and policy backgrounds. This was an important strategy because it gave the report credibility both inside and outside the agency, and helped to ensure that the recommendations would be influential in promoting an agenda of a highly political nature: the reshaping of federal environmental objectives.

In the case of AARC, a broad-based task force was created to prod agency action. The diversity of backgrounds among Task Force members allowed in-depth discussion of a variety of policy options. Together they crafted study recommendations covering a wide range of issues that affect new products, from scientific research priorities to industrial development and business financing. Suzette Dittrich believes the study's credibility relied on a multidisciplinary effort with "small business entrepreneurs, large business men and women, educators, scientists, and others" (Dittrich interview, February 29,1993). Melvin Blase concurs with this assessment and adds that this was "clearly an instance when scientific input into the policy process was important" (Blase, 1993).

HIRING EXPERTS DIRECTLY
There are times when policy studies must be accomplished within time frames that do not allow for the participation of a commission or the methodical work of one of the large research organizations. In the *Spotted Owl* study, Congress hired the scientists directly, a rather unusual approach. Typically, committees seeking expert advice will hold a formal hearing. When they require a formal report on a policy issue, they will usually use one of the many established Washington-based research bodies such as the Congressional Research Service, the OTA, the General Accounting Office (GAO), or the NAS (Lyons & Gordon, 1993). Most of these reports require significant amounts of time for staff to review the subject, interview experts, contract for research, and solicit peer reviews. In this case the two committees agreed that such a lengthy, formal study would not fit the legislative calendar. The hiring of the Gang of Four resulted.

The highly-politicized nature of old-growth forest management required that experts be seen as free from bias. Academics and government scientists met this requirement. Although environmental groups were initially suspicious that the government was using its own scientists, they were pleased with the results (Owens personal communication, March 11, 1993).

They felt that Congress had embarked on a new approach to policy analysis by asking scientists (not politicians) to develop a policy firmly grounded in ecological principles (Owens, 1993). Moreover, because these scientists were not among the regular group that testified on behalf of environmental groups, and were in some cases directly affiliated with government agencies, it was not easy for the timber industry to dismiss their recommendations as biased.

In part because of their credibility as experts, the Gang of Four was able to help restructure the old-growth forest debate. Following the release of the report, "the debate was no longer over trees and acres, but what risk to species do we want to manage for. It gave the politicians some cover: they could say 'based on the best science, I think this is what we should go for.' The debate continues to be affected by the recognition of the importance of ecosystem science" (Owens, 1993).

DOING THE STUDY IN-HOUSE

In order to do a successful policy study "in-house," the sponsor must make adequate resources available, as the EPA did for the *Lead in Gasoline* case. The EPA believed that it was very important that the study be conducted internally. This was necessary because of the volatile and secret nature of the lead issue. Also, implementation of the study recommendations required that EPA staff know the issue inside and out.

By limiting involvement to a small number of people, EPA made sure that the agency would be ready for implementation. The study was conducted by a small team and was approached as an integrated task. Leggett cites the size of the group as "key" and says that too often policy studies are done by dozens of analysts who rarely talk with one another, each person working on a small piece of the study, which is ultimately glued together with other individual pieces at the end of the effort. From beginning to end, the lead study was an integrated project in the hands of no more than six analysts.

Schwartz adds, "We started out with [a] back of the envelope calculation and then we started going in and taking pieces and making them more complicated and less approximate. A big problem with a lot of analyses is that there are different pieces of it, they get done by different people, they get done sequentially, and after you start doing the second one you realize that the first one wasn't done in the way that was needed for the next part. The key for us was always having an integrated system and making incremental improvements without taking it apart or doing things separately.

We always had a spreadsheet which had the final numbers in it for each of the different categories. Then we could refine a category—it was kind of like replaceable units. ... It's not like we built 15 different things and at the last step stuck them together—they were always stuck together" (Schwartz interview, January 29, 1993).

The EPA policy analysts were told that the agency was very interested in their work. Too often, government agencies do not view policy analysis as a critical task, and they directly or indirectly transmit this disinterest to their staffs. Joel Schwartz believes the agency's support of the lead study was crucial. He says, "The person running the agency believed in policy analysis. Since (Alm) had a policy orientation, he thought that was a question that the policy office, not the air office, should be asked" (Schwartz, 1993). Jane Leggett concurs: "A lot of the research of the policy office—whether it's good or bad—doesn't go anywhere because people on a staff level ... are unable to convince people at a higher level that they should undertake it." She concludes that many policy efforts fail because the policy staff are frustrated and not willing to put in the kind of effort invested in the lead study. Leggett argues that staff need to feel that "really important results are coming out and that there is a potential for someone to act on it. Otherwise you're just doing studies and those things tend to drag on over long periods of time" (Leggett, 1993).

EPA provided concrete support to the study as well. The analysts were given high-level administrative support and adequate resources to do their work. Jane Leggett stresses this point: "We were working under ideal conditions—knowing that you had high-level support, feeling like you were getting important findings, knowing that someone would be willing to do something with it, and having enough resources to get the job done." She illustrates her point by noting, "I had the first PC [personal computer] in the agency outside of the Office of Research and Development" (Leggett, 1993).

Even though it was an in-house study, each of the analysts on the lead team had different training, which allowed for a multidisciplinary approach to the policy analysis. Leggett reports, "We had four people with very different expertise and ways of looking at issues. We worked very well as a team together. Having those different points of view, and asking questions of each other by reviewing each other's work, we arrived at different conclusions and covered more of the critical questions than we would have if you had a group with similar expertise and perspective" (Leggett, 1993).

CHOOSING THE RIGHT INSTITUTIONAL AUSPICES

What institutions will lend credibility to the results? There are clearly advantages to sponsoring a formal report prepared by an organization that does not have an immediate stake in the study's outcome. The numerous, seemingly partisan policy studies conducted concerning the continued use of chlorinated organic compounds discussed in Chapter 1 illustrate this point. On the other hand, agencies may opt to produce a report internally. Either of these two approaches may include soliciting the advice of stakeholders. Yet a third approach represented by the FFERDC process involves inviting stakeholders to join in a truly collaborative effort to design solutions. All three approaches can generate influential reports with broad public support.

A variety of institutions participated in the case studies examined. Institutions such as the NAS are well-respected, elite organizations whose publications on science policy are widely read and deemed very credible by policy makers. Similarly, the OTA was created by Congress to research and assess the policy implications of the choices we make about what technologies to use. Until it was dismantled in 1996, it played a valuable role. Environmental policy decisions should be made based upon the best scientific data available. Thus scientists have a role in developing innovative approaches that will assist in the making of new policy decisions. They can also redefine the questions being asked by policy makers and in turn change the terms of the political debate. Science as an institution is both affected by and affects the analysis of environmental policy. The Board on Agriculture, which conducted the *Delaney Paradox* study, is a part of the NAS, a reliable and well-respected research organization.

For the *Spotted Owl* study, scientists were selected from the government and the academic community. Environmentalists were unsure whether the government scientists could produce a study free of policy bias. The highly politicized nature of the old-growth forest issue convinced Congress of the need to find experts who would be seen as free from bias. Both academics and government scientists were able to meet this requirement. Although environmental groups were initially suspicious that the government was using its own scientists, they were pleased with the results (Owens, 1993). Because these scientists did not regularly testify on behalf of environmental groups, and were in some cases directly affiliated with government agencies, it was not easy for the timber industry to dismiss their recommendations as biased.

The *Lead in Gasoline* study was internal. By limiting involvement to a small number of people, EPA made sure that the agency would be ready for implementation. The study was conducted by a small team and approached as an integrated effort. The size of the team and in-house nature of the study were critical factors leading to the success of the study.

The *AARC* study was affiliated and funded by the USDA. The Task Force also included a wide circle of outside representatives from the groups most likely to benefit from a national program to promote new products. Policy studies are often expensive. USDA succeeded in lowering the costs by appointing numerous university and private sector members to the Task Force who were financially supported by their respective institutions. Blase reports that "deans were delighted to pick up the bills," and that the same was true for much of industry (Blase, 1993).

The use of OTA to conduct the *Complex Cleanup* study was particularly important, especially in light of its most important recommendation—institutional reform of DOE. The Senate Armed Forces Committee, by using OTA to conduct the study, recognized that it is often difficult for any agency to consider and then implement fundamental changes in its own management priorities without some external prodding.

The FFERDC process was another results-oriented method. Under the FFERDC method, stakeholders collaborated with DOE in conducting the study and ultimately served as the watchdogs holding DOE responsible for conducting a "fair" study and ensuring "proper" execution of any resulting recommendations selected for implementation. Utilizing institutions such as OTA, and ultimately the FFERDC method, added to the impact of this study.

REVIEWING POLICY STUDY RESULTS

Results of a policy study must be critically appraised before they are released. In some of our cases, evaluation was built into the preparation of the study, as agency personnel or stakeholders were invited to review or comment on the findings as the study progressed. In other instances, there was a formal peer review process once the research was completed. In any review process, differing views on the issues involved must be fully explored, and those most likely to be critical should be given an opportunity to preview the results and provide comments. In general, evaluation helps to ensure that the study has considered important policy questions comprehensively and properly, and thereby enhances the credibility of the study.

The *Delaney Paradox* was subject to the NAS guidelines for peer review. (See Appendix A containing the NAS review guidelines.) The *Lead in Gasoline* report was peer reviewed after it was drafted. Due to time concerns, the *Spotted Owl* report was not subjected to formal peer review, but the scientists did consult with a broad range of experts. However, subsequent research has not substantially challenged the Gang of Four's conclusions (Gordon, 1993). The study provides an example of a "successful" policy study that was generated outside of a formal institutional process, and which never received a peer review.

Complex Cleanup was reviewed by an outside panel of experts put together by OTA. While it existed, OTA provided a useful model of formal report making. To maintain its credibility in the field, OTA followed a standard procedure for all reports. This procedure included:

- Selection of an advisory committee representing diverse interests
- Workshops with a range of experts
- Commissioning consultant research when necessary
- Peer review of the draft report

One of the main goals of the *Reducing Risk* study was to review EPA's report entitled *Unfinished Business*. The SAB, which conducted the review, is affiliated with EPA, but it is also a standing committee of outside experts. Therefore, the report was essentially a peer review of the earlier EPA report, which had been strictly an internal document. Although the Board strongly agreed with the basic concept of using comparative risk to structure environmental priorities, it also had many criticisms. One of the criticisms was that EPA staff did not assess environmental problems in areas that had not historically fallen under EPA's jurisdiction, such as agriculture and transportation. This failing is perhaps symptomatic of a more general problem: It is difficult for agencies to produce internal reports that can analyze problems comprehensively and objectively, free from a set of longstanding preconceptions about current institutional arrangements and policy frameworks.

A second issue that must be addressed during the review process is the form that any policy recommendation should take. One of the most important factors contributing to the success of the *AARC* study was its format. The *Task Force* study is easy to read, attractive, and a "quick study" for busy policy makers. Blase stresses the importance of the format: "As silly as it may sound, I think the red cover on the report made it stand out—people knew it by the color." The report has a very short and carefully worded executive summary that is printed separately from the study itself. Blase says this was important "because a person could pick it up and within 15 minutes get the essence of our recommendations" (Blase, 1993).

Since the study will act as a basis for future laws and regulations, the recommendations must take future technological and knowledge changes into account. The *Lead in Gasoline* study sought to modify existing regulatory criteria concerning permissible lead levels in gasoline. An increased understanding of the risks associated with lead greatly helped to implement the study's recommendations.

Dr. Robert M. White, president emeritus of the National Academy of Engineering (NAE), former vice chairman of the NAS's NRC, and author of the NAE book entitled *Keeping Pace with Science and Engineering*, notes that "legislated criteria are generally an amalgam of scientific knowledge and the value judgments of our representatives in the legislature" (White, 1993). Uncertainties in our understanding of hazards, risks, costs, and benefits are high with respect to environmental laws and regulations because such laws usually address issues at the cutting edge of scientific understanding. Thus finding ways to incorporate future advances in technology as well as knowledge of hazards and risks is an issue of paramount importance.

White provides the following examples of legislation that attempts to incorporate future technological advances: "Some national environmental legislation recognizes the dynamic nature of the technical basis for regulation. In some cases, it mandates research and development programs to improve the data base. In other cases, it provides incentives for the development of new, more cost-effective technology. In still other cases, legislation explicitly includes schedules for reconsidering specific regulatory decisions" (White, 1993). However, even with the above provisions in place, it is still a difficult task to modify environmental laws and regulations.

The *Delaney Paradox* recommended changing existing law. The study demonstrates just how difficult it is to incorporate technological advancements into existing legal frameworks. Due to tremendous improvement in the technology for measuring trace substances in food as well as our improved understanding of the risks associated with these trace elements, the EPA felt confident supporting the recommendations contained in the *Delaney Paradox* (White, 1993). However, since the original clause did not permit any "modification," the recommendations have proven extremely difficult to implement, requiring nearly a decade to be enacted formally.

LEARNING FROM POLICY STUDIES

The ability to use policy studies to catalyze and guide the efforts of federal agencies can be enhanced as individuals and organizations that conduct such studies assimilate the lessons learned fttttttttttttrom experience. Following are three possible approaches to ensure that the appropriate lessons are not lost.

CROSS-CASE ANALYSIS

Cross-case analysis is essential to identify what works and what does not. Some of the many questions that can be analyzed with other, similar cases include: How did decisions about the use of experts and the participation of stakeholders shape the credibility of the final recommendations? How were the new proposals received by the public and the media? How did the design of the study affect implementation? Were the study's recommendations successful in generating policy change in the federal or state arenas? Individuals not involved in any of the cases should be selected to do cross-case analysis.

The notion of comparative risk has evolved over a series of reports and programmatic efforts varying in scale from EPA pilot projects to agencywide reviews and independent, external critiques. In addition, state and local agencies have repeated the "comparative risk reduction" process. These state and local studies usually differ from EPA's national study because they employ a more collaborative approach; a broad range of stakeholders, not just experts, tend to be involved. The *Reducing Risk* study reflected this broader participation. The Relative Risk Reduction Strategies Committee (RRRSC), and particularly the Human Health Subcommittee, found it difficult to rank risk without information on public values. They found that ranking risk goes beyond technical issues to include explicit value considerations. Subsequent efforts have attempted to address this concern by bringing in many different sectors of the affected community. Participants in these comparative risk processes learn from each other, and often become advocates for the process (Manard personal communication, March 16,1993).

The *Spotted Owl* report has provided a successful model for environmental policy studies that can be repeated by different agencies. The approach taken by the scientists encouraged a much broader view of the problem, and one that departed from the normal "problem of the moment" view that often persists on Capitol Hill (Lyons personal communication, February, 1993). The Governor of Georgia has since asked the same four scientists to do a similar study. The House Agriculture and House Merchant Marine Committees are also looking to do analogous studies on northern forests.

The comprehensive "ecosystem management" approach recommended in the *Spotted Owl* report seems to have become a model for resource management. In 1993, the House Natural Resources Committee was still seeking ecosystem management models that could be used across the country (Owens, 1993). The Committee was motivated by the belief that land managers must consider the cumulative impact of activities such as mining and grazing on a variety of species, and on entire watersheds.

DOCUMENTING THE PROCESS OF ANALYSIS IN "REAL TIME"

Decisions and outcomes in each case can only be documented effectively in real time. Documentation enables a research team and sponsors of a study to receive a neutral, critical assessment of the direction and quality of their work right after it is completed. Outside specialists can be brought in to document the study process as it happens (which also will facilitate cross-case analysis after the fact). Such feedback can be used to make substantive or procedural adjustments. These adjustments are more likely to be on target if they are suggested by someone who has followed the study closely.

FINDING OUT WHAT THE RESULTS ARE

Organizations that conduct or sponsor studies should draft follow-up procedural guidelines that incorporate or institutionalize what they have learned. Guidelines can translate the lessons learned from earlier experiences into tangible steps that can help subsequent researchers and sponsors. As they are applied and tested in practice, written guidelines gradually become norms familiar to and recognized by all staff members in an organization. The preparation of written guidelines will also enhance the credibility of a study by ensuring stakeholders (particularly sponsors) that the research underpinning policy recommendations was carried out using "time-tested" procedures.

SETTING THE POLICY RESEARCH AGENDA

Some may think that a discussion of how to decide what policy issues to study should have come earlier in this book. In many ways, though, selecting what to study is an outgrowth of acquired policy experience and thus follows our discussion of how to learn from policy studies.

Often individuals think they know what to study because they are knowledgeable about a topic. More basic questions, however, are even more important. One way of focusing a study effort is known as "futures research." This is particularly relevant when considering environmental issues. As our discussion of the possible regulation of chlorinated organic compounds in Chapter 1 indicated, many environmental issues involve substantial scientific uncertainty.

Futures research, forecasting, and a host of other terms describe techniques for predicting future events, trends, and issues even in the face of enormous uncertainty. Or, as Theodore Gordon describes it, "Futures research is the systematic exploration of what might be" (1992). There are many futures research techniques employed by forecasters. Table 4.1 identifies a selection of them.

Certain futures research techniques are more appropriate for business forecasting, others for macroeconomic forecasting, and still others for technology forecasting. No matter which technique is ultimately selected, certain "axioms" should be kept in mind:

- Forecasting techniques generally assume that the same underlying causal relationship that existed in the past will continue to prevail in the future—in other words, extrapolation is bound to be wrong eventually.
- Forecasts are seldom perfect.
- Forecast accuracy decreases as the range of the forecast increases.
- Forecasts for groups of items tend to be more accurate than forecasts for individual items (Shim, Siegel, & Liew, 1994).

In addition, Gordon (1992) adds:

- Forecasts can be very precise but quite inaccurate.
- Forecasts are incomplete—"The most surprising future is one in which there are no surprises."
- Forecasts can be self-fulfilling or self-defeating.

Futures research has been used in environmental policy making. Indeed, future thinking, principally by environmental policy experts like Lynton K. Caldwell, was the prime factor in the eventual passage of the National Environmental Policy Act (NEPA) in 1970. Since its inception, EPA has targeted prevailing environmental problems. However, the increasing rate of change, primarily in the field of technology, has effectively shrunk the distance between the present and the future. These changes threaten to render present EPA approaches ineffectual in dealing with future environmental problems (Loehr, 1995). The EPA has utilized futures research techniques to chart new directions for the agency for the coming millennium. In 1993, the EPA charged its SAB to

Table 4.1 Forecasting Methods

Qualitative	Quantitative	Indirect
Intuition	Naive Methods	Market Surveys
Expert Opinion	Moving Averages	Input/Output Analysis
Delphi Technique	Exponential Smoothing	Economic Indicators
Scenarios	Trend Analysis	
Assumptions	Decomposition of Time Series	
PERT – Derived	Box-Jenkins	
Simulation	Simple Regression	
Cross-Impact Analysis	Multiple Regression	
Expert Systems Scanning	Econometric Modeling	

- assess different methodologies currently being used to study possible futures and anticipate likely futures events,
- identify some environmental issues that could emerge over the long term (through the year 2025), and
- advise EPA on ways to incorporate futures research into the Agency's activities (U.S. Environmental Protection Agency, 1995).

The SAB formed the Futures Research Committee to address this assignment. In January 1995, after a year and a half, the Committee published its findings in a report entitled *Beyond the Horizon: Using Foresight to Protect the Environmental Future.*

The Environmental Futures Committee, after examining and evaluating the applicability of various futures research techniques, identified three promising approaches: (1) scenarios—or top-down approach, (2) look-out panel—or bottom-up approach, and (3) scanning. Although the Committee considered each method, it did not compare them in detail. Likewise, the Committee declined to state a preference for one technique.

The Committee suggested that as an essential part of its future capabilities, EPA should establish an early-warning system to identify potential future environmental risks. This early warning system should rely on scenarios, a look-out panel, and scanning as input sources. The Committee also recommended that in the longer-term, EPA should focus on five overarching problems:

- Sustainability of terrestrial ecosystems
- Noncancer human health effects
- Total air pollutant loadings
- Nontraditional environmental stressors
- Health of the oceans

Finally, the Committee stressed that EPA, as well as other agencies and organizations, should recognize that global environmental quality is a matter of strategic national interest (Environmental Futures Committee, 1995). Thus, futures research can play a variety of roles in environmental policy studies. First, futures research can help to set the agenda. Second, it offers a way of tracking important structural changes. Third, futures research can help promote better integration of planning and implementation processes (Amara, 1991).

A LOOK IN THE REARVIEW MIRROR

One area not yet discussed relates to the validity of the underlying data, statistics, and other facts gathered for a study. Any inaccuracy or fault in the data, statistics, or facts may skew the results, resulting in erroneous policy recommendations. If policy makers subsequently rely on the recommendations of the study, the results can be disastrous. Such an event would not only erode the credibility of the sponsors, experts, and others involved, but could very well be deleterious to human health and the environment. According to Dr. Robert M. White, president emeritus of the NAE and former vice chairman of the NRC, "When environmental regulatory costs turn out in retrospect to have been unwarranted because regulatory decisions were based on inadequate or inaccurate scientific information, it's only natural to express concern, since costs will have been borne without deriving the projected environmental benefits" (White, 1993).

Factual error can occur in three ways. First, the actual data collected may be improperly recorded. Likewise, a latent error may be present in the statistical data collected. A second type of error can occur when the factual data were themselves accurate but the experts involved interpreted them incorrectly. Dr. White warns that "in environmental … affairs, we frequently are confronted with data for which neither the level of precision nor the level of accuracy is particularly high" (White, 1993). Finally, a third type of error occurs when the factual data are accurate, but the wrong data were utilized in the policy study. This third type of error occurs when data on the wrong parameters are used in the study or when data concerning all of the relevant parameters are not used (Eberstadt, 1995). Closely related to this third type of error is the situation that occurs when the data are accurate by today's technological standards, but may become obsolete with future improvements in data collection.

In many cases it is extremely difficult to identify these errors because they often manifest themselves long after the implemented aims of the policy study recommendations fail to achieve the desired results. However, several steps can be taken to help reduce the incidence of these kinds of errors. Errors in faulty data can be eliminated or at least identified by the experts doing the study. Questionable data may require testing or re-collection to provide the proper baseline data necessary. The first type of error can sometimes be identified and corrected during the peer review process.

The second type of error, erroneous application of accurate data, is more difficult to identify, but can also be detected during the peer review process. "While there may be differences among the parties in their attitudes about what constitutes proper review and evaluation, no one argues that the data ought not to be subject [to rigorous peer review], which presumably results in a body of technical evidence that represents the best that is available at a given time" (White, 1993).

The third type of error, utilizing accurate but inappropriate data, or not considering all the potential parameters, is the most difficult to rectify. A recent study of recycling policies in Pittsburgh, Pennsylvania, unearthed some previously overlooked weaknesses. The study determined that existing recycling policies in Pittsburgh were not economically sound, due in part to the fluctuating market for recyclable materials. Moreover, recycling was not environmentally sound because the negative environmental impacts from the recyclable materials collection process far outweighed the benefits, especially given existing landfill policy and improvements in landfill technology. The Pittsburgh study implies that environmental policy studies need to look at all of the dimensions of a problem and how the parameters interrelate amongst themselves (Hendrickson, Lave, & McMichael, 1995).

A second example of this third type of error was the "mercury in fish" scare. The cause of high mercury in fish was believed to be due to industry discharging heavy metals into the oceans. The scare resulted in a prohibition on the harvest of fish above a certain weight. However, after an examination of museum specimens, researchers discovered that, except in certain isolated incidents, the mercury in ocean fish reflected naturally occurring levels of mercury in the oceans (White, 1993). In this case, early examination of all of the relevant parameters might have anticipated an expensive, but inappropriate reaction.

The problems caused by this third type of error are not insurmountable. Several of the case studies utilized techniques to forestall this third type of error. The *Lead in Gasoline* study looked at a large number of potential effects caused by existing lead levels in gasoline. Although existing lead levels in gasoline had a number of environmental impacts, the study was able to link lead levels in gasoline with the incidence of high blood pressure in human beings. Had the study experts not explored this particular impact on humans, it is quite possible the study would not have resulted in lowering lead levels in gasoline. The study also had a much broader impact in light of more recent determinations indicating that the risks associated with exposure to lead were greater than previously thought (White, 1993).

In the *Spotted Owl* study, the Gang of Four set out to study the differing effects of policy recommendations not only on one species, the spotted owl, but on all of the species in the relevant ecosystem likely to be affected by proposed policy changes. The Gang of Four recognized that a particular policy choice, though beneficial to both industry and the spotted owl, might have unanticipated consequences on other species in the ecosystem. As previously mentioned the Gang of Four consulted experts from over one hundred different fields to make sure they considered all of the relevant factors likely to have an impact on the spotted owl.

CONCLUDING REMARKS

Organizing effective policy studies may appear at the outset to be a daunting task. Although differing approaches can be taken by the many organizations that conduct policy studies, the following organizational tasks can make the difference between an effective and an ineffective policy study:

- Selecting and using experts (researchers, analysts, "doers")
- Shaping the relationship between sponsors and experts
- Choosing the right institutional auspices
- Reviewing policy study results
- Learning from the study
- Setting the policy research agenda

There is a need for precise and accurate data collection when conducting environmental policy studies. Inaccurate data can have devastating consequences for the environment and human health. By looking at data from other areas that relate to the issue at hand, new insights may be achieved as to the most important parameters to be addressed in a policy study.

As the cases demonstrate, there is no single approach that will ensure an effective policy study. Each environmental problem must be addressed in a situationally appropriate way. Issue volatility, political pressure, risk analysis, institutional constraints, and the scientific and technical issues involved are just some of the factors that must be taken into account when organizing a policy study. Using the cases in this volume as a guide should prove beneficial in assessing the factors likely to have an impact in a given policy study.

Arwin van Buuren and Sibout Nooteboom — "Evaluating Strategic Environmental Assessment in The Netherlands: Content, Process and Procedure as Indissoluble Criteria for Effectiveness." *Journal of Impact Assessment and Project Appraisal.*

Reading Commentary

This article focuses on the effectiveness of environmental problem-solving efforts. Using two strategic environmental assessments (SEA) in the Netherlands, which are similar to environmental impact assessments (EIAs) in the United States, the authors argue that the timing and the quality of decision-making influence effectiveness. In particular, the success of an EIA should be based on the content included, the participation of stakeholders and other procedural considerations.

The differences in the two cases indicate how the same process can produce very different outcomes even in the same country. The authors offer a table that compares timing, scope and organization among other factors. They show how a well-structured SEA permits collaborative learning among the various stakeholders. By analyzing alternative policies, designs or solutions, a formal assessment can lead to substantial departures from what was proposed initially.

In the ZZL case, through timely stakeholder engagement, the participants were able to generate a more socially and environmentally sound way of achieving the same economic development goal. This was partly a product of their reliance on joint fact-finding, a way of working together to define the parameters of a study and interpret the results. The increased transparency of the thinking behind the final assessment enhanced the public's trust in the outcome.

The authors also show that SEAs are more successful if they are embedded in a larger planning effort. This increases the chances that an assessment will be linked with other findings and tied more directly to programs and activities related to the implementation of solutions or policies that emerged. In the ZZL case, SEA was combined with CBA, spatial analysis and sustained stakeholder dialogues.

As you read through the two cases, think about the relationship between a good process and a good outcome in environmental problem-solving. Does a good process always lead to a good outcome? Can a good outcome emerge from a bad process?

EVALUATING STRATEGIC ENVIRONMENTAL ASSESSMENT IN THE NETHERLANDS: CONTENT, PROCESS AND PROCEDURE AS INDISSOLUBLE CRITERIA FOR EFFECTIVENESS

Arwin van Buuren and Sibout Nooteboom

Dutch society can be seen as typically post-materialist (Inglehart, 1990), where various interest groups are well organized and numerous stakeholders try to influence spatial planning. Within this society, planning is often both highly controversial and time-consuming. Each spatial function has its advocates, who are well equipped to influence planning decisions and to defend their interests. Despite the fact that the Dutch planning culture is strongly consensus oriented and that stakeholders get many opportunities to participate, actors who feel threatened do not hesitate to go to court to defend their interests, especially in the later stages of decision-making. This tendency has the effect of rendering complex planning processes both unpredictable and time-consuming.

In this context, the knowledge production process often becomes as controversial as the planning process itself. Because impact assessments, cost–benefit analyses and other policy analyses heavily influence the choices made, stakeholders try very hard to influence the research trajectory. They actively question the outcomes of research efforts, especially when the findings are not in line with their own definitions of the problem (Van Buuren and Edelenbos, 2004; Collingridge and Reeve, 1986). Knowledge processes are thus an inherent part of the political struggle surrounding planning decisions. The effectiveness and value of the knowledge production process not only depends on its contribution to the rationalization of political choice, but because this choice is always value laden, it also depends heavily on its contribution to establishing inclusiveness and democracy within the planning process (Van Buuren, 2009; Cashmore, 2004).

The same can be said of strategic environmental assessments (SEAs). SEAs are predefined procedures that structure the process of knowledge production within a spatial planning endeavour. Therivel and Partidário (1992: 19) define an SEA as: 'the formalized, systematic and comprehensive process of evaluating the environmental effects of a policy, plan or programme and its alternatives, including the preparation of a written report on the findings of that evaluation, and the use of these findings in publicly accountable decision-making'. SEAs are meant to ensure that policy options that have significant environmental impacts are weighed duly and deliberately. Often, the effectiveness of an SEA is framed in terms of its contribution to the use of information related to the environmental consequences of a proposed project. An effective SEA is used in decision-making, and ultimately leads to the selection of the most environmentally friendly option and/or the adoption of necessary mitigation measures if the most environmentally friendly option is not selected. However, the effectiveness of an SEA depends not only on the use of the knowledge to enable rational and sustainable policy choices, but also on its contribution to a collaborative dialogue.

In this article, we discuss the various functions an SEA can serve to further the effectiveness of complex and controversial planning processes. In order to establish key SEA effectiveness criteria, it is first necessary to understand the fundamental characteristics and requirements of a planning process that will result in legitimate and effectual outcomes (sections 2 and 3). The features of an effective SEA are explored in the context of two case studies (sections 5 and 6). In section 7, we investigate the conditions that allow for effective SEAs that contribute distinctly to the quality of the planning process. Our findings are discussed in the final section.

CHARACTERISTICS OF AN EFFECTIVE SEA

What makes an SEA effective? The literature is rich in attempts to answer this question (Partidario, 2000; Retief, 2006; Cashmore et al., 2007). Cashmore et al. (2008) defined four effectiveness criteria that determine the transformative potentialities of environmental assessments: learning outcomes (both social and technical); governance outcomes (e.g. stakeholder participation, network development); development outcomes (design choices; consent decisions); and attitudinal and value changes.

In general, an SEA is meant to safeguard environmental interests and ensure that they are given serious consideration in plans and programmes. SEAs were initiated also to further the likelihood that more sustainable policy options are developed and selected (e.g. European Union Directive on Environmental Assessment of Plans and Programmes; see Wallington et al., 2007). As such, the effectiveness of an SEA can be seen as its contribution to the selection of the most sustainable, environmentally friendly planning option. Impact assessments contribute to the body of serviceable knowledge to be considered in choosing between multiple policy options (Jasanoff, 1990). Apart from adding to available information, they may contribute further by framing the argument and longer-term choices.

Despite these obvious benefits, the impact of the SEA on the content of a policy choice is often unclear, as choices tend to evolve over time and the planning process is influenced by additional sources of information as well as the views of stakeholders. Because the process is fluid and influenced by multiple factors, it is often impossible to pinpoint the exact impact of SEA on the final decision. In general, then, impact assessments can be said to contribute to the body of knowledge that is considered when deciding between policy options. They may also augment certain perspectives, legitimize specific choices and provide a rationale for specific spatial functions that are preferred by the authorities.

However, the outcomes of an impact assessment can be highly controversial because they affect to what extent stakeholders can realize their own ambitions. Stakeholders have their own views of problems, their own values and normative frames, and are often capable of mobilizing experts to counter and discredit data collected by planners and policymakers. Frequently, these controversies result in legal action, in which the stakeholders try to convince the court that the impact of spatial development plans have not been sufficiently considered.

It is for this reason that not only the *content* of the SEA but also the *process* of executing an SEA is relevant to the planning process (Pischke and Cashmore, 2006). The impact assessment can both magnify the dispute between stakeholders and minimize it by aiding in the establishment of common ground. Depending on how the SEA is organized, it can certainly contribute to the quality of the collaborative process, and help ensure that stakeholders work together to realize a decision in a consensual manner. By carefully intertwining the process of stakeholder participation and knowledge production, a process of joint fact-finding can be realized in which there occurs a reflexive dialogue and frame reflection between stakeholders with highly diverging perceptions. To fulfil this function, the SEA has to be independent, credible and univocal (Sarewitz, 2004; Twaalfhoven, 1999; Clark and Majone, 1985) and its production process has to be inclusive, democratic and transparent (Van Buuren and Nooteboom, 2009; Woodhouse and Nieusma, 2001; Guston, 2004).

Along with its contributions to the quality of the collaborative process and the ultimate policy choice, an SEA can also contribute to the quality of the decision-making *procedure*. Planning processes surrounding highly controversial public investments are exceptionally difficult to organize and manage. For this and other reasons, the course of planning can be erratic and unpredictable, and the planners often run out of time and budget (Teisman, 2008). Transparent and unambiguous procedures which structure the decision-making process can contribute to their overall quality. The following are the formal steps of SEA in the Dutch planning process:

1. Public announcement of the start of the procedure;
2. Consultation with administrative bodies likely to be involved in the implementation of the plan about the scope and details of the environmental statement;
3. Writing of the environmental statement (termed 'plan-environmental impact assessment');
4. Public display of the environmental statement and draft plan to elicit public feedback and in certain cases consultation of the Netherlands Commission for Environmental Assessment (NCEA);
5. Writing of the final plan based on established environmental impacts and consultations;
6. Publication of the final plan;
7. Evaluation of the impacts of the project after implementation (Ministry of VROM, 2006).

Overall, an SEA adds to the decision-making process only the requirement for a formal statement about the environmental impacts of a course of action, and that the decision to go ahead with a planning project be made after fully considering these impacts. Paradoxically, however, this requirement tends to render the authorities more vulnerable to criticism that environmental information is omitted or undervalued, and can easily add to the 'war of knowledge' which is often fought out in court. Nonetheless, this very risk may instead be viewed as an opportunity to consider more carefully how the detailed planning process is structured within the prevailing legal framework. Those responsible for the SEA may see it as their responsibility to work with the planners so that more key parties agree with the process and final proposal. As such, the SEA can facilitate an ordered, transparent, and timely decision-making process, in which the same questions do not need to be answered again and again, and political support emerges for a well-considered and widely supported final proposal.

CRITERIA FOR SEA EFFECTIVENESS

Based on the previous section, we formulate three criteria with which to evaluate the effectiveness of an SEA undertaken within controversial planning processes:

1. The SEA enables decision-making based on authoritative and undisputed information on the environmental consequences of each alternative choice (content);
2. The SEA contributes to the inclusiveness of the collaborative dialogue, and thus to the realization of support and legitimacy by achieving consensus and frame-reflection (process);
3. As a procedural device, SEA contributes to the timeliness, transparency, and quality of the overall decision-making process (procedure).

If these conditions are met, the SEA would likely have the desired effect on the outcomes of the planning processes. With this interpretation of effectiveness, we move beyond the technical, rational interpretation of the impact of SEAs and broaden our understanding of the elements which determine the contribution of SEA to the effectiveness, legitimacy and overall quality of the decision-making process (Partidário, 2000). We focus on the direct impact of an SEA on the quality of the decision-making process with regard to the quality of its content, stakeholder participation and procedural quality.

Of course, the link between SEA quality and the quality of decision-making is not straightforward. Other factors also influence the quality of decision-making with regard to content, process and procedure. Nonetheless, the above criteria can be said to bring together elements related to various models of science and impact assessment (Cashmore, 2004; Cashmore et al., 2004, 2007). Departing from a critical realist perspective, they integrate a more analytical approach toward the content of the SEA (Thérivel and Minas, 2002) and a participatory approach to the process of SEA drafting and decision-making (see Kørnøv and Thissen, 2000), and incorporate a more governance-oriented approach of decision-making in which procedures can be used to structure and stage a complex decision-making process (see Cashmore et al., 2007). These elements interact continuously which each other and we have come to see them as indissoluble elements of any attempt to determine the effectiveness of an SEA. In the next section, we use two case studies involving SEAs applied in Dutch planning practice to shed light on the conditions under which SEA can realize these ambitions.

METHODOLOGY

Two controversial Dutch planning processes which have received a great deal of political and public attention are compared. These were among the first few planning projects to be conducted in The Netherlands in accordance with the requirements of the SEA Directive, although they pre-empted the formal implementation of SEA requirements. However, environmental impact assessment (EIA) had been applied countless times in The Netherlands, and can be seen as the main precedent to the current SEA system.

Since its introduction to The Netherlands in the 1970s, EIA has been regarded as a supplementary procedure for decisions that require considerable scrutiny in order to secure an environmental permit or planning document. The EIA was supposed to facilitate a more environmentally aware analysis of controversial decisions by explicitly developing alternatives to the proposed plan and comparing their effects with those of the preferred alternative. Such assessments of the most environmentally friendly alternative were made compulsory, along with a formal review of the scoping document and environmental statement by the NCEA. SEA has taken over some of the functions of EIA, and in doing so has eased the process substantially. However, the number of projects expected to make use of an SEA is expected to increase significantly under the current legislation (see also Runhaar and Driessen, 2007).

The first case concerns the Southern Sea Line. The Southern Sea was a large bay within The Netherlands that has given its name to a high-speed rail connection (Zuiderzeelijn, ZZL) that aims to connect the western Netherlands with the northern region. The ZZL proposal came out of negotiations between northern governments and the national government in the mid 1990s in an effort to boost the lagging economy in the north by improving its connection to the economic centre. The national government reserved a significant budget for this high-speed railway connection and much preparatory work was done by the Ministry of Transport. However, at the last moment a Parliamentary Enquiry Committee raised significant doubts about both the need for the railway and the added value of the ZZL. In face of serious delays and cost overruns, the committee demanded a reassessment of the project before they would take a final decision. In the early 2000s the Cabinet assigned a committee to prepare a draft 'structure vision' (a zoning plan), which was a formal strategic spatial assessment that included an SEA, a spatial analysis and a societal cost–benefit analysis (CBA).

The second case concerns the redevelopment of the IJsseldelta Kampen, a large area near the river IJssel. Two spatial investments were proposed to enhance the river's retention capacity in times of high water flow. After a long process of design and dialogue, an SEA was commissioned because adjustments had to be made to regional and local planning documents.

Looking at the two cases, it can be said that they are comparable in that both relate to very controversial planning efforts with multiple stakeholders from opposing domains. In addition, both dealt with large spatial projects (railway infrastructure and a river bypass) and were organized with close interaction between authorities from local and regional levels. Nonetheless, these cases were chosen specifically for the fact that they differ on key factors related to Dutch SEA practice. First, one project was promoted on the basis of a single issue (ZZL), while the other concerns several issues (IJsseldelta Zuid). Second, when considered together, the two projects reflect the application of SEA across a number of domains, namely infrastructure development, regional development, water management, and nature protection. Finally, in the case of the ZZL there was a *national* zoning document, whereas in the case of the IJsseldelta two *regional* zoning documents had to be changed.

The information used in these case studies is based on an extensive analysis of the project histories obtained by examining various policy and research documents and by conducting a series of interviews (about 12 per case) with planners, political authorities, stakeholders and SEA experts. In the case of the ZZL, interviews were conducted as part of an official evaluation commissioned by the Ministry of Transport. During these interviews, considerable attention was paid to the planning process and the function of the SEA in this process, and respondents were asked what they saw as the key contributions of SEA to creating authoritative content, inclusive collaborative dialogue and functional procedures. Although we acknowledge that the examination of just two cases provides a limited basis for generalization, it is important to note that these cases were strategically selected to represent the varied Dutch SEA practice fairly accurately.

CASE STUDIES

SOUTHERN SEA LINE

The task of drafting a structure vision for the Southern Sea Line was given to a project bureau composed of officials from different government ministries. From the start, all formal SEA steps were intertwined with the procedure used to develop the structure vision. This involved multidisciplinary teams of designers, researchers and planners working together to gradually develop the various building blocks that made up the structure vision.

Because the SEA was initiated at the start of the planning process, it had the effect of increasing stakeholder expectations about the way environmental impacts would be assessed and integrated into the vision development process. This effect was triggered by the requirement that the SEA be publicly announced early in the planning process, and that a separate report be produced specifically on environmental impacts. Because of their lack of familiarity with these requirements (which were not yet legally mandated), the project bureau decided to ask the NCEA to advise on the scope of the SEA, as well as to evaluate the quality of the report that was later produced.

A 'starting document' was circulated among authorities and the public, with open invitations for feedback on the desired scope of the SEA report. This included feedback on environmental impacts that had not been considered, as well as alternative ways to reach the project's objectives. Dozens of meetings were organized in several provinces, and hundreds of politicians, officials, stakeholders and citizens attended. Based on the feedback received, a scoping document was prepared and circulated before the actual SEA was written. A social cost–benefit analysis and spatial analysis were also prepared along with the official SEA. Investors in the north were asked to participate in the financing of the project, as the national contributions would not be sufficient to meet the project's needs. A market consultation effort was also organized to further this goal.

These parallel processes were run by both public and private consultancies, and the project bureau coordinated the alternatives and impacts to be assessed. Several alternative technologies and routes for the rail line were developed, assessed, and compared. SEA specialists met with municipalities along all of the railway's routes. A nearly final SEA and draft structure vision were circulated among stakeholders, and the results were taken into consideration (Projectorganisatie Zuiderzeelijn, 2005a, 2005b). The SEA was ready a year after the start of the project bureau, and was submitted to Cabinet (Ministerie van Verkeer en Waterstaat, 2006). However, in April 2006 the Cabinet decided on an alternative course of action called the 'transition alternative' to meet the original purpose of boosting the economy of the north. This alternative had not previously been considered at length and did not entail any major new infrastructure other than a package of economic investments in the north. The Cabinet believed that a high-speed rail line was not economically efficient, and were concerned about its environmental impacts. This draft structure vision was published along with the existing SEA, and stakeholders and the public were again asked for feedback. Public hearings were organized, and hundreds of reactions were received (Ministerie van Verkeer en Waterstaat, *et al.*, 2006; Projectorganisatie Zuiderzeelijn, 2006).

Strikingly, millions of euros had been spent on developing and assessing alternatives for a project that had not been selected. Although apparently wasteful, such efforts were necessary to decide whether the project would solve any problems, or whether it would instead create many new problems. Stakeholder respondents indicated that, in general, they felt that the money had been well spent. In their eyes, the SEA had contributed significantly to the learning process, as had the societal cost–benefit analysis.

SEA enables decision-making based upon authoritative information on the environmental consequences of a wide range of alternatives (content)

ANALYSIS

Summarizing the main effects of the SEA, it is clear that it contributed in many ways to the content of the decision-making, the procedure and the process.

On its second parliamentary consideration in 2006 the SEA was instrumental in bringing about the change of heart with regard to the ZZL. The SEA, together with the CBA and the spatial analysis, made it clear that the original problem definition was inadequate, in that it was based upon the assumption that the distance between the economic centre of The Netherlands and the northern provinces was the reason for the economic problems of the latter. Through the assessment procedure it became clear that the existing economic structure had to be strengthened. This was controversial because many politicians in the northern provinces were outspoken advocates of the ZZL. However, the argument put forth through the SEA was convincing, and it strongly influenced the process. Cabinet broadened the official problem definition for the study to include the so-called 'transition alternative'.

SEA contributes to the quality of the collaborative dialogue and thus to the realization of support and legitimacy by achieving consensus (process)

Interviewees believed that the SEA assisted in developing alternatives that would have been more acceptable for the residents because it elicited suggestions from residents well before the authorities had made a decision.

The SEA process was closely linked to the general planning and assessment process, but it also had a separate consultation track. Unlike what is usually done in The Netherlands, a number of routing alternatives were developed in rough detail, and authorities and residents in potentially affected areas were consulted at several stages of the plan's development. Making use of feedback, and informing people about dilemmas at higher scales of planning, the SEA functioned as a generator of alternatives which opened new avenues for stakeholders to think about other agendas. Although the rail infrastructure proved to be unfeasible in the first phase of the planning process, the northern provinces were asked to consider other ambitions. These were incorporated in the transition alternative, which was then subject to both the CBA and the SEA and assessed as being much more beneficial.

The SEA facilitated a process of learning and frame reflection by delivering undisputed information about the various benefits and disadvantages of different alternatives. The resistance against 'unwelcome facts' was surprisingly moderate, largely because of the transparency and openness of the SEA, the way in which a 'Critical Review Team' safeguarded the quality of the research, and the many opportunities for stakeholders to become involved in the research process.

The railway project was one of the first SEAs undertaken in The Netherlands. The project organization used the SEA procedure voluntarily to structure stakeholder consultation and to organize the assessment process. Although some ministerial officials were sceptical about this instrument, the project organization proceeded to use it and emerged very positive about this decision. They used it to organize the inherently dynamic information search to answer political questions. Although the official requirements of the SEA procedure were minimal, the fact that the SEA was attempted created among residents and interest groups high expectations of being consulted and considered. Hence, they participated more actively than they would normally have done.

As a procedural device, SEA contributes to the timeliness, transparency and quality of the overall decision-making process (procedure)

The SEA team leader was frequently in communication with the director of the project bureau about how to deal with the legal ramifications of the alternatives. The SEA had helped to move the planning process safely away from risks created by environmental laws by seeking information in a timely manner and by looking for alternatives when unexpected impacts emerged in the planning and assessment process.

IJSSEL DELTA SOUTH

High river discharges in 1993 and 1995 not only caused great water damage, but also brought about much societal unrest. In response to this, the Dutch government decided that the discharge capacity of the main rivers in The Netherlands had to grow so that they could handle Rhine river discharges of 16,000 m^3/s at Lobith (near the German border).

This decision had a number of serious implications. A programme called 'Space for the river' was set up which had to result in a concrete programme of measures to be realized before 2015. The capacity of the main rivers in The Netherlands had to be substantially enlarged by this time by measures such as groyne lowering, dyke movements and the creation of inundation areas. To strengthen support, these measures had to be realized in close collaboration with the regional and local authorities and had to take into account their agendas.

In 2003, the decentralized authorities of the two main river regions sent their preferred proposal to the State Secretary of Water Management. Because of a serious bottleneck in the river IJssel near the city of Kampen, expansion measures were seen as necessary and the widening of the riverbed was seen as an adequate short-term solution (2025). At the same time, a bypass of the river south of the Vossemeer was seen as a necessary long-term measure because it was believed that climate change would result in even higher river discharge rates. A spatial reservation was proposed to forestall other spatial developments in this area and to keep the option of a bypass open. In 2003, a parliamentary decision was made at the programme level in favour of the river widening and the creation of a spatial reservation.

However, such a spatial reservation was highly undesirable to officials of the municipality of Kampen and the province of Overijssel, as this area was required to meet growing housing needs. They tried to push forward the bypass as the most effective measure for the short term, and pointed out that other spatial investments were already intended for this area. One of them was a railway called the Hanzeline. Any possible bypass of the river had to cross this Hanzeline twice, and would only be possible if a flyover was part of the route, and if the dimensions of a tunnel under the Vossemeer could be adjusted to the bypass.

In spring 2005 the Ministry of Housing and Spatial Planning decided that the province of Overijssel had to develop an integral 'area development plan' which included the bypass.

At this juncture, a very intensive process was begun in which two possible scenarios were developed by a small intergovernmental project team with minimal interaction with the wider public. This lack of consultation was deemed necessary because it had to be cleared and approved by the central government before the end of 2005, or the line would be built without these adjustments and the bypass would have become prohibitively expensive, if not impossible in the short term.

The process of scenario building was supplemented by a 'voluntary environmental assessment' in which the project team investigated whether their ideas were compatible with the most critical of the environmental objectives formulated by European and national directives. Because of the time pressure, the project director opted for a limited assessment of several key regulations and the SEA was postponed indefinitely.

The project organization presented five scenarios to its stakeholders and citizens, and loud criticism was heard at several information meetings. To counter the criticism, the provincial Deputy invited the citizens to develop their own scenario. Ultimately, this grassroots-led scenario was taken up as the Masterplan.

The Masterplan was elaborated into a formal Intention Agreement between the participating authorities before the formal planning process was begun, and involved the amendment of the provincial and local zoning plans. An SEA was conducted at this stage, and important questions had to be answered in the SEA in relation to the necessity and value of additional housing, the sustainability of the bypass, and the viability of alternative development options for the area (Projectorganisatie IJsseldelta-Zuid, 2007, 2008).

The SEA was necessary to fine-tune the details of the development plan for the area, and to underpin the adjustment of the Provincial Zoning Document. However, as always, the SEA also fuelled a number of new discussions, especially in relation to the way the bypass was to be realized. The SEA writers concluded that a 'blue bypass' with a direct connection between the river and Vossemeer had the most beneficial consequences and the fewest negative external effects (Provincie Overijssel, 2008a, 2008b). However, an open, blue bypass was difficult for the Water Board to accept. They were anxious about the negative hydrological impacts of the bypass and started counter research to support their opinion. In addition, inhabitants and environmental interest groups were anxious about the recreational attraction of a blue bypass and the negative consequences of this for the environment.

The results of the SEA with regard to future population growth were also subject to much debate. Some environmental stakeholders and community associations criticized the assumptions behind these scenarios. Although the governments involved adjusted their plans for the number and location of the houses to be built, much discontent remained because the contrary views and expert opinions gathered by the stakeholder groups were not authoritatively refuted.

ANALYSIS

The SEA for the amendment of the Provincial Zoning Plan to enable project IJsseldelta South was very helpful in framing the reconsideration of the Master-plan, which was initially seen as the preferred alternative. The question was whether a green or a blue bypass would be the better option, and key insights from the SEA caused both the provincial and the municipal governments to rethink the value of a blue bypass. The assessment of different variants allowed for the selection of the option with the most beneficial consequences that added the greatest value to the development of the area as a whole.

> **The SEA enables decision-making based upon authoritative knowledge of the environmental consequences of a wide range of possible alternatives (content)**

Another important insight arising out of the SEA pertained to the unwanted environmental effects of housing beyond the dykes of the bypass. However, this factor was neglected because of strong political pressure to create an attractive housing environment.

Although indeed beneficial, the timing of the SEA did not allow it to contribute meaningfully to the quality of the overall decision to reallocate the whole area. In the eyes of the involved local and regional governments, it helped only to optimize the final planning decision.

> **SEA contributes to the quality of the collaborative dialogue and achieves consensus in the support and legitimacy of the final option (process)**

The SEA did not serve to create an inclusive collaborative process. On the contrary, it functioned as a source of controversy in that it both fuelled the existing debate and initiated new ones. Stakeholders were not satisfied with the way they were involved in the knowledge production process, and some outcomes – for example about the shape of the bypass and housing development outside the dykes – only encouraged further polarized debate.

The way in which the SEA was carried out (with close cooperation between the project organization and the research institute) can be blamed for much of this failure. The amount of interaction and reflection that was facilitated within the project group (which included the various public stakeholders), and within the soundboard group (in which other stakeholder groups came together) was insufficient to facilitate a process of frame reflection and joint learning.

No serious discussion about the added value of the bypass was ever held, and this constituted an important omission in the planning process. The spatial reservation of the area of South Kampen caused the provincial government to decide to speed up a decision about a bypass without securing enough evidence for its necessity. Its value was explained with reference to the extreme river discharges warned of by the Dutch water management authority. The SEA failed to convince all actors because a debate about the added value was lacking.

SEA as a procedural device contributes to the timeliness, transparency and quality of the overall decision-making process (procedure)

The formal procedure of the SEA was used as an argument to postpone environmental assessment to a latter phase of the planning process. Another argument for this decision was the formal status of the Masterplan. There was no formal plan, and therefore no formal SEA was required. Only the next step in the process which was to make a regional planning decision required an official SEA.

Whether the SEA was helpful in organizing the provincial planning procedure is debatable. The formulation of the SEA was mainly an internal matter involving the researchers and the project team. The project manager was highly involved in coupling the outcomes of the SEA to the planning process. The process of adjusting the regional planning document was the lead for the overall process, and the SEA procedure followed the planning procedure in this regard. The same was true of stakeholder involvement, which was organized within the framework of changing the planning document. This arrangement was also used to discuss the research questions, the preliminary results and the final SEA report.

Nevertheless, the SEA was certainly helpful in investigating the negative environmental impacts of the bypass and in designing the adjustments necessary to mitigate these consequences. The SEA revealed that some important habitat types were significantly influenced by the bypass, and thus helped to prevent future delays in the form of legal action on the part of environmental groups.

CASE COMPARISON

In both cases the SEA had an important role in the decision-making process. However, there also are clear differences. Table 1 compares the case studies on a number of crucial factors.

The most important differences between the cases lie in the way in which the SEA was embedded in the decision-making process and when it was executed. Compared to the SEA of the ZZL, the SEA of the IJsseldelta project was executed during a more leisurely phase. The most crucial period of this project was when the Masterplan was drawn up, and this had already passed by the time the SEA was initiated. The SEA in this case was carried out purely to support the formal planning procedure, and to change the provincial zoning plan. This made this SEA much less radical and challenging than the SEA for the ZZL, which was carried out in parallel with the discussion on the need for the whole project.

The ZZL case shows that an SEA can be used to structure stakeholder involvement in such a way that it also contributes to a process of frame reflection and learning. The formal steps of the SEA were combined with the formal steps necessary to design a structure vision. In the case of the IJsseldelta project, perhaps unsurprisingly, the timing of the SEA's implementation caused some earlier debates to be repeated and new debates to be started. The individuals overseeing the SEA process found it difficult to handle this situation, as several crucial decisions had already been made by the time they were involved, and they lacked the authority to offer stakeholders a real say.

An interesting similarity between the cases was the interweaving of the development of the planning document (zoning plan) and the execution of the SEA. Although the planning phases differed, the ways in which the SEA helped to optimize the final planning decisions were highly comparable. We can say that both SEAs helped to improve the search for the most feasible and valuable alternative, as both the planning processes and research processes were intertwined and carried out simultaneously.

Using our definition of effectiveness, we can conclude that the SEA for the ZZL was more effective than that for the IJsseldelta project. First, it was used to organize a serious debate about the necessity and value of the entire project, instead of merely fine-tuning a preselected alternative (as was the case in the IJsseldelta). Second, because large investments were made in combining the stakeholder participation process with the research process, the SEA contributed heavily to frame reflection. It did not simply serve as added fuel for existing controversies not directly addressed by the SEA. Third, in the case of the ZZL, the SEA procedure was used to organize the entire exploration process. In contrast, the SEA in the IJsseldelta case was merely a small player in the procedures put in place to adjust the provincial zoning plan. Because it was introduced very late in the decision-making process, this SEA could not be used to generate viable alternatives or introduce radically new perspectives.

Further, the difference in the way in which the SEA was organized in the two cases was critical. In the ZZL case, the production of the SEA was intertwined with other project organization activities. The SEA experts belonged to the core of the project bureau and were important in developing an explicit strategy for organizing the collaborative process and linking it to larger political decision-making processes. The SEA focused on the groups that were affected because of their location near the possible railway routes, and included the impact on them both in the general CBA and in the generation of alternative proposals. As doubts grew about the feasibility of the line, the SEA focused increasingly on other alternatives. It co-created the alternatives, while respecting and following the main process. In the case of IJsseldelta Zuid, the SEA team was far more distant from the main project team, and the link between the project and the SEA was restricted mainly to the relationship between the project director and the SEA project leader. For these reasons, the SEA in this case could be seen more as a passive information provision tool, with far less influence on the decision-making process.

CONCLUSION AND DISCUSSION

We have analysed two cases involving the application of SEAs to strategic decisions about highly controversial infrastructure projects. These findings are not entirely novel, and reflect many of the findings in previous works related to assessment effectiveness (Thérivel and Minas, 2002; Sheate *et al.*, 2003). In both cases the effectiveness of the SEA was highly dependent on the time of its commissioning, the degree to which it was intertwined with the decision-making process, and the openness of its application. The most visible beneficial effects of the SEA were observed in the ZZL case, where the SEA was instrumental in engaging affected groups. Their input was subsequently used to adjust the planning process. The SEA team worked closely with the general planning team, and this close cooperation allowed environmental information to influence the general planning process. As a result, different alternatives emerged which then were also subject to other assessments. This case demonstrates that the process of conducting an SEA can play a much greater role in determining its ultimate impact than the specific content it generates. The potential of the SEA to create collaborative dialogue and to establish a functional procedure is immense and can exert a strong influence on the quality of the final decision.

The fact that the SEA was initiated early in the discussion on necessity and added value made it far more effective than it was in the IJsseldelta case. The role of timing is indicative of the value of the SEA procedure in structuring and framing the overall planning process and the collaborative dialogue that surrounds it. When the SEA process is adopted midstream (as was the case in IJsseldelta) it cannot serve this structuring function. The project would have developed its own structure and process arrangements, and would not be amenable to the introduction of new procedures. In the ZZL case the project had to be organized from scratch, and the SEA procedure was gratefully embraced as a means of framing the process. This served to embed the SEA far more deeply into the decision-making process, which was an important factor in its ultimate success. This finding mirrors that of Runhaar and Driessen (2007), who similarly argued that the 'synchronization' of SEA and the planning process is a critical factor that determines the impact of an SEA.

In conclusion, the contribution of an SEA to the procedural quality of the planning decision-making process can differ dramatically from case to case. The SEA procedure can be used to structure the larger process, but it can also function as a subordinate procedure with minimal visible structuring impact. More detailed research is necessary to investigate the mechanisms by which SEA makes its procedural contribution to the decision-making processes (see Fischer, 2002).

Table 1 Cross comparison of the case studies

	Southern Sea Line	**Kampen IJsseldelta**
Timing of SEA	Parallel to the discussion about the necessity and added value of the ZZL	In the phase from intention agreement to provincial zoning plan
Scope	Fundamental discussion about ZZL, yes or no	Applied discussion about how to shape the bypass
Organization of SEA	In a very open consultation process with stakeholders	Mainly between experts and project group
Quality checks	NCEA, Critical Review Team	NCEA, contra expertise from the Water Board
Coupling with other tracks	Strong coupling with the CBA, spatial analysis and the zoning plan	Strong coupling with the development of the preferred alternative and the adjustment of the regional and local planning document
Flexibility of scope	Used to incorporate new ideas. The political decision-makers adjusted their perception of the project	Not used: SEA was meant to fill in the necessary knowledge requirements
Contribution to process	Guiding for intensive consultation process with stakeholders and citizens	One of the items that fuelled the debate in the stakeholder process
Contribution to procedure	Guiding in structuring the whole process with regard to the structure plan	Limited role in structuring one specific phase in the process, the provincial zoning plan
Contribution to content	Building block for abandoning ZZL and raising alternatives	Building block for political preference for blue bypass

REFERENCES

Cashmore, M. (2004) The role of science in environmental impact assessment: process and procedure versus purpose in the development theory, *Environmental Impact Assessment Review*, **24**(4), 403–426.

Cashmore, M., R. Gwilliam, R. Morgan, D. Cobb and A. Bond (2004) The interminable issue of effectiveness: substantive purposes, outcomes and research challenges in the advancement of environmental impact assessment theory, *Impact Assessment and Project Appraisal*, **22**(4), 295–310.

Cashmore, M., A. Bond and D. Cobb (2007) The contribution of environmental assessment to sustainable development: toward a richer empirical understanding, *Environmental Management*, 40(3), 516–530.

Cashmore, M., A. Bond and D. Cobb (2008) The role and functioning of environmental assessment: theoretical reflections upon an empirical investigation of causation, *Journal of Environmental Management*, **88**(4), 1233–1248.

Clark, W.C. and G. Majone (1985) The critical appraisal of scientific inquiries with policy implications, *Science, Technology and Human Values*, **10**(3), 6–19.

Collingridge, D. and C. Reeve (1986) *Science speaks to power: the role of experts in policy-making*, London: Frances Pinter Publications.

Fischer, T.B. (2002) Strategic environmental assessment performance criteria. The same requirements for every assessment? *Journal of Environmental Assessment Policy and Management*, **4**(1), 83–99.

Guston, D.H. (2004) Forget politicizing science: let's democratize science, *Issues in Science and Technology*, Fall 2004, 25–28.

Inglehart, R. (1990) *Culture Shift in Advanced Industrial Society*. Princeton, NJ: Princeton University Press.

Jasanoff, S. (1990) *The fifth branch: advisers as policy makers*, Harvard: Harvard University Press.

Kørnøv, L. and W.A.H. Thissen (2000) Rationality in decision – and policy-making: implications for strategic environmental assessment, *Impact Assessment and Project Appraisal*, **18**(3), 191–200.

Ministry of VROM (2006) *Handreiking milieueffectrapportage van plannen (planmer): Europese richtlijn milieubeoordeling van plannen Implementatie in Wet Milieubeheer and Besluit m.e.r. (1994)*, www.vrom. nl/get.asp?file=docs/publicaties/6162.pdf&dn=6162&b=vrom, *last accessed* 25 April 2009.

Ministerie van Verkeer en Waterstaat (2006) *Strategische Milieubeoordeling Structuurvisie Zuiderzeelijn*. http:// www.verkeerenwaterstaat.nl/kennisplein/page_kennisplein.aspx?id= 331127&DossierURI=tcm:195-17272-4, last accessed 25 April 2009.

Ministerie van Verkeer en Waterstaat, Ministerie van Volkshuisvesting, Ruimtelijke Ordening en Milieubeheer en Ministerie van Economische Zaken (2006) *Aanvulling op de Structuurvisie Zuiderzeelijn. Onderzoek naar nut en noodzaak*. http:// www.verkeerenwaterstaat.nl/kennisplein/page_kennisplein.aspx?id=344256&DossierURI=tcm:195-17272-4, last accessed 25 April 2009.

Partidário, M.R. (2000) Elements of an SEA framework: improving the added-value of SEA, *Environmental Impact Assessment Review*, **20**(6), 647–663.

Pischke, F. and M. Cashmore (2006) Decision-oriented environmental assessment: an empirical study of its theory and methods, *Environmental Impact Assessment Review*, **26**(7), 643–662.

Projectorganisatie IJsseldelta-Zuid (2007) *Notitie reikwijdte en detailniveau partiële provinciale planherzieningen IJsseldelta-Zuid*. http:// www.ijsseldelta.info/zuid/?page=cms&sub=list&cid =11&scid=44

Projectorganisatie IJsseldelta-Zuid (2008) *Ontwerp partiële herziening IJsseldelta-Zuid voor de integrale gebiedsontwikkeling, Streekplan Overijssel 2000+*. http:// www.ijsseldelta. info/zuid/?page=cms&sub=list&cid=11&scid=44

Projectorganisatie Zuiderzeelijn 2005a. *Document raadpleging Strategische Milieu Beoordeling*. http://www.verkeerenwaterstaat.nl/kennisplein/page_kennisplein.aspx?id= 324679&DossierURI=tcm:195-17272-4, last accessed 25 April 2009.

Projectorganisatie Zuiderzeelijn 2005b. *Resultaten raadpleging in het kader van de Strategische Milieu Beoordeling (SMB) Zuiderzeelijn*. Den Haag. http:// www.verkeerenwaterstaat.nl/kennisplein/page_kennisplein.aspx?id=324672&DossierURI= tcm:195-17272-4, last accessed 25 April 2009.

Projectorganisatie Zuiderzeelijn (2006) *Aanvulling op de Structuurvisie Zuiderzeelijn*. http://www.verkeerenwaterstaat.nl/kennisplein/page_kennisplein.aspx?id=344326&DossierURI =tcm:195-17272-4, last accessed 25 April 2009.

Provincie Overijssel (2008a) *IJsseldelta-Zuid PlanMER partiële provinciale planherzieningen*. http://www.ijsseldelta.info/zuid/?page=cms&sub=list&cid=11&scid=44, last accessed 25 April 2009.

Provincie Overijssel (2008b) *Reactienota; ingediende adviezen en zienswijzen op de Ontwerp-Partiële Herziening Streekplan Overijssel 2000+ (incl. planMER); IJsseldelta-Zuid*. http://www. ijssel delta.info/zuid/ ?page=cms&sub=list&cid=11&scid=44, last accessed 25 April 2009.

Retief, F. (2006) A performance evaluation of strategic environmental assessment processes within the South African context, *Environmental Impact Assessment Review*, 27(1), 84–100.

Runhaar, H. and P.J. Driessen (2007) What makes strategic environmental assessment successful environmental assessment? The role of context in the contribution of SEA to decision-making, *Impact Assessment and Project Appraisal*, **25**(1), 2–14.

Sarewitz, D. (2004) How science makes environmental controversies worse, *Environmental Science and Policy*, 7, 385–403.

Sheate, W.R., S. Dagg, J. Richardson, R. Aschemann, J. Palerm and U. Steen (2003) Integrating the environment into strategic decision-making: conceptualizing policy SEA, *European Environment*, **13**(1), 1–18.

Teisman, G.R. (2008). Complexity and management of improvement programmes: an evolutionary approach. *Public Management Review*, **10**(3), 341–359.

Thérivel, R. and P. Minas (2002) Measuring SEA effectiveness: ensuring effective sustainability appraisal, *Impact Assessment and Project Appraisal*, **20**(2), 81–91.

Therivel R. and M.R. Partidário (1992). *The practice of strategic environmental assessment*. London: Earthscan.

Twaalfhoven, P. (1999) *The success of policy analysis studies: an actor perspective*, Delft: Eburon.

van Buuren, M.W. (2009) Knowledge for governance, governance of knowledge: inclusive knowledge management in collaborative governance processes, *International Public Management Journal*, in press.

van Buuren, M.W. and J. Edelenbos (2004) Conflicting knowledge: why is knowledge production such a problem, *Science and Public Policy*, **31**(4), 289–299.

van Buuren, M.W. and Nooteboom, S. 2009. The success of SEA in the Dutch planning practice. How formal assessments can contribute to collaborative governance. *Environmental Impact Assessment Review*, in press.

Wallington, T., O. Bina and W. Thissen (2007) Theorising strategic environmental assessment fresh perspectives and future challenges, *Environmental Impact Assessment Review*, **27**(7), 569–584.

Woodhouse, E.J. and D.A. Nieusma (2001) Democratic expertise: integrating knowledge, power, and participation, in M. Hisschemöller, R. Hoppe, W.N. Dunn and J.R. Ravetz (eds) *Knowledge, power, and participation in environmental policy analysis*, Policy Studies Annual Volume **12**, New Brunswick: Transaction Publishers, 73–96.

David Pearce, Giles Atkinson and Susana Mourato — "Executive Summary," "The Stages of Practical Cost-Benefit Analysis" and "Cost-Benefit Analysis and Other Decision-Making Procedures." In *Cost Benefit Analysis and the Environment: Recent Developments.* **Paris: OECD.**

Reading Commentary

Pearce, Atkinson and Mourato trace the theoretical origins of CBA back to infrastructure appraisal in France in the early nineteenth century. They note that it was only formalized as a tool in response to a growing demand for governments to be more "efficient" in their investments after World War II. Since the 1960s, CBA has been used as a tool for policy analysis, most frequently applied to environmental, transportation and health-care policy decisions.

CBA provides policy makers a simplified way of deciding whether or not to adopt a policy (including choosing one version of a policy over its alternatives) based on a calculation of the benefits (i.e., increases in human well-being) versus costs (i.e., reductions in human well-being). The monetary value of many benefits and costs is approximated based on individual willingness to pay (e.g., to pay for benefits or to avoid costs) or willingness to accept compensation (e.g., for losses). These values (approximated at the level of the individual) are then aggregated across populations and over time. The authors point out that the final calculation of value can vary depending on whether willingness to pay (WTP) or willingness to accept (WTA) compensation is used to approximate the value of a cost or benefit, whether a weighting is applied (e.g., greater weight given when these benefits/costs affect disadvantaged groups), or which discount rate was used when trying to calculate future costs and benefits.

Despite its inherent flaws, Pearce, Atkinson and Mourato maintain that CBA remains the most effective evaluative tool for policy-making. Theoretically, as long as a policy's social (i.e., aggregated) benefits outweigh its social costs, it ought to be approved or, in the case of multiple policy options, the option with the greatest net benefit should be selected.

Based on what you will read in the Executive Summary, Box 3.1 and Chapter 18, what do you identify as the potential risks associated with using this type of analysis for policy-making? Are you convinced by the authors' claim that CBA remains the best tool? If not, what other tools or procedures might be used to improve policy-making?

COST-BENEFIT ANALYSIS AND THE ENVIRONMENT RECENT DEVELOPMENTS

David Pearce, Giles Atkinson, and Susana Mourato

INTRODUCTION

The OECD has long championed efficient decision-making using economic analysis. It was, for example, one of the main sponsors of the early manuals in the late 1960s on project evaluation authored by Ian Little and James Mirrlees.[1] Since then, cost-benefit analysis has been widely practised, notably in the fields of environmental policy, transport planning, and healthcare. In the last decade or so, cost-benefit analysis has been substantially developed both in terms of the underlying theory and in terms of sophisticated applications. Many of those developments have been generated by the special challenges that environmental problems and environmental policy pose for cost-benefit analysis. The OECD has therefore returned to the subject in this new and comprehensive volume that brings analysts and decision-makers up to date on the main developments.

HISTORY AND USES OF CBA

The history of cost-benefit analysis (CBA) shows how its theoretical origins date back to issues in infrastructure appraisal in France in the 19th century. The theory of welfare economics developed along with the "marginalist" revolution in microeconomic theory in the later 19th century, culminating in Pigou's *Economics of Welfare* in 1920 which further formalised the notion of the divergence of private and social cost, and the "new welfare economics" of the 1930s which reconstructed welfare economics on the basis of ordinal utility only. Theory and practice remained divergent, however, until the formal requirement that costs and benefits be compared entered into water-related investments in the USA in the late 1930s. After World War II, there was pressure for "efficiency in government" and the search was on for ways to ensure that public funds were efficiently utilised in major public investments. This resulted in the beginnings of the fusion of the new welfare economics, which was essentially cost-benefit analysis, and practical decision-making. Since the 1960s CBA has enjoyed fluctuating fortunes, but is now recognised as the major appraisal technique for public investments and public policy.

THEORETICAL FOUNDATIONS

The essential theoretical foundations of CBA are: benefits are defined as increases in human wellbeing (utility) and costs are defined as reductions in human wellbeing. For a project or policy to qualify on cost-benefit grounds, its social benefits must exceed its social costs. "Society" is simply the sum of individuals. The geographical boundary for CBA is usually the nation but can readily be extended to wider limits. There are two basic aggregation rules. First, aggregating benefits across different social groups or nations involves summing willingness to pay for benefits, or willingness to accept compensation for losses (WTP, WTA respectively), regardless of the circumstances of the beneficiaries or losers. A second aggregation rule requires that higher weights be given to benefits and costs accruing to disadvantaged or low income groups. One rationale for this second rule is that marginal utilities of income will vary, being higher for the low income group. Aggregating over time involves discounting. Discounted future benefits and costs are known as present values. Inflation can result in future benefits and costs appearing to be higher than is really the case. Inflation should be netted out to secure constant price estimates. The notions of WTP and WTA are firmly grounded in the theory of welfare economics and correspond to notions of compensating and equivalent variations. WTP and WTA should not, according to past theory, diverge very much. In practice they appear to diverge, often substantially, and with WTA > WTP. Hence the choice of WTP or WTA may be of importance when conducting CBA.

There are numerous critiques of CBA. Perhaps some of the more important ones are: *a)* the extent to which CBA rests on robust theoretical foundations as portrayed by the Kaldor-Hicks compensation test in welfare economics; *b)* the fact that the underlying "social welfare function" in CBA is one of an arbitrarily large number of such functions on which consensus is unlikely to be achieved; *c)* the extent to which one can make an ethical case for letting individuals' preferences be the (main) determining factor in guiding social decision rules; and *d)* the whole history of neoclassical welfare economics has focused on the extent to which the notion of economic efficiency underlying the Kaldor-Hicks compensation test can or should be separated out from the issue of who gains and loses – the distributional incidence of costs and benefits. CBA has developed procedures for dealing with the last criticism, *e.g.* the use of distributional weights and the presentation of "stakeholder" accounts. Criticisms *a)* and *b)* continue to be debated. Criticism *c)* reflects the "democratic presumption" in CBA, *i.e.* individuals' preference should count.

THE STAGES OF CBA

Conducting a well-executed CBA requires the analyst to follow a logical sequence of steps. The first stage involves asking the relevant questions: what policy or project is being evaluated? What alternatives are there? For an initial screening of the contribution that the project or policy makes to social wellbeing to be acceptable, the present value of benefits must exceed the present value of costs.

Determining "standing" – *i.e.* whose costs and benefits are to count – is a further preliminary stage of CBA, as is the time horizon over which costs and benefits are counted. Since individuals have preferences for when they receive benefits or suffer costs, these "time-preferences" also have to be accounted for through the process of discounting. Similarly, preferences for or against an impact may change through time and this "relative price" effect also has to be accounted for. Costs and benefits are rarely known with certainty so that risk (probabilistic outcomes) and uncertainty (when no probabilities are known) also have to be taken into account. Finally, identifying the distributional incidence of costs and benefits is also important.

DECISION RULES

Various decision rules may be used for comparing costs and benefits. The correct criterion for reducing benefits and costs to a unique value is the net present value (NPV) or "net benefits" criterion. The correct rule is to adopt any project with a positive NPV and to rank projects by their NPVs. When budget constraints exist, however, the criteria become more complex. Single-period constraints – such as capital shortages – can be dealt with by a benefit-cost ratio (B/C) ranking procedure. There is general agreement that the internal rate of return (IRR) should not be used to rank and select mutually exclusive projects. Where a project is the only alternative proposal to the status quo, the issue is whether the IRR provides worthwhile additional information. Views differ in this respect. Some argue that there is little merit in calculating a statistic that is either misleading or subservient to the NPV. Others see a role for the IRR in providing a clear signal as regards the sensitivity of a project's net benefits to the discount rate. Yet, whichever perspective is taken, this does not alter the broad conclusion about the general primacy of the NPV rule.

DEALING WITH COSTS

The cost component is the other part of the basic CBA equation. As far as projects are concerned, it is unwise to assume that because costs may take the form of equipment and capital infrastructure their estimation is more certain than benefits. The experience is that the costs of major projects can be seriously understated. The tendency for policies is for their compliance costs to be overstated. In other words there may be cost pessimism or cost optimism. In light of this it is important to conduct sensitivity analysis, *i.e.* to show how the final net benefit figure changes if costs are increased or decreased by some percentage. Ideally, compliance costs would be estimated using general equilibrium analysis.

Politicians are very sensitive about the effects of regulation on competitiveness. This is why most Regulatory Impact Assessment procedures call for some kind of analysis of these effects. A distinction needs to be made between the competitiveness of nations as whole, and the competitiveness of industries. In the former case it is hard to assign much credibility to the notion of competitiveness impacts. In the latter case two kinds of effects may occur. The first is any impact on the competitive nature of the industry within the country in question – *e.g.* does the policy add to any tendencies for monopoly power? If it does then, technically, there will be welfare losses associated with the change in that monopoly power and these losses should be added to the cost side of the CBA, if they can be estimated. The second impact is on the costs of the industry relative to the costs of competing industries in other countries. Unless the industry is very large, it cannot be assumed that exchange rate movements will cancel out the losses arising from the cost increases. In that case there may be dynamic effects resulting in output losses.

Policies to address one overall goal may have associated effects in other policy areas. Climate change and conventional air pollutants is a case in point. Reductions in climate gases may be associated with reductions in jointly produced air pollutants. Should the two be added and regarded as a benefit of climate change policy? On the face of it they should, but care needs to be taken that the procedure does not result in double counting. To address this it is important to consider the counterfactual, *i.e.* what policies would be in place without the policy of immediate interest. While it is common practice to add the benefits together, some experts have cast doubt on the validity of the procedure.

Finally, employment effects are usually also of interest to politicians and policy-makers. But the extent to which they matter for the CBA depends on the nature of the economy. If there is significant unemployment, the labour should be shadow priced on the basis of its opportunity cost. In turn this may be very low, *i.e.* if not used for the policy or project in question, the labour might otherwise be unemployed. In a fully employed economy, however, this opportunity cost may be such as to leave the full cost of labour being recorded as the correct value.

TOTAL ECONOMIC VALUE

The notion of total economic value (TEV) provides an all-encompassing measure of the *economic value* of any environmental asset. It decomposes into use and non-use (or passive use) values, and further sub-classifications can be provided if needed. TEV does not encompass other kinds of values, such as intrinsic values which are usually defined as values residing "in" the asset and unrelated to human preferences or even human observation. However, apart from the problems of making the notion of intrinsic value operational, it can be argued that some people's willingness to pay for the conservation of an asset, independently of any use they make of it, is influenced by their own judgements about intrinsic value. This may show up especially in notions of "rights to existence" but also as a form of altruism. Any project or policy that destroys or depreciates an environmental asset needs to include in its costs the TEV of the lost asset. Similarly, in any project or policy that enhances an environmental asset, the change in the TEV of the asset needs to be counted as a benefit. For instance, ecosystems produce many services and hence the TEV of any ecosystem tends to be equal to the discounted value of those services.

REVEALED PREFERENCE VALUATION

Economists have developed a range of approaches to estimate the economic value of non-market or intangible impacts. There are several procedures that share the common feature of using market information and behaviour to infer the economic value of an associated non-market impact.

These approaches have different conceptual bases. Methods based on hedonic pricing utilise the fact that some market goods are in fact bundles of characteristics, some of which are intangible goods (or bads). By trading these market goods, consumers are thereby able to express their values for the intangible goods, and these values can be uncovered through the use of statistical techniques. This process can be hindered, however, by the fact that a market good can have several intangible characteristics, and that these can be collinear. It can also be difficult to measure the intangible characteristics in a meaningful way.

Travel cost methods utilise the fact that market and intangible goods can be complements, to the extent that purchase of market goods and services is required to access an intangible good. Specifically, people have to spend time and money travelling to recreational sites, and these costs reveal something of the value of the recreational experience to those people incurring them. The situation is complicated, however, by the fact that travel itself can have value, that the same costs might be incurred to access more than one site, and that some of the costs are themselves intangible (*e.g.* the opportunity costs of time).

Averting behaviour and defensive expenditure approaches are similar to the previous two, but differ to the extent that they refer to individual behaviour to avoid negative intangible impacts. Therefore, people might buy goods such as safety helmets to reduce accident risk, and double-glazing to reduce traffic noise, thereby revealing their valuation of these bads. However, again the situation is complicated by the fact that these market goods might have more benefits than simply that of reducing an intangible bad.

Finally, methods based on cost of illness and lost output calculations are based on the observation that intangible impacts can, through an often complex pathway of successive physical relationships, ultimately have measurable economic impacts on market quantities. Examples include air pollution, which can lead to an increase in medical costs incurred in treating associated health impacts, as well as a loss in wages and profit. The difficulty with these approaches is often the absence of reliable evidence, not on the economic impacts, but on the preceding physical relationships.

STATED PREFERENCE VALUATION: CONTINGENT VALUATION

Stated preference techniques of valuation utilise questionnaires which either directly ask respondents for their willingness to pay (accept), or offer them choices between "bundles" of attributes and from which choices the analysts can infer WTP (WTA).

Stated preference methods more generally offer a direct survey approach to estimating individual or household preferences and more specifically WTP amounts for changes in provision of (non-market) goods, which are related to respondents' underlying preferences in a consistent manner. Hence, this technique is of particular worth when assessing impacts on non-market goods, the value of which cannot be uncovered using revealed preference methods.

This growing interest in stated preference approaches has resulted in a substantial evolution of techniques over the past 10 to 15 years. For example, the favoured choice of elicitation formats for WTP questions in contingent valuation surveys has already passed through a number of distinct stages. This does not mean that uniformity in the design of stated preference surveys can be expected any time soon. Nor is this particularly desirable. Some studies show how, for example, legitimate priorities to minimise respondent strategic bias by always opting for incentive compatible payment mechanisms must be balanced against equally justifiable concerns about the credibility of a payment vehicle. The point is the answer to this problem is likely to vary across different types of project and policy problems.

There remain concerns about the validity and reliability of the findings of contingent valuation studies. Indeed, much of the research in this field has sought to construct rigorous tests of the robustness of the methodology across a variety of policy contexts and non-market goods and services. By and large, one can strike an optimistic note about the use of the contingent valuation to estimate the value of non-market goods. In this interpretation of recent developments, there is a virtuous circle between translating the lessons from tests of validity and reliability into practical guidance for future survey design. Indeed, many of the criticisms of the technique can be said to be imputable to problems at the survey design and implementation stage rather than to some intrinsic methodological flaw. Taken as a whole, the empirical findings largely support the validity and reliability of contingent valuation estimates.

STATED PREFERENCE VALUATION: CHOICE MODELLING

Many types of environmental impact are multidimensional in character. Hence an environmental asset that is affected by a proposed project or policy often will give rise to changes in component attributes each of which command distinct valuations. The application of choice modelling (CM) approaches to valuing multidimensional environmental problems has been growing steadily in recent years. CM is now routinely discussed alongside the arguably better-known contingent valuation method in state-of-the-art manuals regarding the design, analysis and use of stated preference studies. While there are a number of different approaches under the CM umbrella, it is arguably the *choice experiment* variant (and to some extent, *contingent ranking*) that has become the dominant CM approach with regard to applications to environmental goods. In a choice experiment, respondents are asked to choose their most preferred option from a choice set of at least two options, one of which is the status quo or current situation. It is this CM approach that can be interpreted in standard welfare economic terms, an obvious strength where consistency with the theory of cost-benefit analysis is a desirable criterion.

Much of the discussion about, for example, validity and reliability issues in the context of contingent valuation (CV) studies applies in the context of the CM. While it is likely that on some criteria, CM is likely to perform better than CV – and *vice versa* – the evidence for such assertions is largely lacking at present. While those few studies that have sought to compare the findings of CM and CV appear to find that the total value of changes in the provision of the same environmental good in the former exceeds that of the latter, the reasons for this are not altogether clear. However, whether the two methods should be seen as always competing against one another – in the sense of say CM being a more general and thereby superior method – is debatable. Both approaches are likely to have their role in cost-benefit appraisals and a useful contribution of any future research would also be to aid understanding of when one approach should be used rather than the other.

OPTION VALUE

The notion of *quasi option value* was introduced in the environmental economics literature some three decades ago. In parallel, financial economists developed the notion of "*option value*". QOV is not a separate category of economic value. Rather it is the difference between the net benefits of making an optimal decision and one that is not optimal because it ignores the gains that may be made by delaying a decision and learning during the period of delay. Usually, QOV arises in the context of irreversibility. But it can only emerge if there is uncertainty which can be resolved by learning. If the potential to learn is not there, QOV cannot arise.

Can QOV make a significant difference to decision-making? Potentially, yes. It is there to remind us that decisions should be made on the basis of maximum feasible information about the costs and benefits involved, and that includes "knowing that we do not know". If this ignorance cannot be resolved then nothing is to be gained by delay. But if information can resolve it, then delay can improve the quality of the decision. How large the gain is from this process is essentially an empirical question since QOV is the difference in the net benefits of an optimal decision and a less than optimal one.

WTP VERSUS WTA?

Traditionally, economists have been fairly indifferent about the welfare measure to be used for economic valuation: willingness to pay (WTP) and willingness to accept compensation (WTA) have both been acceptable. By and large, the literature has focused on WTP. However, the development of stated preference studies has, fairly repeatedly, discovered divergences, sometimes substantial ones, between WTA and WTP. These differences still would not matter if the nature of property rights regimes were always clear. WTP in the context of a potential improvement is clearly linked to rights to the status quo. Similarly, if the context is one of losing the status quo, then WTA for that loss is the relevant measure. By and large, environmental policy tends to deal with improvements rather than deliberate degradation of the environment, so there is a presumption that WTP is the right measure. The problems arise when individuals can be thought of as having some right to a future state of the environment. If that right exists, their WTP to secure that right seems inappropriate as a measure of welfare change, whereas their WTA to forego that improvement seems more relevant. In practice, the policy context may well be one of a mixture of rights, *e.g.* a right to an improvement attenuated by the rights of others not to pay "too much" for that improvement.

Finding out why, empirically, WTA and WTP differ also matters. If there are legitimate reasons to explain the difference then the preceding arguments apply and one would have to recommend that CBA should always try to find both values. The CBA result would then be shown under both assumptions. But if the observed differences between WTA and WTP are artefacts of questionnaire design, there is far less reason to be concerned at the difference between them. The fallback position of their approximate equality could be assumed. Unfortunately, the literature is undecided as to why the values differ. This again suggests showing the CBA results under both assumptions about the right concept of value.

VALUING ECOSYSTEM SERVICES

Research is now being conducted on the value of ecosystem services. The aim is to estimate the total economic value (TEV) of ecosystem change. The problems with valuing changes in ecosystem services arise from the interaction of ecosystem products and services, and from the often extensive uncertainty about how ecosystems function internally, and what they do in terms of life support functions. Considerable efforts have been made to value specific services, such as the provision of genetic information for pharmaceutical purposes. The debate on that issue usually shows how complex valuing ecosystem services can be. But even that literature is still developing, and it does not address the interactive nature of ecosystem products and services.

Once it is acknowledged that ecosystem functioning may be characterised by extensive uncertainty, by irreversibility and by non-linearities that generate potentially large negative effects from ecosystem loss or degradation, the focus shifts to how to behave in the face of this combination of features. The short answer is that decision-making favours precaution. But just what precaution means is itself a further debate.

DISCOUNTING

Some advances have been prompted by the alleged "tyranny of discounting" – the fact that discounting has a theoretical rationale in the underlying welfare economics of CBA, but with consequences that many seem to find morally unacceptable. This unacceptability arises from the fact that distant future costs and benefits may appear as insignificant present values when discounting is practised. In turn, this appears to be inconsistent with notions of intergenerational fairness. Current activities imposing large costs on future generations may appear insignificant in a cost-benefit analysis. Similarly, actions now that will benefit future generations may not be undertaken in light of a cost-benefit analysis.

The weakness of the conventional approach, which assumes that one positive discount rate is applied for all time, is that it neither incorporates uncertainty about the future nor attempts to resolve the tyranny problem. Additionally, the assumption of a constant discount rate is exactly that – an assumption. The "escapes" from the tyranny problems centre on several approaches.

First, many studies find that very often (but not always), people actually discount "hyperbolically", *i.e.* people actually do use time-declining discount rates. If what people do reflects their preferences, and if preferences are paramount, there is a justification for adopting time-declining discount rates.

Second, there is also uncertainty about future interest rates: here it can be shown that uncertainty about the temporal weights – *i.e.* the *discount factor* – is consistent with a time-declining certainty equivalent *discount rate*. Introducing uncertainty about the state of the economy more generally can be shown also to generate time-declining rates, if certain conditions are met.

Third, by positing the "tyranny" problem as a social choice problem in which neither the present nor the future dictates outcomes, and adopting reasonable ethical axioms can be shown to produce time-declining rates.

In terms of the uncertainty and social choice approaches, the time-path of discount rates could be very similar with long term rates declining to the "lowest possible" rates of, say, 1%.

But time-consistency problems remain and some experts would regard any time-declining discount rate as being unacceptable because of such problems. Others would argue that the idea of a long-run optimising government that never revises its "optimal" plan is itself an unrealistic requirement for the derivation of an optimal discount rate.

219

VALUING HEALTH AND LIFE

Considerable strides have been made in recent years in terms of clarifying both the meaning and size of the "value of a statistical life" (VOSL). One of the main issues has been how to "transfer" VOSLs taken from non-environmental contexts to environmental contexts. Non-environmental contexts tend to be associated with immediate risks such as accidents. In contrast, environmental contexts are associated with both immediate and future risks. The futurity of risk may arise because the individual in question is not at immediate risk from *e.g.* current levels of pollution but is at risk in the future when there is greater vulnerability to risk. Or futurity may arise because the risk is latent as with diseases such as asbestosis or arsenicosis. All this suggests *a)* that valuations of immediate risk might be transferred to environmental immediate risk contexts (provided that the perception of the risk is the same) but *b)* future risks need to be valued separately.

In terms of practical guidelines, the age of the respondent who is valuing the risk matters. Age may or may not be relevant in valuing immediate risks – the literature is ambiguous. The general rule, then, is to ensure that age is controlled for in any primary valuation study. For "benefits transfer" the rule might be one of adopting a default position in which immediate risks are valued the same regardless of age (*i.e.* the VOSL does not vary with age), with sensitivity analysis being used to test the effects of lower VOSLs being relevant for older age groups. Age is very relevant for valuing future risks. Thus a policy which lowers the general level of exposure to pollution should be evaluated in terms of the (lower than immediate VOSL) valuations associated with younger people's valuations of future risks, plus older persons' valuation of that risk as an immediate risk.

Some environmental risks fall disproportionately on the very young and the very old. A complex issue arises with valuing risks to children. The calculus of willingness to pay now seems to break down since children may have no income to allocate between goods, including risk reduction, may be ill-informed about or be unaware of risks, and may be too young to articulate preferences anyway. The result is that adults' valuations of the risks *on behalf of* children need to be estimated. The literature on which to base such judgements is only now coming into existence. Preliminary findings suggest that the resulting values of WTP may be higher for adults valuing on behalf of children than they are for adults speaking on behalf of themselves. The safest conclusion at this stage is that bringing the effects on children into the domain of CBA is potentially important, with a default position being to use the adult valuations of "own" life risks for the risks faced by children.

EQUITY

One important issue is equity or the distributional incidence of costs and benefits. Incorporating distributional concerns implies initially identifying and then possibly weighting the costs and benefits of individuals and groups on the basis of differences in some characteristic of interest (such as income or wealth). First, there is the relatively straightforward but possibly arduous task of assembling and organising raw (*i.e.* unadjusted) data on the distribution of project costs and benefits. Second, these data could then be used to ask what weight or distributional adjustment would need to placed on the net benefits (net costs) of a societal group of interest for a given project proposal to pass (fail) a distributional cost-benefit test. Third, explicit weights reflecting judgement about society's preferences towards distributional concerns can be assigned and net benefits re-estimated on this basis.

A crucial question then is where should cost-benefit analysts locate themselves upon this hierarchy? Given that cost-benefit appraisals are sometimes criticised for ignoring distributional consequences altogether then the apparently simplest option of cataloguing how costs and benefits are distributed could offer valuable and additional insights. This suggests that, at a minimum, cost-benefit appraisals arguably should routinely provide these data. Whether more ambitious proposals should be adopted is a matter of deliberating about whether: first, the gains in terms of being able to scrutinise the (weighted) net benefits of projects in the light of societal concerns about both efficiency and equity outweighs; second, the losses arising from the need for informed guesswork in interpreting the empirical evidence with regards to the treatment of the latter.

On the one hand, empirical evidence about the "correct" magnitudes of distributional weights can be usefully employed in distributional CBA as its application to the case of climate change illustrates. On the other hand, even apparently small changes in assumptions about the size of distributional weights – indicated by the range of values in available empirical studies – can have significant implications for recommendations about a project's social worth. This finding should not be a surprise for it primarily reflects the complexity involved in trying to disentangle society's distributional preferences. As a practical matter, the danger is whether the most ambitious proposals for distributional CBA generate more heat than light. While it would worthwhile for research to seek further understanding of these preferences – perhaps making greater use of stated preference methods – in the interim, estimating implicit weights might be the most useful step beyond the necessary task of cataloguing the distribution of project cost and benefits.

SUSTAINABILITY AND CBA

While there remains a debate about what it means for development to be sustainable, there is now a coherent body of academic work that has sought to understand what a sustainable development path might look like, how this path can be achieved and how progress towards it might be measured. Much of this work considers the pursuit of sustainable development to be an aggregate or macroeconomic goal. Comparatively little attention has been paid to the implications of notions of sustainability for CBA. However, a handful of recommendations do exist with regards to how cost-benefit appraisals can be extended to take account of recent concerns about sustainable development.

According to one perspective there is an obvious role for appraising projects in the light of these concerns. This notion of strong sustainability starts from the assertion that certain natural assets are so important or critical (for future, and perhaps current, generations) so as to warrant protection at current or above some other target level. If individual preferences cannot be counted on to fully reflect this importance, there is a paternal role for decision-makers in providing this protection. With regards to the relevance of this approach to cost-benefit appraisals, a handful of contributions have suggested that sustainability is applicable to the management of a *portfolio* of projects. This has resulted in the idea of a shadow or compensating project. For example, this could be interpreted as meaning that projects that cause environmental damage are "covered off" by projects that result in environmental improvements. The overall consequence is that projects in the portfolio, on balance, maintain the environmental status quo.

There are further ways of viewing the problem of sustainable development. Whether these alternatives – usually characterised under the heading "weak sustainability" – are complementary or rivals has been a subject of debate. This debate would largely dissolve if it could be determined which assets were critical. As this latter issue is itself a considerable source of uncertainty, the debate continues. However, the so-called "weak" approach to sustainable development is useful for a number of reasons. While it has primarily be viewed as a guide to constructing green national accounts (*i.e.* better measures of income, saving and wealth), the focus on assets and asset management has a counterpart in thinking about project appraisal. For example, this might emphasise the need for an "asset check". That is, what the stocks of assets are before the project intervention and what they are likely to be after the intervention? It might also add another reason for the tradition in cost-benefit analysis of giving greater weight to projects which generate economic resources for saving and investment in economies where it is reckoned that too little net wealth (per capita) is being passed on to future generations.

BENEFITS TRANSFER

Benefits or value transfer involves taking economic values from one context and applying them to another. Transfer studies are the bedrock of practical policy analysis in that only infrequently are policy analysts afforded the luxury of designing and implementing original studies. In general then, analysts must fall back on the information that can be gleaned from past studies. This is likely to be no less true in the case of borrowing or transferring WTP values to policy questions involving environmental or related impacts. Almost inevitably, benefits transfer introduces subjectivity and greater uncertainty into appraisals in that analysts must make a number of *additional* assumptions and judgements to those contained in original studies. The key question is whether the added subjectivity and uncertainty surrounding the transfer is acceptable and whether the transfer is still, on balance, informative.

Surprisingly given its potentially central role in environmental decision-making, there are no generally accepted practical transfer protocols to guide analysts. However, a number of elements of what might constitute best practice in benefits transfer might include the following. First, the studies included in the analysis must themselves be sound. Initial but crucial steps of any transfer are very much a matter of carefully scrutinising the accuracy and quality of the original studies. Second, in conducting a benefits transfer, the study and policy sites must be similar in terms of population and population characteristics. If not then differences in population, and their implications for WTP values, need to be taken into account. Just as importantly, the change in the provision of the good being valued at the two sites also should be similar.

The holy grail of benefits transfer is the consolidation of data on non-market values in emerging transfer databases. Yet, while databases are to be welcomed and encouraged, these developments still need to be treated with some caution. Thus, there is a widely acknowledged need for more research to secure a better understanding of when transfers work and when they do not as well as developing methods that might lead to transfer accuracy being improved.

However, a competent application of transfer methods demands informed judgement and expertise and sometimes, according to more demanding critics, as advanced technical skills as those required for original research. At the very least, it suggests that practitioners should be explicit in their analysis about important caveats regarding a proposed transfer exercise as well as take account of the sensitivity of their recommendations to changes in assumptions about economic values based on these transfers.

CBA AND OTHER DECISION-MAKING GUIDANCE

A significant array of decision-guiding procedures are available and include cost-effectiveness analysis (CEA) and multi-criteria analysis (MCA). These procedures vary in their degree of comprehensiveness, where this is defined as the extent to which all costs and benefits are incorporated. In general, only MCA is as comprehensive as CBA and may be more comprehensive once goals beyond efficiency and distributional incidence are considered. All the remaining procedures either deliberately narrow the focus on benefits, *e.g.* to health or environment, or ignore cost. Procedures also vary in the way they treat time. Environmental Impact Assessment and Life Cycle Analysis are essential inputs into a CBA, although the way these impacts are dealt with in "physical terms" may not be the same in a CBA. They are not decision-making procedures in their own right. Risk assessments tend to be focused on human health only but ecological risk assessments are also fairly common. Once again, neither enables a comprehensive decision to be made.

SOME POLITICAL ECONOMY

Political economy, or "political economics", seeks to explain why the economics of the textbook is rarely embodied in actual decision-making. CBA is very much a set of procedures derived from an analytical framework that is as theoretically "correct" as possible. Unsurprisingly, actual decisions may be made on very different bases to this analytical approach. The reasons lie in the role played by "political" welfare functions rather than the social welfare functions of economics, distrust about or disbelief in monetisation, the capture of political processes by those not trained in economics, beliefs that economics is actually "common sense" and easily understood, and, of course, genuine mistrust of CBA and its theoretical foundations based on the debates that continue within the CBA community and outside it. But explaining the gap between actual and theoretical design is not to justify the gap. Theoretical economists need a far better understanding of the pressures that affect actual decisions, but those who make actual decisions perhaps also need a far better understanding of economics.

Box 3.1. Achieving air quality targets in Europe

The European Union has set air quality targets for the year 2010 with respect to various pollutants. The ones selected here are NO_2, SO_2 and PM. The ambient concentration levels of these pollutants associated with the future standards are compared to the projected concentrations in a "reference scenario", *i.e.* the ambient air quality that would prevail if the standards were not mandated. The standards are consistent with a 10% reduction in emissions of sulphur oxides, 8% for nitrogen oxides and 50% for particulate matter, all relative to the reference scenario. Using epidemiological information about dose-response relationships, the change in ambient concentration of each pollutants can be linked to various health end-states. Those chosen are short-term (or "acute") effects: reduced mortality, reduced hospital admissions, reduced respiratory symptoms in children (PM only), and restricted activity days for adults. Long term ("chronic") effects include reduced mortality and reduced respiratory illness. Values for these impacts were taken from the literature showing how individuals are willing to pay for reductions in these health end states including "values of statistical life" taken from studies looking at willingness to pay to reduce life risks.

A second category of effects is related to material damage, *e.g.* acidic corrosion of buildings. It is easy to see that reduced corrosion avoids cleaning and repair costs. The study in question did assess these benefits in money terms for sulphur oxides only. Other impacts, notably, ecosystem effects (forests, wetlands, soil etc.), reduced damage to crops, and changes in visibility were not quantified or valued. Hence total benefits will be understated to the extent that these effects are excluded. Significant uncertainties surround some of the dose-response functions and more uncertainty attaches to the valuation estimates, especially those relating to life risk reduction.

The resulting benefits and costs are shown below. They all relate to cities since rural areas were found to comply with the standards without specific action.

Several observations can be made. First, morbidity and material damage reduction are unimportant. Second, acute health effects are significantly less important than chronic health effects, and acute effects may in fact have a negligible value due to the very brief periods of life that are "saved" by reduced episodes of acute pollution. Third, the range of cost values is very wide, by an order of magnitude. The range for benefits is similarly very large, again by an order of magnitude. The explanation for this wide range lies in the dominant effect of mortality reductions on the estimates, and the fact that a wide range of values of reduced mortality riskare used (Euro 0.36 to 10 million per "statistical life"). Fourth, but not shown here, some 90% of the benefit arises from PM reduction. This is explained by the fact that the standards require the biggest reduction in PM (50%) and the fact that PM is implicated in the biggest amount of health damage. Fifth, the results are shown in a somewhat unusual fashion, *i.e.* Comparing annual benefits and annual costs rather than present values. The study omits any mention of a discount rate, but it is extremely unlikely that undiscounted benefits and costs are the same for each year. It is possible that the authors believed they could avoid dealing with the choice of a discount rate, an often controversial feature of CBA. But discounting is still relevant – see Annex 3.A1. This is an unsatisfactory feature of the study. Finally, benefits appear to exceed costs again by an order of magnitude, whether low or high estimates are taken.

Costs per annum		Benefits per annum	
SO$_2$	EUR 4 to 48 million	Short-term mortality	EUR 0 to 8 153 million
NOX	EUR 5 to 285 million	Hospital admissions	EUR 2 to 6 million
PM	EUR 50 to 300 million	Long-term mortality	EUR 5 438 to 58 149 million
		Other morbidity	EUR 2 million
		Materials	EUR 58 million
Total	**EUR 59 to 633 million**	**Total**	**EUR 5 500 to 66 368 million**

Since this study there has been a substantial debate about the validity of applying risk values to chronic mortality in the manner shown, and it seems likely that the consensus now would be that the health benefits in this study are significantly lower. Recall, however, that the study omits several other kinds of benefit.

Source: Olsthoorn et al. 1999.

3.7. ACCOUNTING FOR RISING RELATIVE VALUATIONS

It is not unusual to find discounting, inflation and relative price changes being confused in a CBA. They are three very different things. Discounting arises because of the underlying value judgement in CBA, and taken from welfare economics, that individuals' preferences count. As long as individuals prefer now to later, this value judgement must be applied to time. The discounting of future benefits and costs is thus determined by the rate at which individuals express this "time preference." Inflation, as was noted, is simply a rise in the *general* price level. While it does not matter strictly which year's prices are used in a CBA, it is important to select just one year and to net out all future inflation. Typically, the "base year" is chosen and all costs and benefits are valued at the prices ruling in that year. Suppose this year is Year 1 and that the price level has an index of 100 in that year. Inflation might run at, say, 3% per annum, so that a benefit in year 10 might appear to be $(1.03)^{10} = 1.34$ times higher than the same benefit in year 1. CBA proceeds by dividing the benefit in year 10, valued at year 10 prices, by 1.34 to express it in year 1 prices. The basic rule is simple: net out all general price changes.

A relative price change is different again. What this says is that some benefits and costs attract a higher valuation over time *relative to the general level of prices*. This might be because the benefit or cost in question has a positive income elasticity of willingness to pay, perhaps because it is simply valued more at higher incomes. It can be important to include this rising (or falling) relative valuation in a CBA, and it is especially important for environmental impacts. For example, it may be surmised that, as the overall stock of environmental assets diminishes over time, each unit of the environment will attract a higher "price". This reflects a positive *income elasticity of willingness to pay for the environment*. Annex 3.A1 shows in more detail how this is accounted for.

Pearce (2003a) surveys the evidence on the income elasticity of WTP for environmental improvements. The empirical estimates suggest that the income elasticity of WTP for environmental change is less than unity, and numbers like 0.3–0.7 seem about right.

Chapter 18

COST-BENEFIT ANALYSIS AND OTHER DECISION-MAKING PROCEDURES

CBA is often contrasted with other decision-making aids such as cost-effectiveness analysis (CEA) and multi-criteria analysis (MCA). But the assumption that these aids are substitutable is not valid and great care is needed in defining the question to be asked and in determining which technique is most relevant to helping with the decision. This chapter provides an overview of various techniques. In addition to CEA and MCA it looks at risk assessment, environmental impact assessment, strategic environmental assessment, risk-benefit, risk-risk, and health-health analysis. Each of these approaches reveals insights into features of good decision-making, but CBA tends to have a more comprehensive approach.

18.1. A GALLERY OF PROCEDURES

This volume is concerned with recent developments in cost-benefit analysis (CBA). In Chapter 19 we look at some of the reasons that some decision-makers are distrustful of CBA. This distrust is one reason (and it is important to understand it may not be the dominant reason) that some people look for alternatives to CBA. Other reasons for looking for alternatives include:

- A desire to have decision-aiding procedures that are not so demanding in informational terms.
- A desire to have procedures that can be widely understood and which are not reliant on expert.
- A desire to have "rapid" procedures given that political decisions cannot always wait for the results of a CBA.

Over the years, various techniques of appraisal have emerged in the environmental field. We list these as:

- Environmental Impact Assessment (EIA) or Environmental Assessment (EA).
- Strategic Environmental Assessment (SEA).
- Life Cycle Analysis (LCA).
- Risk Assessment (RA).
- Comparative Risk assessment (CRA).
- Risk-Benefit Analysis (RBA).
- Risk-Risk Analysis (RRA).
- Health-Health Analysis (HHA)
- Cost-Effectiveness Analysis (CEA).
- Multi-Criteria Analysis (MCA).

In the remainder of this chapter we look very briefly at each of these procedures. Space forbids a detailed assessment which can be found, for example, in EFTEC (1998). The idea is simply to "locate" CBA in this range of procedures. It is important to understand that the procedures vary significantly in their comprehensiveness and that it cannot be assumed that each is a substitute for the other.

18.2. ENVIRONMENTAL IMPACT ASSESSMENT (EIA)

EIA is a systematic procedure for collecting information about the environmental impacts of a project or policy, and for measuring those impacts. It will immediately be obvious that EIA is not a comprehensive evaluation procedure. It ignores non-environmental impacts and it ignores costs. Less obviously, it may not account in a detailed way for the ways in which impacts vary with time. Nonetheless, EIA is an essential part of any evaluative procedure. If we use the benchmark of CBA, then EIA is an essential input to CBA. CBA covers the other impacts of projects and policies, and it goes one stage further than EIA by attempting to put money values on the environmental impacts. Most EIAs do make an effort, however, to assess the significance of environmental impacts. Some may go further and give the impacts a score (the extent of the impact) and a weight (its importance). Weights might be derived from public surveys but more usually are determined by the analyst in question. Unlike CBA, EIA has no formal decision rule attached to it (*e.g.* benefits must exceed costs), but analysts would typically argue that its purpose is to look at alternative means of minimising the environmental impacts without altering the benefits of the project or policy.

In general, then:

- EIA is an essential input to any decision-making procedure.
- Impacts may be scored and weighted, or they become inputs into a CBA.
- EIA would generally look for ways to minimise environmental impacts without changing (significantly anyway) the benefits or costs of the project or policy.

18.3. STRATEGIC ENVIRONMENTAL ASSESSMENT (SEA)

SEA is similar to EIA but tends to operate at a "higher" level of decision-making. Instead or single projects or policies, SEA would consider entire programmes of investments or policies.

The goal is to look for the synergies between individual policies and projects and to evaluate alternatives in a more comprehensive manner. An SEA is more likely than an EIA to consider issues like: is the policy or project needed at all; and, if it is, what are the alternative options available? In this sense, SEA is seen to be more pro-active than EIA which tends to be reactive. Proactive here means that more opportunity exists for programmes to be better designed (from an environmental perspective) rather than accepting that a specific option is chosen and the task is to minimise environmental impacts from that option. Again, while it encompasses more issues of concern, SEA remains non-comprehensive as a decision-guiding procedure. Issues of time, cost and non-environmental costs and benefits do not figure prominently. Relative to the benchmark of CBA, SEA goes some way to considering the kinds of issues that would be relevant in a CBA – *e.g.* the "with/without" principle and consideration of alternatives.

18.4. LIFE CYCLE ANALYSIS (LCA)

LCA is similar to EIA in that it identifies the environmental impacts of a policy or project and tries to measure them. It may or may not measure the impacts in the same units, any more than EIA tries to do this. Typically, when attempts are made to adopt the same units they do not include money, although some LCAs have done this. The chief difference between EIA and LCA is that LCA looks not just at the impacts directly arising from a project or policy, but at the whole "life cycle" of impacts. For example, suppose the policy problem is one of choosing between the "best" forms of packaging for a product, say fruit juice. The alternatives might be cartons, bottles and cans. LCA would look at the environmental impacts of each option but going right back to the materials needed for manufacturing of the container (*e.g.* timber and plastics, glass, metals) and the ways in which they will be disposed of once consumers have consumed the juice. Included in the analysis would be the environmental impacts of primary resource extraction and the impacts from landfill, incineration etc. LCAs proceed by establishing an inventory of impacts and then the impacts are subjected to an assessment to establish the extent of impact and the weight to be attached to it. Relative to the benchmark of CBA, LCA is essentially the physical counterpart to the kind of environmental impact analysis that is required by a CBA. In itself LCA offers no obvious decision rule for policies or projects. Though widely advocated as a comprehensive decision-guidance, LCA does not (usually) consider non-environmental costs and benefits. Hence it is not a comprehensive decision-guide. However, if the choice context is one where one of several options has to be chosen (we must have cans or bottles or cartons, but not none of these), then, provided other things are equal, LCA operates like a cost-effectiveness criterion (see below).

18.5. RISK ASSESSMENT (RA)

Risk assessment involves assessing either the health or environmental risks (or both) attached to a product, process, policy or project. Risk assessments may be expressed in various ways:

- As the probability of some defined health or ecosystem effect occurring, *e.g.* a 1 in 100 000 chance of mortality from continued exposure to some chemical.
- As a number of incidences across a defined population, *e.g.* 10 000 premature deaths per annum out of some population.

- As a defined incidence per unit of exposure, *e.g.* X% increase in premature mortality per unit air pollution.
- As a "no effect" level of exposure, *e.g.* below one microgram per cubic metre there may be no health effect.

Risk assessments may not translate into decision rules very easily. One way they may do this is if the actual or estimated risk level is compared to an "acceptable" level which in turn may be the result of some expert judgement or the result of a public survey. A common threshold is to look at unavoidable "everyday" risks and to judge whether people "live with" such a risk. This may make it acceptable. Other procedures tend to be more common and may define the acceptable level as a no-risk level, or even a non-risk level with a sizeable margin or error. Procedures establishing "no effect" levels, *e.g.* of chemicals, define the origin of what the economist would call a "damage function" but cannot inform decision-making unless the goal is in fact to secure that level of risk. Put another way, "no effect" points contain no information about the "damage function".

18.6. COMPARATIVE RISK ASSESSMENT (CRA)

CRA involves analysing risks but for several alternative projects or policies. The issue is then which option should be chosen and the answer offered by CRA is that the option with the lowest risk should be chosen. Efforts are made to "normalise" the analysis so that like is compared to like. For example, one might want to choose between nuclear energy and coal-fired electricity. One approach would be to normalise the risks of one kilowatt hour of electricity and compute, say, deaths per kWh. The option with the lowest "death rate" would then be chosen. However, in this case, the normalisation process does not extend to cost, so that CRA may want to add a further dimension, the money cost of generating one kWh. Once this is done, the focus tends to shift to cost-effectiveness analysis – see below. A further problem concerns the nature of risk. "One fatality" appears to be a homogenous unit, but if people are not indifferent to the manner of death or whether it is voluntarily or involuntarily borne, then, in effect, the normalisation has failed. Once again, one can see that CRA is not a comprehensive decision-guide since the way it treats costs (if at all) may not be all-embracing. Nor would CRA deal with benefits.

18.7. RISK-BENEFIT ANALYSIS (RBA)

RBA tends to take two forms, each of which is reducible to another form of decision rule. In other words, RBA is not a separate procedure. The first meaning relates to benefits, costs and risks, where risks are treated as costs and valued in money terms. In that case the formula for accepting a project or policy would be:

[Benefits – Costs – Risks] > 0

This is no different to a CBA rule.

In the second case the RBA rule reduces to CRA. Benefits might be standardised, *e.g.* to "passenger kilometres" and the risk element might be fatalities. "Fatalities per passenger kilometre" might then be the thing that should be minimised. As with CRA, cost may or may not enter the picture. If it does, then RBA tends to result in CBA or cost-effectiveness analysis.

18.8. RISK-RISK ANALYSIS (RRA)

RRA tends to focus on health risks and asks what would happen to health risks if some policy was adopted and what would happen if it was not adopted. The "with/without" focus is familiar in CBA. The novelty tends to be the fact that not undertaking a policy may itself impose costs in terms of lives or morbidity. For example, a policy of banning or lowering consumption of saccharin might have a justification in reducing health risks from its consumption. But the with-policy option may result in consumers switching to sugar in place of the banned saccharin, thus increasing morbidity by that route. The advantage of RRA is that it forces decision-makers to look at the behavioural responses to regulations. Once again, however, all other components in a CBA equation are ignored, so the procedure is not comprehensive.

18.9. HEALTH-HEALTH ANALYSIS (HHA)

HHA is similar to RRA but instead of comparing the risks with and without the behavioural reaction to a policy, it compares the change in risks from a policy with the risks associated with the *expenditure* on the policy. As such, it offers a subtle focus on policy that is easily overlooked. Since policies costs money, the money has to come from somewhere and, ultimately, the source is the taxpayer. But if taxpayers pay part of their taxes for lifesaving policies, their incomes are reduced. Some of that reduced income would have been spent on lifesaving or health-enhancing activities. Hence the taxation actually increases life risks. HHA compares the anticipated saving in lives from a policy with the lives lost because of the cost of the policy. In principle, policies costing more lives than they save are not desirable. HHA proceeds by estimating the costs of a life-saving policy and the number of lives saved. It then allocates the policy costs to households. Life risks are related to household incomes through regression analysis, so that it is possible to estimate lives lost due to income reductions. Once again, the procedure is not comprehensive: policies could fail an HHA test but pass a CBA test, and *vice versa*.

18.10. COST-EFFECTIVENESS ANALYSIS (CEA)

The easiest way to think about CEA is to assume that there is a single indicator of effectiveness, E, and this is to be compared to a cost of C. Suppose there is now just a single project or policy to be appraised. CEA would require that E be compared to C. The usual procedure is to produce a cost-effectiveness ratio (CER):

$$CER = \frac{E}{C}.$$ [18.1]

Notice that E is in some environmental unit and C is in money units. The fact that they are in different units has an important implication which is, unfortunately, widely disregarded in the literature. A moment's inspection of [18.1] shows that the ratio is perfectly meaningful – *e.g.* it might be read as US Dollars per hectare of land conserved. But the ratio says nothing at all as to whether the conservation policy in question is worth undertaking. In other words, CEA cannot help with the issue of whether or not to undertake any conservation. It should be immediately obvious that this question cannot be answered unless E and C are in the same units.

CEA can only offer guidance on which of several alternative policies (or projects) to select, given that one has to select one. By extension, CEA can *rank* any set of policies, all of which could be undertaken, but given that at least some of them must be undertaken. To see the limitation of CEA, equation [18.1] should be sufficient to show that an entire list of policies, ranked by their cost-effectiveness, could be adopted without any assurance that any one of them is actually worth doing. The notion of "worth doing" only has meaning if one can compare costs and benefits in a manner that enables one to say costs are greater (smaller) than benefits. In turn, that requires that costs and benefits have a common *numeraire* which, in principle, could be anything. In CBA the numeraire is money.

If we suppose that there are $i = 1...n$ potential policies, with corresponding costs Ci and effectiveness Ei then CEA requires that we rank the policies according to

$$CER_i = \frac{E_i}{C_i}$$
[18.2]

This ranking can be used to select as many projects as fit the available budget, *i.e.*:

$$Rank \quad by \quad CER_i \quad s.t. \quad \sum_i C_i = \overline{C}$$
[18.3]

A further issue with CEA is the process of selecting the effectiveness measure. In CBA the principle is that benefits are measured by individuals' preferences as revealed by their willingness to pay for them. The underlying value judgement in CBA is "consumer" or "citizen sovereignty". This amounts to saying that individuals are the best judges of their own well-being. Technically, the same value judgement could be used in CEA, *i.e.* the measure of effectiveness could be based on some attitude survey of a random sample of individuals. In practice, CEA tends to proceed with indicators of effectiveness chosen by experts. Rationales for using expert choices are *a)* that experts are better informed than individuals, especially on issues such as habitat conservation, landscape, etc., and *b)* that securing indicators from experts is quicker and cheaper than eliciting individuals' attitudes.

18.11. MULTI-CRITERIA ANALYSIS (MCA)

MCA is similar in many respects to CEA but involves multiple indicators of effectiveness. Technically, CEA also works with multiple indicators but increasingly resembles simple models of MCA since different effectiveness indicators, measured in different units, have to be normalised by converting them to scores and then aggregated via a weighting procedure. Like CEA, policy or scheme cost in an MCA is always (or should always be) one of the indicators chosen. The steps in an MCA are as follows:

- The goals or objectives of the policy or investment are stated.
- These objectives are not pre-ordained, nor are they singular (as they are in CBA which adopts increases in economic efficiency as the primary objective) and are selected by "decision-makers".
- Generally, decision-makers will be civil servants whose choices can be argued to reflect political concerns.
- MCA then tends to work with experts' preferences. Public preferences may or may not be involved.
- "Criteria" or, sometimes, "attributes" which help achieve the objectives are then selected. Sometimes, objectives and criteria tend to be fused, making the distinction difficult to observe. However, criteria will generally be those features of a good that achieve the objective.
- Such criteria may or may not be measured in monetary terms, but MCA differs from CBA in that not all criteria will be monetised.
- Each option (alternative means of securing the objective) is then given a score and a weight. Pursuing the above example, a policy might score 6 out of 10 for one effect, 2 out of 10 for another effect, and 7 out of 10 for yet another. In turn, experts may regard the first effect as being twice as important as the second but only half as important as the third. The weights would then be 2, 1 and 4 respectively.
- In the simplest of MCAs, the final outcome is a weighted average of the scores, with the option providing the highest weighted score being the one that is "best". More sophisticated techniques might be used for more complex decisions.
- To overcome issues relating to the need for criteria to be independent of each other (*i.e.* experts' preferences based on one criterion should be independent of their preferences for that option based on another criterion), more sophisticated techniques might be used, notably "multi-attribute utility theory" (MAUT). MAUT tends to be over-sophisticated for most practical decision-making.

The formula for the final score for a project or policy using the most simple form of MCA is:

$$S_i = \sum_j m_j . S_j \qquad [18.4]$$

where i is the ith option, j is the jth criterion, m is the weight, and S is the score.

MCA offers a broader interpretation of CEA since it openly countenances the existence of multiple objectives. Issues relating to MCA and which are the subject of debate are as follows:

- As with CEA, when effectiveness is compared to cost in ratio form MCA cannot say anything about whether or not it is worth adopting any project or policy at all (but see Annex

18.A1). Its domain is restricted to choices between alternatives in a portfolio of options some of which must be undertaken. Both MCA and CEA are therefore "efficient" in the sense of seeking to secure maximum effectiveness for a given unit of cost, but may be "inefficient" in the sense of economic efficiency. Annex 18.A1 illustrates the problem further and shows that MCA produces the same result as a CBA only when *a)* the scores on the attributes are the same, *b)* the weights in the MCA correspond to shadow prices in the CBA, and *c)*, which follows from *b)*, the weight on cost is unity.

- MCA generally proceeds by adopting scores and weights chosen by experts. To this extent MCA is not as "accountable" as CBA where the money units reflect individuals' preferences rather than expert preferences. Put another way, the raw material of CBA is a set of individuals' votes, albeit votes weighted by income, whereas experts are unelected and may not be accountable to individual voters.
- MCA tends to be more "transparent" than CBA since objectives and criteria are usually clearly stated, rather than assumed. Because of its adoption of multiple objectives, however, MCA tends to be less transparent than CEA with a single objective.
- It is unclear how far MCA deals with issues of time discounting and changing relative valuations.
- Distributional implications are usually chosen as one of the objectives in an MCA and hence distributional impacts should be clearly accommodated in an MCA.

18.12. SUMMARY AND GUIDANCE FOR DECISION-MAKERS

A significant array of decision-guiding procedures are available. This chapter shows that they vary in the degree of comprehensiveness where this is defined as the extent to which all costs and benefits are incorporated. In general, only MCA is as comprehensive as CBA and may be more comprehensive once goals beyond efficiency and distributional incidence are considered. All the remaining procedures either deliberately narrow the focus on benefits, *e.g.* to health or environment, or ignore cost. Procedures also vary in the way they treat time. EIA and LCA are essential inputs into a CBA, although the way these impacts are dealt with in "physical terms" may not be the same in a CBA. Risk assessments, of which HHA and RRA are also variants, tend to be focused on human health only. The essential message is that the procedures are not substitutes for each other.

ANNEX 18.A1

MULTI-CRITERIA ANALYSIS AND THE "DO NOTHING" OPTION

For the "do nothing" option to be included correctly in an evaluation it is necessary for costs and benefits to be measured in the same units. When MCA adopts the form of cost-effectiveness, with the multiple criteria of effectiveness being compared *in ratio form* to cost, then MCA cannot evaluate the "do nothing" option. This is because the units of effectiveness are weighted scores whilst the measure of cost is money. Numerator and denominator are not in the same units. The "escape" from this problem is for costs to be given a score (usually the absolute level of money cost) and a weight. If we think of the weighted scores as "utils" (or any other unit of account) then MCA can handle the "do nothing" option. If the ratio of benefits to costs is less than unity, the "do something" option is rejected. Similarly, if utils of benefits minus utils of costs is negative, the do something option would also be rejected.

In this way, MCA can be modified to handle the do nothing option. However, it can easily be shown that MCA will give the same result as CBA under very limited conditions.

Table 18.A1.1 shows the procedure adopted in a simple MCA. Let the score for E1 be 10, E2 = 5 and E3 = 30. The scores are multiplied by chosen weights, assumed to be W1 = 4, W2 = 6, W3 = 10. Cost is weighted at unity. The sum of the weighted scores shows that "do something" is a "correct" choice. If the weights W1 … W3 are prices, then Table 18.A1.1 would appear as a CBA, *i.e.* MCA and CBA would produce formally identical results.

Table 18.A1.1 shows that the selection of weights is important. An "unweighted" approach (which means raw scores are weighted at unity) would reject the policy but the weighted approach would accept it. As long as the weights in Table 18.A1.1 correspond to the prices in a CBA, however, then CBA and MCA would generate the same result.

Finally, if we assume shadow prices and MCA weights are the same, but that the weight applied to cost in the MCA is, say, 8, then weighted cost would appear as –400 in Table 18. A1.1 and weighted MCA would reject the do something option.

We can summarise the conditions for CBA and MCA to generate the same result:

a) Attribute scores must be the same.
b) MCA weights must correspond to shadow prices and, in particular.
c) Costs must be weighted at unity.

Robert Costanza, Rudolf de Groot, Paul Sutton, Sander van der Ploeg, Sharolyn J. Anderson, Ida Kubiszewski, Stephen Farber, R. Kerry Turner —"Changes in the Global Value of Ecosystem Services." *Journal of Global Environmental Change.*

Reading Commentary

Ecosystem services, or the benefits people derive from ecosystems, have become a new focus for environmental problem-solving and policy-making. The total value of global ecosystem services in 2011 was $125 trillion. Between 1997 and 2011, Costanza et al. say we lost $2.7 trillion per year in ecosystem services. While these estimates are probably low, they demonstrate the extremely high value that ought to be attached to environmental protection. Of course, there are strong arguments against placing a monetary value on nature. Many environmentalists would argue that nature has what they would call intrinsic value. They worry that efforts to calculate the economic worth of nature will lead to reductions in environmental protection because those with sufficient wealth will just "buy the right" to pollute or destroy natural resources. Others have problems with the methods being used to calculate the value of ecosystem services. How can we definitively know what a mangrove forest in a particular location provides by way of benefits to a nearby coastal community?

The authors counter these criticisms by pointing out that valuation is already underway but hidden from view. In their opinion, the only way to protect nature is to make such valuations transparent. They see their work as a contribution that parallels the calculation of gross domestic product. Since CBA is used all the time to justify policy decisions, they want to be sure that the benefits of ecosystem services are not overlooked. While valuation methods are not perfect, the authors argue that they have been improved over time, and will continue to be improved as we develop more sophisticated valuation techniques.

Do you agree with the authors' assertion that we must improve our ability to measure the value of global ecosystem services?

CHANGES IN THE GLOBAL VALUE OF ECOSYSTEM SERVICES

Robert Costanza, Rudolf de Groot, Paul Sutton
Sander van der Ploeg, Sharolyn J. Anderson, Ida
Kubiszewski, Stephen Farber, R. Kerry Turner

1. INTRODUCTION

Ecosystems provide a range of services that are of fundamental importance to human well-being, health, livelihoods, and survival (Costanza et al., 1997; Millennium Ecosystem Assessment (MEA), 2005; TEEB Foundations, 2010; TEEB Synthesis, 2010). Interest in ecosystem services in both the research and policy communities has grown rapidly (Braat and de Groot, 2012; Costanza and Kubiszewski, 2012). In 1997, the value of global ecosystem services was estimated to be around US$ 33 trillion peryear (in 1995 $US), a figure significantly larger than global gross domestic product (GDP) at the time. This admittedly crude underestimate of the welfare benefits of natural capital, and a few other early studies (Daily, 1997; de Groot, 1987; Ehrlich and Ehrlich, 1981; Ehrlich and Mooney, 1983; Odum, 1971; Westman, 1977) stimulated a huge surge in interest in this topic.

In 2005, the concept of ecosystem services gained broader attention when the United Nations published its Millennium Ecosystem Assessment (MEA). The MEA was a four-year, 1300-scientist study for policymakers. Between 2007 and 2010, a second international initiative was undertaken by the UN Environment Programme, called the Economics of Ecosystems and Biodiversity (TEEB) (TEEB Foundations, 2010). The TEEB report was picked up extensively by the mass media, bringing ecosystem services to a broader audience. Ecosystem services have now also entered the consciousness of mainstream media and business. The World Business Council for Sustainable Development has actively supported and developed the concept (WBCSD, 2011, 2012). Hundreds of projects and groups are currently working toward R. Costanza et al./Global Environmental Change 26 (2014) better

Table 18.A1.1 Weighted input data for an MCA: cost weighted at unity

	Do something: raw scores	Do something: weighted scores
Cost	−50	−50
E1	+10	+40
E2	+5	+30
E3	+30	+300
Sum of (weighted) scores	−5	+320

understanding, modeling, valuation, and management of ecosystem services and natural capital. It would be impossible to list all of them here, but emerging regional, national, and global networks, like the Ecosystem Services Partnership (ESP), are doing just that and are coordinating their efforts (Braat and de Groot, 2012; de Groot et al., 2011).

Probably the most important contribution of the widespread recognition of ecosystem services is that it reframes the relationship between humans and the rest of nature. A better understanding of the role of ecosystem services emphasizes our natural assets as critical components of inclusive wealth, well-being, and sustainability. Sustaining and enhancing human well-being requires a balance of all of our assets—individual people, society, the built economy, and ecosystems. This reframing of the way we look at "nature" is essential to solving the problem of how to build a sustainable and desirable future for humanity.

Estimating the relative magnitude of the contributions of ecosystem services has been an important part of changing this framing. There has been an on-going debate about what some see as the "commodification" of nature that this approach supposedly implies (Costanza, 2006; McCauley, 2006) and what others see as the flawed methods and questionable wisdom of aggregating ecosystem services values to larger scales (Chaisson, 2002). We think that these critiques are largely misplaced once one understands the context and multiple potential uses of ecosystem services valuation, as we explain further on.

In this paper we (1) update estimates of the value of global ecosystem services based on new data from the TEEB study (de Groot et al. 2012, 2010a,b); (2) compare those results with earlier estimates (Costanza et al,, 1997) and with alternative methods (Boumans et al., 2002); (3) estimate the global changes in ecosystem service values from land use change over the period 1997–2011; and (4) review some of the objections to aggregate ecosystem services value estimates and provide some responses (Howarth and Farber, 2002).

We do not claim that these estimates are the only, or even the best way, to understand the value of ecosystem services. Quite the contrary, we advocate pluralism based on a broad range of approaches at multiple scales. However, within this range of approaches, estimates of aggregate accounting value for ecosystem services in monetary units have a critical role to play in heightening awareness and estimating the overall level of importance of ecosystem services relative to and in combination with other contributors to sustainable human well-being (Luisetti et al., 2013).

2. WHAT IS VALUATION?

Valuation is about assessing trade-offs toward achieving a goal (Farber et al., 2002). All decisions that involve trade-offs involve valuation, either implicitly or explicitly (Costanza et al., 2011). When assessing trade-offs, one must be clear about the goal. Ecosystem services are defined as the benefits people derive from ecosystems – the support of sustainable human well-being that ecosystems provide (Costanza et al., 1997; Millennium Ecosystem Assessment (MEA), 2005). The value of ecosystem services is therefore the *relative* contribution of ecosystems to that goal. There are multiple ways to assess this contribution, some of which are based on individual's perceptions of the benefits they derive. But the support of sustainable human well-being is a much larger goal (Costanza, 2000) and individual's perceptions are limited and often biased (Kahneman, 2011). Therefore, we also need to include methods to assess benefits

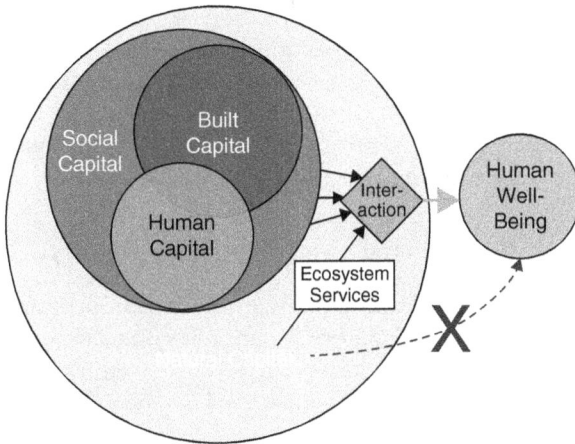

Figure 1 Interaction between built, social, human and natural capital required to produce human well-being. Built and human capital (the economy) are embedded in society which is embedded in the rest of nature. Ecosystem services are the relative contribution of natural capital to human well-being, they do not flow directly. It is therefore essential to adopt a broad, transdisciplinary perspective in order to address ecosystem services.

to individuals that are not well perceived, benefits to whole communities, and benefits to sustainability (Costanza, 2000). This is an on-going challenge in ecosystem services valuation, but even some of the existing valuation methods like avoided and replacement cost estimates are not dependent on individual perceptions of value. For example, estimating the storm protection value of coastal wetlands requires information on historical damage, storm tracks and probability, wetland area and location, built infrastructure location, population distribution, etc. (Costanza et al., 2008). It would be unrealistic to think that the general public understands this complex connection, so one must bring in much additional information not connected with perceptions to arrive at an estimate of the value. Of course, there is ultimately the link to built infrastructure, which people perceive as a benefit and value, but the link is complex and not dependent on the general public's understanding of or perception of the link.

It is also important to note that ecosystems cannot provide any benefits to people without the presence of people (human capital), their communities (social capital), and their built environment (built capital). This interaction is shown in Fig. 1. Ecosystem services do not flow directly from natural capital to human well-being – it is only through interaction with the other three forms of capital that natural capital can provide benefits. This is also the conceptual valuation framework for the recent UK National Ecosystem Assessment (http://uknea.unep-wcmc.org) and the Intergovernmental Platform on Biodiversity and Ecosystem Services (IPBES – http://www.ipbes.net). The challenge in ecosystem services valuation is to assess the relative contribution of the natural capital stock in this interaction and to balance our assets to enhance sustainable human well-being.

The relative contribution of ecosystem services can be expressed in multiple units – in essence any of the contributors to the production of benefits can be used as the "denominator" and other contributors expressed in terms of it. Since built capital in the economy, expressed in monetary units, is one of the required contributors, and most people understand values expressed in monetary units, this is often a convenient denominator for expressing the relative contributions of the other forms of capital, including natural capital. But other units are certainly possible (i.e. land, energy, time, etc.) – the choice is largely about which units communicate best to different audiences in a given decisionmaking context.

3. VALUATION IS NOT PRIVATIZATION

It is a misconception to assume that valuing ecosystem services in monetary units is the same as privatizing them or commodifying them for trade in private markets (Costanza, 2006; Costanza et al., 2012; McCauley, 2006; Monbiot, 2012). Most ecosystem services are public goods (non-rival and non-excludable) or common pool resources (rival but non-excludable), which means that privatization and conventional markets work poorly, if at all. In addition, the non-market values estimated for these ecosystem services often relate more to *use* or *non-use* values rather than *exchange* values (Daly, 1998). Nevertheless, knowing the value of ecosystem services is helpful for their effective management, which in some cases can include economic incentives, such as those used in successful systems of payment for these services (Farley and Costanza, 2010). In addition, it is important to note that valuation is unavoidable. We already value ecosystems and their services every time we make a decision involving trade-offs concerning them. The problem is that the valuation is implicit in the decision and hidden from view. Improved transparency about the valuation of ecosystem services (while recognizing the uncertainties and limitations) can only help to make better decisions.

It is also incorrect to suggest (McCauley, 2006) that conservation based on protecting ecosystem services is betting against human ingenuity. Recognizing and measuring natural capital and ecosystem services in terms of stocks and flows is a prime example of enlightened human ingenuity. The study of ecosystem services has merely identified the limitations and costs of 'hard' engineering solutions to problems that in many cases can be more efficiently solved by natural systems. Pointing out that the 'horizontal levees' of coastal marshes are more cost-effective protectors against hurricanes than constructed vertical levees (Costanza et al., 2008) and that they also store carbon that would otherwise be emitted into the atmosphere (Luisetti et al., 2011) implies that restoring or recreating them for this and other benefits is only using our intelligence and ingenuity, not betting against it.

The ecosystem services concept makes it abundantly clear that the choice of "the environment versus the economy" is a false choice. If nature contributes significantly to human well-being, then it is a major contributor to the *real* economy (Costanza et al., 1997), and the choice becomes how to manage all our assets, including natural and human-made capital, more effectively and sustainably (Costanza et al., 2000).

4. USES OF VALUATION OF ECOSYSTEM SERVICES

The valuation of ecosystem services can have many potential uses, at multiple time and space scales. Confusion can arise, however, if one is not clear about the distinctions between

Table 1 Range of uses for ecosystem service valuation

Use of valuation	Appropriate values	Appropriate spatial scales	Precision needed
Raising awareness and interest	Total values, macro aggregates	Regional to global	Low
National income and well-being accounts	Total values by sector and macro aggregates	National	Medium
Specific policy analyses	Changes by policy	Multiple depending on policy	Medium to high
Urban and regional land use planning	Changes by land use scenario	Regional	Low to medium
Payment for ecosystem services	Changes by actions due payment	Multiple depending on system	Medium to high
Full cost accounting	Total values by business, product, or activity and changes by business, product, or activity	Regional to global, given the scale of international corporations	Medium to high
Common asset trusts	Totals to assess capital and changes to assess income and loss	Regional to global	Medium

these uses. Table 1 lists some of the potential uses of ecosystem services valuation, ranging from simply raising awareness to detailed analysis of various policy choices and scenarios. For example, Costanza et al. (1997) was clearly an awareness raising exercise with no specific policy or decision in mind. As its citation history verifies, it was very successful for this purpose. It also pointed out that ecosystem service values could be useful for several of the other purposes listed in Table 1, and it stimulated subsequent research and application in these areas. There have been thousands of subsequent studies addressing the full range of uses listed in Table 1.

5. AGGREGATING VALUES

Ecosystem services are often assessed and valued at specific sites for specific services. However some uses require aggregate values over larger spatial and temporal scales (Table 1). Producing such aggregates suffers from many of the same problems as producing any aggregate estimate, including macroeconomic aggregates such as GDP. Table 2 lists a range of possible approaches for aggregating ecosystem service values (Kubiszewski et al., 2013a).

Basic benefit transfer, the technique used in Costanza et al. (1997) assumes a constant unit value per hectare of ecosystem type and multiplies that value by the area of each type to arrive at aggregate totals. This can be improved somewhat by adjusting values using expert opinion of local conditions (Batker et al., 2008). Benefit transfer is analogous to the approach taken in GDP accounting, which aggregates value by multiplying price times quantity for each sector of the economy. Our aggregate is an accounting measure of the quantity of ecosystem services (Howarth and Farber, 2002). In this accounting dimension the measure is based on virtual non-market prices and incomes, not real prices and incomes. We return to this point later when we examine some of the criticisms of the original 1997 study.

While simple and easy, this approach obviously glosses over many of the complexities involved. This degree of approximation is appropriate for some uses (Table 1) but ultimately a more spatially explicit and dynamic approach would be preferable or essential for some other uses. These approaches are beginning to be implemented (Bateman et al., 2013; Boumans et al., 2002; Burkhard et al., 2013; Costanza et al., 2008; Costanza and Voinov, 2003; Crossman et al., 2012; Goldstein et al., 2012; Nelson et al., 2009) and this represents the cutting edge of research in this field.

Regional aggregates are useful for assessing land use change scenarios. National aggregates are useful for revising national income accounts. Global aggregates are useful for raising awareness and emphasizing the importance of ecosystem services relative to other contributors to human well-being. In this paper, we provide some updated global estimates, recognizing that this is only one among many potential uses for ecosystem services valuation, and that this use has special requirements, limitations, and interpretations.

6. ESTIMATES OF GLOBAL VALUE

Costanza et al. (1997) estimated the value of 17 ecosystem services for 16 biomes and an aggregate global value expressed in monetary units. This estimate was based on a simple benefit transfer method described above.

Notwithstanding the limitations and restrictions in benefit transfer techniques (Brouwer, 2000; Defra, 2010; Johnston and Rosenberger, 2010) it is an attractive option for researchers and policy-makers facing time and budget constraints. Value transfer has been used for valuation of environmental resources in many instances. Nelson and Kennedy (2009) provide a critical overview of 140 meta-analyses.

de Groot et al. (2012) estimated the value of ecosystem services in monetary units provided by 10 main biomes (Open oceans, Coral reefs, Coastal systems, Coastal wetlands, Inland wetlands, Lakes, Tropical forests, Temperate forests, Woodlands, and Grasslands) based on local case studies across the world. These studies covered a large number of ecosystems, types of landscapes, different definitions of services, different areas, different levels of scale, time and complexity and different valuation methods. In total, approximately 320 publications were screened and more than 1350 data-points from over 300 case study locations were stored in the Ecosystem Services Value Database (ESVD) (http://www.fsd.nl/esp/80763/5/0/50). A selection of 665 of these value data points were used for the analysis. Values were expressed in terms of 2007 'International' $/ha/year, i.e. translated into US$ values on the basis of Purchasing Power Parity (PPP) and contains site-, study-, and context-specific information from the case studies. We added some additional estimates for this paper, notably for urban and cropland systems (see Supporting Material for details).

A detailed description of the ESVD is given in van der Ploeg et al. (2010). de Groot et al. (2012) provides details of the results. Below, we provide a comparison of the de Groot et al. (2012) results with the Costanza et al. (1997) results in order to estimate the changes in the flow of ecosystem services over this time period.

After some consolidation of the typologies used in the two studies we can compare the de Groot et al. (2012) estimates per service and per biome with the Costanza et al. (1997) estimates in Table 3, and in more detail in Supporting Material, Table SI. Table SI lists the mean value for each service and biome for both 1997 and 2011. Table 4 is a summary of the number of estimates, mean, standard deviation, median, and minimum and maximum values used in de Groot et al. (2012). All values are in international $/ha/yr and were derived from the ESV database. Note that there is a wide range of the number of studies for each biome, ranging from 14 for open ocean to 168 for inland wetlands. This is a significantly larger number of studies than were available for the Costanza et al. study (less than 100). One can also note the wide variation and high standard deviation for several of the biomes. For example, values for coral reefs varied from a low of 36,794 $/ha/yr to a high of 2,129,122 $/ha/yr. Given a sufficient number of studies, some of this variation can be explained by other variables. For example, De Groot et al. performed a meta-regression analysis for inland wetlands using 16 independent variables in a model with an adjusted R^2 of 0.442. Variables that were significant in explaining the value of inland wetlands included the area of the study site, the type of inland wetland, GDP/capita, and population of the country in which the wetland occurred, the proximity of other wetlands,

Table 2 Four levels of ecosystem service value aggregation (Kubiszewski et al., 2013a,b)

Aggregation method	Assumptions/ approach	Examples
1. Basic value transfer	Assumes values constant over ecosystem types	Costanza et al. (1997), wLiu et al. (2010a,b)
2. Expert modified value transfer	Adjusts values for local ecosystem conditions using expert opinion surveys	Batker et al. (2008)
3. Statistical value transfer	Builds statistical model of spatial and other dependencies	de Groot et al. (2012)
4. Spatially explicit functional modeling	Builds spatially explicit statistical or dynamic systems models incorporating valuation	Boumans et al. (2002), Costanza et al. (2008), Nelson et al. (2009)

Table 3 Changes in area, unit values and aggregate global flow values from 1997 to 2011 (green are values that have increased, red are values that have decreased)

Biome	Area (e6 ha)			Unit values 2007$/ha/yr			Aggregate Global Flow Value el 2 2007$/yr				Change in Value el 2 2007$/yr	
							A. Original	B. Change unit values only	C. Change area only	D. Change both unit values and area	2011–1997	
							Assuming 1997 area and 1997 unit values	Assuming 1997 area and 2011 unit values	Assuming 2011 area and 1997 unit values	Assuming 2011 area and 2011 unit values	E. Column C–Column A	F. Column D–Column B
	1997	2011	Change 2011–1997	1997	2011	Change 2011–1997	1997	2011	2011	2011	A	B
Marine	36,302	36,302	0	796	1,368	572	28.9	60.5	29.5	49.7	0.6	(10.9)
Open Ocean	33,200	33,200	0	348	660	312	11.6	21.9	11.6	21.9	-	-
Coastal	3,102	3,102	0	5,592	8,944	3,352	17.3	38.6	18.0	27.7	0.6	(10.9)
Estuaries	180	180	0	31,509	28,916	-2,593	5.7	5.2	5.7	5.2	-	-
Seagrass/Algae Beds	200	234	34	26,226	28,916	2,690	5.2	5.8	6.1	6.8	0.9	1.0
Coral Reefs	62	28	-34	8,384	352,249	343,865	0.5	21.7	0.2	9.9	(0.3)	(11.9)
Shelf	2,660	2,660	0	2,222	2,222	0	5.9	5.9	5.9	5.9	:	:

Terrestrial	15,323	15,323	0	1,109	4,901	3,792	17.0	84.5	12.1	75.1	(4.9)	(9.4)
Forest	4,855	4,261	-594	1,338	3,800	2,462	6.5	19.5	4.7	16.2	(1.8)	(3.3)
Tropical	1,900	1,258	-642	2,769	5,382	2,613	5.3	10.2	3.5	6.8	(1.8)	(3.5)
Temperate/Boreal	2,955	3,003	48	417	3,137	2,720	1.2	9.3	1.3	9.4	0.0	0.2
Grass/Rangelands	3,898	4,418	520	321	4,166	3,845	1.2	16.2	1.4	18.4	0.2	2.2
Wetlands	330	188	-142	20,404	140,174	119,770	6.7	36.2	3.4	26.4	(3.3)	(9.9)
Tidal Marsh/Mangroves	165	128	-37	13,786	193,843	180,057	2.3	32.0	1.8	24.8	(0.5)	(7.2)
Swamps/Floodplains	165	60	-105	27,021	25,681	-1,340	4.5	4.2	1.6	1.5	(2.8)	(2.7)
Lakes/Rivers	200	200	0	11,727	12,512	785	2.3	2.5	2.3	2.5	-	-
Desert	1,925	2,159	234	-	-	0w		-	-	-	-	-
Tundra	743	433	-310	-	-	0		-	-	-	-	-
Ice/Rock	1,640	1,640	0	-	-	0		-	-	-	-	-
Cropland	1,400	1,672	272	126	5,567	5,441	0.2	7.8	0.2	9.3	0.0	1.5
Urban	332	352	20	-	6,661	6,661		2.2	-	2.3	-	0.1
Total	51,625	51,625	0				45.9	145.0	41.6	124.8	(4.3)	(20.2)

Table 4 Summary of the number of estimates, mean, standard deviation, median, minimum and maximum values used in de Groot et al. (2012). Values are in international $/ha/yr, derived from the ESV database

	No. of estimates	Total of service means (TEV)	Total of St. Dev. of means	Total of median values	Total of minimum values	Total of maximum values
Open oceans	14	491	762	135	85	1664
Coral reefs	94	352,915	668,639	197,900	36,794	2129,122
Coastal systems	28	28,917	5045	26,760	26,167	42,063
Coastal wetlands	139	193,845	384,192	12,163	300	887,828
Inland wetlands	168	25,682	36,585	16,534	3018	104,924
Rivers and lakes	15	4267	2771	3938	1446	7757
Tropical forest	96	5264	6526	2355	1581	20,851
Temperate forest	58	3013	5437	1127	278	16,406
Woodlands	21	1588	317	1522	1373	2188
Grasslands	32	2871	3860	2698	124	5930

and the valuation method used for the study. If this number of studies were available for the other biomes in our global assessment, we could use this type of meta-regression to produce more accurate estimates. However, for the current estimate, we must continue to rely on global averages.

Global averages per ha may vary between the two time periods we are comparing for three distinct reasons: (1) new (and generally more numerous and complete) estimates of the unit values of ecosystem services per ha; (2) changes in the average functionality of ecosystem per ha; and (3) changes in value per ha due to changes in human, social, or built capital. The actual estimates conflate these causes and we see no way of disentangling them at this point. However, since global population only increased by 16% between 1997 and 2011 (from 5.83 to 7 billion), and, if anything, ecosystems are becoming more stressed and less functional, we can attribute most of the increase in unit values to more comprehensive, value estimates available in 2011 than in 1997.

Table 3 shows that values per ha estimated by de Groot et al. (2012) are an average of 8 times higher than the equivalent estimates from Costanza et al. (1997) (both converted into $2007). Only inland wetlands and estuaries did not show a significant increase in estimated value per ha, but these were among the best studied biomes in 1997. Some biomes showed significant increases in value. For example, tidal marsh/mangroves increased from abound 14,000 to around 194,000 $/ha/yr. This is largely due to new studies of the storm protection, erosion control, and waste treatment values of these systems. Coral reefs also increased tremendously in estimated value from around 8000 to around 352,000 $/ha/yr due to additional studies of storm protection, erosion protection, and recreation. Cropland and urban system also increased dramatically, largely because there were almost no studies of these systems in 1997 and there have subsequently been several new studies (Wratten et al., 2013).

Table 3 also shows the aggregate global annual value of services, estimated by multiplying the land area of each biome by the unit values. Column A uses the original values from Costanza et al. (1997) converted to 2007 dollars (total = $45.9 trillion/yr). If we assume that land areas did not change between the two time periods, the new estimate, shown in column B is $145 trillion/yr, are more than 3 times larger than the original estimate. This is due solely to updated unit values. However, land use *has* changed significantly between the two years, changing the supply (the flow) of ecosystem services. If we use the new land use estimates shown in Table 3 (see Supporting Material for details) and the 1997 unit values, we get the estimates in column C – a total of $41.6 trillion/ yr. Column E is the change in value due to land use change using the 1997 unit values. Marine systems show a slight increase in value, while terrestrial systems show a large decrease. This decrease is largely due to decreases in the area of high value per ha biomes (tropical forests, wetlands, and coral reefs – shown in red in Table 3) and increases in low value per ha biomes. The total net decrease is estimated to be $4.3 trillion/yr. It is almost certain that the functionality of ecosystems per ha has also declined in many cases so the supply effects are surely greater than this. Column D shows the combined effects of both changes in land areas and updated unit values. The net effect yields an estimate of $124.8 trillion/yr – 2.7 times the original estimate. For comparison, global GDP was approximately 46.3 trillion/yr in 1997 and $75.2 trillion/yr in 2011 (in $2007).

The difference between columns D and B is the estimated loss of ecosystem services based on land use changes and using the 2011 unit value estimates. This is shown in column F. In this case marine systems show a large loss (\$10.9 trillion/yr), due mainly to a decrease in coral reef area and the substantially larger unit value for coral reef using the 2011 unit values. Terrestrial systems also show a large loss, dominated by tropical forests and wetlands, but countered by small increases in the value of grasslands, cropland, and urban systems. Overall, the total net decrease is estimated to be \$20.2 trillion in annual services since 1997. Given the more comprehensive unit values employed in the 2011 estimates, this is a better approximation than using the 1997 unit values, but certainly still a conservative estimate. The present value of the discounted flow of ecosystem services consumed would represent part of the stock of inclusive wealth lost/gained over time (UNU-IHDP, 2012).

As we have previously noted, basic value transfer is a crude first approximation at best. We could put ranges on these numbers based on the standard deviations shown in Table 4, but there are other sources of error and caveats as well, as described in Costanza et al. including errors in estimating land use changes. However, we think that solving these problems will most likely lead to even larger estimates. For example, one problem is the limited number of valuation studies available and we expected that as more studies became available from 1997 to 2011 the unit value estimates would increase, and they did.

We also anticipate that more sophisticated techniques for estimating value will lead to larger estimates. For example, more sophisticated integrated dynamic and spatially explicit modeling techniques have been developed and applied at regional scales (Barbier, 2007; Bateman et al., 2013; Bateman and Jones, 2003; Costanza and Voinov, 2003; Goldstein et al., 2012; Nelson et al., 2009). However, few have been applied at the global scale. One example is the Global Unified Metamodel of the Biosphere (GUMBO) that was developed specifically to simulate the integrated earth system and assess the dynamics and values of ecosystem services (Boumans et al., 2002). GUMBO is a 'metamoder in that it represents a synthesis and simplification of several existing dynamic global models in both the natural and social sciences at an intermediate level of complexity. It includes dynamic feedbacks among human technology, economic production, human welfare, and ecosystem goods and services within and across 11 biomes. The dynamics of eleven major ecosystem goods and services for each of the biomes have been simulated and evaluated. A range of future scenarios representing different assumptions about future technological change, investment strategies and other factors, have been simulated. The relative value of ecosystem services in terms of their contribution to supporting both conventional economic production and human well-being more broadly defined were estimated under each scenario. The value of global ecosystem services was estimated to be about 4.5 times the value of Gross World Product (GWP) in the year 2000 using this approach. For a current global GDP of \$75 trillion/yr this would be about \$347 trillion/yr, or almost three times the column D estimate in Table 3. This is to be expected since the dynamic simulation can include a more comprehensive picture of the complex interdependencies involved. It is also important to note that this type of model is the only way to potentially assess more than marginal changes in ecosystem services, including irreversible thresholds and tipping points (Rockstrom et al., 2009; Turner et al., 2003).

7. CAVEATS AND MISCONCEPTIONS

We want to make clear that expressing the value of ecosystem services in monetary units does not mean that they should be treated as private commodities that can be traded in private markets. Many ecosystem services are public goods or the product of common assets that cannot (or should not) be privatized (Wood, 2014). Even if fish and other provisioning services enter the market as private goods, the ecosystems that produce them (i.e. coastal systems and oceans) are common assets. Their value in monetary units is an estimate of their benefits to society expressed in units that communicate with a broad audience. This can help to raise awareness of the importance of ecosystem services to society and serve as a powerful and essential communication tool to inform better, more balanced decisions regarding trade-offs with policies that enhance GDP but damage ecosystem services.

Some have argued that estimating the global value of ecosystem services is meaningless, because if we lost all ecosystem services human life would end, so their value must be infinite (Chaisson, 2002). While this is certainly true, as was clearly pointed out in the 1997 paper (Costanza et al., 1997), it is a simple misinterpretation of what our estimate refers to. Our estimate is more analogous to estimating the total value of agriculture in national income accounting. Whatever the fraction of GDP that agriculture contributes now, it is clear that if all agriculture were to stop, economies would collapse to near zero. What the estimates are referring to, in both cases, is the *relative* contribution, expressed in monetary units, of the assets or activities at the current point in time. Referring to Fig. 1, human well-being comes from the interaction of the four basic types of capital shown. GDP picks up only a fraction of this total contribution (Costanza et al., 2014; Kubiszewski et al., 2013b). What we have estimated is the relative contribution of natural capital now, with the current balance of asset types. Some of this contribution is already included in GDP, embedded in the contribution of natural capital to marketed goods and services. But much of it is not captured in GDP because it is embedded in services that are not marketed or not fully captured in marketed products and services. Our estimate shows that these services (i.e. storm protection, climate regulation, etc.) are much larger in relative magnitude right now than the sum of marketed goods and services (GDP). Some have argued that this result is impossible, wrongly assuming that all of our value estimates are based on willingness-to-pay and that that cannot exceed aggregate ability-to-pay (i.e. GDP). But for it to be impossible, one would have to argue that *all* human benefits are marketed and captured in GDP. This is obviously not the case. Another example is the many other types of goods and services traded on "black markets" that in some countries far exceed GDP. Moreover, our estimate is an accounting measure based on virtual not real prices and incomes and it is these virtual total expenditures that should not be exceeded (Costanza et al., 1998; Howarth and Farber, 2002). It is also important for policy to evaluate gains/losses in stocks and consequent service flows (analogous to net GDP). The discounted present value of such stock/flow changes is a measure of a component of inclusive wealth or wellbeing.

8. CONCLUSIONS

The concepts of ecosystem services flows and natural capital stocks are increasingly useful ways to highlight, measure, and value the degree of interdependence between humans and the rest of nature. This approach is complementary with other approaches to nature conservation, but provides conceptual and empirical tools that the others lack and it communicates with different audiences for different purposes. Estimates of the global accounting value of ecosystem services expressed in monetary units, like those in this paper, are mainly useful to raise awareness about the magnitude of these services relative to other services provided by human-built capital at the current point in time. Our estimates show that global land use changes between 1997 and 2011 have resulted in a loss of ecosystem services of between $4.3 and $20.2 trillion/yr, and we believe that these estimates are conservative. One should not underestimate the importance of the change in awareness and worldview that these global estimates can facilitate – it is a necessary precursor to practical application of the concept using changes in the flows of services for decisionmaking at multiple scales. It allows us to build a more comprehensive and balanced picture of the assets that support human well-being and human's interdependence with the wellbeing of all life on the planet.

REFERENCES

Barbier, E.B., 2007. Valuing ecosystem services as productive inputs. Econ. Policy 22, 177–229.

Bateman, I.J., Harwood, A.R., Mace, G.M., Watson, R.T., Abson, D.J., Andrews, B., Binner, A., Crowe, A., Day, B.H., Dugdale, S., Fezzi, C., Foden, J., Hadley, D., Haines-Young, R., Hulme, M., Kontoleon, A., Lovett, A.A., Munday, P., Pascual, U., Paterson, J., Perino, G., Sen, A., Siriwardena, G., van Soest, D., Termansen, M., 2013. Bringing ecosystem services into economic decision-making: land use in the United Kingdom. Science 341, 45–50.

Bateman, I.J., Jones, A.P., 2003. Contrasting conventional with multi-level modelling approaches to meta-analysis: expectation consistency in UK woodland recreation values. Land Econ. 79, 235–258.

Batker, D., Swedeen, P., Costanza, R., de la Torre, I., Boumans, R., Bagstad, K., 2008. A New View of the Puget Sound Economy: The Economic Value of Nature's Services in the Puget Sound Basin. Earth Economics, Tacoma, WA.

Boumans, R., Costanza, R., Farley, J., Wilson, M.A., Portela, R., Rotmans, J., Villa, F., Grasso, M., 2002. Modeling the dynamics of the integrated earth system and the value of global ecosystem services using the GUMBO model. Ecol. Econ. 41, 529–560.

Braat, L., de Groot, R., 2012. The ecosystem services agenda: bridging the worlds of natural science and economics, conservation and development, and public and private policy. Ecosyst. Serv. 1, 4–15.

Brouwer, R., 2000. Environmental value transfer: state of the art and future prospects. Ecol. Econ. 32, 137–152.

Burkhard, B., Crossman, N., Nedkov, S., Petz, K., Alkemade, R., 2013. Mapping and modelling ecosystem services for science, policy and practice. Ecosyst. Serv. 4, 1–146.

Chaisson, E.J., 2002. Cosmic Evolution: The Rise of Complexity in Nature. Harvard University Press, Cambridge, MA.

Costanza, R., 2000. Social goals and the valuation of ecosystem services. Ecosystems 3, 4–10.

Costanza, R., 2006. Nature: ecosystems without commodifying them. Nature 443, 749.

Costanza, R., d'Arge, R., de Groot, R., Farber, S., Grasso, M., Hannon, B., Limburg, K., Naeem, S., O'Neill, R.V., Paruelo, J., Raskin, R.G., Sutton, P., van den Belt, M., 1998. The value of ecosystem services: putting the issues in perspective. Ecol. Econ. 25, 67–72.

Costanza, R., Daly, M., Folke, C., Hawken, P., Holling, C.S., McMichael, A.J., Pimentel, D., Rapport, D., 2000. Managing our environmental portfolio. Bioscience 50, 149–155.

Costanza, R., dArge, R., de Groot, R., Farber, S., Grasso, M., Hannon, B., Limburg, K., Naeem, S., Oneill, R.V., Paruelo, J., Raskin, R.G., Sutton, P., van den Belt, M., 1997. The value of the world's ecosystem services and natural capital. Nature 387, 253–260.

Costanza, R., Kubiszewski, I., 2012. The authorship structure of ecosystem services as a transdisciplinary field of scholarship. Ecosyst. Serv. 1, 16–25.

Costanza, R., Kubiszewski, I., Ervin, D., Bluffstone, R., Boyd, J., Brown, D., Chang, H., Dujon, V., Granek, E., Polasky, S., Shandas, V., Yeakley, A., 2011. Valuing ecological systems and services. FI 000 Biol. Rep. 3, 14.

Costanza, R., Kubiszewski, I., Giovannini, E., Lovins, H., McGlade, J., Pickett, K.E., Ragnarsdottir, K.V., Roberts, D., De Vogli, R., Wilkinson, R., 2014. Time to leave GDP behind. Nature 505, 283–285.

Costanza, R., Pérez-Maqueo, O., Martinez, M.L., Sutton, P., Anderson, S.J., Mulder, K., 2008. The value of coastal wetlands for hurricane protection. AMBIO: J. Hum. Environ. 37, 241–248.

Costanza, R., Quatrini, S., Øystese, S., 2012. Response to George Monbiot: The Valuation of Nature and Ecosystem Services is Not Privatization. Responding to Climate Change (RTCC).

Costanza, R., Voinov, A., 2003. Landscape Simulation Modeling: A Spatially Explicit, Dynamic Approach. Springer, New York.

Crossman, N., Burkhard, B., Nedkov, S., 2012. Quantifying and mapping ecosystem services. J. Biodivers. Sci. Ecosyst. Serv. Manage. 8, 1–185.

Daily, G.C., 1997. Nature's Services: Societal Dependence on Natural Ecosystems. Island Press, Washington, DC.

Daly, H.E., 1998. The return of Lauderdale's paradox. Ecol. Econ. 25, 21–23.

de Groot, R., 1987. Environmental functions as a unifying concept for ecology and economics. Environmentalist Summer 7, 105–109.

de Groot, R., Brander, L., van der Ploeg, S., Costanza, R., Bernard, F., Braat, L., Christie, M., Crossman, N., Ghermandi, A., Hein, L., Hussain, S., Kumar, P., McVittie, A., Portela, R., Rodriguez, L.C., ten Brink, P., van Beukering, P., 2012. Global estimates of the value of ecosystems and their services in monetary units. Ecosyst. Serv. 1, 50–61.

de Groot, R., Costanza, R., Broeck, D.V.D., Aronson, J., Burkhard, B., Gomez-Bag – gethun, E., Haines-Young, R., Kubiszewski, I., Muller, F., Petrosillo, I., Potschin, M., Ploeg, S.V.D., Zurlini, G., 2011. A global partnership for ecosystem services. Solutions 2, 42–43.

de Groot, R.S., Fisher, B., Christie, M., Aronson, J., Braat, L., Haines-Young, R., Gowdy, J., Maltby, E., Neuville, A., Polasky, S., Portela, R., Ring, I., 2010. Integrating the ecological and economic dimensions in biodiversity and ecosystem service valuation. In: Kumar, P. (Ed.), The Economics of Ecosystems and Biodiversity: Ecological and Economic Foundations. Earthscan, London.

de Groot, R.S., Kumar, P., van der Ploeg, S., Sukhdev, P., 2010. Estimates of monetary values of ecosystem services. In: Kumar, P. (Ed.), The Economics of Ecosystems and Biodiversity: Ecological and Economic Foundations. Earthscan, London.

Defra, 2010. Improving the Use of Environmental Valuation in Policy Appraisal: A Value Transfer Strategy. Defra, London.

Ehrlich, P., Ehrlich, A., 1981. Extinction: The Causes and Consequences of the Disappearance of Species. Random House, New York.

Ehrlich, P.R., Mooney, H.A., 1983. Extinction, substitution, and ecosystem services. Bioscience 33, 248–254.

Farber, S.C., Costanza, R., Wilson, M.A., 2002. Economic and ecological concepts for valuing ecosystem services. Ecol. Econ. 41, 375–392.

Farley, J., Costanza, R., 2010. Payments for ecosystem services: from local to global. Ecol. Econ. 69, 2060–2068.

Goldstein, J.H., Caldarone, G., Duarte, T.K., Ennaanay, D., Hannahs, N., Mendoza, G., Polasky, S., Wolny, S., Daily, G.C., 2012. Integrating ecosystem-service tradeoffs into land-use decisions. Proc. Natl. Acad. Sci. U.S.A. 109, 7565–7570.

Howarth, R.B., Farber, S., 2002. Accounting for the value of ecosystem services. Ecol. Econ. 41, 421–429.

Johnston, R.J., Rosenberger, R.S., 2010. Methods, trends and controversies in contemporary benefit transfer. J. Econ. Surv. 24, 479–510.

Kahneman, D., 2011. Thinking Fast and Slow. Farrar, Straus and Giroux, New York.

Kubiszewski, I., Costanza, R., Dorji, P., Thoennes, P., Tshering, K., 2013. An initial estimate of the value of ecosystem services in Bhutan. Ecosyst. Serv. 3, e11–e21.

Kubiszewski, I., Costanza, R., Franco, C., Lawn, P., Talberth, J., Jackson, T., Aylmer, C., 2013. Beyond GDP: measuring and achieving global genuine progress. Ecol. Econ. 93, 57–68.

Liu, S., Costanza, R., Farber, S., Troy, A., 2010a. Valuing ecosystem services: theory, practice and the need for a trans-disciplinary synthesis Ecological Economics Reviews. Book Series: Annals of the New York Academy of Sciences, vol. 1185., pp. 54–78.

Liu, S., Costanza, R., Troy, A., D'Aagostino, J., Mates, W., 2010b. Valuing New Jersey's ecosystem services and natural capital: a spatially explicit benefit transfer approach. Environ. Manage. 45, 1271–1285.

Luisetti, T., Bateman, I.J., Turner, R.K., 2011. Testing the fundamental assumption of choice experiments. Land Econ. 87, 284–296.

Luisetti, T., Jackson, E.L., Turner, R.K., 2013. Valuing the European coastal blue carbon storage benefit. Mar. Pollut. Bull. 71, 101–106.

McCauley, D.J., 2006. Selling out on nature. Nature 443, 27–28.

Millennium Ecosystem Assessment (MEA), 2005. Ecosystems and Human Well-Being: Synthesis. Island Press, Washington, DC.

Monbiot, G., 2012. Putting a price on the rivers and rain diminishes us all. The Guardian.

Nelson, E., Mendoza, G., Regetz, J., Polasky, S., Tallis, H., Cameron, D.R., Chan, K.M.A., Dailey, G.C., Goldstein, J., Dareiva, P.M., Lansdorf, E., Naidoo, R., Ricketts, T.H., Shaw, M.R., 2009. Modeling multiple ecosystem services, biodiversity conservation, commodity production, and tradeoffs at landscape scales. Front. Ecol. Environ. 7, 4–11.

Nelson, J.P., Kennedy, P.E., 2009. The use (and abuse) of meta-analysis in environmental and natural resource economics: an assessment. Environ. Resour. Econ. 42, 345–377.

Odum, H.T., 1971. Environment, Power and Society. John Wiley, New York.

Rockstrom, J., Steffen, W., Noone, K., Persson, A., Chapin, F.S., Lambin, E.F., Lenton, T.M., Scheffer, M., Folke, C., Schellnhuber, H.J., Nykvist, B., de Wit, C.A., Hughes, T., van der Leeuw, S., Rodhe, H., Sorlin, S., Snyder, P.K., Costanza, R., Svedin, U., Falkenmark, M., Karlberg, L., Corell, R.W., Fabry, V.J., Hansen, J., Walker, B., Liverman, D., Richardson, K., Crutzen, P., Foley, J.A., 2009. A safe operating space for humanity. Nature 461, 472–475.

TEEB Foundations, 2010. The Economics of Ecosystems and Biodiversity: Ecological and Economic Foundations. Earthscan, London and Washington.

TEEB Synthesis, 2010. Mainstreaming the Economics of Nature: A Synthesis of the Approach, Conclusions and Recommendations of TEEB. Earthscan, London and Washington.

Turner, R.K., Paavola, J., Cooper, P., Farber, S., Jessamy, V., Georgiou, S., 2003. Valuing nature: lessons learned and future research directions. Ecol. Econ. 46, 493–510.

UNU-IHDP, 2012. Inclusive Wealth Report: Measuring Progress Toward Sustainability. UNU-IHDP, Bonn.

van der Ploeg, S., De Groot, R.S., Wang, Y., 2010. In: Development, F.F.S. (Ed.), The TEEB Valuation Database: Overview of Structure, Data, and Results. Wagenin – gen, The Netherlands.

WBCSD, 2011. Guide to Corporate Ecosystem Valuation: A Framework for Improving Corporate Decision-Making. Geneva, Switzerland.

WBCSD, 2012. Biodiversity and Ecosystem Services: Scaling Up Business Solutions. Company Case Studies that Help Achieve Global Biodiversity Targets. Geneva, Switzerland.

Westman, W.E., 1977. How much are nature's services worth? Science 197, 960–964.

Wood, M.C., 2014. Nature's Trust: Environmental Law for a New Ecological Age. Cambridge University Press, Cambridge, UK.

Wratten, S., Sandhu, H., Cullen, R., Costanza, R., 2013. Ecosystem Services in Agricultural and Urban Landscapes. Wiley-Blackwell, Oxford, UK.

Donald Ludwig — "Limitations of Economic Valuation of Ecosystems." *Ecosystems Journal*.

Reading Commentary

Ludwig offers a far-reaching critique of economic valuation methods frequently used to assess environmental choices. His main argument is that we are not always aware enough of the extent to which typical valuation methods are based on assumptions rather than empirical evidence. He points out that we generally do not have enough information or sufficient understanding to make anything more than vague qualitative predictions about ecosystem dynamics. For example, Ludwig argues that a contingent valuation analysis may be inaccurate because "it asks people to give prices to things that are not ordinarily priced". Because of this, it is quite common to underestimate the value of affected goods and services. In turn, inaccurate estimates lead to inaccurate policies that fail to protect ecosystem services to the extent they should.

Ludwig's concerns regarding the difficulties of using economic strategies to value environmental systems are convincingly presented but beg the following question: how then do we move forward when we have to make a decision about whether and how to proceed on a project with environmental implications?

Ludwig offers some clues to the answer, though it is clear that the path is not simple or even fully developed. He mentions that we need to work harder to understand the error bounds on estimates. Continuing to use existing valuation models, or relying on their underlying assumptions, may be an exercise in tail chasing. Instead, if we strive to understand the built in "errors," we may be better positioned to try to escape the purely utilitarian approach to valuation that is likely holding us back.

For example, while risk assessment focuses on possible risks to humans, ecosystem services analysis focuses on the valuation of services provided by our environment. The tool reframes the relationship between humans and the rest of nature, a step often considered as essential to moving toward solving the problem of how to build a sustainable and desirable future (Costanza 1998). The logic is clear: by giving us a more comprehensive and balanced picture of the environmental assets that support human well-being, we can better form policies to protect natural resources.

Ludwig, however, does not stop there. He asks us to consider whether "irrational responses," if they were treated differently, would be more influential. In other words, do we have the capacity to see the moral, cultural and other noneconomic values of a project or the environment itself? Can our tools capture it?

LIMITATIONS OF ECONOMIC VALUATION OF ECOSYSTEMS

Donald Ludwig

INTRODUCTION

Valuation methods have been prominent in recent discussions because they are being used in legal efforts to protect and restore ecosystems (Portney 1994). Such methods also seem to be a promising way to include ecological values when various public policies and projects are under consideration. Economic valuations may have quite perverse and pernicious effects unless they are applied with a careful regard for their limitations. I believe that, in fact, the proper domain of application of such methods is quite limited. Such methods work well for small projects of minor importance, or possibly for fine-tuning of larger projects. But these methods—and economic methods in general—are inappropriate and harmful when used to determine important public policies.

There is a large literature on economics and the environment. The following are some of my favorite works. They offer some discussion of the underlying issues rather than attempting to offer a theory or methods that must be accepted holus-bolus. Goulder and Kennedy (1997) examine some of the philosophical issues that underlie application of economic methods. A deeper and more devastating critique was given by Gunnar Myrdal (1953). I urge everyone to read these works. Some of the ideology that underlies much economic thought is discussed by Bromley (1990) and Galbraith (1992). Comprehensive expositions of the issues involved in environmental economics are provided by Cooper (1981) and Common (1995). Both of these authors go to great pains to explain economic assumptions and discuss their validity. More polemical works are Sagoff (1988) and Brown (1991). Portney (1994) is an economist writing for other economists about the significance of the issue of economic valuation. This work is quite accessible.

My main objections to valuation are:

1. Economic values are generally of tertiary importance: personal and social values are of a higher order and are generally incompatible with economic values.
2. Economic theories require a host of simplifying assumptions that have dubious validity. The effect of these simplifications is seldom if ever assessed.
3. Market measures or survey measures are inappropriate for decisions that involve important ecological questions.
4. The methods customarily used for valuation of ecosystems have such severe internal flaws that the results have little significance.

After considering each of these points in more detail, I offer a few hints for the wary who may be about to cooperate in a valuation exercise.

HIGHER DOMAINS OF VALUE

Scitovsky (1976) is a good reference for the relative importance of economic and other values, but few of us require much convincing that economic values belong to a different sphere than personal and social values. It is commonplace for victims to be granted monetary compensation for their injuries. Often the reaction of the victim is that, although it helps to have some money, no amount of money can make up for the trauma. In some cases of abuse, the victims simply want an apology. Lawyers may advise against such apologies on the grounds that it may increase liability. Such a policy aggravates the original injury. The money and the injury have to do with different domains of value.

The higher domains of value involve our sense of personal integrity and dignity. They often are combined with religious, ethical, and social values, or perhaps a sense of place. Love and friendship cannot be bought. Those who subordinate personal and social values to monetary values often are considered to be sociopaths. Human sociopaths generally are confined or otherwise repressed, but corporate sociopathy seems to be a norm. It is illuminating to consider altruists (Monroe 1996). Monroe's conclusion after many intensive interviews with people who have committed acts of great heroism and altruism—a typical example is a person who endangered both himself and his family to rescue Jews from Nazi Germany—is that these people felt that they could not do otherwise. Not to help would be to deny their common humanity. There was no cost-benefit calculation: the decision was taken instantly. We honor such people and wish that we could be more like them. They are in touch with a higher domain of value.

ECONOMIC ASSUMPTIONS ARE OFTEN INAPPROPRIATE

The objectives of economic theory stem from the utilitarian goal of finding consistent rules for human behavior and decision making. The issue immediately arises whether we shall look for a descriptive (positive) theory or a prescriptive (normative) theory (Daston 1988). Shall we describe how people *actually* behave, or shall we give rules that describe how they *ought* to behave? In case of collective decisions, the distinction is crucial. This distinction is systematically confused in economic theory and applications (Myrdal 1953). Since nobody can be opposed to rational and objective public decision, it is taken to mean assent to the idea that decisions can be made by toting up the costs and the benefits and seeing which comes out ahead. In contrast, A. Sen (1970), writing on Collective Choice and Social Welfare, states in his preface "Indeed the premise of this book is based on the belief that the problem cannot be satisfactorily discussed within the confines of economics." Sen recently was awarded a Nobel prize for his work.

When applying economic theory, a difficulty immediately arises that the costs and benefits may be experienced by different groups of people. This problem is finessed by assuming that side payments from one group to the other will compensate. The fact that such side payments are seldom or never made does not seem to impede the practical application of the idea. It also fails to address the question that arises in my first example: suppose the injury and the compensation are incommensurable? (Common 1995, p. 155). This difficulty is disposed of in economic theory by postulating that everybody at all times has a "utility function" that he/

she tries to maximize. Lest we collapse into gales of laughter at such an assumption, think of the parallel assumption in behavioral ecology that everybody is at all times striving to increase his/her inclusive fitness. Ecologists come out ahead in this comparison because they actually perform experiments to check the validity of the theory. McFadden (1998) provides a wealth of detail that challenges the economic assumption.

The price to be paid for a "rational" economic theory is adoption of a number of sweeping assumptions. The term rational appears in quotes because it is actually a shorthand for the assumption that everybody has a utility function. Economists say in effect that you are only rational if you always act to maximize a utility function that they can cope with. Such assumptions are made not on the basis of empirical evidence, but because they seem to be required for a workable theory (Common 1995, p. 127). Among the most important of these expedient assumptions is one that various goods can be substituted for one another. That is, deficiencies in one good can be compensated by supplying other goods in greater abundance.

Such assumptions make it possible to speak of the price of a good rather than assigning value only to certain combinations of goods. A parallel assumption in ecology would be that if nutrients are lacking, they can be compensated by increasing the water supply or solar radiation. When considering problems of growth in the face of limited resources, economists often have used a "production function" that describes the output as the product of a number of factors, some of which are in principle unbounded. The analogy in ecology would express plant growth as given by the product of factors depending upon nutrients, solar input, water, etc. Any deficiency in water supply could be compensated by increasing either the solar input or the nutrient supply. The assumption is absurd in ecology, and it is for the most part unsupported by evidence in economics. The fallacy is not in using a theory that involves compensation, but rather in extrapolating relationships that exhibit compensation far beyond the domain for which evidence is available.

As an example of such extrapolation, when the "limits to growth" issue arose in public discourse, economists were able to dismiss fears about running out of material resources on the grounds that unbounded increases in capital and technology would compensate, based upon the production function argument (Dasgupta and Heal 1979; Solow 1974, p. 11). For some empirical evidence to the contrary, see Kopp and Smith (1982). Another notable triumph of economic theory over common sense was the report of the NAS committee on global climate change (NAS 1991). The conclusion of the assembled savants was that since we expect economic growth at the rate of, for example, 5% per year, any deleterious effects of global warming can easily be compensated from a portion of the additional income.

The danger that we risk when adopting economic methods for valuation of ecosystems is that these methods cannot be divorced from the underlying assumptions and theory. If you accept the methods, you are adopting those assumptions, unless perhaps you are in a position to follow the whole chain of reasoning. To illustrate how arcane the chain of reasoning can be, consider the issue of uncertainty. This is one of the great problems in any discussion of the possible or plausible effects of various actions that are under consideration. We generally do not have enough information and understanding to make anything more than a vague qualitative prediction about responses of ecosystems. We ignore our uncertainty at our peril, or the peril of future generations. How does economic theory deal with uncertainty? There are a number of methods, based upon additional assumptions. One of the most egregious of these is "rational expectations" (Starrett 1988, p. 100–3). This is the assumption that we all can predict the mean future outcomes. Hahn (1985, p. 17–8) and Hahn and Solow(1995, pp. 1, 140) have dissenting opinions, and there is evidence to the contrary, but the assumption is vital for some simple theories, so it is retained.

Another assumption relating to uncertainty is essential for the fundamental theorems of welfare economics, which attempt a proof of the validity of Adam Smith's concept of the "invisible hand." According to this theory, if everybody goes about maximizing their own personal utility, then the result of the collective actions will be optimal—in the sense that we arrive at a state where nobody can be made better off except by making someone else worse off. This theory is used to justify the idea that free markets are always superior to planning. It rests on an assumption that there is a "complete set of markets." That is, we have not only a spot market for each possible commodity, but also a market for all possible future deliveries of the commodity: a kind of world commodities exchange. In reality, there is no complete set of markets, nor can there ever be for environmental goods. That is why other measures are used to determine their economic value. The worry about the effects of uncertainty about the future is finessed by assuming that the complete set of markets includes a futures market for every conceivable future state of the world. At time zero, all of these markets clear, that is, binding agreements are made about the prices of these commodities in every possible future state of the world (Hahn 1985). In fact, there is no need for money in such a world. If you believe that these are reasonable assumptions, then you believe in the invisible hand. When you apply standard economic methods to valuation, you are subscribing to assumptions like these (Common 1995, p. 125).

It might be argued that inconsistencies and unwarranted assumptions are not necessarily fatal objections. Indeed, economic methods are valuable if we keep their limitations in mind. As an example where these limitations may cause severe problems for proper valuation of ecological services, consider the matter of discounting of future benefits. In essence, these arguments are based upon substitutability of present benefits for future ones. A standard justification relies on the "opportunity cost" of capital. That is, if we devote resources to a given project, that forecloses the possibility of devoting them to some other project. Instead of investing in a restoration project, we might invest an equivalent sum in government bonds. The rate of return on the bonds is fixed, so we may be sure that a sum of $1 invested this year will be worth $(1 + r)$ in a year. Hence $1 received next year is worth $1/(1 + r)$ now: the future return is discounted. If we believe that ecological and monetary values can be substituted for each other, it follows that ecological services received next year are diminished by the same fraction. Is this a reasonable conclusion? What if the services include provision of fresh water, or moderation of the effects of flooding or droughts? Some believe that ecosystem services should not be discounted, and there is some evidence that people do not apply the usual sort of discounting to benefits that are received in the far future (Heal 1997).

More generally, one might ponder whether a single measure of value is adequate to capture the many complexities of behavior and services provided by ecosystems. Perhaps the economic assumptions are adequate under some circumstances, but how do we determine what those circumstances are? Economics, like any discipline that purports to deal with important practical matters, is susceptible to inconsistencies, fallacies, and blunders on a massive scale. Harm is done when those making use of the results are unable to judge the validity of the assumptions and reasoning that underlie the predictions and prescriptions.

MARKET MEASURES OR SURVEYS ARE INAPPROPRIATE FOR DECISIONS THAT INVOLVE IMPORTANT ECOLOGICAL QUESTIONS

On this question, I can do no better than to quote Sagoff (1988, 44):

> A market or quasi-market approach to arithmetic, for example, is plainly inadequate. No matter how much people are willing to pay, three will never be the square root of six. Similarly, segregation is a national curse, and if we are willing to pay for it, that does not make it better but only makes us worse. Similarly, the case for abortion rights must stand on the merits; it cannot be priced at the margin.

> Our failures to make the right decisions in these matters are failures in arithmetic, failures in wisdom, failures in taste, failures in morality—but they are not market failures. There are no relevant markets to have failed.

> What separates these questions from those for which markets are appropriate is this: They involve matters of knowledge, wisdom, morality, and taste that admit of better or worse, right or wrong, true or false—and these concepts differ from economic optimality. Surely environmental questions—the protection of wilderness, habitats, water, land, and air as well as policy towards environmental safety and health—involve moral and aesthetic principles and not just economic ones. This is consistent, of course, with cost-effective strategies for implementing our environmental goals and with a recognition of the importance of personal freedoms and economic constraints.

The fundamental fallacy in application of economic criteria to determine our policies is the substitution of techniques for judgement. As I have pointed out above, these techniques are burdened with numerous concessions to expediency. They are unaccompanied by any estimates of the magnitudes of the distortions that are thereby introduced and therefore at best can only serve to raise questions for further investigation. Here is a supporting quotation from Common (1995, p. 5):

> Economics, that is, largely ignores: history; the material laws of nature; and the study of the nature of man. Despite this, economics has much to say about the means of achieving sustainability-related social goals that is important and useful. but which is often ignored in the process of policy formation. The point is not that economists have nothing interesting to say, but rather, that their expert status relates to questions about means rather than ends. With regard to the debate on the ultimate goals of policy, economists should not be accorded the privileged status that they frequently appear to claim for themselves.

THE METHODS USED FOR ECONOMIC
VALUATION ARE FLAWED

The contingent valuation approach may at first sight appear as a clever finesse of many of the issues I have raised. To find out what something is worth, ask a representative sample of those involved what they would be willing to pay for it. A comprehensive critique of this idea is presented by Diamond and Hausman (1994). I shall mention only a few points. One of the main assumptions that surveyers make is that the responses are actually responses to the questions propounded by the interviewer. Responses such as zero or infinite value are disregarded on the grounds that they are unreasonable. In many cases, such responses are a form of protest about the question (Sagoff 1988). Other answers are in the category of the "warm glow," which might be a vague feeling that we should do something to protect the environment. This latter effect is present when survey responses assign the same value to five lakes as to a single lake in the group. Inconsistencies like these are common.

An additional difficulty is that people are asked to place prices on things that are not ordinarily priced. For some commodities, we form an opinion about a suitable price from long experience in a market. If there is no such experience and no such market, there may be little consistency among responses and little validity in inferences drawn from the responses. The situation is much worse when people with limited experience with an ecosystem are asked to come up with a value. In our society, complicated technical issues are not usually determined by such surveys. For instance, standards for environmental pollutants are determined by an administrative process that includes expert inputs. Cooper (1981, p. 46) says, "it is silly to believe in evaluations in situations of scientific ignorance." McFadden (1998) provides a detailed critique of contingent valuation methods.

Diamond and Hausmann conclude as follows:

> We believe that contingent valuation is a deeply flawed methodology for measuring non-use values, one that does not measure what its proponents claim to be measuring. The absence of direct market parallels affects both the ability to judge the quality of contingent valuation responses and the ability to calibrate responses to have usable numbers. It is precisely the lack of experience both in markets and in the consequences of such decision that makes contingent valuation questions so hard to answer and the responses so suspect.

> We have argued that internal consistency tests (particularly adding-up tests) are required to assess the reliability and validity of such surveys. When these tests have been done, contingent valuation has come up short.

BLUFF AND BLUSTER

In spite of being convinced of the inappropriateness of the standard valuation methods, some of us may find ourselves on a committee to advise on or perform valuation. Good questions to ask in such situations are: Why am I here? Whose interests does it serve for me to be here? If I am unable to modify the scope or methods or the terms of reference of the investigation, what purpose do I serve?

259

When it come to discussion of the methods and the validity thereof, one may be subjected to tactics of bluff and intimidation. In such circumstances, it helps to have a few questions at the ready. A good question always is "What are the error bounds on your estimates?" Do not be fobbed off by some answer that computes the inverse square root of the sample size and uses it in a statistical bound. Sampling deviations (which such methods calculate) are often tiny in comparison with structural defects in the model. For example, how many responses were modified or thrown out because they were "irrational"? If they were treated differently, how influential would they be? Have the consistency checks discussed by Diamond and Hausman (1994) been applied?

If someone invokes the fundamental theorems of welfare economics (the invisible hand argument), ask about the assumptions behind such theorems. Do they assume a complete set of markets? What does that mean? Are there any examples of complete sets of markets? How do they deal with uncertainty? Do they assume rational expectations? What does that mean? If everybody has rational expectations, why does the stock market go up and down so much? Do they invoke the assumptions of consumer sovereignty and revealed preferences? What do those concepts mean? If any of these assumptions are satisfied only approximately, how large and important are the corresponding errors? Precisely what does economic efficiency mean? Does it have any connection with equity or justice? If not, why not?

If we can have questions such as these answered as part of the valuation process, we can much better understand and appreciate the results.

REFERENCES

Bromley DW. 1990. The ideology of efficiency: searching for a theory of policy analysis. J. Environ Econ Manage 19:86–107.
Brown P. 1991. Why climate change is not a cost/benefit problem. In: White JC, editor. Global climate change. New York: Elsevier, p 33–44.
Common M. 1995. Sustainability and policy: limits to economics. Cambridge, UK: Cambridge University Press.
Cooper C. 1981. Economic evaluation and the environment. London: Hodder and Staughton.
Dasgupta P, Heal G. 1979. Economic theory and exhaustible resources. Cambridge, UK: Cambridge University Press.
Daston L. 1988. Classical probability in the enlightenment. Princeton, NJ: Princeton University Press.
Diamond P, Hausman J. 1994. Contingent valuation: is some number better than no number? J Econ Perspect 8(4):45–64.
Galbraith JK. 1992. The culture of contentment. Boston: Houghton Mifflin Co.
Goulder LH, Kennedy D. 1997. Valuing ecosystem services: philosophical bases and empirical methods. In: Dailey G, editor. Nature's services. Washington, DC: Island Press, p 23–47.
Hahn, F. 1985. In praise of economic theory, pp 10–28 in F. Hahn Money, growth and stability. Oxford: Basil Blackwell
Hahn, F. and Solow R. 1995. A critical essay on modern macroeconomic theory. Cambridge MA: MIT Press.
Heal G. 1997. Discounting and climate change. Climatic Change 37:335–43.
Kopp RJ, Smith VK. 1982. Neoclassical measurement of ex ante resource substitution. Adv Econ Res Energy 4:183–98.

McFadden D. 1998. Rationality for economists, to appear in Journal of Risk and Uncertainty. Available at http://www. santafe.edu/sfi/publications/98wplist.html.

Monroe KR. 1996. The heart of altruism. Princeton, NJ: Princeton University Press.

Myrdal G. 1953. The political element in the development of economic thought. London: Routledge Paul.

Portney PR. 1994. The contingent valuation debate: why economists should care. J Econ Perspect 8(4):3–17.

Sagoff M. 1988. The economy of the earth: philosophy, law, and the environment. Cambridge, UK: Cambridge University Press.

Scitovsky T. 1976. The joyless economy: an inquiry into human satisfaction and consumer dis-satisfaction. New York: Oxford University Press.

Sen A. 1970. Collective choice and social welfare. San Francisco, CA: Holden-Day.

Solow R. 1974. The economics of resources or the resources of economics. Am Econ Rev May: 1–14.

Starrett D. 1988. Foundations of public economics. Cambridge, UK: Cambridge University Press.

Howard Kunreuther and Paul Slovic — "Challenges in Risk Assessment and Risk Management." *Annals of the American Academy of Political and Social Science.*

Reading Commentary

Risk assessment is yet another tool that is often used in environmental policy-making. Risk is usually framed as the probability of a hazard occurring times the potential impact or losses that will result. Kunreuther and Slovic summarize several of the challenges related to risk assessment in this introduction to a special volume of the *Annals of the American Academy of Political and Social Science* focused on risk management. They point to a worrying trend: the use of tools like risk assessment is becoming "much more politicized and contentious." Environmental evaluation tools—including EIAs, CBA and risk assessment—are used to help policy makers decide whether to adopt or to change environmental policies and regulations. While they are often touted as being "objective," "scientific" and "apolitical," the uncertainty inherent in any of these kinds of analyses makes them vulnerable to manipulation in support of a favorite policy (rather than as honest evaluations of the gains and losses associated with alternative courses of action).

The insights summarized by Kunreuther and Slovic also raise several important questions that extend to the rest of the environmental policy-making process. Namely, how should we handle the uncertainty (in risk, valuation etc.) inherent to environmental policy-making? What role should tools like RA play in policy-making given the inherent uncertainty involved? How can we improve decision-making under conditions of uncertainty? How can we address the growing divide between the scientific experts who wield these tools and the communities that are affected by decisions that are based on these tools? What is the role of the government in risk management? What is the role of the community? And finally, what are some ways in which we can make the underlying assumptions (and nonobjective judgments) inherent in these tools more transparent?

CHALLENGES IN RISK ASSESSMENT AND RISK MANAGEMENT

Howard Kunreuther and Paul Slovic

During the past two decades, our society has grown healthier and safer on average and has spent billions of dollars and immense effort to become so. Nevertheless, the public has become more, rather than less, concerned about risk. We have come to see ourselves as being exposed to more serious risks than we faced in the past, and we believe that the worst is yet to come.

A second dramatic trend is that risk assessment and risk management, like many other facets of our society, have become much more politicized and contentious. Polarized views, controversy, and overt conflict have become pervasive. The public has lost faith in the ability of science, industry, and government to manage the risks from many important technologies, such as nuclear power and chemicals and their wastes. In addition, the conflict is exacerbated by sharp differences between people as to who should and does benefit or lose as a result of specific decisions—for example, siting a noxious facility.

Difficulties in managing risks from technology are compounded by the fact that there is often great uncertainty associated with estimates of those risks. This uncertainty is sometimes due to a sparse database from which to derive risk estimates. Knowledge of the ways in which accidents, illnesses, or other forms of harm result from exposure to a technology may also be lacking. If the hazard is latent, then it may be even more difficult to determine the potential harmful consequences. In addition, there may be many factors triggering a particular disease (such as cancer), making it difficult to determine the responsibility of a particular technology or event (for example, groundwater contamination).

In addition, each party concerned with a particular problem involving risks to health, safety, or the environment has its own goals and agenda, often framed around how risky a particular activity is likely to be. Scientific experts frequently disagree on the nature of the risks, so that each interested party can typically find someone to support its position. Given the lack of adequate theoretical models and data for many of these risks, it is often difficult to evaluate the differences between these estimates.

The U.S. Congress has also been concerned with risk assessment and risk management. Several recent bills have proposed detailed risk assessments and benefit-cost analyses as a basis for determining appropriate regulations and standards. It is not easy to forecast the impact that this proposed legislation might have on the treatment of risk in our society.

THE NEED FOR A NEW PERSPECTIVE

The conflicts and controversies surrounding risk are not due to public irrationality or ignorance but, instead, may be seen as a side effect of our remarkable form of participatory democratic government, amplified by certain powerful technological and social changes. The technological change allows the electronic and print media to inform us of bad (trust-destroying) news from all over the world, often right as it happens. Special interest groups, well funded and sophisticated in using their own experts and the media, effectively communicate their viewpoints and influence risk policy debates and decisions. All of this is blended into an adversarial legal system that pits expert against expert, contradicting each other's risk assessments and further destroying the public trust.

The young science of risk assessment and the techniques of benefit-cost analysis and decision analysis face numerous challenges in addressing questions about how society should manage its health, safety, and environmental risks. Scientific analysis of risks cannot allay our fears of major accidents or delayed cancers unless we trust the system.

It is essential that we understand the complex psychological, social, cultural, and political forces that dictate success and failure in risk assessment and risk management. Understanding the root causes of conflict and ineffective action gives hope for improvement. As we come to understand the complexity of the root causes of risk conflict, we also recognize the need for radically different approaches to risk management.

ORGANIZATION OF THIS VOLUME

The articles in this volume of *The Annals* provide a perspective from leading scholars on the challenges in risk assessment and risk management. Each of the articles addresses a set of key topics in the authors' field of expertise. The questions raised provide a blueprint for action as well as an agenda for future research.

UNCERTAINTY AND RISK ASSESSMENT

John Graham and Lorenz Rhomberg's article reviews the types of scientific data that can be collected to assess risks. Such data range from real-world observations to controlled clinical trials. The power and limitations of these different approaches are illustrated with empirical examples. Graham and Rhomberg conclude with a set of cautionary notes on the difficulties of settling value differences with risk assessments due to uncertainties in the estimates and the possibility that the assessment may exacerbate conflicts between potential winners and losers.

The principal message from Robert Poliak's article is that risk assessment today cannot be viewed as scientific because too little is known about the relationship between exposure to certain substances and contraction of diseases. Hence he is somewhat skeptical of proposals that attempt to achieve uniformity of risk assessments across federal agencies. He concludes his article by pointing out that the government needs to be concerned with issues of trust, public perception of risk, and value differences in developing risk regulations.

Dale Jamieson's article focuses on the high level of uncertainty associated with many societal problems and suggests that people can thus find or create risk data to support their own agendas. He points out that uncertainty is more than just an analysis of scientific data, being determined also by the degree of trust that people have in the institution providing the data and the ways in which people are likely to behave in the future. He thus concludes that we need to bring science into closer contact with the public and our policymakers in order to improve the decision-making process under conditions of uncertainty.

William Freudenburg also focuses on the challenges in using risk assessment by focusing on the difficulties that scientists and technologists have had in dealing with the public. He argues that risk assessors are often overconfident, frequently overlooking or underestimating the unknowns. He introduces the word "recreancy" to depict our reliance on scientific experts who we fear will not do their job. The inability of scientists to recognize or admit their blind spots has created a sense of distrust in experts on the part of the public.

VALUATION AND RISK

Robin Gregory, Thomas Brown, and Jack Knetsch point out that valuations of environmental risks are often contentious because people focus on different dimensions of the risk and weigh them differently. Gregory and his colleagues indicate some of the challenges in obtaining valuations such as loss/gain differentials and varying rates of time preferences by individuals. The absence of prices makes assessment of environmental values difficult and may require the use of new approaches such as value trees and citizen juries.

The importance of understanding the sociocultural system associated with any problem is the principal theme of Roy Rappaport's article. Using the example of an oil spill's impact on fish and other marine life, he points out that for some individuals, the loss may be viewed as economic, while for others it is seen as a threat to their culture. Many values such as integrity and equity cannot be expressed in quantitative terms but need to be recognized when evaluating specific risks and strategies for dealing with them. The article contends that one must take into account these apprehensions of the affected public even if the experts view the risks as much lower.

Baruch Fischhoff addresses the issues of eliciting values from the public and using them to establish trust. He indicates that one can use focus groups, citizen juries, and public opinion surveys to do this. Fischhoff also makes the case that people are anxious to understand the models being used to estimate risk and to know how good they are. By sharing their knowledge with the public, it may be possible for scientists to become trusted by citizens.

RISK COMMUNICATION

William Leiss addresses the questions of how to improve the dialogue on risk between experts and the public and how to achieve a higher degree of social consensus. He first reviews two earlier phases in the evolution of risk communication, where quantitative risk estimates and comparative risk estimates determined priorities (Phase I) and where an emphasis was placed on characteristics of successful communication (Phase II). He feels that the current phase, which emphasizes social interrelations between players in the risk management game, has a better chance of successfully addressing his questions than did the other two phases. Two case studies illustrate how this third phase has been operationalized.

The importance of risk communication in the amplification and attenuation of risk is highlighted in the article by Roger and Jeanne Kasperson. They trace the different phases of social amplification from the time that an event is channeled through individuals and networks, through the communication process with its potential ripple effects, including the creation of stigma. A case study from Goiânia, Brazil, illustrates the impact of the media on tourism in the aftermath of a radiation accident in the region. The media's coverage produced a decline in tourism, resulting in social and economic consequences far greater than one would expect given the limited degree of radiation exposure and risk from the event. The Kaspersons conclude that one needs to create political regimes and institutions for risk containment and risk reduction.

Kip Viscusi and Richard Zeckhauser examine the role of hazard warnings in communicating information on risk. They point out that warnings can facilitate decentralized risk-taking decisions when there are heterogeneous preferences. In issuing warnings, one needs to consider the length, format, and type of warning as well as who receives it. It is particularly important to focus on how individuals process information, so one can determine how likely a particular warning will be understood. The article concludes by posing a set of challenges for communicating information on risks.

THE PROCESS OF RISK MANAGEMENT

Howard Kunreuther and Paul Slovic contend that risk and its management should be viewed as a game in which the rules must be socially negotiated within the context of a specific problem. This contextualist view provides insight into why technical approaches to risk management often fail. It also highlights the need to understand the values and goals of the interested parties, thus emphasizing the importance of institutional, procedural, and societal processes. The siting of hazardous facilities is used to illustrate this importance of process as well as outcome measures in developing risk management strategies.

Ralph Keeney points out that the role of values in the risk management process requires making trade-offs between costs and reductions in specific risks such as the chances of death. He argues that there are differences in how a person should evaluate a particular strategy if it involves just that person as an individual (for example, "What should I pay for an air bag?") or if it involves society (whether cars should be required to have air bags). In order to determine a person's values, one needs to specify that person's objectives and the attributes that define them. Two case studies illustrate how this methodology can be used for making trade-offs between costs and benefits.

Robin Cantor examines the role that risk management will play in the policymaking process in the coming years. Some people feel that congressional reform of the risk management process will lead to a more rational process, with peer review and scientific consensus. Others feel it will paralyze the federal role in risk management. A key question that needs to be addressed is whether reforms in the risk assessment process will enable one to reach some type of consistency in guidelines and consensus across different groups. Cantor concludes the article by indicating the importance of considering equity in developing risk management strategies and having an open and participatory process rather than relying solely on experts.

RISK MANAGEMENT STRATEGIES

In their article, Richard Zeckhauser and Kip Viscusi point out that it is necessary for government to supplement market processes due to public misperceptions of risk and the negative externalities resulting from particular activities, such as water and air pollution. They suggest that one appropriate role for government is to reassure the public about certain risk levels in products through testing, as performed by agencies such as the Food and Drug Administration. Another role is for the government to insure certain risks against which the private sector cannot offer protection, such as catastrophic losses from natural disasters. A third role is setting standards for certain risks that the public cannot easily measure or that produce negative externalities, such as the number of asbestos fibers in the air.

Patrick Field, Howard Raiffa, and Lawrence Susskind explore strategies for siting hazard-ous facilities in poor neighborhoods. Through a hypothetical brainstorming session between community and state representatives, the interested parties conclude that the facility will be acceptable only if it is part of a broader development program jointly created by the citizens in the poor neighborhoods and other interested parties. The community needs to feel it is better off with the facility than without it with respect to its economic well-being as well as its health and safety. The article concludes with a set of principles for effective siting.

Roger Noll suggests that one of the reasons that it is difficult to determine whether risk regulation is too stringent or not stringent enough is the difficulty that the public has in identifying health, safety, and environmental risks and effective responses to them. Due to the uncertain nature of these risks, the regulators are highly dependent on the type of infor-mation provided to them. Hence the stringency of each regulation reflects the intensities of political support and opposition in a particular situation that are partially determined by public perception. Noll feels that one needs to recognize that risk regulations are often based on dreaded consequences that are beyond an individuals control. The challenge is how to design better institutional arrangements and public education to construct regulations for coping with these risks.

The final article in this volume, by James Krier, explores how the legal system can utilize information on risk effectively. He points out that the legal system includes both legislative and administrative bodies, both of which often play a more significant role than the courts in dealing with risk issues. He stresses the importance of the legal system in making sure that administrative agencies fulfill their regulatory requirements and in enabling victims to seek compensation from a risk producer through the law of torts. Krier suggests that cost-benefit analysis be used to determine how much risk is acceptable. Today experts feel we overregulate due to irrational fears on the part of the public concerning the dangers of new technologies. It is an open question as to how public perceptions of risk can be incorporated in benefit – cost analyses.

John Sterman — "A Skeptic's Guide to Computer Models."
In *Managing a Nation: The Microcomputer Software Catalog.*
Boulder, CO: Westview Press.

Reading Commentary

Models have become ubiquitous in environmental problem-solving and policy-making. Sterman presents three types—optimization, simulation and econometric—of models. Understanding their strengths and shortcomings is important. Being able to do so allows you to choose wisely. Optimization models are hard to use because they require specifying the overall goal. Simulation models address some of the shortcomings of optimization models, but have their own limitations. Once boundaries are set for purposes of building a simulation model, these boundaries can become an obstacle to effective problem-solving. Econometric models are limited by their primary assumptions, which follow from basic economic theory.

As Sterman points out, models are only as good as the assumptions they are based on. By changing key assumptions, you can end up with very different results. For example, economic models assume that people act rationally and have all the information they need to make smart decisions. This leads to unrealistic results, however, since most people have to make decisions without complete information. Modelers make assumptions based on their worldview. For example, they may assume that economic growth can and should continue forever. These same modelers are not likely to emphasize assumptions about sustainability.

Environmental factors are often left out of or discounted in economic models. This is one of the reasons that ecosystem services (discussed in the Costanza et al. reading) have become a more common way of valuing environmental costs and benefits in policy-making and problem-solving. Even given all these shortcomings, Sterman still advocates using models but always with caution. Modeling results alone are not enough to justify final decisions. They should be combined within a common sense and other considerations. As Sterman states, "[modeling] is a tool for improving judgment and intuition". For this to happen, assumptions and biases must always be made clear to those who intend to rely on model outputs.

The questions at the end of Sterman's classic article are important when we survey environmental challenges in our communities: What is the problem being addressed? How is it bounded (or framed)? What time horizon is being set? Are people assumed to act rationally? Does the model assume that key actors have full information? These same questions should probably be asked about any tool we rely upon for environmental analysis.

A SKEPTIC'S GUIDE TO
COMPUTER MODELS

John Sterman

But Mousie, thou art no they lane In proving foresight may be vain;
The best-laid schemes o' mice an' men Gang aft a-gley,
An lea'e us nought but grief an' pain,
For promis'd joy!

THE INEVITABILITY OF USING MODELS

Computer modeling of social and economic systems is only about three decades old. Yet in that time, computer models have been used to analyze everything from inventory management in corporations to the performance of national economies, from the optimal distribution of fire stations in New York City to the interplay of global population, resources, food, and pollution. Certain computer models, such as *The Limits to Growth* (Meadows et al. 1972), have been front page news. In the US, some have been the subject of numerous congressional hearings and have influenced the fate of legislation. Computer modeling has become an important industry, generating hundreds of millions of dollars of revenues annually.

As computers have become faster, cheaper, and more widely available, computer models have become commonplace in forecasting and public policy analysis, especially in economics, energy and resources, demographics, and other crucial areas. As computers continue to proliferate, more and more policy debates—both in government and the private sector—will involve the results of models. Though not all of us are going to be model builders, we all are becoming model consumers, regardless of whether we know it (or like it). The ability to understand and evaluate computer models is fast becoming a prerequisite for the policy maker, legislator, lobbyist, and citizen alike.

During our lives, each of us will be faced with the result of models and will have to make judgments about their relevance and validity. Most people, unfortunately, cannot make these decisions in an intelligent and informed manner, since for them computer models are *black boxes*: devices that operate in completely mysterious ways. Because computer models are so poorly understood by most people, it is easy for them to be misused, accidentally or intentionally. Thus there have been many cases in which computer models have been used to justify decisions already made and actions already taken, to provide a scapegoat when a forecast turned out wrong, or to lend specious authority to an argument.

If these misuses are to stop and if modeling is to become a rational tool of the general public, rather than remaining the special magic of a technical priesthood, a basic understanding of models must become more widespread. This paper takes a step toward this goal by offering model consumers a peek inside the black boxes. The computer models it describes are the kinds used in foresight and policy analysis (rather than physical system models such as NASA uses to test the space shuttle). The characteristics and capabilities of the models, their advantages and disadvantages, uses and misuses are all addressed. The fundamental assumptions of the major modeling techniques are discussed, as is the appropriateness of these techniques for foresight and policy analysis. Consideration is also given to the crucial questions a model user should ask when evaluating the appropriateness and validity of a model.

MENTAL AND COMPUTER MODELS

Fortunately, everyone is already familiar with models. People use models—mental models—every day. Our decisions and actions are based not on the real world, but on our mental images of that world, of the relationships among its parts, and of the influence our actions have on it.

Mental models have some powerful advantages. A mental model is flexible; it can take into account a wider range of information than just numerical data; it can be adapted to new situations and be modified as new information becomes available. Mental models are the filters through which we interpret our experiences, evaluate plans, and choose among possible courses of action.

The great systems of philosophy, politics, and literature are, in a sense, mental models.

But mental models have their drawbacks also. They are not easily understood by others; interpretations of them differ. The assumptions on which they are based are usually difficult to examine, so ambiguities and contradictions within them can go undetected, unchallenged, and unresolved.

That we have trouble grasping other people's mental models may seem natural. More surprising, we are not very good at constructing and understanding our own mental models or using them for decision making. Psychologists have shown that we can take only a few factors into account in making decisions (Hogarth 1980; Kahneman, Slovic, and Tversky 1982). In other words, the mental models we use to make decisions are usually extremely simple. Often these models are also flawed, since we frequently make errors in deducing the consequences of the assumptions on which they are based.

Our failure to use rational mental models in our decision-making has been well demonstrated by research on the behavior of people in organizations (e.g., families, businesses, the government). This research shows that decisions are not made by rational consideration of objectives, options, and consequences. Instead, they often are made by rote, using standard operating procedures that evolve out of tradition and adjust only slowly to changing conditions (Simon 1947, 1979). These procedures are determined by the role of the decision makers within the organization, the amount of time they have to make decisions, and the information available to them.

But the individual perspectives of the decision-makers may be parochial, the time they have to weigh alternatives insufficient, and the information available to them dated, biased, or incomplete. Furthermore, their decisions can be strongly influenced by authority relations, organizational context, peer pressure, cultural perspective, and selfish motives. Psychologists and organizational observers have identified dozens of different biases that creep into human decision making because of cognitive limitations and organizational pressures (Hogarth

1980; Kahneman, Slovic, and Tversky 1982). As a result, many decisions turn out to be incorrect; choosing the best course of action is just too complicated and difficult a puzzle.

Hamlet exclaims (perhaps ironically) "What a piece of work is a man, how noble in reason, how infinite in faculties…!" But it seems that we, like Hamlet himself, are simply not capable of making error-free decisions that are based on rational models and are uninfluenced by societal and emotional pressures.

Enter the computer model. In theory, computer models offer improvements over mental models in several respects:

They are explicit; their assumptions are stated in the written documentation and open to all for review.
They infallibly compute the logical consequences of the modeler's assumptions.
They are comprehensive and able to interrelate many factors simultaneously.

A computer model that actually has these characteristics has powerful advantages over a mental model. In practice, however, computer models are often less than ideal:

They are so poorly documented and complex that no one can examine their assumptions. They are black boxes.
They are so complicated that the user has no confidence in the consistency or correctness of the assumptions.
They are unable to deal with relationships and factors that are difficult to quantify, for which numerical data do not exist, or that lie outside the expertise of the specialists who built the model.

Because of these possible flaws, computer models need to be examined carefully by potential users. But on what basis should models be judged? How does one know whether a model is well or badly designed, whether its results will be valid or not? How can a prospective user decide whether a type of modeling or a specific model is suitable for the problem at hand? How can misuses of models be recognized and prevented? There is no single comprehensive answer, but some useful guidelines are given on the following pages.

THE IMPORTANCE OF PURPOSE

A model must have a clear purpose, and that purpose should be to solve a particular problem. A clear purpose is the single most important ingredient for a successful modeling study. Of course, a model with a clear purpose can still be incorrect, overly large, or difficult to understand. But a clear purpose allows model users to ask questions that reveal whether a model is useful for solving the problem under consideration.

Beware the analyst who proposes to model an entire social or economic system rather than a problem. Every model is a representation of a system—a group of functionally interrelated elements forming a complex whole. But for the model to be useful, it must address a specific problem and must simplify rather than attempting to mirror in detail an entire system.

What is the difference? A model designed to understand how the business cycle can be stabilized is a model of a problem. It deals with a part of the overall economic system.

A model designed to understand how the economy can make a smooth transition from oil to alternative energy sources is also a model of a problem; it too addresses only a limited system within the larger economy. A model that claims to be a representation of the entire economy is a model of a whole system. Why does it matter? The usefulness of models lies in the fact that they simplify reality, putting it into a form that we can comprehend. But a truly comprehensive model of a complete system would be just as complex as that system and just as inscrutable. The map is not the territory—and a map as detailed as the territory would be of no use (as well as being hard to fold).

The art of model building is knowing what to cut out, and the purpose of the model acts as the logical knife. It provides the criterion about what will be cut, so that only the essential features necessary to fulfill the purpose are left. In the example above, since the purpose of the comprehensive model would be to represent the entire economic system, few factors could be excluded. In order to answer all questions about the economy, the model would have to include an immense range of long-term and shortterm variables. Because of its size, its underlying assumptions would be difficult to examine. The model builders—not to mention the intended consumers—would probably not understand its behavior, and its validity would be largely a matter of faith.

A model designed to examine just the business cycle or the energy transition would be much smaller, since it would be limited to those factors believed to be relevant to the question at hand. For example, the business cycle model need not include long-term trends in population growth and resource depletion. The energy transition model could exclude short-term changes related to interest, employment, and inventories. The resulting models would be simple enough so that their assumptions could be examined. The relation of these assumptions to the most important theories regarding the business cycle and resource economics could then be assessed to determine how useful the models were for their intended purposes.

TWO KINDS OF MODELS: OPTIMIZATION VERSUS SIMULATION AND ECONOMETRICS

There are many types of models, and they can be classified in many ways. Models can be static or dynamic, mathematical or physical, stochastic or deterministic. One of the most useful classifications, however, divides models into those that optimize versus those that simulate. The distinction between optimization and simulation models is particularly important since these types of models are suited for fundamentally different purposes.

OPTIMIZATION

The Oxford English Dictionary defines *optimize* as "to make the best of most of; to develop to the utmost." The output of an optimization model is a statement of the best way to accomplish some goal. Optimization models do not tell you what will happen in a certain situation. Instead they tell you what to do in order to make the best of the situation; they are normative or prescriptive models.

Let us take two examples. A nutritionist would like to know how to design meals that fulfill certain dietary requirements but cost as little as possible. A salesperson must visit certain cities and would like to know how to make the trip as quickly as possible, taking into account the available flights between the cities. Rather than relying on trial and error, the nutritionist and the salesperson could use optimization models to determine the best solutions to these problems.

An optimization model typically includes three parts. The *objective function* specifies the goal or objective. For the nutritionist, the objective is to minimize the cost of the meals. For the salesperson, it is to minimize the time spent on the trip. The *decision variables* are the choices to be made. In our examples, these would be the food to serve at each meal and the order in which to visit the cities. The *constraints* restrict the choices of the decision variables to those that are acceptable and possible. In the diet problem, one constraint would specify that daily consumption of each nutrient must equal or exceed the minimum requirement. Another might restrict the number of times a particular food is served during each week. The constraints in the travel problem would specify that each city must be visited at least once and would restrict the selection of routes to actually available connections.

An optimization model takes as inputs these three pieces of information—the goals to be met, the choices to be made, and the constraints to be satisfied. It yields as its output the best solution, i.e., the optimal decisions given the assumptions of the model. In the case of our examples, the models would provide the best set of menus and the most efficient itinerary.

LIMITATIONS OF OPTIMIZATION

Many optimization models have a variety of limitations and problems that a potential user should bear in mind. These problems are: difficulties with the specification of the objective function, unrealistic linearity, lack of feedback, and lack of dynamics.

Specification of the Objective Function: Whose Values? The first difficulty with optimization models is the problem of specifying the objective function, the goal that the model user is trying to reach. In our earlier examples, it was fairly easy to identify the objective functions of the nutritionist and the salesperson, but what would be the objective function for the mayor of New York? To provide adequate city services for minimal taxes? To encourage the arts? To improve traffic conditions? The answer depends, of course, on the perspective of the person you ask.

The objective function embodies values and preferences, but which values, whose preferences? How can intangibles be incorporated into the objective function? How can the conflicting goals of various groups be identified and balanced? These are hard questions, but they are not insurmountable. Intangibles often can be quantified, at least roughly, by breaking them into measurable components. For example, the quality of life in a city might be represented as depending on the rate of unemployment, air pollution levels, crime rate, and so forth. There are also techniques available for extracting information about preferences from interviews and other impressionistic data.

Just the attempt to make values explicit is a worthwhile exercise in any study and may have enormous value for the clients of a modeling project.

It is important that potential users keep in mind the question of values when they examine optimization models. The objective function and the constraints should always be scrutinized to determine what values they embody, both explicitly and by omission. Imagine that a government employee, given responsibility for the placement of sewage treatment plants along a river, decides to use an optimization model in making the decision. The model has as its objective function the cheapest arrangement of plants; a constraint specifies that the arrangement must result in water quality standards being met. It would be important for the user to ask how the model takes into account the impacts the plants will have on fishing, recreation, wild species, and development potential in the areas where they are placed. Unless these considerations are explicitly incorporated into the model, they are implicitly held to be of no value.

Linearity. Another problem, and one that can seriously undermine the verisimilitude of optimization models, is their linearity. Because a typical optimization problem is very complex, involving hundreds or thousands of variables and constraints, the mathematical problem of finding the optimum is extremely difficult. To render such problems tractable, modelers commonly introduce a number of simplifications.

Among these is the assumption that the relationships in the system are linear. In fact, the most popular optimization technique, linear programming, requires that the objective function and all constraints be linear.

Linearity is mathematically convenient, but in reality it is almost always invalid. Consider, for example, a model of a firm's inventory distribution policies. The model contains a specific relationship between inventory and shipments—if the inventory of goods in the warehouse is 10 percent below normal, shipment may be reduced by, say, 2 percent since certain items will be out of stock. If the model requires this relationship to be linear, then a 20 percent shortfall will reduce shipments by 4 percent, a 30 percent shortfall by 6 percent, and so on. And when the shortfall is 100 percent? According to the model, shipments will still be 80 percent of normal. But obviously, when the warehouse is empty, no shipments are possible. The linear relationship within the model leads to an absurdity.

The warehouse model may seem trivial, but the importance of non-linearity is well demonstrated by the sorry fate of the passenger pigeon, *Ectopistes migratorius.* When Europeans first colonized North America, passenger pigeons were extremely abundant. Huge flocks of the migrating birds would darken the skies for days. They often caused damage to crops and were hunted both as a pest and as food. For years, hunting had little apparent impact on the population; the prolific birds seemed to reproduce fast enough to offset most losses. Then the number of pigeons began to decline—slowly at first, then rapidly. By 1914, the passenger pigeon was extinct.

The disappearance of the passenger pigeons resulted from the non-linear relationship between their population density and their fertility. In large flocks they could reproduce at high rates, but in smaller flocks their fertility dropped precipitously. Thus, when hunting pressure was great enough to reduce the size of a flock somewhat, the fertility in that flock also fell. The lower fertility lead to a further decrease in the population size, and the lower population density resulted in yet lower birth rates, and so forth, in a vicious cycle.

Unfortunately, the vast majority of optimizations models assume that the world is linear. There are, however, techniques available for solving certain non-linear optimization problems, and research is continuing.

Lack of Feedback. Complex systems in the real world are highly interconnected, having a high degree of feedback among sectors. The results of decisions feed back through physical, economic, and social channels to alter the conditions on which the decisions were originally made. Some models do not reflect this reality, however. Consider an optimization model that computes the best size of sewage treatment plants to build in an area. The model will probably assume that the amount of sewage needing treatment will remain the same, or that it will grow at a certain rate. But if water quality improves because of sewage treatment, the area will become more attractive and development will increase, ultimately leading to a sewage load greater than expected.

Models that ignore feedback effects must rely on *exogenous variables* and are said to have a narrow boundary. Exogenous variables are ones that influence other variables in the model but are not calculated by the model. They are simply given by a set of numerical values over time, and they do not change in response to feedback. The values of exogenous variables may come from other models but are most likely the product of an unexaminable mental model. The *endogenous variables,* on the other hand, are calculated by the model itself. They are the variables explained by the structure of the model, the ones for which the modeler has an explicit theory, the ones that respond to feedback.

Ignoring feedback can result in policies that generate unanticipated side effects or are diluted, delayed, or defeated by the system (Meadows 1982). An example is the construction of freeways in the 1950s and 1960s to alleviate traffic congestion in major US cities. In Boston it used to take half an hour to drive from the city neighborhood of Dorchester to the downtown area, a journey of only a few miles. Then a limited access highway network was built around the city, and travel time between Dorchester and downtown dropped substantially.

But there's more to the story. Highway construction led to changes that fed back into the system, causing unexpected side effects. Due to the reduction in traffic congestion and commuting time, living in outlying communities became a more attractive option. Farmland was turned into housing developments or paved over to provide yet more roads. The population of the suburbs soared, as people moved out of the center city. Many city stores followed their customers or were squeezed out by competition from the new suburban shopping malls. The inner city began to decay, but many people still worked in the downtown area—and they got there via the new highways. The result? Boston has more congestion and air pollution than before the highways were constructed, and the rush – hour journey from Dorchester to downtown takes half an hour, again.

In theory, feedback can be incorporated into optimization models, but the resulting complexity and non-linearity usually render the problem insoluble. Many optimization models therefore ignore most feedback effects. Potential users should be aware of this when they look at a model. They should ask to what degree important feedbacks have been excluded and how those exclusions might alter the assumptions and invalidate the results of the model.

Lack of Dynamics. Many optimization models are static. They determine the optimal solution for a particular moment in time without regard for how the optimal state is reached or how the system will evolve in the future. An example is the linear programming model constructed in the late 1970s by the US Forest Service, with the objective of optimizing the use of government lands. The model was enormous, with thousands of decision variables and tens of thousands of constraints, and it took months just to correct the typographical errors in the model's huge database. When the completed model was finally run, finding the solution required full use of a mainframe computer for days.

Despite the gigantic effort, the model prescribed the optimal use of forest resources for only a single moment in time. It did not take into account how harvesting a given area would affect its future ecological development. It did not consider how land – use needs or lumber prices might change in the future. It did not examine how long it would take for new trees to grow to maturity in the harvested areas, or what the economic and recreational value of the areas would be during the regrowth period. The model just provided the optimal decisions for a single year, ignoring the fact that those decisions would continue to influence the development of forest resources for decades.

Not all optimization models are static. The MARKAL model, for example, is a large linear programming model designed to determine the optimal choice of energy technologies. Developed at the Brookhaven National Laboratory in the US, the model produces as its output the best (least-cost) mix of coal, oil, gas, and other energy sources well into the next century. It requires various exogenous inputs, such as energy demands, future fuel prices, and construction and operating costs of different energy technologies. (Note that the model ignores feedbacks from energy supply to prices and demand.) The model is dynamic in the sense that it produces a "snapshot" of the optimal state of the system at five-year intervals.

The Brookhaven model is not completely dynamic, however, because it ignores delays. It assumes that people, seeing what the optimal mix is for some future year, begin planning far enough in advance so that this mix can actually be used. Thus the model does not, for example, incorporate construction delays for energy production facilities. In reality, of course, it takes time— often much longer than five years—to build power plants, invent new technologies, build equipment, develop waste management techniques, and find and transport necessary raw materials.

Indeed, delays are pervasive in the real world. The delays found in complex systems are especially important because they are a major source of system instability. The lag time required to carry out a decision or to perceive its effects may cause overreaction or may prevent timely intervention. Acid rain provides a good example. Although there is already evidence that damage to the forests of New England, the Appalachians, and Bavaria is caused by acid rain, many scientists suspect it will take years to determine exactly how acid rain is formed and how it affects the forests. Until scientific and then political consensus emerges, legislative action to curb pollution is not likely to be strong. Pollution control programs, once passed, will take years to implement. Existing power plants and other pollution sources will continue to operate for their functional lifetimes, which are measured in decades. It will require even longer to change settlement patterns and lifestyles dependent on the automobile. By the time sulfur and nitrogen oxide emissions are sufficiently reduced, it may be too late for the forests.

Delays are a crucial component of the dynamic behavior of systems, but—like nonlinearity—they are difficult to incorporate into optimization models. A common simplification is to assume that all delays in the model are of the same fixed length. The results of such models are of questionable value. Policy makers who use them in an effort to find an optimal course of action may discover, like the proverbial American tourist on the back roads of Maine, that "you can't get there from here."

WHEN TO USE OPTIMIZATION

Despite the limitations discussed above, optimization techniques can be extremely useful. But they must be used for the proper problems. Optimization has substantially improved the quality of decisions in many areas, including computer design, airline scheduling, factory siting, and oil refinery operation. Whenever the problem to be solved is one of choosing the best from among a well-defined set of alternatives, optimization should be considered. If the meaning of *best* is also well defined, and if the system to be optimized is relatively static and free of feedback, optimization may well be the best technique to use. Unfortunately, these latter conditions are rarely true for the social, economic, and ecological systems that are frequently of concern to decision makers.

Look out for optimization models that purport to forecast actual behavior. The output of an optimization model is a statement of the best way to accomplish a goal. To interpret the results as a prediction of actual behavior is to assume that people in the real system will in fact make the optimal choices. It is one thing to say, "in order to maximize profits, people should make the following decisions," and quite another to say "people will succeed in maximizing profits, because they will make the following decisions." The former is a prescriptive statement of what to do, the latter a descriptive statement of what will actually happen.

Optimization models are valid for making prescriptive statements. They are valid for forecasting only if people do in fact optimize, do make the best possible decisions. It may seem reasonable to expect people to behave optimally—after all, wouldn't it be irrational to take second best when you could have the best? But the evidence on this score is conclusive: real people do not behave like optimization models. As discussed above, we humans make decisions with simple and incomplete mental models, models that are often based on faulty assumptions or that lead erroneously from sound assumptions to flawed solutions. As Herbert Simon puts it,

> The capacity of the human mind for formulating and solving complex problems is very small compared with the size of the problem whose solution is required for objectively rational behavior in the real world or even for a reasonable approximation to such objective rationality. (Simon 1957, p. 198)

Optimization models augment the limited capacity of the human mind to determine the objectively rational course of action. It should be remembered, however, that even optimization models must make simplifying assumptions in order to be tractable, so the most we can hope from them is an approximation of how people ought to behave. To model how people actually behave requires a very different set of modeling techniques, which will be discussed now.

SIMULATION

The Latin verb *simulare* means to imitate or mimic. The purpose of a simulation model is to mimic the real system so that its behavior can be studied. The model is a laboratory replica of the real system, a *microworld* (Morecroft 1988). By creating a representation of the system in the laboratory, a modeler can perform experiments that are impossible, unethical, or prohibitively expensive in the real world.

Simulations of physical systems are commonplace and range from wind tunnel tests of aircraft design to simulation of weather patterns and the depletion of oil reserves. Economists and social scientists also have used simulation to understand how energy prices affect the economy, how corporations mature, how cities evolve and respond to urban renewal policies, and how population growth interacts with food supply, resources, and the environment. There are many different simulation techniques, including stochastic modeling, system dynamics, discrete simulation, and role-playing games. Despite the differences among them, all simulation techniques share a common approach to modeling.

Optimization models are prescriptive, but simulation models are descriptive. A simulation model does not calculate what should be done to reach a particular goal, but clarifies what would happen in a given situation. The purpose of simulations may be *foresight* (predicting how systems might behave in the future under assumed conditions) or *policy design* (designing new decision-making strategies or organizational structures and evaluating their effects on the behavior of the system).

In other words, simulation models are "what if" tools. Often such "what if" information is more important than knowledge of the optimal decision. For example, during the 1978 debate in the US over natural gas deregulation, President Carter's original proposal was modified dozens of times by Congress before a final compromise was passed. During the congressional debate, the Department of Energy evaluated each version of the bill using a system dynamics model (Department of Energy 1979). The model did not indicate what ought to be done to maximize the economic benefits of natural gas to the nation. Congress already had its own ideas on that score. But by providing an assessment of how each proposal would affect gas prices, supplies, and demands, the model generated ammunition that the Carter administration could use in lobbying for its proposals.

Every simulation model has two main components. First it must include a representation of the physical world relevant to the problem under study. Consider for example a model that was built for the purpose of understanding why America's large cities have continued to decay despite massive amounts of aid and numerous renewal programs (Forrester 1969). The model had to include a representation of the physical components of the city—the size and quality of the infrastructure, including the stock of housing and commercial structures; the attributes of the population, such as its size and composition and the mix of skills and incomes among the people; flows (of people, materials, money, etc.) into and out of the city; and other factors that characterize the physical and institutional setting.

How much detail a model requires about the physical structure of the system will, of course, depend on the specific problem being addressed. The urban model mentioned above required only an aggregate representation of the features common to large American cities. On the other hand, a model designed to improve the location and deployment of fire fighting resources in New York City had to include a detailed representation of the streets and traffic patterns (Greenberger, Crenson, and Crissey 1976).

In addition to reflecting the physical structure of the system, a simulation model must portray the behavior of the actors in the system. In this context, behavior means the way in which people respond to different situations, how they make decisions. The behavioral component is put into the model in the form of decision-making rules, which are determined by direct observation of the actual decision-making procedures in the system.

Given the physical structure of the system and the decision-making rules, the simulation model then plays the role of the decision makers, mimicking their decisions. In the model, as in the real world, the nature and quality of the information available to decision makers will depend on the state of the system. The output of the model will be a description of expected decisions. The validity of the model's assumptions can be checked by comparing the output with the decisions made in the real system.

An example is provided by the pioneering simulation study of corporate behavior carried out by Cyert and March (1963). Their field research showed that department stores used a very simple decision rule to determine the floor price of goods. That rule was basically to mark up the wholesale cost of the items by a fixed percentage, with the value of the markup determined by tradition. They also noted, however, that through time the traditional markup adjusted very slowly, bringing it closer to the actual markup realized on goods when they were sold. The actual markup could vary from the normal markup as the result of several other decision rules: when excess inventory piled up on the shelves, a sale was held and the price was gradually reduced until the goods were sold; if sales goals were exceeded, prices were boosted. Prices were also adjusted toward those of competitors.

Cyert and March built a simulation model of the pricing system, basing it on these decision-making rules. The output of the model was a description of expected prices for goods. When this output was compared with real store data, it was found that the model reproduced quite well the actual pricing decisions of the floor managers.

LIMITATIONS OF SIMULATION

Any model is only as good as its assumptions. In the case of simulation models, the assumptions consist of the descriptions of the physical system and the decision rules. Adequately representing the physical system is usually not a problem; the physical environment can be portrayed with whatever detail and accuracy is needed for the model purpose. Also, simulation models can easily incorporate feedback effects, nonlinearities, and dynamics; they are not rigidly determined in their structure by mathematical limitations as optimization models often are. Indeed, one of the main uses of simulation is to identify how feedback, nonlinearity, and delays interact to produce troubling dynamics that persistently resist solution. (For examples see Sterman 1985, Morecroft 1983, and Forrester 1969.)

Simulation models do have their weak points, however. Most problems occur in the description of the decision rules, the quantification of soft variables, and the choice of the model boundary.

Accuracy of the Decision Rules. The description of the decision rules is one potential trouble spot in a simulation model. The model must accurately represent how the actors in the system make their decisions, even if their decision rules are less than optimal. The model should respond to change in the same way the real actors would. But it will do this only if the model's assumptions faithfully describe the decision rules that are used under different circumstances. The model therefore must reflect the actual decision-making strategies used by the people in the system being modeled, including the limitations and errors of those strategies.

Unfortunately, discovering decision rules is often difficult. They cannot be determined from aggregate statistical data, but must be investigated first hand. Primary data on the behavior of the actors can be acquired through observation of actual decision making in the field, that is, in the boardroom, on the factory floor, along the sales route, in the house-hold. The modeler must discover what information is available to each actor, examine the timeliness and accuracy of that information, and infer how it is processed to yield a decision. Modelers often require the skills of the anthropologist and the ethnographer. One can also learn about decision making through laboratory experiments in which managers operate simulated corporations (Sterman 1989). The best simulation modeling draws on extensive knowledge of decision making that has been developed in many disciplines, including psychology, sociology, and behavioral science.

Soft Variables. The majority of data are soft variables. That is, most of what we know about the world is descriptive, qualitative, difficult to quantify, and has never been recorded. Such information is crucial for understanding and modeling complex systems. Yet in describing decision making, some modelers limit themselves to hard variables, ones that can be measured directly and can be expressed as numerical data. They may defend the rejection of soft variables as being more scientific than "making up" the values of parameters and relationships for which no numerical data are available. How, they ask, can the accuracy of estimates about soft variables be tested? How can statistical tests be performed without numerical data?

Actually, there are no limitations on the inclusion of soft variables in models, and many simulation models do include them. After all, the point of simulation models is to portray decision making as it really is, and soft variables—including intangibles such as desires, product quality, reputation, expectations, and optimism – are often of critical importance in decision making. Imagine, for example, trying to run a school, factory, or city solely on the basis of the available numerical data. Without qualitative knowledge about factors such as operating procedures, organizational structure, political subtleties, and individual motivations, the result would be chaos. Leaving such variables out of models just because of a lack of hard numerical data is certainly less "scientific" than including them and making reasonable estimates of their values. Ignoring a relationship implies that it has a value of zero—probably the only value known to be wrong! (Forrester 1980)

Of course, all relationships and parameters in models, whether based on soft or hard variables, are imprecise and uncertain to some degree. Reasonable people may disagree as to the importance of different factors. Modelers must therefore perform sensitivity analysis to consider how their conclusions might change if other plausible assumptions were made. Sensitivity analysis should not be restricted to uncertainty in parameter values, but should also consider the sensitivity of conclusions to alternative structural assumptions and choices of model boundary.

Sensitivity analysis is no less a responsibility for those modelers who ignore soft variables. Apparently hard data such as economic and demographic statistics are often subject to large measurement errors, biases, distortions, and revisions. Unfortunately, sensitivity analysis is not performed or reported often enough. Many modelers have been embarrassed when third parties, attempting to replicate the results of a model, have found that reasonable alternative assumptions produce radically different conclusions. (See the discussion below of the experiment conducted by the Joint Economic Committee with three leading econometric models.)

Model Boundary. The definition of a reasonable model boundary is another challenge for the builders of simulation models. Which factors will be exogenous, which will be endogenous? What feedbacks will be incorporated into the model? In theory, one of the great strengths of simulation models is the capacity to reflect the important feedback relationships that shape the behavior of the system and its response to policies. In practice, however, many simulation models have very narrow boundaries. They ignore factors outside the expertise of the model builder or the interests of the sponsor, and in doing so they exclude important feedbacks.

The consequences of omitting feedback can be serious. An excellent example is provided by the Project Independence Evaluation System (PIES) model, used in the 1970s by the US Federal Energy Administration and later by the US Department of Energy. As described by the FEA, the purpose of the model was to evaluate different energy strategies according to these criteria: their impact on the development of alternative energy sources, their impact on economic growth, inflation, and unemployment; their regional and social impacts; their vulnerability to import disruptions; and their environmental effects (Federal Energy Administration 1974, P- D-

Surprisingly, considering the stated purpose, the PIES model treated the economy as exogenous. The economy—including economic growth, interest rates, inflation, world oil prices, and the costs of unconventional fuels—was completely unaffected by the US domestic energy situation—including prices, policies, and production. The way the model was constructed, even a full embargo of imported oil or a doubling of oil prices would have no impact on the economy.

Its exogenous treatment of the economy made the PIES model inherently contradictory. The model showed that the investment needs of the energy sector would increase markedly as depletion raised the cost of getting oil out of the ground and synthetic fuels were developed. But at the same time, the model assumed that higher investment needs in the energy sector could be satisfied without reducing investment or consumption in the rest of the economy and without raising interest rates or inflation. In effect, the model let the economy have its pie and eat it too.

In part because it ignored the feedbacks between the energy sector and the rest of the economy, the PIES model consistently proved to be overoptimistic. In 1974 the model projected that by 1985 the US would be well on the way to energy independence: energy imports would be only 3.3 million barrels per day, and production of shale oil would be 250,000 barrels per day. Furthermore, these developments would be accompanied by oil prices of about $22 per barrel (1984 dollars) and by vigorous economic growth. It didn't happen. In fact, at the time this paper is being written (1988), oil imports are about 5.5. million per day, and the shale oil industry remains a dream. This situation prevails despite the huge reductions in oil demand that have resulted from oil prices of over $30 per barrel and from the most serious economic recession since the Great Depression.

A broad model boundary that includes important feedback effects is more important than a great amount of detail in the specification of individual components. It is worth noting that the PIES model provided a breakdown of supply, demand, and price for dozens of fuels in each region of the country. Yet its aggregate projections for 1985 weren't even close. One can legitimately ask what purpose was served by the effort devoted to forecasting the demand for jet fuel or naphtha in the Pacific Northwest when the basic assumptions were so palpably inadequate and the main results so woefully erroneous.

In fairness it must be said that the PIES model is not unique in the magnitude of its errors. Nearly all energy models of all types have consistently been wrong about energy production, consumption, and prices. The evidence shows clearly that energy forecasts actually lag behind the available information, reflecting the past rather than anticipating the future (Department of Energy 1983). A good discussion of the limitations of PIES and other energy models is available in the appendix of Stobaugh and Yergin (1979).

Overly narrow model boundaries are not just a problem in energy analysis. *The Global 2000 Report to the President* (Barney 1980) showed that most of the models used by US government agencies relied significantly on exogenous variables. Population models assumed food production was exogenous. Agriculture models assumed that energy prices and other input prices were exogenous. Energy models assumed that economic growth and environmental conditions were exogenous. Economic models assumed that population and energy prices were exogenous. And so on. Because they ignored important intersectoral feedbacks, the models produced inconsistent results.

ECONOMETRICS

Strictly speaking, econometrics is a simulation technique, but it deserves separate discussion for several reasons. First, it evolved out of economics and statistics, while most other simulation methods emerged from operations research or engineering. The difference in pedigree leads to large differences in purpose and practice. Second, econometrics is one of the most widely used formal modeling techniques. Pioneered by Nobel Prize-winning economists Jan Tinbergen and Lawrence Klein, econometrics is now taught in nearly all business and economics programs. Econometric forecasts are regularly reported in the media, and ready-to use statistical routines for econometric modeling are now available for many personal computers. And third, the well-publicized failure of econometric models to predict the future has eroded the credibility of all types of computer models, including those built for very different purposes and using completely different modeling techniques.

Econometrics is literally the measurement of economic relations, and it originally involved statistical analysis of economic data. As commonly practiced today, econometric modeling includes three stages – specification, estimation, and forecasting. First the structure of the system is specified by a set of equations. Then the values of the parameters (coefficients relating changes in one variable to changes in another) are estimated on the basis of historical data. Finally, the resulting output is used to make forecasts about the future performance of the system.

Specification. The model specification is the description of the model's structure. This structure consists of the relationships among variables, both those that describe the physical setting and those that describe behavior. The relationships are expressed as equations, and a large econometric model may have hundreds or even thousands of equations reflecting the many interrelationships among the variables.

For example, an econometric model of the macroeconomy typically will contain equations specifying the relationship between GNP and consumption, investment, government activity, and international trade.

It also will include behavioral equations that describe how these individual quantities are determined. The modeler may expect, for instance, that high unemployment reduces inflation and vice versa, a relationship known as the Phillips curve. One of the equations in the model will therefore express the Phillips curve, specifying that the rate of inflation depends on the amount of unemployment. Another equation may relate unemployment to the demand for goods, the wage level, and worker productivity. Still other equations may explain wage level in terms of yet other factors.

Not surprisingly, econometrics draws on economic theory to guide the specification of its models. The validity of the models thus depends on the validity of the underlying economic theories. Though there are many flavors of economics, a small set of basic assumptions about human behavior are common to most theories, including modem neoclassical theory and the "rational expectations" school. These assumptions are: optimization, perfect information, and equilibrium.

In econometrics, people (economic agents, in the jargon), are assumed to be concerned with just one thing—maximizing their profits. Consumers are assumed to optimize the "utility" they derive from their resources. Decisions about how much to produce, what goods to purchase, whether to save or borrow, are assumed to be the result of optimization by individual decision makers. Non-economic considerations (defined as any behavior that diverges from profit or utility maximization) are ignored or treated as local aberrations and special cases.

Of course, to optimize, economic agents would need accurate information about the world. The required information would go beyond the current state of affairs; it also would include complete knowledge about available options and their consequences. In most econometric models, such knowledge is assumed to be freely available and accurately known.

Take, for example, an econometric model simulating the operation of a firm that is using an optimal mix of energy, labor, machines, and other inputs in its production process. The model will assume that the firm knows not only the wages of workers and the prices of machines and other inputs, but also the production attainable with different combinations of people and machines, even if those combinations have never been tried. Rational expectation models go so far as to assume that the firm knows future prices, technologies, and possibilities, and that it can perfectly anticipate the consequences of its own actions and those of competitors.

The third assumption is that the economy is in or near equilibrium nearly all of the time. If disturbed, it is usually assumed to return to equilibrium rapidly and in a smooth and stable manner. The prevalence of static thinking is the intellectual legacy of the pioneers of mathematics and economics. During the late nineteenth century, before computers or modem cybernetic theory, the crucial questions of economic theory involved the nature of the equilibrium state for different situations. Given human preferences and the technological possibilities for producing goods, at what prices will commodities be traded, and in what quantities? What will wages be? What will profits be? How will a tax or monopoly influence the equilibrium?

These questions proved difficult enough without tackling the more difficult problem of dynamics, of the behavior of a system in flux. As a result, dynamic economic theory— including the recurrent fluctuations of inflation, of the business cycle, of the growth and decline of industries and nations – remained primarily descriptive and qualitative long after equilibrium theory was expressed mathematically. Even now, dynamic behavior in economics tends to be seen as a transition from one equilibrium to another, and the transition is usually assumed to be stable.

The rich heritage of static theory in economics left a legacy of equilibrium for econometrics. Many econometric models assume that markets are in equilibrium at all times. When adjustment dynamics are modeled, variables are usually assumed to adjust in a smooth and stable manner toward the optimal, equilibrium value, and the lags are nearly always fixed in length. For example, most macroeconometric models assume that capital stocks of firms in the economy adjust to the optimal, profit – maximizing level, with a fixed lag of several years. The lag is the same whether the industries that supply investment goods have the capacity to meet the demand or not. (See, for example, Eckstein 1983 and Jorgenson 1963).

Yet clearly, when the supplying industries have excess capacity, orders can be filled rapidly; when capacity is strained, customers must wait in line for delivery. Whether the dynamic nature of the lag is expressed in a model does make a difference. Models that explicitly include the determinants of the investment delay will yield predictions significantly different from models that assume a fixed investment lag regardless of the physical capability of the economy to fill the demand (Senge 1980). In general, models that explicitly portray delays and their determinants will yield different results from models that simply assume smooth adjustments from one optional state to another.

Estimation. The second stage in econometric modeling is statistical estimation of the parameters of the model. The parameters determine the precise strengths of the relationships specified in the model structure. In the case of the Phillips curve, for example, the modeler would use past data to estimate precisely how strong the relationship between inflation and unemployment has been. Estimating the parameters involves statistical regression routines that are, in essence, fancy curve-fitting techniques. Statistical parameter estimates characterize the degree of correlation among the variables. They use historical data to determine parameter values that best match the data themselves.

All modeling methods must specify the structure of the system and estimate parameters. The use of statistical procedures to derive the parameters of the model is the hallmark of econometrics and distinguishes it from other forms of simulation. It gives econometricians an insatiable appetite for numerical data, for without numerical data they cannot carry out the statistical procedures used to estimate the models. It is no accident that the rise of econometrics went hand in hand with the quantification of economic life. The development of the national income and produce accounts by Simon Kuznets in the 1930s was a major advance in the codification of economic data, permitting consistent measures of economic activity at the national level for the first time. To this day all major macroeconometric models rely heavily on the national accounts data, and indeed macroeconomic theory itself has adapted to the national accounts framework.

Forecasting. The third step in econometric modeling is forecasting, making predictions about how the real system will behave in the future. In this step, the modeler provides estimates of the future values of the exogenous variables, that is, those variables that influence the other variables in the model but aren't themselves influenced by the model. An econometric model may have dozens of exogenous variables, and each must be forecast before the model can be used to predict.

LIMITATIONS OF ECONOMETRIC MODELING

The chief weak spots in econometric models stem from the assumptions of the underlying economic theory on which they rest: assumptions about the rationality of human behavior, about the availability of information that real decision makers do not have, and about equilibrium. Many economists acknowledge the idealization and abstraction of these assumptions, but at the same time point to the powerful results that have been derived from them. However, a growing number of prominent economists now argue that these assumptions are not just abstract— they are false. In his presidential address to the British Royal Economics Society, E. H. Phelps-Brown said:

> The trouble here is not that the behavior of these economic chessmen has been simplified, for simplification seems to be part of all understanding. The trouble is that the behavior posited is not known to be what obtains in the actual economy. (Phelps-Brown 1972, p. 4)

Nicholas Kaldor of Cambridge University is even more blunt:

> ...in my view, the prevailing theory of value – what I called, in a shorthand way, "equilibrium economics"—is barren and irrelevant as an apparatus of thought... (Kaldor 1972, p. 1237)

As mentioned earlier, a vast body of empirical research in psychology and organizational studies has shown that people do not optimize or act as if they optimize, that they don't have the mental capabilities to optimize their decisions, that even if they had the computational power necessary, they lack the information needed to optimize. Instead, they try to satisfy a variety of personal and organizational goals, use standard operating procedures to routinize decision making, and ignore much of the available information to reduce the complexity of the problems they face. Herbert Simon, in his acceptance speech for the 1978 Nobel Prize in economics, concludes:

> There can no longer be any doubt that the micro assumptions of the theory—the assumptions of perfect rationality—are contrary to fact. It is not a question of approximation; they do not even remotely describe the processes that human beings use for making decisions in complex situations (Simon 1979, p. 510).

Econometrics also contains inherent statistical limitations. The regression procedures used to estimate parameters yield unbiased estimates only under certain conditions.

These conditions are known *as maintained hypotheses* because they are assumptions that must be made in order to use the statistical technique. The maintained hypotheses can never be verified, even in principle, but must be taken as a matter of faith. In the most common regression technique, ordinary least squares, the maintained hypotheses include the unlikely assumptions that the variables are all measured perfectly, that the model being estimated corresponds perfectly to the real world, and the random errors in the variables from one time period to another are completely independent. More sophisticated techniques do not impose such restrictive assumptions, but they always involve other a priori hypotheses that cannot be validated.

Another problem is that econometrics fails to distinguish between correlations and causal relationships. Simulation models must portray the causal relationships in a system if they are to mimic its behavior, especially its behavior in new situations. But the statistical techniques used to estimate parameters in econometric models don't prove whether a relationship is causal. They only reveal the degree of past correlation between the variables, and these correlations may change or shift as the system evolves. The prominent economist Robert Lucas (1976) makes the same point in a different context.

Consider the Phillips curve as an example. Though economists often interpreted the Phillips curve as a causal relationship—a policy trade-off between inflation and unemployment—it never did represent the causal forces that determine inflation or wage increases. Rather, the Phillips curve was simply a way of restating the past behavior of the system. In the past, Phillips said, low unemployment had tended to occur at the same time inflation was high, and vice-versa. Then, sometime in the early 1970s, the Phillips curve stopped working; inflation rose while unemployment worsened. Among the explanations given by economists was that the structure of the system had changed. But a modeler's appeal to "structural change" usually means that the inadequate structure of the model has to be altered because it failed to anticipate the behavior of the real system!

What actually occurred in the 1970s was that, when inflation swept prices to levels unprecedented in the industrial era, people learned to expect continuing increases. As a result of the adaptive feedback process of learning, they learned to deal with high inflation through indexing, COLAs, inflation – adjusting accounting, and other adjustments. The structure, the causal relationships of the system, did not change. Instead, causal relationships that had been present all along (but were dormant in an era of low inflation) gradually became active determinants of behavior as inflation worsened. In particular, the ability of people to adapt to continuing inflation existed all along but wasn't tested until inflation became high enough and persistent enough. Then the behavior of the system changed, and the historical correlation between inflation and unemployment broke down.

The reliance of econometric estimation on numerical data is another of its weaknesses. The narrow focus on hard data blinds modelers to less tangible but no less important factors. They ignore both potentially observable quantities that haven't been measured yet and ones for which no numerical data exist. (Alternatively, they may express an unmeasured factor with a proxy variable for which data already exists, even though the relationship between the two is tenuous—as when educational expenditure per capita is used as a proxy for the literacy of a population.)

Among the factors excluded from econometric models because of the hard data focus are many important determinants of decision making, including desires, goals, and perceptions. Numerical data may measure the results of human decision making, but numbers don't explain how or why people made particular decisions. As a result, econometric models cannot be used to anticipate how people would react to a change in decision-making circumstances.

Similarly, econometric models are unable to provide a guide to performance under conditions that have not been experienced previously. Econometricians assume that the correlations indicated by the historical data will remain valid in the future. In reality, those data usually span a limited range and provide no guidance outside historical experience. As a result, econometric models are often less than robust: faced with new policies or conditions, the models break down and lead to inconsistent results.

An example is the model used by Data Resources, Inc. in 1979 to test policies aimed at eliminating oil imports. On the basis of historical numerical data, the model assumed that the response of oil demand to the price of oil was rather weak—a 10 percent increase in oil price caused a reduction of oil demand of only 2 percent, even in the long run. According to the model, for consumption to be reduced by 50 percent (enough to cut imports to zero at the time), oil would have to rise to $800 per barrel. Yet at that price, the annual oil bill for the remaining 50 percent would have exceeded the total GNP for that year, an impossibility (Sterman 1981). The model's reliance on historical data led to inconsistencies. (Today, with the benefit of hindsight, economists agree that oil demand is much more responsive to price than was earlier believed. Yet considering the robustness of the model under extreme conditions could have revealed the problem much earlier.)

Validation is another problem area in econometric modeling. The dominant criterion used by econometric modelers to determine the validity of an equation or a model is the degree to which it fits the data. Many econometrics texts (e.g., Pindyck and Rubinfeld 1976) teach that the statistical significance of the estimated parameters in an equation is an indicator of the correctness of the relationship. Such views are mistaken. Statistical significance indicates how well an equation fits the observed data; it does not indicate whether a relationship is a correct or true characterization of the way the world works. A statistically significant relationship between variables in an equation shows that they are highly correlated and that the apparent correlation is not likely to have been the result of mere chance. But it does not indicate that the relationship is causal at all.

Using statistical significance as the test of model validity can lead modelers to mistake historical correlations for causal relationships. It also can cause them to reject valid equations describing important relationships. They may, for example, exclude an equation as statistically insignificant simply because there are few data about the variables, or because the data don't contain enough information to allow the application of statistical procedures.

Ironically, a lack of statistical significance does not necessarily lead econometric modelers to the conclusion that the model or the equation is invalid. When an assumed relationship fails to be statistically significant, the modeler may try another specification for the equation, hoping to get a better statistical fit. Without recourse to descriptive, microlevel data, the resulting equations may be ad hoc and bear only slight resemblance to either economic theory or actual behavior. Alternatively, the modelers may attempt to explain the discrepancy between the model and the behavior of the real system by blaming it on faulty data collection, exogenous influences, or other factors.

The Phillips curve again provides an example. When it broke down, numerous revisions of the equations were made. These attempts to find a better statistical fit met with limited success. Some analysts took another tack, pointing to the oil price shock, Russian wheat deal, or other one-of-a-kind events as the explanation for the change. Still others argued that there had been structural changes that caused the Phillips curve to shift out to higher levels of unemployment for any given inflation rate.

These flaws in econometrics have generated serious criticism from within the economic profession. Phelps-Brown notes that because controlled experiments are generally impossible in economics "running regressions between time series is only likely to deceive" (Phelps-Brown 1972, p. 6). Lester Thurow notes that econometrics has failed as a method for testing theories and is now used primarily as a "showcase for exhibiting theories." Yet as a device for advocacy, econometrics imposes few constraints on the prejudices of the modeler. Thurow concludes:

> By simple random search, the analyst looks for the set of variables and functional forms that give the best equations. In this context the best equation is going to depend heavily upon the prior beliefs of the analyst. If the analyst believes that interest rates do not affect the velocity of money, he finds a 'best' equation that validates his particular prior belief. If the analyst believes that interest rates do affect the velocity of money, he finds a 'best' equation that validates this prior belief. (Thurow 1983, pp. 107–8)

But the harshest assessment of all comes from Nobel laureate Wassily Leontief:

> Year after year economic theorists continue to produce scores of mathematical models and to explore in great detail their formal properties; and the econometricians fit algebraic functions of all possible shapes to essentially the same sets of data without being able to advance, in any perceptible way, a systematic understanding of the structure and the operations of a real economic system. (Leontief 1982, p. 107; see also Leontief 1971.)

But surely such theoretical problems matter little if the econometric models provide accurate predictions. After all, the prime purpose of econometric models is short-term prediction of the exact future state of the economy, and most of the attributes of econometrics (including the use of regression techniques to pick the "best" parameters from the available numerical data and the extensive reliance on exogenous variables) have evolved in response to this predictive purpose.

Unfortunately, econometrics fails on this score also; in practice, econometric models do not predict very well. The predictive power of econometric models, even over the short-term (one to four years), is poor and virtually indistinguishable from that of other forecasting methods. There are several reasons for this failure to predict accurately.

As noted earlier, in order to forecast, the modeler must provide estimates of the future values of the exogenous variables, and an econometric model may have dozens of these variables. The source of the forecasts for these variables may be other models but usually is the intuition and judgment of the modeler. Forecasting the exogenous variables consistently, much less correctly, is difficult.

Not surprisingly, the forecasts produced by econometric models often don't square with the modeler's intuition. When they feel the model output is wrong, many modelers, including those at the "big three" econometric forecasting firms—Chase Econometrics, Wharton Econometric Forecasting Associates, and Data Resources—simply adjust their forecasts. This fudging, or add factoring as they call it, is routine and extensive. The late Otto Eckstein of Data Resources admitted that their forecasts were 60 percent model and 40 percent judgment ("Forecasters Overhaul Models of Economy in Wake of 1982 Errors," *Wall Street*

Journal, 77 February 1983). *Business Week* ("Where Big Econometric Models Go Wrong," 30 March 1981) quotes an economist who points out that there is no way of knowing where the Wharton model ends and the model's developer, Larry Klein, takes over. Of course, the adjustments made by add factoring are strongly colored by the personalities and political philosophies of the modelers. In the article cited above, the *Wall Street Journal* quotes Otto Eckstein as conceding that his forecasts sometimes reflect an optimistic view: "Data Resources is the most influential forecasting firm in the country…If it were in the hands of a doom-and-gloomer, it would be bad for the country."

In a revealing experiment, the Joint Economic Committee of Congress (through the politically neutral General Accounting Office) asked these three econometric forecasting firms (DRI, Chase, and Wharton) to make a series of simulations with their models, running the models under different assumptions about monetary policy. One set of forecasts was "managed" or add factored by the forecasters at each firm. The other set consisted of pure forecasts, made by the GAO using the untainted results of the models. As an illustration of the inconsistencies revealed by the experiment, consider the following: when the money supply was assumed to be fixed, the DRI model forecast that after ten years the interest rate would be 34 percent, a result totally contrary to both economic theory and historical experience. The forecast was then add factored down to a more reasonable 7 percent. The other models fared little better, revealing both the inability of the pure models to yield meaningful results and the extensive ad hoc adjustments made by the forecasters to render the results palatable (Joint Economic Committee 1982).

Add factoring has been criticized by other economists on the grounds that it is unscientific. They point out that, although the mental models used to add factor are the mental models of seasoned experts, these experts are subject to the same cognitive limitations other people face. And whether good or bad, the assumptions behind add factoring are always unexaminable.

The failure of econometric models have not gone unnoticed. A representative sampling of articles in the business press on the topic of econometric forecasting include the following headlines:

- "1980: The Year The Forecasters Really Blew It." (Business Week, 14 July 1980).
- "Where The Big Econometric Models Go Wrong." (Business Week, 30 March 1981).
- "Forecasters Overhaul Models of Economy in Wake of 1982 Errors." (Wall Street Journal, 17 February, 1983).
- "Business Forecasters Find Demand Is Weak in Their Own Business: Bad Predictions Are Factor." (Wall Street Journal, 7 September 1984).
- "Economists Missing The Mark: More Tools, Bigger Errors." (New York Times, 12 December 1984).

The result of these failures has been an erosion of credibility regarding all computer models no matter what their purpose, not just econometric models designed for prediction. This is unfortunate. Econometric models are poor *forecasting* tools, but well-designed simulation models can be valuable tools for *foresight* and *policy design*. Foresight is the ability to anticipate how the system will behave if and when certain changes occur. It is not forecasting, and it does not depend on the ability to predict. In fact, there is substantial agreement among modelers of global problems that exact, point prediction of the future is neither possible nor necessary:

> ...at present we are far from being able to predict social-system behavior except perhaps for carefully selected systems in the very short term. Effort spent on attempts at precise prediction is almost surely wasted, and results that purport to be such predictions are certainly misleading. On the other hand, much can be learned from models in the form of broad, qualitative, conditional understanding—and this kind of understanding is useful (and typically the only basis) for policy formulation. If your doctor tells you that you will have a heart attack if you do not stop smoking, this advice is helpful, even if it does not tell you exactly when a heart attack will occur or how bad it will be. (Meadows, Richardson, and Bruckmann 1982, p. 279)

Of course, policy evaluation and foresight depend on an accurate knowledge of the history and current state of the world, and econometrics has been a valuable stimulus to the development of much-needed data gathering and measurement by governments and private companies. But econometric models do not seem well-suited to the types of problems of concern in policy analysis and foresight. Though these models purport to simulate human behavior, they in fact rely on unrealistic assumptions about the motivations of real people and the information available to them. Though the models must represent the physical world, they commonly ignore dynamic processes, disequilibrium, and the physical basis for delays between actions and results. Though they may incorporate hundreds of variables, they often ignore soft variables and unmeasured quantities. In real systems the feedback relationships between environmental, demographic, and social factors are usually as important as economic influences, but econometric models often omit these because numerical data are not available. Furthermore, econometrics usually deals with the short term, while foresight takes a longer view. Over the time span that is of concern in foresight, real systems are likely to deviate from their past recorded behaviors, making unreliable the historical correlations on which econometric models are based.

CHECKLIST FOR THE MODEL CONSUMER

The preceding discussion has focused on the limitations of various modeling approaches in order to provide potential model consumers with a sense of what to look out for when choosing a model. Despite the limitations of modeling, there is no doubt that computer models can be and have been extremely useful foresight tools. Well-built models offer significant advantages over the often faulty mental models currently in use.

The following checklist provides further assistance to decision makers who are potential model users. It outlines some of the key questions that should be asked to evaluate the validity of a model and its appropriateness as a tool for solving a specific problem.

• What is the problem at hand? What is the problem addressed by the model?
• What is the boundary of the model? What factors are endogenous? Exogenous? Excluded? Are soft variables included? Are feedback effects properly taken into account? Does the model capture possible side effects, both harmful and beneficial?

- What is the time horizon relevant to the problem? Does the model include as endogenous components those factors that may change significantly over the time horizon?
- Are people assumed to act rationally and to optimize their performance? Does the model take non-economic behavior (organizational realities, non-economic motives, political factors, cognitive limitations) into account?
- Does the model assume people have perfect information about the future and about the way the system works, or does it take into account the limitations, delays, and errors in acquiring information that plague decision makers in the real world?
- Are appropriate time delays, constraints, and possible bottlenecks taken into account?
- Is the model robust in the face of extreme variations in input assumptions?
- Are the policy recommendations derived from the model sensitive to plausible variations in its assumptions?
- Are the results of the model reproducible? Or are they adjusted (add factored) by the model builder?
- Is the model currently operated by the team that built it? How long does it take for the model team to evaluate a new situation, modify the model, and incorporate new data?
- Is the model documented? Is the documentation publicly available? Can third parties use the model and run their own analyses with it?

CONCLUSIONS

The inherent strengths and weaknesses of computer models have crucial implications for their application in foresight and policy analysis. Intelligent decision-making requires the appropriate use of many different models designed for specific purposes—not reliance on a single, comprehensive model of the world. To repeat a dictum offered above, "Beware the analyst who proposes to model an entire social or economic system rather than a problem." It is simply not possible to build a single, integrated model of the world, into which mathematical inputs can be inserted and out of which will flow a coherent and useful understanding of world trends.

To be used responsibly, models must be subjected to debate. A cross-disciplinary approach is needed; models designed by experts in different fields and for different purposes must be compared, contrasted, and criticized. The foresight process should foster such review.

The history of global modeling provides a good example. The initial global modeling efforts, published in *World Dynamics* (Forrester 1971) and *The Limits To Growth* (Meadows et al. 1972) provoked a storm of controversy. A number of critiques appeared, and other global models were soon developed. Over a period of ten years, the International Institute for Applied Systems Analysis (IIASA) conducted a program of analysis and critical review in which the designers of global models were brought together. Six major symposia were held, and eight important global models were examined and discussed. These models had different purposes, used a range of modeling techniques, and were built by persons with widely varying backgrounds. Even after the IIASA conferences, there remain large areas of methodological and substantive disagreement among the modelers. Yet despite these differences, consensus did emerge on a number of crucial issues (Meadows, Richardson, and Bruckmann 1982), including the following:

- Physical and technical resources exist to satisfy the basic needs of all the world's people into the foreseeable future.
- Population and material growth cannot continue forever on a finite planet.
- Continuing "business as usual" policies in the next decades will not result in a desirable future nor even in the satisfaction of basic human needs.
- Technical solutions alone are not sufficient to satisfy basic needs or create a desirable future.

The IIASA program on global modeling represents the most comprehensive effort to date to use computer models as a way to improve human understanding of social issues. The debate about the models created agreement on crucial issues where none had existed. The program helped to guide further research and provided a standard for the effective conduct of foresight in both the public and private sectors.

At the moment, model-based analyses usually take the form of studies commissioned by policy makers. The clients sit and wait for the final reports, largely ignorant of the methods, assumptions, and biases that the modelers put into the models. The policy makers are thus placed in the role of supplicants awaiting the prophecies of an oracle. When the report finally arrives, they may, like King Croesus before the Oracle at Delphi, interpret the results in accordance with their own preconceptions. If the results are unfavorable, they may simply ignore them. Policy makers who use models as black boxes, who accept them without scrutinizing their assumptions, who do not examine the sensitivity of the conclusions to variations in premises, who do not engage the model builders in dialogue, are little different from the Delphic supplicants or the patrons of astrologers. And these policy makers justly alarm critics, who worry that black box modeling abdicates to the modelers and the computer a fundamental human responsibility (Weizenbaum 1976).

No one can (or should) make decisions on the basis of computer model results that are simply presented, "take 'em or leave 'em." In fact, the primary function of model building should be educational rather than predictive. Models should not be used as a substitute for critical thought, but as a tool for improving judgment and intuition. Promising efforts in corporations, universities, and public education are described in Senge 1989; Graham, Senge, Sterman, and Morecroft 1989; Kim 1989; and Richmond 1987.

Towards that end, the role of computer models in policy making needs to be redefined. What is the point of computer modeling? It should be remembered that we all use models of some sort to make decisions and to solve problems. Most of the pressing issues with which public policy is concerned are currently being handled solely with mental models, and those mental models are failing to resolve the problems. The alternative to continued reliance on mental models is computer modeling. But why turn to computer models if they too are far from perfect?

The value in computer models derives from the differences between them and mental models. When the conflicting results of a mental and a computer model are analyzed, when the underlying causes of the differences are identified, both of the models can be improved.

Computer modeling is thus an essential part of the educational process rather than a technology for producing answers. The success of this dialectic depends on our ability to create and learn from shared understandings of our models, both mental and computer.

Properly used, computer models can improve the mental models upon which decisions are actually based and contribute to the solution of the pressing problems we face.

REFERENCES

Barney, Gerald O., ed. 1980. *The Global 2000 Report to the President.* 3 vols. Washington, DC: US Government Printing Office.

Business Forecasters Find Demand Is Weak in Their Own Business: Bad Predictions Are Factor. *Wall Street Journal, 7* September 1984.

Cyert, R., and March, J. 1963. A *Behavioral Theory of the Firm.* Englewood Cliffs, NJ: Prentice Hall.

Department of Energy. 1979. *National Energy Plan II.* DOE/TIC-10203. Washington, DC: Department of Energy.

———. 1983. *Energy Projections to the Year 2000. Washington, DC: Department of Energy, Office of Policy, Planning, and Analysis.*

Eckstein, O. *1983. The DRI Model of the US Economy.* New York, McGraw Hill.

Economists Missing the Mark: More Tools, Bigger Errors. *New York Times,* 12 December 1984.

Federal Energy Administration. 1974. *Project Independence Report.* Washington, DC: Federal Energy Administration.

Forecasters Overhaul Models of Economy in Wake of 1982 Errors. *Wall Street Journal, 17* February, 1983.

Forrester, Jay W. 1969. *Urban Dynamics.* Cambridge, Mass.: MIT Press.

———. 1971. *World Dynamics. Cambridge, Mass.: MIT Press.*

———. 1980. Information Sources for Modeling the National Economy. *Journal of the American Statistical Association* 75(371):555–574.

Graham, Alan K.; Senge, Peter M.; Sterman, John D.; and Morecroft, John D. W. 1989. Computer Based Case Studies in Management Education and Research. In *Computer-Based Management of Complex Systems,* eds. P. Milling and E. Zahn, pp. 317–326. Berlin: Springer Verlag.

Grant, Lindsey, ed. 1988. *Foresight and National Decisions: The Horseman and the Bureaucrat.* Lanham, Md.: University Press of America.

Greenberger, M., Crenson, M. A., and Crissey, B. L. 1976. *Models in the Policy Process.* New York: Russell Sage Foundation.

Hogarth, R. M. 1980. *Judgment and Choice.* New York: Wiley.

Joint Economic Committee. 1982. *Three Large Scale Model Simulations of Four Money Growth Scenarios.* Prepared for subcommittee on Monetary and Fiscal Policy, 97th Congress 2nd Session, Washington, DC

Jorgenson, D. W. 1963. Capital Theory and Investment Behavior. *American Economic Review* 53:247–259.

Kahneman, D., Slovic, P., and Tversky, A. 1982. *Judgment Under Uncertainty:* Heuristics and Biases. Cambridge: Cambridge University Press.

Kaldor, Nicholas. 1972. The Irrelevance of Equilibrium Economics. *The Economic Journals:* 1237–55.

Kim, D. 1989. Learning Laboratories: Designing a Reflective Learning Environment. In *Computer-Based Management of Complex Systems, P.* Milling and E. Zahn, eds., pp. 327–334. Berlin: Springer Verlag

Leontief, Wassily. 1971. Theoretical Assumptions and Nonobserved Facts. *American Economic Review* 61(1): 1–7.

———. 1982. Academic Economics. *Science,* 217: 104–107.

Lucas, R. 1976. Econometric Policy Evaluation: A Critique. In *The Phillips Curve and Labor Markets,* K. Brunner and A. Meltzer, eds. Amsterdam: North-Holland.

Meadows, Donella H.; Meadows, Dennis L.; Randers, Jorgen.; and Behrens, William W. 1972. *The Limits to Growth.* New York: Universe Books.

Meadows, Donella H. 1982. Whole Earth Models and Systems. *CoEvolution Quarterly,* Summer 1982, pp. 98–108.

Meadows, Donella H.; Richardson, John; and Bruckmann, Gerhart 1982. *Groping in The Dark.* Somerset, NJ: Wiley.

Morecroft, John D. W. 1983. System Dynamics: Portraying Bounded Rationality. *Omega II:* 131–142.

———. 1988. System Dynamics and Microworlds for Policy Makers. *European Journal of Operational Research* 35(5):301–320.

———. 1980: The Year The Forecasters Really Blew It *Business Week,* 14 July 1980.

Phelps-Brown, E. H. 1972. The Underdevelopment of Economics. *The* Economic Journal 82:1–10.

Pindyck, R., and Rubinfeld, D. 1976. *Econometric Models and Economic Forecasts.* New York: McGraw Hill.

Richmond, B. 1987. *The Strategic Forum.* High Performance Systems, Inc., 45 Lyme Road, Hanover, NH 03755, USA.

Senge, Peter M. 1980. A System Dynamics Approach to Investment Function Formulation and Testing. *Socioeconomic Planning Sciences* 14:269–280.

———. 1989. Catalyzing Systems Thinking Within Organizations. In *Advances in Organization Development,* F. Masaryk, ed., forthcoming.

Simon, Herbert. 1947. *Administrative Behavior.* New York: MacMillan.

———. 1957. *Models of Man.* New York: Wiley.

———. 1979. Rational Decisionmaking in Business Organizations. *American Economic Review* 69:493–513.

Sterman, John D. 1981. The Energy Transition and the Economy: A System Dynamics Approach. Ph.D. dissertation, Massachusetts Institute of Technology, Cambridge.

———. 1985. A Behavioral Model of the Economic Long Wave. *Journal of Economic Behavior and Organization* 6(1): 17–53.

———. 1989. Modeling Managerial Behavior: Misperceptions of Feedback in a Dynamic Decision Making Experiment. *Management Science* 35(3):321–339.

Stobaugh, Robert and Yergin, Daniel. 1979. *Energy Future.* New York: Random House.

Thurow, Lester. 1983. *Dangerous Currents.* New York: Random House.

Weizenbaum, J. 1976. *Computer Power and Human Reason: from Judgment to Calculation.* San Francisco: W. H. Freeman.

Where The *Big* Econometric Models Go Wrong. *Business Week,* 30 March 1981.

The author wishes to acknowledge that many of the ideas expressed here emerged from discussions with or were first formulated by, among others, Jay Forrester, George Richardson, Peter Senge, and especially Donella Meadows, whose book Groping in the Dark (Meadows, Richardson, and Bruckmann 1982), was particularly helpful. Robert Burns, "To a Mouse".

"Scenario Planning," "Scenario Methodology" and "Driving Forces." In *Puget Sound Future Scenarios*. UW Urban Ecology Research Lab, University of Washington.

Reading Commentary

You are asked to create a management plan for a national park for 2050. How do you plan so far ahead? Do you assume that everything will be the same? Or, do you plan for change? What kind? How do you decide? Scenario planning is a systematic method of planning given a great deal of uncertainty and a complex future. By identifying key drivers of change and how they interact, scenario planning makes it possible to compare various possible strategies in light of alternative futures.

While most of the tools we have presented in this chapter wish away most of the uncertainty surrounding environmental problem-solving, scenario planning embraces the fact that the future is uncertain. There is no right way to plan for 2050. There are just too many unknowns. Imagining multiple futures, however, lets planners take a wide range of possibilities into account. It is possible to plan without specifying a single version of the future.

Scenario planning also accepts that multiple factors may have a major effect on the environment going forward, and that these factors can influence or impact each other. In the Puget Sound Case, the scenario planners identified two key drivers—climate change and changes in human behavior. They also postulated four trajectories based on different assumptions about the rate of growth, shifting socioeconomic characteristics, modifications in governance arrangements and different levels of regional investments. They assumed that each trajectory would influence the key drivers differently. Overall, the scenario-planning approach allowed the participants to sort out future possibilities without having to commit to one "best estimate" of the future.

The process of scenario planning incorporates multiple viewpoints and many different types of information. Usually, this requires that many different people with a wide range of viewpoints need to be involved. This collaborative approach tends to generate ideas that might otherwise be overlooked. For example, 152 experts from 88 agencies (from a variety of fields) participated in the Puget Sound scenario-planning effort.

This reading is an excerpt from a longer report. The full study produced six scenarios (named "forward," "order," "innovation," "barriers," "collapse" and "adaptation") and compared them in order to suggest how a large-scale restoration effort ought to proceed. You can take a look at the full report online if you would like to see the complete results of the scenario-planning effort.

CHAPTER 3
SCENARIO PLANNINGUW URBAN ECOLOGY RESEARCH LAB

WHY SCENARIOS?

Our ability to predict future conditions is critical to inform management and planning decisions. However, long–range futures are difficult to predict due to the complex interactions and uncertainty of important driving forces. Thus our assumptions about predictability and uncertainty ultimately influence our ability to conduct effective future assessments. While past observations are important to inform our expectations of the future, the interactions among uncertain driving forces may create novel conditions that fall outside the expected uncertainty range. Scenarios provide an effective approach for planning and decision making by addressing the expanded range of uncertainty. Scenarios illuminate opportunities and risks by exploring the most divergent and relevant plausible future conditions that can emerge due to the interactions of multiple uncertain driving forces.

Visioning and forecasts are traditionally used in strategic planning as tools for envisioning desirable futures and projecting current trajectories. Scenarios go beyond visions and predictions by looking at a complete suite of plausible, divergent and compelling futures. Table 3.1 describes the main differences among forecasts, visions and scenarios in terms of their future views, assumptions and approaches. Scenarios are appropriate to tackle long time frames and complex environments. When the future is dominated by uncertainty, scenario planning allows decision makers to consider multiple plausible futures generated by the interactions of uncertain driving forces.

WHAT IS SCENARIO PLANNING?

Scenario Planning is a strategic planning approach for making long-term decisions. Rather than focusing on the predictions of a single outcome, scenarios examine the interactions of various key uncertain factors that may create alternative plausible futures (Figure 3.1). Scenarios are hypotheses of alternative plausible futures designed to highlight the risks and opportunities and assess the effectiveness of alternative strategies. The suite of scenarios incorporates the most divergent future conditions in order to rigorously and systematically test the efficacy of alternative strategies. Simply put, scenarios help us ask: If the future turns out differently than originally anticipated, will our strategy still work?

Scenarios are useful when the uncertainty is high and the risk associated with forecasting the wrong trajectory is great. Scenarios start by identifying the drivers of change influencing the issue of interest (focal issue), and providing insight about the direction for the assessment process, i.e. what questions should managers be asking. Then, by considering the uncertainty of key driving forces, scenarios reveal the implications of potential future trajectories. Scenarios challenge managers' assumptions about the future in a way that a single forecast cannot. Scenario planning is based on the premise that by exploring the most divergent

Table 3.1 Forecasts, visions, and scenarios

Future View	Their beliefs	Their Approach	Tool
Extrapolators	Believe that the future will represent a logical extension of the past.	Identify past trends and extrapolate them in a reasoned and logical manner.	Forecasts
Goal Analysts	'The future will be determined by the beliefs and actions of various individuals, organizations and institutions, and therefore the future is modifiable by these entities.	Project the future by examining the stated and implied goals of various decisions makers and trendsetters.	Visions
Scenario Planners	The uncertainty and complexity of future conditions is largely controlled by the interaction of the most important and uncertain key driving forces affecting a focal issue.	Develop multiple divergent, relevant and plausible scenarios to illustrate risk and opportunities of strategies	Scenarios

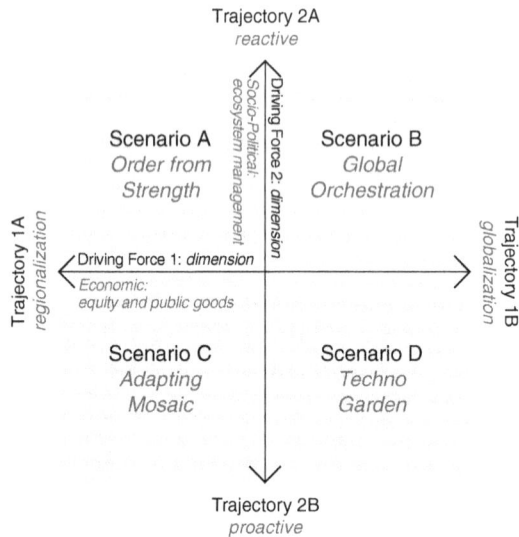

Figure 3.1 Anatomy of a scenario
(Millenium Ecosystem Assessment, 2003)

plausible future conditions, managers can illuminate options and risks that would otherwise be hidden or dismissed. While the scenario development process is significantly more complex and resource intensive than a singular forecast, the benefit gained is the ability to assess the robustness of alternative strategies under plausible future conditions.

Scenario planning emerged in WWII when the US Air Force needed to anticipate what its opponents would do (Lindgren et al, 2003). A decade after the war, Herman Kahn brought scenario planning into the business world through the Hudson Institute (Schwartz, 1991). Perhaps the most well known use of scenarios comes from Pierre Wack's work with Royal Dutch Shell. Wack helped his managers believe and prepare for a world where oil prices would increase dramatically (Schwartz, 1991). Soon after Shell's success scenarios became a common strategic tool. However, while the free form process of early scenario planning led to original thinking, it kept the connections to decision making loose (Ringland, 1998). Only during the last two decades has a more structured approach emerged. The Puget Sound Future Scenarios are based on the Intuitive Logics School (ILS) method and practiced by the Stanford Research Institute, reflecting a structured approach to scenario writing true to the original intent (Ringland, 1998).

While scenarios have been most heavily used within the business environment, recent ecosystem management projects have been utilizing scenario planning's ability to portray complex information and an uncertain future. Some of the most impressive projects include the Millennium Ecosystem Assessment (MEA), the Intergovernmental Panel on Climate Change (IPCC), California's Water Plan 2005, and the Northern Highland Lakes District research (MEA 2003; IPCC 2000; Department of Water Resources 2005; Peterson et al 2003). Published reports point towards an intensive and demanding process, a collaboration with a diverse set of experts, and an ability to creatively see risks and opportunities that were not obvious at the onset of the study (MEA, 2003; Peterson et al, 2003).

HOW TO DEVELOP SCENARIOS

At first, scenarios may seem like stories that are creatively written, much like novels. But while there are several variations on how to conduct scenario planning, scenarios differ from fictional stories by being structurally and explicitly grounded in scientific knowledge. Scenarios focus on key drivers, complex interactions and irreducible uncertainties in order to generate the futures within which we can assess alternative strategies. According to the ILS method, scenario planning generally involves eight key steps (Schwartz 1991; Peterson et al 2003; Lindgren et al, 2003) (Appendix A: 8 Steps of Scenario Building):

1. Identify focal issue or decision
2. Identify driving forces
3. Rank importance & uncertainty
4. Select the scenario logics
5. Develop the scenarios
6. Select metrics for monitoring
7. Assess impacts for different scenarios
8. Evaluate alternative strategies

DEALING WITH UNCERTAINTY

Scenarios allow planners and managers to fully consider the uncertainty associated with key drivers of change into the assessment process. The methodology for integrating uncertainty distinguishes scenarios from other future assessment approaches. The two main differences include the interactions among multiple uncertainties and the selection of drivers beyond the manager's control.

Predictive models are developed based on previous observations. Uncertainty is estimated based on known probability distributions of key drivers. Models work very well when the future's uncertainty can be described by past trajectories and associated fluctuations. Model predictions are limited when the future is highly uncertain and when driving forces exhibit non-linearities, discontinuities, thresholds and emergent properties. Forecasting future trajectories is further complicated when multiple driving forces simultaneously change in unpredictable ways. While we can estimate the uncertainty associated with one driver when we know its probability distribution, the way uncertain factors interact may generate unpredictable outcomes and surprises. Coupled human-natural systems are highly uncertain. Furthermore ecosystem functions are highly context driven, and the uncertainty associated with future trajectories can create future conditions that do not resemble past occurrences.

Scenarios expand the assumptions of forecasting by considering hypothetical boundary conditions generated by interactions of driving forces and their range of uncertainties, and testing these hypotheses with expert knowledge. For example, we can consider the full spectrum of future conditions under alternative climate change scenarios and their interactions with alternative technological futures. We may not be able to accurately predict what will happen when we simultaneously consider the full spectrum of alternative rates of technological change and magnitude of climate impact, but we can isolate alternative trajectories and ask experts to hypothesize about the potential outcomes of those interactions. Scenarios function by combining scientific knowledge with expert judgment to help decision makers look at novel interactions of uncertain drivers. The key benefit of the alternative scenarios comes from anticipating impacts that lie beyond the probable estimates based on past observations alone (Figure 3.2).

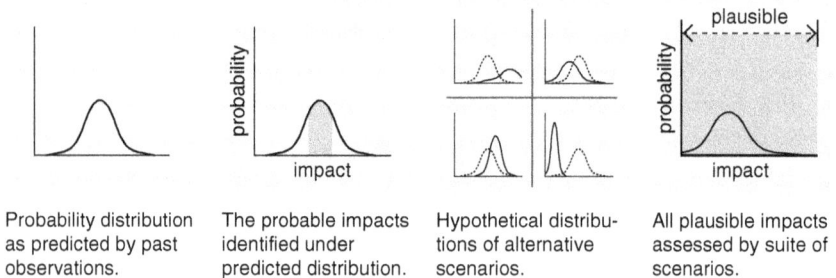

| Probability distribution as predicted by past observations. | The probable impacts identified under predicted distribution. | Hypothetical distributions of alternative scenarios. | All plausible impacts assessed by suite of scenarios. |

Figure 3.2 Probability distributions and scenario building

Scenarios look at a different type of uncertainty than looked at in traditional future assessments because they focus on drivers that lie outside the immediate control of decision makers. Several 'alternative futures' studies explore different projections by focusing on drivers that reflect the strategic decision being made. For example, many planning studies ask how the region may sprawl under alternative growth regulations (USGS, 2004; NASA, 2004). Scenarios instead focus on drivers that would alter the efficacy of the strategy. For example, global climate impacts, a collapsing economy or natural hazards could directly and indirectly affect the nearshore ecosystem condition, but cannot be controlled by the Puget Sound Nearshore Partnership. While other future assessments simplify uncertain outcomes of drivers outside their realm of influence, scenario planners specifically integrate this type of uncertainty in order to robustly test the efficacy of their decisions. The problem with simplifying these uncertainties is that it prevents decision makers from being able to test their strategies against factors that may be most influential in affecting the outcome of their plan. The trick is identifying the most relevant uncertainties to focus on.

THE BENEFITS AND LIMITATIONS OF SCENARIOS

Scenario planning is not an alternative to other planning approaches such as visioning and extrapolation. They each have different objectives. For example, when the future is relatively certain extrapolation can be a much more efficient method. Scenarios work best when the future is highly uncertain and are helpful when there are multiple key uncertain drivers simultaneously impacting the future. Scenarios can also be a useful communication tool, especially in bridging the gap between scientists, policy makers and the general public.

While the benefits of scenarios are great, so are the investments. In addition to a considerable investment of time and energy, scenarios pose challenges to managers who are more comfortable with traditional strategic planning. The amount of time required to conduct the various steps can be frustrating to experts who believe forecasting can provide a sufficient range of conditions in a more efficient manner. Further, scenarios combine facts with expert values (Schwartz, 1991). The incorporation of values can be uncomfortable to scientists who generally rely on verifiable facts. Finally, scenarios depend on knowledge that currently lies at the fringe, as their purpose is to transform our knowledge into new perspectives. This transformation requires planners to suspend their judgment long enough to appreciate new insights. When this transformation occurs it "often generates a heartfelt 'Aha' ", but more often than not, scenarios fail at achieving this goal (Schwartz, 1991, p100).

Successful scenarios are easy to communicate, interpret and learn from: they're provocative, pushing the reader to think about the 'unthinkable'; they're plausible, making use of real world facts and models to construct futures that could actually happen; they're internally consistent, looking at the context and examining changes across a wide spectrum of concerns; they're divergent, acknowledging different possibilities; and they're open, allowing readers who are not widely involved with the issue to think about their own choices and plans within each future (Cascio, 2004).

Scenarios are particularly effective when they surprise and challenge participants in the scenario process, enabling them to 'think the unthinkable' (Schwartz, 1991, p100 quoting Herman Kahn, 1965). Scenarios have been developed for over fifty years, and numerous organizations and publications have attempted to define and refine the structure and process of scenario planning. Many have produced insightful alternatives that allowed practitioners to anticipate future uncertainties and plan accordingly. However, there have also been many scenarios that merely reconfirmed preconceptions and prejudices (Ratcliffe, 2000). Despite the structured methodology, the development of scenarios remains more an art than a science. In the end, the success of scenarios lies in the hands of its participants.
(Millenium Ecosystem Assessment, 2003)

CHAPTER 4
SCENARIO METHODOLOGY

PROCESS

Scenario development for the Puget Sound Future Scenarios has been an on-going process for the last two years. The steps identified in this process generally follow a classic 8-step scenario development technique developed by Stanford Research Institute and more recently applied by the Millennium Ecosystem Assessment. In the first year the UERL and PSNP teams focused on laying out the scenario parameters including the focal issue, time scale, driving forces and scenario logics. The second year has involved developing the final scenario narratives. The methodology presented below describes an integrative, iterative, and systematic process for ensuring the final scenarios are relevant, plausible, divergent and internally consistent.

PARTICIPATING EXPERTS

The development of scenarios requires the input and collaboration of a diverse set of expertise. It is our assumption that no one person or panel of scientists can effectively create scenarios on its own. The most fundamental emphasis of scenarios is to challenge initial assumptions formulated by past observation. Therefore the integration of different experts, representing a multitude of disciplines and backgrounds, and having the opportunity to both inform and reflect on the process is critical to the success of the final scenarios. Further, scenario development requires the active involvement of a specific type of expert–experts who have knowledge of key driving forces that are powerfully influencing this region's future; who are simultaneously comfortable with accurate scientific data and a high level of uncertainty associated with a long term outlook; and who are able to communicate across disciplinary boundaries in order to capture the interactions among key driving forces over a dynamic array of spatial and temporal scales.

To develop the Puget Sound Scenarios, the UERL has involved 152 experts, representing more than 88 agencies and bringing together disciplines ranging from atmospheric science to economics and filmmaking. The final scenarios reflect three iterations of input, synthesis and feedback from this group of experts. Each expert has contributed hours of input, from preparing for and attending meetings to providing follow-up feedback on materials.

INTERVIEWS

The most intensive aspect of scenario development is conducting interviews, providing the chance to incorporate diverse scientific and policy backgrounds, and to test assumptions about the relationships among key drivers. Overall the scenario development process has involved 30 interviews, including phone interviews, individual meetings and panel discussions. Two phases of interviews with the experts allowed the UERL to formulate initial hypotheses about the trajectories of multiple drivers and their interactions, develop the scenarios and refine the hypotheses. The first phase of interviews identified the key drivers of change influencing this region's future with a focus on the nearshore ecosystem. The second phase of interviews, organized in panels of experts representing the final set of selected drivers, constituted the basis for developing the final scenarios by identifying key dimensions and trajectories for each driver and interactions between the drivers.

The objective of the first phase of interviews was to capture the breadth of issues influencing the nearshore ecosystem. The final selection of experts was organized around their disciplinary background to ensure effective exchange of ideas and a focused discussion. Groups included physical scientists, biological scientists, social and behavioral scientists, planners, the private sector, nonprofit organizations, public agencies and advocacy groups for minority populations. Interview questions primarily focused on drivers and changes affecting the state of the Puget Sound region in 50 years, and laying out the assumptions behind those relationships. Interview notes were coded to cluster keywords of changes and drivers into a manageable set of key driving forces, while still maintaining the multiple dimensions of each cluster.

The second phase of interviews was comprised of panel discussions intended to refine the scenario logics and finalize the scenario narratives. Overall the UERL identified over 200 experts, contacted over 100 different agencies, and personally interviewed 53 regional experts representing 12 expert panel teams. Expert teams represented the key driving forces identified by the first phase of interviews. Within this set, a 'core team' joined the two panels of the 'key driving forces' directing the scenario logics. The core team met both at the beginning and end of the second phase. Their initial responsibilities included refining the scenario logics in order to depict the most relevant and divergent alternative futures and formulating the scenario hypotheses for each scenario. At the end of phase 2 the core team provided a final check to ensure the internal consistency of the scenarios narratives. Supporting panel discussions representing the remaining driving forces identified by the initial interviews met to define each driving force, identify critical dimensions of the driving force, and compare future trajectories of each dimension under the scenario hypotheses developed by the core team. In addition two panel discussion involved agency and communication experts whose objective was to ensure the final scenario's usability (Figure 4.1).

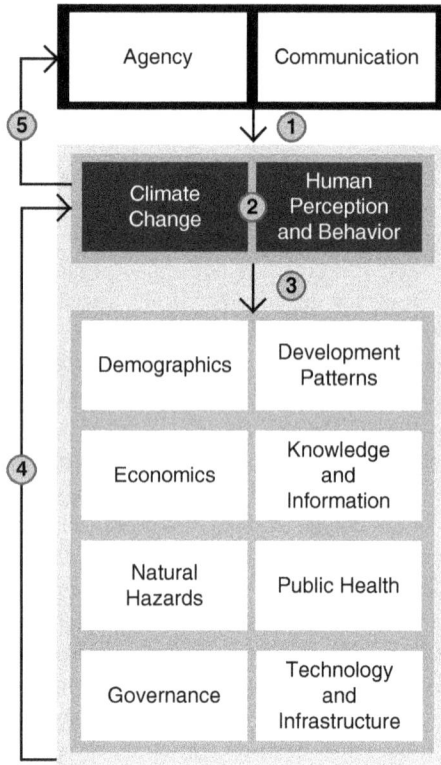

Figure 4.1 Panel discussions

WORKSHOP

In the fall of 2006 the Urban Ecology Research Lab led a one-day workshop to develop the initial scenario logics. Thirty-eight people attended the workshop, including representatives from academia, public agencies, the private sector and non-governmental organizations. Several academic disciplines were represented including geomorphology, geography, climatology, oceanography, ecology, biology, urban planning, business and economic development. The primary objective of the workshop was to select the most important and uncertain driving forces that would define the scenario logics. This step of the scenario process represents not only the most critical decision influencing the relevance and divergence of the final scenarios, but is also a decision that cannot be made independently by the diverse groups. The workshop provided a forum for participants to exchange knowledge about identified driving forces, to discuss potential merits of how the driving forces are ranked, to develop a shared agreement of which two drivers are the most important and uncertain, and to begin to develop hypotheses about the implications of alternative future trajectories.

SYNTHESIS

Between each step in the scenario development process, input from the experts was synthesized into communicable products leading to feedback from experts and a movement to the next step. The synthesis was largely conducted by the Urban Ecology Research Team including Dr. Marina Alberti and Michal Russo. The major synthesis tasks included coding and interpreting interview data. The first phase interviews were synthesized to identify the driving forces, clustering over 300 keywords into 10 aggregated drivers and developing a fact sheet to communicate the essential importance of each one. The second phase of interviews, or panel discussions, were synthesized to develop internally consistent scenario narratives integrating feedback from various teams. Based on expert input, a relational database was developed to summarize the relationships among selected dimensions of the key drivers, such that trajectories of indicators of each dimension could be developed under each scenario.

SCENARIO DEVELOPMENT

The ILS process generally includes eight steps. This report focuses on steps 1–5, which aim at the development of the scenarios.

FOCAL ISSUE

The focal issue represents the question about the future that an organization is confronting. An effective focal issue must be collectively agreed upon by the leading organization and participating experts (Peterson et al 2003) and must include the articulation of both the temporal and spatial extent of the project (Schwartz, 1991). A common problem with large-scale assessments comes from losing sight of the objective of the project as the development of the analysis proceeds (Lingren et al 2003). Articulating the focal issue at the onset brings clarity to project goals and helps the process stays on track. The Future Without Workgroup (FWW) provided the initial focal issue for the project: The 'Future Without' focal issue is to assess the future of the Puget Sound nearshore ecosystem assuming a large scale restoration project does not occur. A series of round-table discussion between the FWW and the UERL helped refine the focus of the project. The first phase of interviews with regional experts provided feedback on the focal issue, and helped refine the selection of relevant questions.

DRIVING FORCES

Driving forces represent key variables that influence a phenomenon or issue that constitutes the focus of a decision, the focal issue. Some common driving forces include demographics, economics, and science and technology. The final selection of driving forces ultimately defines the relevant parameters with which to describe the scenarios. Identifying the driving forces represents the most research-intensive aspect of scenario planning, requiring both a thorough investigation of the literature and dialogue with various experts. For the Puget Sound Scenarios, interviews with regional experts identified a wide spectrum of changes and drivers influencing this region's future. Interview notes were coded to cluster keywords into a final set of driving forces. For each driving force a definitional fact sheet was developed by the UERL team. At the workshop, participants received the factsheets and provided feedback on the selection and definition of the final set.

UNCERTAINTY AND IMPORTANCE

The final selection of driving forces is ranked in terms of each driver's level of uncertainty and importance in relation to the focal issue. This step formulates the direction of the final scenarios by identifying the most divergent and relevant conditions influencing alternative futures. Importance can be defined as the magnitude and extent of impact a driver has on the focal issue, or the cascading effect of the driving force on other drivers. The uncertainty of a driving force can be defined as having low predictability and a wide range of possible outcomes. Further, within the scenario-planning framework, uncertain driving forces generally represent drivers that are not readily controlled by the leading organization. The ultimate selection of 2 key driving forces in the Puget Sound scenarios relied on the ranking and agreement of workshop participants. Initially, interview experts discussed the importance and uncertainty of driving forces and their input was shared with workshop attendants. The final agreement over the selection of the two drivers came from individual, team and, lastly, a whole room ranking of each driving force. The driving forces with lower importance and uncertainty rankings continued to support the development of the final scenarios.[6]

SCENARIO LOGICS

The interaction among the key driving forces creates the frame for the scenario logics. For each driving force scenario planners select one or more relevant dimension and identify the key uncertainties that characterize its trajectories. For example, assuming demography is one of the two most uncertain and important driving forces, a key dimension could be population growth, and two alternative trajectories could be fast or slow growth. The logics are initially defined by crossing the axes representing the selected dimensions such that four frames are created, each representing a divergent yet plausible scenario for which defining parameters are controlled by the interaction of the two drivers' future trajectories. However, if more than two key driving forces or more than one dimension of each driving force are identified as having a critical influence on isolating alternative future conditions, than the process ends up having more than four scenarios. In theory the scenario development process can produce a large number of scenarios. A key to a successful process is to iteratively refine the scenario logics in order to ensure that the final set of scenarios represents the most divergent and relevant alternative future conditions.

The final scenario logics for the Puget Sound scenarios underwent a series of iterations incorporating critical expert feedback. The initial logics were developed during the workshop. Nine teams developed nine sets of logics by crossing the same key driving forces. After the workshop the UERL looked for redundancy and inconsistencies within the logics and synthesized the 9 sets of into one set, incorporating two dimensions for each driver. Lastly, the core team, representing experts from the disciplinary fields of the key driving forces refined the synthesized set. This team of 14 regional experts re-defined each driver's dimensions, selected the most plausible interactions among the drivers, and identified divergent future trajectories of each dimension in order to develop the final scenario logics.

SCENARIO NARRATIVES

The scenario narratives represent the final plot of the scenarios, containing detailed information on the future condition. Each scenario is developed by exploring the implications created by the interaction of the two key driving forces. Further, once the initial defining conditions are described, scenario planners can begin to articulate the trajectories of the other driving forces characterizing each scenario. The major goal of developing the scenario narratives is writing internally consistent and compelling stories that inform manager's view of future opportunities and challenges in relation to the focal issue. The Puget Sound scenarios relied on four main elements to develop the final narratives: key drivers, trajectories of the supporting drivers, storylines and alternative system states. The next section describes the objective and approach underlying each element.

THE FOUR ELEMENTS USED TO SYNTHESIZE THE FINAL SCENARIOS

KEY DRIVERS

The key drivers represent the most important and uncertain driving forces influencing the focal issue. Scenario planners develop hypotheses about how each key driver (or more accurately a specific dimension of that driver) affects the future trajectory of each supporting key driving forces, either directly or indirectly. Key drivers were originally selected during the workshop and then refined by a core team of experts representing the two key driving forces.

SUPPORTING TRAJECTORIES

Additional driving forces were used to describe the scenarios. Participating experts within panel discussions identified critical dimensions of their respective driving force, and a set of potential indicators to describe the trajectories of each diver under each scenario. Specifically, experts were asked to identify dimensions that have been previously analyzed (such that information is available to assess current status and future trends); that are relevant (to the scenarios and the nearshore ecosystem); that are uncertain (at least two alternative trajectories exist for this dimension for this region over the next fifty years); and that are not-highly correlated (such that the three dimensions measure very different attributes of the driving force).

The final future trajectory for each indicator under each alternative scenario was extrapolated from current trends. Overarching principles guiding the relationship among multiple future trajectories emerged from the panel discussion. These overarching trajectories formed a strong feedback loop to the original hypotheses, and helped shape the narratives of each scenario. Future trajectories are hypothesized based primarily on the influence of the key drivers, secondly by the trajectories of the other supporting indicators, and thirdly by the other dimensions of that driving force. These trajectories provide essential details filling the plot of each scenario, and ultimately leading to internally consistent and compelling scenarios.

STORYLINES

A third element of the scenario development came from the storylines. Scenario development is rooted in a common set of archetypal future worlds: worlds that evolve gradually, worlds influenced by a strong push for sustainability goals, worlds that exploit nature, and worlds where new human values and forms emerge (MEA, 2003). These archetypes represent the storylines for each scenario, i.e., the underlying theme or plot that synthesizes our assumptions about the future and reflects the basic premise behind each narrative. The Puget Sound Future Scenarios' storylines draw heavily from the past scenario precedence, including the Millennium Ecosystem Assessment, Peter Schwartz' The Art of the Long View (1993), Robert Costanza's Visions of Alternative Futures and their Use in Policy Analysis (2000) and the Northern Highland Lake District Scenarios (Peterson, 2003). The storylines are created by combining two essential pieces of the plot, our assumption about the worldview of society within the scenario and whether we are optimistic or skeptical about the validity of the particular worldview. Together, these pieces of the storyline help strengthen the linkages between the initial hypotheses of the key drivers and the assumptions of the supporting driver's future trajectories.

The storylines are described using four key elements: worldviews, societal emphasis, human-nature relationships, and future outlooks. Worldviews reflect the general beliefs of society about how the world works. Worldviews are dynamic, changing from one culture to another and from one time period to another. Some well known past and current worldviews are that the world is flat, that god will punish us if we sin, that humans evolved from apes, and that mass can neither be created nor destroyed. For these scenarios we combined global worldviews about the main emphasis in society (MEA, 2003) with specific human to nature relationships (Holling et al, 2002). The main societal emphasis reflects the approach society relies on more heavily in order to solve problems, including technological innovation, free market enterprise, collaborations, policy and regulations or new knowledge. The human-nature relationship can be described as how society (in aggregate) views humans' relationship to nature, whether functions are interdependent or independent, whether humans can (re)produce nature, whether nature is there for humans to consume, or humans are intended to steward nature. Future outlook on the other hand is the implicit decision in scenario building about whether or not that worldview is actually 'true'. If the optimists are right, and the worldview is correct, the scenario will reflect a future where society will prevail and positive changes are on the horizon. If the skeptics are right, and the worldview is incorrect, the scenario will likely reach a crisis.

SYSTEM STATE

The last element, the system state, communicates the level of resilience and types of pressure within each scenario. This element helps assess the extent to which the final scenarios are divergent in terms of the opportunities and challenges they pose for nearshore ecosystem restoration. Resilience[7] is defined as the capacity of an ecosystem to withstand disturbances without shifting to a qualitatively different state (Carpenter, 2001). The more resilient a system, the easier it can bounce back and rebuild itself after a perturbation. Sources of pressures refer to the magnitude and type of pressures that the region may experience under the conditions of each scenario. A pressure is generally a form of distress that focuses the attention and consequent financial investment and regulatory emphasis within a society. Some examples of pressure can include natural or man-made hazards, economic decline and wealth distribution, health and resource availability, corrupt governance, inadequate reform and infrastructure failure. Each type of pressure will inevitably have a different influence on the regulatory emphasis and financial investments, and thereby each may have significantly different implications for restoration management in this region. For example, an economic recession may distract political attention from ecological issues, and may actually lead to loosened regulations attempting to draw in economic growth. On the other hand a public health epidemic may heighten societal awareness of ecological implications on human welfare and lead to larger investments in ecological restoration efforts. System state variables were gathered from interviews with participating experts (in both phases of the project) and from relevant research publications.

CHAPTER 5
DRIVING FORCES

OVERVIEW

Driving forces are the main ingredients for scenario planning, describing factors or phenomena which alter the future trajectory in significant ways. Assumptions we make about the future often reflect changes we see within our environment, though these changes represent only the tip of an iceberg. A driving force represents the whole iceberg. By identifying and discussing the driving forces significantly influencing future conditions, scenario planners can make explicit the assumptions behind the scenario narratives. This section describes ten driving forces identified by participating experts, including two key drivers: climate change and human perceptions and behavior, and eight supporting driving forces: demography, development patterns, economy, governance, public health, natural hazards, knowledge and information, and infrastructure and technology.

Each of the ten driving forces combines the input from the first phase of interviews into a multi-faceted compendium, containing a clustering of similar drivers and the changes associated with them. In the second phase of panel discussion each driving force was further refined by selecting three to four critical dimensions and consequent indicators of that driver (Table 5.1). The following section includes the definition of each driving force and its relevancy to the focal issue as well as descriptions for each of its dimensions in terms of the selected indicator (in parenthesis), its current status and projected future trend.

Table 5.1 Driving forces dimensions and indicators

Driving Force		Dimension	Indicator
Key Drivers	Climate change	Temperature	Temperature
		Precipitation	Precipitation
		Fluctuation	Fluctuation
	Human Perceptions and Behavior	Individualism / Collectivism	Group scale of sharing
			Level of congruence
		Future Valuation	Discount rate
			Long-term public investments
Supporting Drivers	Demography	Population growth	Rate of growth
		Age Structure	Age distribution at 5-year intervals
		Migration	In-migration as % of pop
	Development Patterns	Intensity of Development	Number of people per impervious area
		Configuration	Forest Aggregation Index
		Diversity and Fit	Walkability
		New Development Growth Rate	Number of housing permits added each year
	Economy	Economic Growth	GDP growth rate
		Economic Inequality	Gini index and Lorenz curve
		Stability of Economy	Percentage of industry sector contribution
		Trade Dependence	Import / Export Dollars

Category	Indicator	Description
Governance	Leadership (Strength & Effectiveness)	Number of bills introduced and passed into law
	Locus of Power	Number of decision makers; type of interactions
	Types of Partnerships	Influence of public, private, non-profit, academia partnerships
Knowledge and Information	Educational Attainment	% of population 25+ with a high school (HS) degree; a Bachelors or higher (BA+)
	Investment in Education	Spending per capita for K-12 and higher
	Accessibility	Access to knowledge and information
Natural Hazards	Vulnerability	Spatial distribution of natural hazards
	Magnitude of Events	Cost of Natural Hazards
	Frequency	# of hydrologic disasters per year
Public Health	Health Status	%self assessed poor or fair health
	Resource Distribution	% without health insurance
	Resource Abundance	Acres of shellfish growing area and farmland
Technology and Infrastructure	Connectivity	Connectivity of transportation, energy, waste (water and solid), and water (drinking and storm) infrastructure
	Investments	Expenditure on highways, transit, electric, gas, waste, sewer, and water
	Type of Infrastructure	Dominance of rigid vs. adaptive technology

Scenario Assignment:
Cost-Benefit Analysis

You are on the governor's personal staff. You are there because you have superb credentials as a policy analyst, including a joint doctoral degree in environmental planning and resource economics. There is a controversy over the long-delayed cleanup of a highly polluted Superfund site. At one time, there was a foundry at an inner suburban site. Long-abandoned underground storage tanks at the site leaked into nearby wetlands. No one disputes that. While there are few residential neighbors who have been adversely affected, the city is eager to reuse the site for a new mall. Efforts have been made to pin financial responsibility for the cleanup (which is likely to cost $35 million or more) on various successor corporations, but this strategy has failed. The company that ran the foundry went out of business, and there are not even any clients of the foundry around anymore who can be held accountable for the cost of the cleanup. The state government has no funds to pay for the cleanup beyond the $3 million it spent more than a decade ago to cap the site and sequester the polluted water and soil.

The town government that is trying to push forward with plans for a new mall wants the state (or the federal government) to fund an arterial road linking the cleaned-up site to the nearby interstate. This would make the mall more accessible, and perhaps justify additional state or federal investment in the cleanup.

The governor is getting pressure from many sides. Her transportation secretary thinks the mall developer should pay for the cleanup and the new road. Her secretary of environment thinks that the cleanup is a good idea, but is not prepared to specify what share of the cost should be borne by the state. Environmental groups charge that there are probably long-term health effects experienced by water users in the area that might be caused by dangerous materials buried at the site leaching into nearby aquifers. Local officials argue that the region as a whole will share the economic benefits of the redevelopment of the site, so it is not unreasonable to ask the state to cover the cost of the cleanup. The governor has asked you to outline a comprehensive CBA that can be done to help answer all these questions:

1. How should benefits and costs in this case be defined?
2. What, in your view, are the most important costs and benefits to focus on in the study?
3. What should be the time frame and the geographic scope of the study?
4. How should the governor handle the question of the inequitable distribution of benefits and costs associated with reuse of the site?

Role-Play Exercise Assignment:
Negotiating Societal Risk Assessment

This assignment is a negotiation exercise, known as role-play simulation, derived from research conducted at the Program on Negotiation (PON), an interdisciplinary, multi-university research center based at Harvard Law School. Professor Lawrence Susskind wrote the role-play scenario, which is enamed "Humboldt." The scenario is based on a famous case that occurred in Europe. The focus is to practice negotiation and conflict resolution tactics, as it pertains to the evaluation and implementation of policies with uncertainty regarding their associated societal risks. Within this exercise, a group of seven parties have to decide whether they will come to an agreement, with the assistance of a mediator. The meeting is occurring under the sponsorship of the elected regional governors of Humboldt, who will not be in attendance. Significant issues in economic development and environmental protection are at stake. The governors have indicated that they would prefer unanimity, but have pledged to support an agreement that at least five of the seven parties endorse.

The participants are encouraged to follow closely the General and Confidential Instructions (a sample for one party is presented below) that they are given before they begin the exercise. However, as in real life, the participants must make sure to go beyond and interpret them according to their own interests and strategies as well as fill in the many blanks with creativity. Will they be able to overcome lies and deception, and instead engage in joint fact-finding and build relationships of trust?

In this negotiation, the seven negotiating parties are representing, overall, two broad sets of interests.

- On one side, there is the (1) mayor of Milltown; (2) director of economic development for the Humboldt Region; (3) lawyer representing Imports Inc.; and (4) coordinator of the Humboldt Business Association.
- On the other side, there is the (1) head of the Dairy Cooperative; (2) head of the Regional Environmental Action League; and (3) representative of the Arcadia Environmental Quality Agency.

The participants' responsibility is to explore if they can build effective coalitions and persuasive arguments relying upon the tools they have explored in the book, and figure out how to create and claim value both within and outside the realms of their coalition. Will they be able to benefit the communities they represent?

General Instructions

The Humboldt region has always been economically depressed. South of Humboldt River there are several towns, none with more than 10,000 inhabitants. The southern part of the Humboldt region is mostly given over to family-run dairy farming. North of the river is the largest city in the region—Milltown. It has almost 100 inhabitants, most of whom work in the steel mill. Unfortunately, the mill is scheduled to close in about a year, and all efforts to reverse that decision have failed.

For the past few years regional government officials have been trying diligently to diversify the industrial base in Humboldt, but they have not been successful. Their only new prospect is a local manufacturing plant—Smith's Rollers, a firm that produces specialized rollers used in paper production. Smith's has operated successfully in Humboldt for decades. A multinational corporation, Imports, Inc., that wants to expand the plant, recently purchased it. Imports, Inc., is considering a site on the Humboldt River next to a former steel mill. The new plant, if it is built, could provide almost 500 new jobs within a year (not including new construction jobs) and might expand further. The site on the river is attractive for two reasons. First, roller production involves a plastic lamination process that generates a chemical byproduct, known as "lamina." That by-product has typically been treated, then fed into a river or stream. Thus the Humboldt River site has a distinct advantage for Imports, Inc. Second, the production of the plastic lamination generates noxious gases. Since the plant is in an industrial area that is already quite polluted, the odors will not be especially troublesome.

The population of Milltown is quite excited about the prospect of the new Smith's Rollers plant. They recently elected a new mayor who promised to do "everything possible" to get the plant built. The *Milltown Journal*, a weekly newspaper, has supported the new mayor and editorialized about the need for the new plant. However, in a recent editorial the *Milltown Journal* noted that the Environmental Action League was already "making trouble," raising questions about the long-term implications (both economic and environmental) of polluting the Humboldt River further and about the immediate health impacts that the new plant might bring.

Farmers south of the river recently held a meeting to discuss the possible impacts the new plant might have on dairy production. Rumor has it that the air pollution likely to be caused by the plant could adversely affect the herd. In the past there have been periodic problems caused by pollutants dumped into the river upstream. The dairy farmers are worried that the new plant could further contaminate the river, put them out of business and reduce the value of their land.

The Humboldt region is in the southwestern corner of Arcadia—a small democratic country bordered by Austria, Switzerland, Germany and Italy. Arcadia has only recently adopted its first environmental protection laws. These were enacted primarily to please Arcadia's neighbors in the European Union (EU). Arcadia still does not have specific air pollution or water pollution standards. It does, however, have an environmental impact assessment (EIA) law that requires regional governments to take all reasonable steps to minimize adverse environmental impacts of major new industrial projects. It is hard to say whether the Smith's Rollers plant is a major new project. Specific regulations pursuant to the EIA law have not yet been adopted. And there is substantial debate in the Arcadia Parliament about how best to define "major projects."

The elected regional governors in Humboldt have decided to invite all the parties with a stake in the Smith's Rollers plant decision to participate in an informal discussion. From the governors' standpoint, it would be best if everyone could agree on how to proceed. The governors have invited the following people:

- The mayor of Milltown
- The head of the Dairy Cooperative of Southern Humboldt
- The head of the regional Environmental Action League
- The lawyer representing Imports, Inc. (the multinational corporation that recently purchased Smith's Rollers)
- The director of economic development for the Humboldt region
- A representative of the Arcadia Environmental Quality Agency (EQA)
- The coordinator of the Humboldt Business Association (an official organization of business leaders)

Two smaller preliminary meetings will be held before the large meeting. At one, the Dairy Cooperative, the Environmental Action League and the Arcadia EQA will meet at the Action League's offices to discuss their concerns about the proposed project. At the other, the mayor, Imports, Inc., the economic development director, and the coordinator of the Business Association will meet (in the mayor's office) to discuss their mutual concerns.

The governors, in an attempt to help build consensus, have identified a well-known professional mediator to assist the parties. The governors have pledged to support any agreement that at least five of the seven participants endorse. They have also indicated that they would prefer unanimity. The governors have publicly pledged to cover all "reasonable" costs associated with impact studies or other aspects of an acceptable agreement.

Confidential Instructions to the Head of the Dairy Cooperative

Your task is to protect the herd and to ensure that dairy farming can continue in the southern portion of the Humboldt region. You do not consider yourself an environmentalist, but you are pleased that the Environmental Action League has gotten people stirred up about the possible impact of the new plant.

You are not interested in having the Environmental Quality Agency (EQA) come in and tell Humboldt what to do. You have enough problems with the National Dairy Commission, which is always telling the farmers what they can and what they cannot do.

The regional governors have never been especially sympathetic to the dairy farmers, so it comes as something of a surprise that they have even invited the cooperative to participate in these discussions. You intend to make the most of it. Your people want to count on you to get the important following points across:

1. No new plant if it poses any threat to dairy farming south of the river.
2. A guarantee that if the plant is built and pollution problems arise, the regional governors will cover the cost of replacing any portion of the herd that is lost (the current value of the total herd is estimated at $25 million).
3. The farmers do not want taxes increased to cover the cost of subsidizing a new plant that they do not even want.

End of Unit III Written Assignment:
Environmental Assessment

Most environmental planners presume that policy decisions regarding the use of natural resources and patterns of development can be enhanced through the application of various analytical tools. (1) Explain why and how you agree or disagree with this, with reference to each of the tools described in Unit III. (2) What do you think are the relative strengths and limitations of each of the analytical tools?

First Example Response to Assignment:
Strengths, Weaknesses and Policy Implications of Environmental
Analysis Tools

Analytical tools from impact assessments to scenario planning are intended to inform environmental policies on natural resource and land use. This essay analyzes the strengths and weaknesses of these approaches, and their capacity to enhance policy-making, suggesting that while all offer benefits, none serves as a panacea for facilitating clear policy solutions.

A. Impact Assessments

Environmental Impact Assessments (EIA), Strategic Environmental Assessments (SEA) and Social Impact Assessments (SIA)

- **Strengths:** Impact assessments offer systematic, quantitative approaches to measuring deleterious impacts of future projects or policies (Van Buuren and Nooteboom, 2009). They heighten attention to environmental (EIAs/ SEAs) or social (SIAs) concerns, and can shift values and social norms over time (O'Faircheallaigh, 2009; Cashmore, Bond and Cobb, 2007; Jha-Thakur et al. 2009). Developers and policy makers thus assume greater accountability (Cashmore, Bond and Cobb, 2007). Knowing that EIAs are mandated in the United States prevents highly hazardous projects from entering the pipeline (Susskind, 2014). Impact assessments increase transparency—facilitating public scrutiny and access to data—and create a platform for collaborative dialogue (Van Buuren and Nooteboom, 2009; Cashmore, Bond and Cobb, 2007; Jha-Thakur et al. 2009).
- **Weaknesses:** Impact assessments require significant time and personnel, and their parameters are often limited by data availability (Jha-Thakur et al. 2009). Because parameters vary in scope and units, cardinal values are translated into interval numbers—a vast oversimplification to cross compare impact (Susskind, 2014). Without aggregating, however, the assessment remains an unruly database with no clear method for weighing trade-offs or for taking political action. The process entails numerous subjective choices—from weighting variables to defining time horizons—and because in the US project *proponents* are responsible for contracting the assessment, impact is frequently underestimated (Therivel et al., 2009).
- **Can impact assessments enhance policy decisions?** Impact assessments often create false perceptions of due diligence (Cashmore, Bond and Cobb, 2007), and bias undermines the rigor of the assessment and thus usefulness for policy-making. The process would be more effective if we set

universal standards, but conditions are too context specific to do so effectively. Nevertheless, through a postpositivist lens, impact assessments with stakeholder participation can provide significant opportunity for collective learning—a process that enables changing perceptions and institutional reform for better governance (Cashmore, Bond and Cobb, 2007; Jha-Thakur et al. 2009).

B. Cost-Benefit Analysis

- **Strengths:** CBAs are comprehensive. Unlike EIAs, they weigh environmental, economic, social and health costs simultaneously (Pearce, Atkinson and Mourato 2006). The tool provides a rationalized, formalized and quantitative model that prescribes whether or not to act, and which can also compare alternatives and optimal scales (ibid.).
- **Weaknesses:** CBAs attempt to monetize health, life, and ethical or cultural values—all of which arguably could have infinite value. By solely relying on individuals' willingness to pay, we may run the risk of disregarding societal interests like "national pride" or "collective self-respect" (Sagoff 1988). The optimal geographical and temporal scope is unclear, and weighing benefits/costs in the future proves problematic. Theorists argue that discounting fails to represent future generations' interests (Pearce, Atkinson and Mourato 2006). Without complete information of all benefits and costs, a CBA is ineffective (ibid.).
- **Can CBA enhance policy decisions?** CBA should not supplant moral judgment but can augment decision-making where costs and benefits are not initially clear. CBA can help determine priorities. In assessing air quality, the EU identified that reducing particulate matter (PM) accrued 90% of the benefits—strong reasoning for prioritizing PM mitigation policies (Pearce, Atkinson and Mourato 2006). Whereas CBA can *suggest* whether particular costs are worth paying and *can* consider equity, the tool can also produce egregious outcomes, so it is critical to remain mindful of results.

C. Ecosystem Services Analysis

- **Strengths:** Rather than treating ecosystems as single entities—as with mitigation banking where ecosystems are "replaced" by the acre—ESAs' identify the *processes* that must be replaced to maintain ecosystem vitality (Polasky and Segerson 2009). By monetizing ecoservices, complex systems that are otherwise overlooked—such as water filtration or carbon sequestration—are converted into salient and significant values (Costanza et al. 2014; Gómez-Baggethun et al. 2010).

- **Weaknesses:** ESAs' underlying economic assumptions are contested: goods are not substitutable to the degree assumed, and discounting inappropriately diminishes the value of future ecosystems (Ludwig 2000). Assessing value also proves problematic: people rarely understand ecosystem processes fully and have no price point for comparison for nonmarket goods (Ludwig 2000). Finally, many argue that nature is a personal and social value, with no monetary equivalent. Nature's commodification risks inducing competition and undermining moral incentives by emphasizing self-interest (Gómez-Baggethun et al. 2010; Ludwig 2000).
- **Can valuating ecosystem services enhance policy decisions?** Economic arguments consistently catalyze political agency. Thus, despite gross assumptions, illustrating the value of ecoservices to humans provides powerful incentive for instating protection policies (Costanza et al. 2014). For example, New York chose to restore a wetland for $1 billion in order to ensure effective water filtration in the Catskills. In comparison, a new filtration plant would have cost $6 billion—a salient argument for preservation policies (Sagoff 2002).

D. Risk Assessment

- **Strengths:** By relying solely on market processes, humans underestimate risk and negative externalities. RA provides a viable mechanism for elucidating risks and costs (Kunreuther et al. 1996; EEA 2008). Like CBA, RA provides a comprehensive approach to comparing variables across "different languages" and for prioritizing action and further research (EEA 2008).
- **Weaknesses:** Risk always exists, so the question becomes how much is *acceptable*, which requires value judgment rather than quantitative analysis (Kunreuther et al. 1996; Lowrence 1980) (e.g., *Should we remove all toxic soil from a site, or can we cover it with fresh soil?*). Varying values and perceptions of risk foster contradictions among experts, lead to public distrust and convolute the ability to take action (Pearce, Atkinson and Mourato 2006). Many argue that overregulation limits beneficial technology development—for example, genetically modified organisms (Kunreuther et al. 1996). Nevertheless, trends show both an overreliance and exaggerated confidence in RA, resulting in "premature conclusions based on elementary science indicators" (EEA 1998; Lowrence 1980). Despite potentially high levels of liability, RA does not delineate responsibility for consequences (Lowrence, 1980).
- **Can risk assessment enhance policy decisions?** RA could be instrumental if it considers cultural, political and psychological forces, if it employs a participatory process that fosters consensus building between

experts and laypeople, and if it defines the collective goals from the start (Kunreuther et al. 1996). These elements are necessary for the "perception of risk" to be communal, and for the results to clearly identify mutual and best courses for action.

E. Modeling

Dynamic Modeling, Agent-Based Modeling, Optimization, Simulation and Econometrics

- **Strengths:** Modeling proves useful for complex situations where nonlinearity, uncertainty, and lags in time and space significantly influence future events (Costanza and Ruth 1998). It interrelates many factors simultaneously, linking cause and effect closely within space and time (Costanza and Ruth 1998; Sterman 1991). The interactive quality allows non-modelers to become more "intellectually and emotionally involved" in the model (Costanza and Ruth 1998, 185). Its manner of visualizing outcomes can thus facilitate learning and consensus building arguably better than aforementioned decision aids (Costanza and Ruth 1998; Law 2005). Where certain models lack, others compensate: optimization provides strong prescriptive recommendations, and simulation can contribute descriptive foresight to policy design, whereas, agent-based models best account for heterogeneity between agents (Sterman 1991).
- **Weaknesses:** Models risk becoming "black boxes." Assumptions are critical for assessing validity of the outcomes, and yet frequently they are inaccessibly complex or left unexplained (Sterman 1991). Both econometrics and optimization receive criticism for their assumptions. Econometrics overgeneralizes human rationality and excludes consideration for desires and goals integral to decision-making. Meanwhile, optimization determines an "optimal" solution based on one set of values and assumes a false linearity, failing to consider feedback mechanisms (Sterman 1991).
- **Can modeling enhance policy decisions?** Because of their inherent subjectivity, models can become "self-fulfilling prophecies," leveraged as a means to legitimize rather than inform policy (Costanza and Ruth 1998). They also, however, present opportunity for collaborative, cross-disciplinary discussion—particularly through structured stakeholder walk-throughs—which could counteract bias (Law 2005). To be most influential in policy-making, analysts should interact with decision makers in the process, and expand business-as-usual predictions to show the potential effects of policy decisions (Sterman 1991). As demonstrated through the International Institute for Applied Systems Analysis program on global modeling,

comparing multiple models can serve as a powerful tool for education and consensus building around complex problems like population growth (Sterman 1991).

F. Scenario Planning

Scenario Planning (SP) and Anticipatory Governance (AG)

- **Strengths:** SP helps decision-making when there are multiple, compounding uncertainties and potentially high risk in future outcomes (UW Urban Ecology Research Lab 2008). Unlike forecasting, SP does not rely on historical data—a potentially improper indicator—and develops hypotheses for not one, but numerous potential "futures" (UW Urban Ecology Research Lab 2008). By doing so, SP considers the long term (and future generations) more thoroughly than methods like RA or CBA. It also builds a shared language and framework for discussing the problem, serving as a provocative communication tool to achieve alignment and collective learning among scientists, policy makers, and public (UW Urban Ecology Research Lab 2008; Aldrich 2011).
- **Weaknesses:** SP is time and energy intensive. If the future is relatively certain, extrapolation or forecasting tends to be more efficient (UW Urban Ecology Research Lab 2008). Theorists also critique that experts are limited by their past experiences and do not consider rare events (Quay 2010). Often scenarios fail to reach a point where new insight is created, and instead "reconfirm preconditions and prejudices" (UW Urban Ecology Research Lab 2008, 10). Consequently, SP has been described as an "art" rather than a consistent methodology for achieving effective results (UW Urban Ecology Research Lab 2008; Quay 2010).
- **Can scenario planning enhance policy decisions?** SP does not replace any of the tools previously discussed, but can serve as an effective framework for these tools. Through the process, key choices are rationalized and ultimately legitimized, facilitating consensus around policy decisions (Susskind 2014). SP improves the "predict-and-plan" approach by producing multiple considerations and hypotheses—many previously unconsidered—as well as strategies for alternative outcomes.

Consequently, SP increases flexibility and strengthens communities' adaptive capacities (Aldrich 2011). Anticipatory governance illustrates how SP can help foster climate change preparedness. For example, to enhance its adaptability, the city of Phoenix has prepared for multiple alternative futures shaped by growth and climate-driven water shortages (Quay 2010).

In summary, it is key to remember that these decision aids are truly *aids*. They must be seen as knowledge-production processes rather than prescriptive solution-generating exercises. The decisions determined and the effectiveness of the subsequent policy will ultimately be based on how the results of these tools are interpreted and further acted upon. No analysis is neutral. Analyses embed values, personal perceptions, and aspects like "integrity" that are difficult to quantify. Making these analyses as transparent as possible, engaging a range of stakeholders who can critically assess assumptions and integrating means for flexibility in the face of uncertain futures will be the best possible tools for effective policy-making.

References

Aldrich, S. C. 2011. *Communities Adapting to Climate Change—Scenario Planning*. Cambridge, MA: Lincoln Institute of Land Policy.

Cashmore, M., A. Bond and D. Cobb. 2007. "The Contribution of Environmental Assessment to Sustainable Development: Toward a Richer Empirical Understanding." *Environmental Management* **40**: 516–30.

Costanza, R., and M. Ruth. 1998. "Using Dynamic Modeling to Scope Environmental Problems and Build Consensus." *Environmental Management* **22**: 183–95.

Costanza, R. et al. 2014. "Changes in the Global Value of Ecosystem Services." *Global Environmental Change* **26**: 152–8.

EEA. 1998. *Environmental Risk Assessment: Approaches, Experiences and Information Sources.* Luxembourg: European Environmental Agency. https://www.eea.europa.eu/publications/GH-07-97-595-EN-C2.

Gómez-Baggethun, E., R. de Groot, P. L. Lomas and C. Montes, C. 2010. "The History of Ecosystem Services in Economic Theory and Practice: From Early Notions to Markets and Payment Schemes." *Ecological Economics* **69**: 1209–18.

Jha-Thakur, U., P. Gazzola, D. Peel, T. B. Fischer and S. Kidd, S. 2009. "Effectiveness of Strategic Environmental Assessment—the Significance of Learning." *Impact Assessment Project Appraisal* **27**: 133–44.

Kunreuther, H., P. Slovic, A. W. Heston and N. A. Weine. 1996. "Challenges in Risk Assessment and Risk Management." *Annals of the American Academy of Political and Social Science* **545**: 8–13.

Law, A. M. 2005. In *Proceeding of the 2005 Winter Simulation Conference*, edited by M. E. Kuhl, N. M. Steiger, F. B. Armstrong and J. A. Joines, 24–32.

Lowrence, W. W. 1980. In *Societal Risk Assessment*, edited by R. Schwing and W. Albers. New York: Plenum, 1–62.

Ludwig, D. 2000. "Limitations of Economic Valuation of Ecosystems." *Ecosystems* **3**: 31–5.

O'Faircheallaigh, C. 2009. "Effectiveness in Social Impact Assessment: Aboriginal Peoples and Resource Development in Australia." *Impact Assessment and Project Appraisal* **27**: 95–110.

Pearce, D., G. Atkinson and S. Mourato. 2006. *Cost-Benefit Analysis and the Environment: Recent Developments*. Paris: OECD Publishing, 15–28, 59–60 and 269–77.

Polasky, S., and K. Segerson. 2009. "Integrating Ecology and Economics in the Study of Ecosystem Services: Some Lessons Learned." *Annual Review of Resource Economics* **1**: 409–34.

UW Urban Ecology Research Lab. 2008. *Puget Sound Future Scenarios*. University of Washington. http://urbaneco.washington.edu/wp/wpcontent/uploads/2012/09/scenarios_report.pdf.

Quay, R. 2010. "Anticipatory Governance." *Journal of the American Planning Association* **76**: 496–511.

Sagoff, M. 1988. *The Economy of the Earth: Philosophy, Law, and the Environment*. Cambridge: Cambridge University Press, 24–49.

———. 2002. "On the Value of Natural Ecosystems." *Politics and the Life Sciences* **21**: 19–25.

Sterman, J. D. 1991. "A Skeptic's Guide to Computer Models." In *Managing a Nation: The Microcomputer Software Catalog*, edited by Gerald O. Barney, W. Brian Kreutzer and Martha J. Garrett. Boulder, CO: Westview Press. 209–29.

Susskind, L. 2014. Personal communication. November 4.

———. 2014. Personal communication. October 16.

Therivel, R. et al. 2009. "Sustainability-focused Impact Assessment: English Experiences." *Impact Assessment Project Appraisal* **27**: 155–68.

Van Buuren, A., and S. Nooteboom. 2009. "Evaluating Strategic Environmental Assessment in the Netherlands: Content, Process and Procedure as Indissoluble Criteria for Effectiveness." *Impact Assessment and Project Appraisal* **27**: 145–54.

Second Example Response to Assignment: Environmental Management Can Be Enhanced through Analytical Tools

There exists a plethora of analytical tools to support environmental decision-making. I broadly categorize them as tools to assess trade-offs that rank options or provide decision rules, and tools to generate and increase knowledge. In order to assess whether these tools improve decision-making, let us first examine a few baseline scenarios without these tools:

1. A single decision-maker decides based on her mental model of what should be.
2. Numerous decision-makers collectively organize their individual and collective mental models and values to decide what ought to happen, through a referendum, for example.
3. In a scarcity-based market system, socially optimal decisions are made by aggregating individual market transactions that maximize personal utility through individual choice.

In a homogeneous society with a shared, rule-based system of ethics, the first and second approaches may be valid, with one or multiple decision-makers agreeing, "yes, we shall not do X, and must do Y," and that decision would be clear and accepted. The third scenario presumes perfect information in the market: that producers provide honest and accurate information, and that consumers can honestly reconcile prices, information and their values.

In a world of heterogeneous publics where decision makers have imperfect information about an uncertain future, past and present, additional tools are required to understand trade-offs, to clarify the complex world and to make decisions. But does this lead to unequivocally better decisions than those made using simpler models and heuristics? No.

Planners must bear in mind epistemic hubris: the arrogance of thinking we know. We must be cognizant of the limits of our knowledge. Decisions should be made acknowledging all forms of uncertainty and with contingencies to protect from serious consequences of human error. The table below lists the different tools and their strengths and limitations, which is followed by a discussion of the degree to which each enhances decision-making.

The Environmental Impact Assessment (EIA) requires projects or policies to explicitly quantify and assess a range of possible impacts on the natural and human environment. While designed to provide the public and decision-makers with greater information on the process, the process does not have an explicit decision rule beyond whether the EIA was performed and published

Table 1 Strengths and Limitations of Analytical Tools

Analytical Tool	Strengths	Limitations
Environmental Impact Assessment	Requires information be provided to public	Aggregation of impacts Decision rule
Cost-Benefit Analysis	Tangibly accounts impacts of policy Explicit decision rule	Intergenerational equity Distributional impacts Incorporating moral values Oversimplification
Ecosystem Analysis	Demonstrates linkage between economy and environment Raises importance of ecosystem services	Implied fungibility of ecosystems Assessing static value
Risk Assessment	Highlights consequences of failure	Managing public perception
Simulation and Modeling	Makes models and assumptions explicit Explains complex systems	Limits to modeling behavior Measurable inputs bias
Scenario Planning	Explicitly addresses uncertainty	Start-up time and cost

with due diligence. Some have argued that the greatest value of such a process comes from the threat of transparency, which has halted truly awful projects from starting the public process.

That being said, the EIA process can lead to bad outcomes by providing a veneer of legitimacy to potentially flawed analysis. I am concerned that over time, project proponents can learn to subvert the process. For example, they can produce dense documentation of such unnecessary length and complexity that it can tax the resources of reviewers. This is particularly a problem in the United States, where proponents are responsible for conducting studies. Though agencies in the United Kingdom and Europe tend to perform analyses for project proponents, they face an agency problem when they self-regulate their policies (Cashmore, Bond and Cobb 2007; Therivel et al. 2009).

Cost-Benefit Analysis (CBA) is an added layer to the EIA process that provides an explicit decision rule through the valuation of costs and benefits: which option maximizes societal utility? By monetizing impacts it can aggregate different units into one that is tangible to the layperson: money. I was told at MIT by an instructor with substantial government experience that he had never seen an unfavorable government CBA of a government-sponsored project. Agencies will find ways to "make the numbers work" for projects that have political support. Though a common criticism is that CBA has a bias against the poor, assessing costs and benefits can highlight distributional issues regarding who receives benefits and who suffers impacts (Pearce, Atkinson and Mourato 2006).

Ecosystems analysis is an input to CBA that quantifies the economic value of natural systems. It is important to render tangible to the public the invisible services provided by ecosystems as it demonstrates that the economy cannot be divorced from the environment. This awareness raising of the trends in ecosystem health is paramount. For example, the technique can be extremely useful to communicate to a larger public how specific projects of restoration or preservation will overwhelmingly benefit the economy, as was the case of New York City's drinking water supply. Instead of commissioning a new water filtration plant, New York City restored Catskill ecosystems and thus achieved a rate of return of an astonishing 90%–170% (Chichilnisky and Heal 1998).

In the United States, without major regulatory shifts, I think that project appraisal will always undervalue ecosystem services, given the strong tendency for projects to be approved through litigation or with its threat in mind. It is common for judges to make decisions based on Kaldor-Hicks efficiency (maximizing societal utility), and in a case of economic interests versus ecosystem services, I think that ecosystem services will be systematically undervalued. The application of ecosystem analysis thus presents a vulnerability for policy and projects to this systemic bias. Further, the monetization implies a fungibility that does not exist. As with distributional economic impacts, location matters. But one cannot compensate an ecosystem that has been reduced to near-zero value with some equivalently manufactured one at a different site.

Models and simulations are a broad category of tools that can be inputs to other analytical processes such as CBA or ecosystem analysis. I think that physical models are useful to explain natural phenomena and predict changes over time. Of less use, and potentially dangerous, are optimization and models predicting human behavior. While they perform well for average cases, they may miss the devastating effects of the improbable or how individuals may behave in unfortunate circumstances. Precisely what the policy maker should guard against. Doran's (2001) failed attempt to model the political complexity of watershed management of the Fraser River inspires a plea to advance the state of the art, but I would argue it demonstrates the futility of trying to model complex social systems.

Modeling resources might be put to better use building dynamic models for building consensus among actors (Costanza and Ruth 1998), though this presumes that consensus on the models themselves is achievable. It is vital to be cognizant of the limits of models, including their bias toward the easily quantifiable (Sterman 1991). Decisions based on them must be made with full acknowledgment of their inherent risk and uncertainty. That is why the US military has added an additional step to the model-building process. After validation, someone must take ownership of the model and the errors that may arise (Law 2009).

For projects and policies with potential negative consequences, it is important to evaluate vulnerabilities and how systems can fail through *risk assessment*. However, the difference between acceptable risk and estimated risk (the probability of a hazard occurring multiplied by its impact) is highly sensitive to public acceptance and probability estimation. The problem with low-probability, high-impact events is that they are wickedly difficult to predict from past data. The effort expended on precisely calculating these probabilities would be better used in mapping out how systems can fail and how to limit exposure to failure or ensure a safe failure.

Decision-making in uncertainty can be helped by *scenario planning*, a collaborative creative visioning of future exogenous states and subsequent assessment of plan performance. The process should humbly acknowledge the variability of the future and require increased flexibility of plans, leading to adaptable institutions. However, I question the long-term feasibility of this approach in the context of the US executive agency. On the one hand, scenario planning was first developed for large corporations, where changes in management are endogenous to the performance of the corporation.

On the other hand, executive agencies tend to have a change of leadership every political cycle. Thus, I wonder how stable the culture change brought about by the process, which requires a long preparation time with little immediate benefit, is to such frequent changes. Two of the three agencies studied by Quay (2010) are water commissions that appear to be arm's length to government, while the process in New York City was initiated under one mayor and seems to have been created as its own organization.

With a healthy dose of skepticism, environmental policy-making can benefit from analytical tools. It is important, however, in order not to make harmful mistakes, to remember the limits of these tools.

References

Cashmore, M., A. Bond and D. Cobb. 2007. "The Contribution of Environmental Assessment to Sustainable Development: Toward a Richer Empirical Understanding." *Environmental Management* 40, no. 3: 516–30.

Chichilnisky, G., and G. Heal. 1998. "Economic Returns from the Biosphere." *Nature* 391: 629–30.

Costanza, R., and M. Ruth. 1998. "Using Dynamic Modeling to Scope Environmental Problems and Build Consensus." *Environmental Management* 22, no. 2: 183–95.

Doran, J. 2001. "Agent-Based Modelling of Ecosystems for Sustainable Resource Management." In *Multi-Agent Systems and Applications*, edited by Michael Luck, Vladimir Mařík, Olga Štěpánková and Robert Trappl. Berlin and Heidelberg: Springer, 383–403.

Law, A. M. 2009. "How To Build Valid and Credible Simulation Models." In *Simulation Conference, 2009: Proceedings of the Winter*, edited by M. D. Rossetti, R. R. Hill, B. Johansson, A. Durkin and R.G. Ingalls, 24–33.

Pearce, D., G. Atkinson and S. Mourato. 2006. *Cost-benefit Analysis and the Environment: Recent Developments*. Paris: OECD Publishing.

Quay, R. 2010. "Anticipatory Governance." *Journal of the American Planning Association* 76, no. 4: 496–511.

Sterman, J. D. 1991. "A Skeptic's Guide to Computer Models." In *Managing a Nation: The Microcomputer Software Catalog*, edited by Gerald O. Barney, W. Brian Kreutzer, Martha J. Garrett. Boulder, CO: Westview Press, 209–29.

Therivel, Riki, Gemma Christian, Claire Craig, Russell Grinham, David Mackins, James Smith, Terry Sneller, Richard Turner, Dee Walker and Motoko Yamane A. 2009. "Sustainability-focused Impact Assessment: English Experiences." *Impact Assessment and Project Appraisal* 27, no. 2: 155–68.

1 Arwin van Buuren and Sibout Nooteboom are at Erasmus University Rotterdam, Department of Public Administration, Room M8–31, PO Box 1738, 3000 DR Rotterdam, The Netherlands, Tel. + 31 10 4082635, e-mail: vanbuuren@fsw.eur.nl

Unit IV

COLLECTIVE ACTION TO SOLVE ENVIRONMENTAL PROBLEMS

Introduction

Unit IV highlights the vital connection between environmental problem-solving and democratic decision-making. While individual consumers and citizens can certainly make environmental improvements on their own (e.g., by recycling household waste or reducing their personal carbon footprint), most environmental problem-solving requires collective action—either by whole communities, whole countries or even all the nations of the world acting together. Unfortunately, it is not easy to generate agreement at any of these levels. Stakeholders often start with conflicting philosophies. Additionally, a high likelihood that any proposed solution will affect them differently. Since self-interest is an important motivation, some way of resolving these underlying conflicts and making sure that multiple interests are met simultaneously must be found.

Groups with different interests often define problems in ways that reflect their biases. They see what they want to see and hear what they expect to hear. They advocate methods of analysis that will yield results that support the outcomes they favor. This brings us back to the claim we made at the outset of this book: environmental problem-solving will inevitably involve both science and politics. It requires science, because a popular agreement that does not actually solve a problem is of little use. It requires politics, because almost all collective action requires some form of effective government involvement.

The chapter begins with an excerpt from *Breaking Robert's Rules*. In it, Lawrence Susskind and Jeffrey Cruikshank advocate for a consensus-building approach to mediating conflicting interests. Although consensus building is oftentimes more time and resource intensive, the authors argue that it is more likely to lead to fairer, wiser, more efficient and more stable outcomes. In other words, they advocate for a more democratic and inclusive approach to decision-making.

Ian Shapiro's excerpt summarizes what political philosophers know about collective decision-making in democratic settings. He also discusses whether citizens have a right to participate in decisions that affect them. You will see that we hold to the belief that majority rule, the usual way in which citizens make decisions in a democracy, is not good enough. We would rather see contending groups attempt to reach agreement whenever they engage in environmental problem-solving. Why should they not try? The worst case is that they fall back to majority rule. However, if they seek consensus from the outset, they have an opportunity to deliberate and try to find mutually agreeable trades and outcomes.

Oftentimes, in a consensus-building approach, it helps to have a professional "neutral" or mediator who helps manage the process. Lawrence Susskind and Connie Ozawa emphasize the role of urban and regional planners as mediators whose contribution it is to foster a fair process in which stakeholders with conflicting interests can reach a negotiated agreement.

The Mark Sagoff reading points out that not all political disagreements stem from conflicting economic self-interest. Emotional, spiritual and intellectual commitments can prove even more important. He explains that even though sometimes we act as consumers motivated by private economic concerns, many other times we act as citizens hoping to ensure that fairness or justice is achieved (not just for ourselves). In this volume, we take the distinction between our economic interests as consumers and our political interests as citizens very seriously.

The goal of this last unit of the book is to demonstrate that collective decision-making can be a means of balancing science and politics in environmental problem-solving, and that the way to do this is to involve interested stakeholders directly. One of the critiques of consensus-building or collaborative approaches to decision-making is that they are not always genuine attempts to engage all stakeholders in decision-making. Thomas C. Beierle (1998) identifies a series of social goals that can be used to assess different forms of public participation. The main point we are trying to make by sharing this perspective is that public participation in and of itself is not inherently good. Instead, the question of who is asked to participate and at what point in the decision-making process they are asked to participate is integral to developing a meaningful participatory approach.

We have also included the excerpt by Eugene A. Rosa, Ortwin Renn and Aaron McCright on risk governance because, as we discussed in the introduction to Unit III, evaluating risk as well as developing scenarios and a plan to manage risk is complicated by the inherent uncertainty. The authors propose collective decision-making in relation to risk management as a way of ensuring that any risk is not unfairly borne by one subset of the population.

The last two readings are crucial. The reading by Garrett Hardin ("Tragedy of the Commons") and the reading by Elinor Ostrom (from *Governing the Commons*) are iconic environmental policy references that often appear in collections like this. Although many professors would start their course on environmental policy-making with these texts, we want you to pursue them with "fresh eyes" after reading about democratic theory and environmental problem-solving first.

We also include scenarios. The first explores how to engage stakeholders in regional efforts to manage dwindling water supplies. The second asks you to recommend ways of involving the public in formulating a municipal land-use plan in the face of racial tension. The third empowers you to reflect on how you would work out differences among stakeholders with radically different values (not interests) to secure a resilient agreement. The final written assignment gives you a chance to contrast alternative methods of democratic decision-making in environmental problem-solving situations. MIT student responses to the assignment are included too.

Commentaries and Reading Excerpts

Lawrence Susskind and Jeffrey Cruikshank — "Convening," "Assigning Roles and Responsibilities," "Facilitating Group Problem Solving," "Reaching Agreement" and "Holding Parties to Their Commitments." In *Breaking Robert's Rules*. Oxford: Oxford University Press.

Reading Commentary

The tyranny of the majority is a growing worry in democratic societies. We see this at the city or metro level where citywide interests are willing to proceed with development projects that severely disadvantage small, often marginal, groups. Even if elected officials represented by a majority of residents decide to do something, the interests of minorities ought to be of concern. In the public policy world, the assumption that a project or a policy ought to go ahead if the "gains to the gainers outweigh the losses to the losers" is continually contested by those with progressive views. Conscious of the need to redress these concerns while fostering effective problem-solving, Susskind and Cruikshank offer the consensus-building approach, a battle-tested, collaborative decision-making process that engages the widest possible array of stakeholder representatives in face-to-face problem-solving, as an efficient check on the tyranny of the majority.

Hard political bargaining usually proceeds on the assumption that to achieve one's goals, other parties must not get what they want. This premise engenders "win-at-all-cost" tactics including bluffing, lying (or what is now called "strategic misrepresentation"), dirty tricks, false news and other techniques for undermining the credibility of one's opponents in the public eye. We have more than enough experience to know that these tactics are counterproductive; parties are unlikely to be convinced to support something that is not in their best interest, especially when others are clearly out to hurt them. Instead, the consensus-building approach proceeds on an entirely different assumption: what if the best way to satisfy one's interests is to find a low-cost way to meet the most important interests of the other parties, working collaboratively to find a "package" that maximizes the total value of the outcome for everyone?

The consensus-building approach, whose five core steps are outlined in detail in the authors' excerpt, seeks fairer, wiser, more efficient and more stable results by maximizing the stakeholders' joint gains through reachable, durable and implementable agreements that foster actionable shifts in policy and political alignments. Through careful joint fact-finding, often in the presence of trained neutrals, parties are empowered to inform their decisions by working together, relying on jointly-selected analysts rather than deferring to partisan experts, and crafting packages that try to meet the most important interests on all sides. Engaging in a consensus-building approach is thus an endeavor that does not ask anyone to trade away their interests or compromise their identity. Rather, the goal is to generate agreements that allow all parties to satisfy their most important interests while finding a space to reconcile conflicting values.

The consensus-building approach may also provide another key benefit: a persistent structure that can bolster a project as stakeholders move through inevitable post-agreement challenges. Conflicts in the public realm are increasing in number and complexity. In the past, what usually signified the settlement of a political dispute was the literal or figurative signing of an official agreement by all stakeholders, or a court decision. What was seen as a final deal, and the end of the negotiation, may be different now. Instead, the final product often includes the creation of an oversight committee or organization to monitor "postsignature actions" or, depending on what happens, to modify, rescope or restructure the agreement. With continuously evolving problems comes a need for continuously evolving solutions.

In today's fast-paced, information-rich world, even the best-laid plan may not work out as envisioned. Given a wide range of uncertainties, it may not make sense to formulate a permanent agreement. It may be more important for the implementers of a plan or project to be nimble, adaptable and capable of absorbing and responding to inevitable disruptions. The trusting relationships formed among parties through the consensus-building approach are a crucially valuable outcome of an initial negotiation, because these enable the parties to come back later, together, to strengthen or revise their agreement to effectively mirror the changing environmental problem-solving context around them.

Note: In the upcoming text, "CBA" refers to "consensus-building approach."

Step 1
CONVENING

Lawrence Susskind and Jeffrey Cruikshank

1.1 INITIATE DISCUSSION WITH POTENTIAL ORGANIZER(S)

Anyone can initiate a consensus-building process by raising the idea with the right individual(s) or official(s) in charge. The critical step at the outset is to identify a potential convener—an elected or appointed official or a senior official in the private sector with the formal authority to take action. This person does not need to know very much about CBA in order to play a convening role. At the outset, it is crucial to discuss the advantages and disadvantages of using consensus building, rather than relying exclusively on whatever the usual way of making group decisions might be.

Give the potential convener a copy of *Breaking Robert's Rules*. Explain that they are not being asked to give up any of their formal authority. Point out that the primary reason to use CBA is because it has been proven to be a better way of getting group agreement; in essence, the total value (when generated by consensus) is likely to be greater than the sum of the parts (i.e., the value that the individuals can create on their own). It may take a little longer to get CBA off the ground (i.e., to ensure that all the relevant stakeholders are represented, to get agreement on ground rules, and so on), but the chances are that the overall time it takes to get to an agreement that can be implemented effectively will be less.

1.2 INITIATE AN ASSESSMENT

If the convener is willing to explore the idea further, the next step is to initiate the preparation of an **assessment** (sometimes called a conflict assessment). The convener either can tap someone she knows to do this (as long as that person is viewed as nonpartisan or neutral by all the key stakeholders), or the convener can seek help from a professional neutral (i.e., a trained facilitator or mediator). There are professional neutrals listed in almost every telephone book.

Preparation of an assessment involves interviewing the obvious stakeholders (i.e., the first circle) privately and confidentially. This round of interviewees usually suggests a second circle of people to talk to, as well. In the public arena, the fact that an assessment is being prepared should be made public, so people who feel they have an important contribution to make can also ask to be interviewed (a third circle of interviewees). An assessment doesn't need to take very long or cost very much. Once the interviews are done, the assessor needs to write a short synthesis "mapping the conflict"—that is, laying out the major categories of stakeholders and their views on each possible agenda item. This short summary, and a one-page matrix (like the one Connie prepared for Bill on page 58) summarizing the results of the interviews, should be sent to everyone interviewed. Their comments and corrections should be incorporated by the assessor into a final version of the assessment (in which no one who was interviewed is mentioned by name).

1.3 USE THE ASSESSMENT TO IDENTIFY APPROPRIATE STAKEHOLDER REPRESENTATIVES

The assessment should help the assessor spell out for to the convener who the **appropriate people** are to "invite to the table," assuming CBA goes forward. The assessor has to suggest a way to make sure that all relevant stakeholder groups are represented. This is hard in some cases, because some stakeholders may not already be organized. It is almost always possible, however, to find some individual willing to stand in for a hard-to-represent group.

1.4 FINALIZE COMMITMENTS TO INVOLVE APPROPRIATE STAKEHOLDER REPRESENTATIVES IF A CONSENSUS-BUILDING PROCESS GOES FORWARD

Once all the potential invitees have been identified (by the convener) and the interviewees have had a chance to comment on the draft assessment, the assessor should formulate a potential agenda, work plan, timetable, and budget, and suggest ground rules. These need not be more than one page each in length. (See suggested ground rules in appendix B.) These, too, should be sent in draft to all the stakeholders in preliminary form for their review. They should be asked directly by the assessor if they would agree to participate in a CBA process, if one were initiated by the convener. The attached materials should make it clear what the participants would be asked to accept.

1.5 DECIDE WHETHER TO COMMIT TO CBA

Based on the reactions of the key stakeholders, the assessor should be in a position to make a recommendation to the convener about whether to go forward. If the assessor believes that the most important stakeholders will participate, at least in an organizational session, and based on the confidential interviews that—regardless of what may have been said in the newspapers or in other public settings—there is a possibility of finding common ground, then the assessor should tell this to the convener. The convener can decide (either on his own, or with appropriate co-conveners) whether to send a letter of invitation to all relevant stakeholders, urging them to participate.

With the suggested agenda, timetable, budget, and ground rules attached to the invitation, each participant should be able to see what is expected of them. Such invitations usually ask the recipients to commit to attend an organizational meeting at which the convener will asks the parties if they are prepared to work with the person he has suggested as the facilitator. (It could be the same person who prepared the assessment, or someone new.) The parties must then make their own decisions: whether to go forward, who the facilitator should be, and whether to accept or modify the agenda, ground rules, timetable, budget, and even the list of invitees.

1.6 MAKE SURE THAT THOSE IN POSITIONS OF AUTHORITY AGREE TO THE PROCESS

Often the convener needs to contact other elected or appointed officials or members of organizations (and sometimes heads of nongovernmental or business groups) who, while they may not be direct stakeholders, could become involved later on, depending on what the CBA generates by way of proposals. It is usually a good idea to brief those individuals before the process begins to make sure they are willing to support the effort. Final implementation of whatever agreement is produced often requires formal action by such related officials.

Step 2

ASSIGNING ROLES AND RESPONSIBILITIES

2.1 SPECIFY WHO WILL TAKE RESPONSIBILITY FOR CONVENING, FACILITATING, RECORDING, MODERATING OR CHAIRING MEETINGS, REPRESENTING KEY STAKEHOLDER GROUPS, AND PROVIDING EXPERT ADVICE

At the first meeting, the group as a whole should review its roles and responsibilities. Presumably, all the participants "represent" some larger set of interested stakeholders or constituents. They may also bring special expertise to the discussion. To the extent that one of the participants is unwilling or unable to meet the expectations of the others, with regard to her representational responsibilities, it may be necessary to find a substitute. "Representation" may consist of nothing more than agreeing to serve as a go-between or a two-way channel of information.

In small groups like the Blaine Centennial Committee, it is possible to handle the assignments of these responsibilities in a less formal way, relying on members of the group to take on their assignments periodically. The convener may want to name a member of the group as moderator or chair (for purposes of providing the "outside world" with a point of contact). That role can also be assigned (by the group) to the facilitator.

It is not necessary for the convener to attend every meeting, although she might want to name someone to sit in for her on a regular basis. Technical advisors or experts must be approved by the group as a whole. Often, they are relegated to task forces or subcommittees named by the full group to generate proposals or explore specific topics in advance of full group meetings. Meetings are usually managed by the facilitator. Either a member of the facilitation staff or one of the participants should serve as recorder for each session— keeping a visible written record of key points discussed and preparing a short summary of items of agreement and disagreement to distribute to all the participants soon after each meeting.

2.2 SET RULES REGARDING THE INVOLVEMENT OF ALTERNATES AND OBSERVERS

The credibility of almost any group decision-making process is enhanced when the interactions among the participants are transparent. On the other hand, the work of any deliberative body or team can be made more difficult by the presence of observers. It is important for each CBA group to set clear ground rules regarding which, if any, of its sessions will be open to observers and whether or not the observers will be recognized to speak. In a highly public context, when observers are permitted, it is a good idea to require them to initial written ground rules regarding acceptable behavior and the "rules of engagement" before they are admitted. (This is usually less important inside an organization or a firm.)

Some groups permit all participants to name (at the outset) an alternate empowered to participate on their behalf if they cannot attend a meeting. When this procedure is in effect, it is the participant's obligation to ensure that his alternate is up-to-speed on what has happened thus far, as well as what will be covered at the meeting.

2.3 FINALIZE THE AGENDA, GROUND RULES, WORK PLAN, AND BUDGET IN WRITTEN FORM (FOR PUBLIC OR ORGANIZATIONAL REVIEW)

At the first meeting, all the participants should take part in a thorough review of the proposed agenda, ground rules, work plan, and budget for the CBA process. Whatever revisions are proposed, a written version should be circulated prior to the second meeting, so that all participants can review them with their relevant constituents or outside actors prior to adopting them at the beginning of the second meeting. There should be a provision in the ground rules allowing any participant to suggest changes in the ground rules, work plan, agenda, or budget at any time with the concurrence of the full group. At the beginning of the second meeting, all participants should be prepared to adopt a final set of ground rules and a work plan.

2.4 ASSESS OPTIONS FOR COMMUNICATING WITH THE CONSTITUENCIES REPRESENTED AS WELL AS WITH THE COMMUNITY-AT-LARGE

Once the discussions are underway, some means of communicating with the broader array of constituents will be needed. How this will be done should be determined at the first meeting. The recorder should send meeting summaries to the participants in a form that they, in turn, can easily distribute to other interested stakeholders. A website may be appropriate in certain circumstances. Written meeting summaries distributed by mail to a continuously-updated mailing list may be most appropriate. The final report or recommendations of the group may need to be distributed in still other forms to ensure the most widespread review possible.

Step 3
FACILITATING GROUP PROBLEM SOLVING

3.1 STRIVE FOR TRANSPARENCY (DISTRIBUTE WRITTEN SUMMARIES OF ALL MEETINGS)

CBA involves not only a set of tools and techniques (like assessment and facilitation) but also a commitment to joint problem solving. Members of any organization or community are not likely to view a joint problem-solving effort as legitimate, however, unless they understand exactly who is meeting, and how well they have handled their assignments. Thus, the legitimacy of any consensus-building effort hinges in large part on the way the process is perceived by those likely to be affected by what is decided.

The legitimacy of both the process and the outcome, in turn, depends on the transparency of the effort. Not only are the written agenda, ground rules, work plan, and budget crucial to the spirit of transparency but so are the written summaries of regular meetings. These should be easily available to anyone interested, along with a draft of the final report, in time for nonparticipants to offer comments before final decisions are made. Unlike traditional minutes, meetings summaries should *not* mention who said what. Rather, they should highlight points of agreement and disagreement, and summarize the evidence and arguments that carried weight with the group.

3.2 SEEK EXPERT INPUT WHEN JOINT FACT-FINDING MIGHT BE HELPFUL

A great many consensus-building processes involve decisions that hinge, at least in part, on scientific or technical judgments of one kind or another. When technical considerations come into play, all participants should have equal access to the best possible (nonpartisan) advice. It is almost always desirable to avoid the kinds of "dueling experts" (i.e., technical advisors selected by each side to reinforce their interpretation of technical inputs and undermine the technical arguments of others) so common in the courtroom.

Joint fact-finding hinges on the participants working together to (1) spell out the technical matters or questions on which they want advice, (2) select a range of experts to advise the group as a whole, (3) interact with those experts before they begin their work to discuss how they intend to address or answer the group's concerns, and (4) interact with those experts, once they have generated preliminary findings, to discuss the policy implications of their work. A knowledgeable facilitator can often serve as an "interlocutor" between the experts and participants with less technical expertise.

3.3 CREATE WORKING SUBCOMMITTEES IF APPROPRIATE

A consensus-building process built around a very full agenda can sometimes be enhanced by appointing subcommittees of interested participants and technical advisors to explore individual or clusters of agenda items, prior to the full group's first effort to address them. Subcommittees should not be given decision-making responsibilities. Rather, they should be offered a setting within which a great deal of technical material can be explored and "homework" on related efforts elsewhere can be reviewed. The goal of subcommittees should be to make the work of the full group easier and to provide a starting point for informed discussion of one or a cluster of agenda items.

3.4 SEEK TO MAXIMIZE JOINT GAINS THROUGH THE BRAINSTORMING OF PACKAGES

One of the strongest arguments for the use of CBA—regardless of the scope of the decision-making problem—is that it aims to maximize joint gains in order to create as much value as possible. In other words, it aims to find the best possible way of responding to the conflicting concerns or needs of the stakeholders. The harder the group works to expand the proverbial "pie," the larger the "slice" each is likely to get. The most efficient way to expand the pie is to look for mutually beneficial trades, or what are called **packages,** that give each group more of what is most important to them, in exchange for granting others what they need. This is best accomplished by playing the game of "what-if." One group offers what it thinks will be attractive to the others on a what-if basis, as long as it is guaranteed that what it finds most valuable from the others will be provided in return. Sometimes, the process of packaging is best handled by having the facilitator meet privately with key stakeholders before or between meetings (or during breaks in actual meetings) and then crafting a bundle of proposals, or a package, without saying who specifically offered what to whom.

3.5 SEPARATE INVENTING FROM COMMITTING

As our colleagues Roger Fisher, Bill Ury, and Bruce Patton have explained in their classic book, *Getting to Yes*, it is hard to get people to brainstorm or play the game of what-if when they are worried that everything they say may be construed as a firm commitment, even before they are ready to make such promises. One of the ground rules in most consensus-building processes, therefore, is that no one should be asked to make firm commitments to a package or a proposal until they are absolutely ready to do so. This ground rule can be summarized in the phrase, "separate inventing from committing," meaning that nothing said during brainstorming can later be thrown back at someone as a promise made earlier.

3.6 USE THE HELP OF A SKILLED FACILITATOR

The steps in the CBA process can get very complicated, especially when there are many parties at the table. Coalitions or alliances sometimes form away from the table. Interpersonal difficulties can impede communication. Outside pressures (usually in the form of unrealistic time-tables) can make everyone defensive.

The best way to head off or resolve such difficulties is to have somebody manage the dialogue, someone whose only concern is to make sure that everyone is treated fairly and that the group remains true to its agenda, timetable, and work plan. It is often best to have a professional facilitator play this role, although there are many instances in which someone from inside one of the relevant organizations is trusted enough by all the participants to take on this role. There are growing numbers of accredited professional facilitators (sometimes called dispute resolvers, mediators, or neutrals) who can provide these services on a very reasonable fee-for-service basis.

3.7 USE A SINGLE-TEXT PROCEDURE

As a CBA process moves along, more and more paper accumulates. Meeting summaries, subcommittee reports, and proposals from individual participants keep appearing. At some point, these all need to be folded into a single-text that everyone can support. While a group may tentatively finish discussing an important item on its agenda in week two, there can be no agreement on anything until the full package is agreed upon by everyone (or almost everyone) at the conclusion of the process. Very often, a statement to this effect is included in the ground rules.

3.8 MODIFY THE AGENDA, GROUND RULES, AND DEADLINES AS YOU GO

As the process of brainstorming proceeds and new packages emerge, it may become dear that the group should consult with additional individuals who were not thought to be central when the process began. Indeed, it may even be appropriate to add additional stakeholder representatives late in the process, depending on whether new issues have emerged. Should this become necessary, the group should revisit its agenda (work plan) and proposed deadlines. It may even need to reconsider some of its ground rules. This is perfectly OK, as long as all such decisions are made collectively and in a transparent way. New members, if they are added, need to be given a chance to review everything that transpired prior to their arrival and to raise any questions that seem important to them. While the group as a whole is unlikely to go back over everything that has emerged at that point—and is contained in the current version of the single-text—they might, in light of the concerns expressed by a new member, be willing to reconsider something that has already been discussed. Remember, no one has made a firm or final commitment until they have reviewed the full package and checked it with their constituents.

Step 4
REACHING AGREEMENT

4.1 SEEK UNANIMITY ON A WRITTEN PACKAGE OF COMMITMENTS

Straw polls are a good device for determining how close the participants are to reaching agreement on a single-text (presented by the facilitator or the group leader). Often, it is necessary to add items to the emerging agreement to provide additional benefits to parties that have not yet indicated their approval. This does not mean that everyone involved needs to find the proposed agreement entirely to their liking. Rather, participants should be asked to consider the "full package" in relation to estimated gains and losses they are most likely to get if there is no agreement. While this undoubtedly involves estimates or forecasts of what is likely to happen (rather than something that can be calculated with certainty), it provides the most appropriate point of comparison for each participant. Thus, the objective of CBA is to seek unanimity on a written package that offers all stakeholders something more valuable than what they can expect in the absence of an agreement.

4.2 USE CONTINGENT COMMITMENTS, IF APPROPRIATE, TO DEAL WITH UNCERTAINTY OR RISK

Sometimes, there are too many unknowns for a participant in a CBA process to estimate the value to them of the package being proposed. Contingent options can be used to reduce or eliminate such uncertainties. For instance, one participant may only be able to agree with a package if a particular (but unlikely) circumstance can be eliminated entirely as a future possibility. While everyone else in the group is convinced that the concern of that stakeholder is highly unlikely (e.g., there is less than a 1-percent chance that the thing they are worried about will occur), the participant for whom it is the worst of all possibilities may need further assurance before she can sign an agreement. This can be accomplished by including a contingent option in the agreement indicating that if this highly unlikely event does occur, the group will have to get together again and revise the agreement (i.e., all bets would be off).

Sometimes, a table of contingent options can be added to an agreement to win the support of remaining holdouts. Such a table would spell out a wide range of if–then obligations, such as: "if interest rates increase by 5 percent, 6 percent, 7 percent, or more over the next three years, the financial contribution of each participant indicated in the package will be revised as follows." Instead of arguing that interest rates are not likely to increase by 5 percent in the next few years (and reaching no agreement because someone can not be convinced), the group would spell out the revised terms of their agreement under each of these increasingly unlikely circumstances and incorporate the relevant table into the agreement itself.

4.3 ADHERE TO AGREED-UPON DECISION-MAKING PROCEDURES

It is important that all participants in a CBA process stick to the procedures they prescribe at the outset. This is important for ensuring the perceived legitimacy of the process in the eyes of the at-large. The only changes allowed in the process of decision making should be those made **by the fall group**, using a method of amending the ground rules made explicit in the original ground rules.

4.3.1 ASK WHO CAN'T LIVE WITH THE PACKAGE

When the facilitator or dialogue leader thinks the stakeholders are close to overwhelming consensus, she should ask, "Who can't live with the package of proposals detailed in the most recent version of the single text?" Anyone who indicates that he can't live with such a package is obliged **under the terms of the ground rules** to explain why they object (i.e., what interests of theirs aren't being met).

4.3.2 ASK THOSE WHO OBJECT TO SUGGEST IMPROVEMENTS THAT WOULD MAKE THE PACKAGE ACCEPTABLE TO THEM WITHOUT MAKING IT UNACCEPTABLE TO OTHERS

It is then the obligation of those who object to offer specific "improvements" that would make the agreement acceptable to them without losing the support of others. If a person indicating unhappiness with the package cannot think of any way to do this, others in the group should be encouraged to help. If no one can think of any way of responding to the concerns of the holdout(s), the rest of the group must decide whether to continue with its deliberations or to conclude them, having reached consensus (i.e., seeking unanimity, but settling for over-whelming agreement once it is clear that no one can think of any way to respond to the remaining concerns of the holdout(s) without leaving others worse off).

4.4 KEEP A WRITTEN RECORD OF ALL AGREEMENTS

The final written version of the package should indicate who could not live with the draft agreement and why. It should also include in a footnote detailing the ideas or offers considered at the last minute in an effort to respond to the remaining holdout(s).

4.5 MAINTAIN COMMUNICATION WITH ALL RELEVANT CONSTITUENTS AND THE COMMUNITY-AT-LARGE

The CBA process is not complete until the final draft of the package, approved by those at the meeting, is circulated by the participants to their constituents and to the convener or convening agencies prior to the final meeting of the stakeholders.

Step 5
HOLDING PARTIES TO THEIR COMMITMENTS

5.1 SEEK RATIFICATION OF THE DRAFT AGREEMENT BY CHECKING BACK WITH ALL RELEVANT CONSTITUENCIES

The CBA process is complete when the stakeholders return one last time to meet face-to-face to review the comments received when the participants took the penultimate (i.e., "next-to-final") draft of the agreement out for review. It may be necessary to modify the package one last time to ensure the support of constituents that the participants are supposed to represent. If major changes to the package are made, however, it may be necessary to go through the final step once again.

5.2 AT A FINAL MEETING, ASK ALL THE STAKEHOLDER REPRESENTATIVES TO INDICATE THEIR PERSONAL SUPPORT FOR THE PACKAGE BY SIGNING THE AGREEMENT

While not everyone at the final meeting will have the authority to speak in a legal sense, for the constituents or stakeholders they supposedly represent, it is nevertheless appropriate to ask each individual involved to sign the agreement. Usually, participants are asked to sign a statement that indicates their personal support for the package and their personal promise to work to implement the agreement (and to follow through on any commitment they have made). Sometimes such statements include a sentence indicating that the signatories have, indeed, taken the final draft of the agreement to their constituents for review.

5.3 PRESENT THE RECOMMENDED PACKAGE OF PROPOSALS TO THOSE WITH THE FORMAL AUTHORITY TO ACT

The facilitator and/or the group leader (if one has been designated) should meet with the convener or convening agencies to present the final signed package produced by the CBA process. They should explain how the group arrived at its recommendations and make themselves available to answer questions about both the content of the agreement and the procedures that were followed. They should ask those with the authority to act to indicate their reaction to the agreement and to describe the steps they intend to take to follow through on what has been proposed.

5.3.1 LOOK FOR WAYS TO MAKE INFORMALLY NEGOTIATED AGREEMENTS BINDING

Often it is possible for the convener or convening agencies to transform the proposals from the CBA group into binding mechanisms that will hold everyone to his or her commitments. For example, elected officials can vote on a proposed package and incorporate it into the terms of permits, licenses, regulations, formal policy statements, or legislation. In the corporate context, negotiated agreements can take a contractual form that legally binds the relevant decision makers to adhere to the terms of the agreement. In other situations, the officers of an organization can vote to enact the recommendations produced by the CBA process they set in motion. Including a dispute resolution clause in the agreement is a good way to make sure that unexpected problems do not cause an agreement to unravel but instead trigger a dispute resolution procedure that has been agreed upon ahead of time.

5.4 RECONVENE THE PARTIES IT THOSE IN AUTHORITY CAN NOT LIVE WITH THE PACKAGE TO SEE WHAT CHANGES MIGHT BE POSSIBLE

If those in positions of authority determine there are aspects of the package they can not support, it may be desirable to reconvene the stakeholders to discuss further modifications to the agreement, rather than let it languish. Ideally, the draft of the final agreement reviewed by all the constituents should be circulated informally to the relevant convener or convening bodies for their comments, avoiding this kind of impasse.

5.5 MONITOR CHANGING CIRCUMSTANCES DURING IMPLEMENTATION AND RECONVENE IF NECESSARY

Sometimes, negotiated agreements call for periodic review by stakeholders to assess progress toward implementation. This may involve reconvening the full group on the anniversary of their final meeting, or periodic review by a monitoring subcommittee selected by the full group and named in the package. The facilitator is usually designated as the person to bring the participants together to monitor progress or to prepare a revised proposal that would allow the group to amend its final agreement and take account of unexpected events.

Ian Shapiro — "Aggregation, Deliberation, and the Common Good." In _State of Democratic Theory._ Princeton: Princeton University Press.

Reading Commentary

How do we arrive at "the common good," or outcomes that are likely to benefit the majority of society? In this chapter, Shapiro compares aggregative democratic approaches (i.e., summing individual votes to reveal the preferences of the majority) to deliberative democratic approaches (i.e., "manufacturing" a shared conception of the "common good" by transforming people's preferences through deliberation).

Shapiro points out that aggregative democratic approaches (i.e., majority rule) may lead to the "tyranny of the majority" (i.e., minority tyranny or arbitrary outcomes). He compares this to deliberative democracy as described by Amy Gutmann and Dennis Thompson (1996) and Bruce Ackerman and James Fishkin (2002). He argues that both models, but particularly the deliberative model, fail to take into account the role of power in policy-making. In the case of aggregative democracy, powerful actors can determine what is on the agenda and the order of voting, thereby controlling the outcome. In the case of deliberative democracy, he uses the example of national health-care reform under the Clinton administration to demonstrate how powerful actors set the agenda (e.g., through political contributions, media ads, lobbying, etc.) and excluded options (e.g., single-payer health care), thereby limiting the range of outcomes that could be reached through deliberation.

Shapiro's main thesis is that power cannot be removed from politics. He reminds us to ask who decides which issues are presented for discussion and, possibly, decision? Who sets the agenda? Who chooses the "experts" who provide input into the policy-making process or decide which criteria should be used to make sure their input is "scientific" or "balanced"?

This book has been designed to introduce you to a variety of ways of evaluating environmental policy and environmental policy-making. It is also intended to encourage you to question your personal assumptions about how natural resources should be distributed and how these allocation decisions should be made. Advocates of both approaches to democratic decision-making argue that democracy ensures political legitimacy. They also assert that political legitimacy is crucial to sustainable collective action. As you read this excerpt and the rest of the chapter, we hope you will continue to reflect on the ways in which both the process and outcomes of environmental problem-solving can be enhanced.

CHAPTER 1
AGGREGATION, DELIBERATION, AND THE COMMON GOOD

Ian Shapiro

Underlying the normative literature on democracy is a series of debates about rationality. They revolve around the question whether the classic democratic notions of a "will of the people" or "common good" have any coherent meaning. The idea that democracy does or should converge on a rationally identifiable common good finds its locus classicus in Jean-Jacques Rousseau's *Social Contract*, and in particular in his contention that decision procedures should reveal a general will that embodies the common good. Rousseau ([1762] 1968: 72) famously, if vaguely, characterized this by saying that we start with "the sum of individual desires" and subtract "the plusses and minuses which cancel each other out"; then "the sum of the difference is the general will." Attempts to make sense of this formulation have spawned two literatures, an aggregative literature, which has been geared to finding out just how we are supposed to do the relevant math, and a deliberative literature, which has been partly motivated by impatience with the aggregative one. Deliberative theorists are concerned with getting people to converge on the common good where this is understood more robustly than as a totting up of exogenously fixed preferences.

In §1.1 I make the case that proponents of aggregative conceptions of the common good are right that Rousseau's formulation of the problem cannot be solved, but that considerably less turns on this failure, for democratic politics, than they often suppose. The expectation that a general will should be discoverable rests upon implausible expectations of collective rationality and a misconstrual of what stable democratic politics requires. Turning to proponents of deliberative conceptions in §1.2, I argue that they share a touching faith in deliberation's capacity to get people to converge on the common good. Sometimes people's interests are irreducibly at odds, precluding this possibility. Moreover, in the world of actual politics people confront one another in massively unequal power contexts—in the United States most obviously owing to the role of money in politics. Deliberation theorists tend to confuse problems associated with the unequal power contexts in which deliberation occurs with a deliberative deficit, mistaking the doughnut for the hole. Some contend that what is of interest for democratic politics is what deliberation would produce under ideal, noncoercive, conditions. I doubt that we can answer such questions. In any case I argue that, if such conditions prevailed, the facts about politics that lead people to call for deliberation would no longer obtain.

1.1 AGGREGATIVE CONCEPTIONS OF THE COMMON GOOD

At least since Plato's time, political theorists have warned that democracy fosters mob rule rather than the common good. As the franchise expanded over the course of the nineteenth century, Alexis de Tocqueville ([1835] 1969: 246–61) and John Stuart Mill ([1859]

1978: 4) also cautioned against democracy's propensity to lead to the "tyranny of the majority." In this they echoed Rousseau's concern that a majority might satisfy its members' interests at the expense of the minority ([1762] 1968: 73) and Madison's (Hamilton, Madison, and Jay [1788] 1966: 122–28) discussion of the dangers presented by "majority factions" in *Federalist* No. 10.

1.1.1 DEMOCRACY'S ALLEGED IRRATIONALITY

Modern social choice theorists have held that the problem is worse than these classical authors realized in that majority rule can lead to arbitrary outcomes and even to minority tyranny. They agree that the goal of democratic decision procedures should be to discover something like a general will, referred to in the modern idiom as a social welfare function, but, following Kenneth Arrow (1951), they argue that this is impossible. Extending an old insight of the Marquis de Condorcet ([1785] 1972), Arrow showed that under some exceedingly weak assumptions, majority rule leads to outcomes that are opposed by a majority of the population. For instance, if voter I's ranked preferences are A > B > C, voter II's are C > A > B, and voter Ill's are B > C > A, then there is a majority for A over B (voters I and II), a majority for B over C (voters I and III), and a majority for C over A (voters II and III). This outcome, known as a voting cycle, violates the principle of transitivity, generally taken to be an essential feature of rationality because it permits a self-contradictory ranking of societal preferences. Moreover, it opens up the possibility that whoever controls the order of voting can determine the outcome, provided she knows the preferences of the voters. Even if outcomes are not consciously manipulated, they might nonetheless be arbitrary in the sense that, had the alternatives been voted on in some order other than they actually were, the result would have been different. In short, democracy might lead to tyranny of the majority, but it might also lead to tyranny of a strategically well placed minority or to tyranny of irrational arbitrariness.[1]

Notice that power relations enter into aggregative conceptions of the common good only implicitly, and then in an implausible fashion. The presumption behind trying to render the Rousseauist project coherent is that if it cannot be done, then collective decisions that *are* taken amount to illicit impositions masquerading as democratic decisions. And since there is a wide consensus in the literature that the project cannot be rendered coherent, the libertarian implication, that collective action of all kinds should be limited as much as possible, is held by libertarians such as Riker, Weingast, Buchanan, and Tullock to follow from this result (see Shapiro 1996: 30–42).

Fear of tyranny by majority factions led Madison and the Federalists to devise a political system composed of multiple vetoes in order to make majority political action difficult: a separation-of-powers system in which "ambition will be made to counteract ambition" (Hamilton, Madison, and Jay [1788] 1966: 318). This included an independent court with the power to declare legislation unconstitutional and a president whose election and hence legitimacy are independent of the legislature; strong bicameralism in which legislation must pass both houses and in which two-thirds majorities in both houses can override the president's veto power; and a federal system in which there is constant jurisdictional tension between federal and state governments. The findings in the post-Arrovian social choice literature have led commentators such as Riker and Weingast to endorse this multiplication of institutional veto points on the possibility of governmental action, arguing that courts should hem in legislatures as much as possible, lest they compromise individual rights, particularly property rights, with irrational and perhaps manipulated collective decisions.[2]

Yet we can grant Arrow his victory over Rousseau without being persuaded of the merits of these ossifying institutional arrangements. The decisive question, after all, is compared to what? Arrow's finding deals not merely with majority rule. His theorem shows that, given the diversity of preferences he postulates, his modest institutional conditions, and his unexceptionable constraints on rationality, *no* mechanism will produce a rational collective decision. Libertarians suppose that the alternative is to minimize governmental action as much as possible, but that is inadequate for two reasons. First, it is mistaken to suppose that making governmental action difficult limits collective action. Perhaps owing to their proclivity for thinking in a social contract idiom, libertarian commentators often write as if "not having" collective action is a coherent option in societies that nonetheless have private property, enforcement of contracts, and the standard panoply of negative freedoms. The recent experience of postcommunist countries such as Russia should remind us that these are costly institutions requiring continual collective enforcement (see Holmes and Sunstein 1999). The libertarian constitutional scheme is a collective action regime maintained by the state, one that is disproportionately financed by implicit taxes on those who would prefer an alternative regime. The more appropriate question, then, is not "whether-or-not collective action?" but rather "what sort of collective action?"

Second, liberal constitutionalists like Riker and Weingast (1988) tend to focus on potential institutional pathologies of legislatures, while ignoring those of the institutions that they would have curb legislative action. At least in the United States, courts are themselves majoritarian institutions. There is every reason to believe they would be at least as vulnerable to cycles as legislatures, and possibly even more susceptible to manipulation. Chief justices, who have considerable control over court agendas and the order in which issues are taken up, know a good deal about their colleagues' preferences, owing to the fact that they decide many closely related cases, and personnel turnover is incremental and slow. In theory judges are constrained by doctrines and precedents, but, as Oliver Wendell Holmes insisted, much of the time these can be rendered consistent with any outcome by a sufficiently enterprising judge.[3] Indeed, it may be that less of the information pertinent to manipulation is available in a Senate of 100, a third of whom are up for election every two years, or a House of Representatives of 435, all of whom are up for reelection every two years—not to mention the population at large. True, high incumbent re – election rates slow down turnover of legislators, and much of their work is done in smaller committees. Granting this, there is still no reason to believe them more susceptible than courts to the potential for arbitrary or manipulated outcomes identified by Arrow.

1.1.2 COMPETING VIEWS OF RATIONAL COLLECTIVE DECISION

More important, perhaps, than these weaknesses in the liberal constitutionalist critique of democracy are the expectations about what would be a nonarbitrary decision-making outcome. Arrow may have established that often there may be no such thing as a Rousseauian general will, but why should we be troubled by that? Transitivity may well be a reasonable property of individual rationality, but it is far from clear that it makes sense to require it of collective decisions. If the New York Giants beat the Dallas Cowboys, who in turn beat the Washington Redskins, no one suggests that the Redskins should not play the Giants lest the principle of transitivity be violated. Deadlocked committees sometimes make decisions by the toss of a coin—arbitrary, perhaps, but necessary for collective life to go on. In such circumstances it matters more that each contest or decision mechanism was perceived to be fair than that a different outcome might not have occurred on a different day (see Mueller 1989: 390–92).

If we abandon the expectation that there is a Rousseauian general will or social welfare function waiting out there to be discovered like a Platonic form floating in metaphysical space, we might nonetheless be persuaded of the merits of majority rule for decision making in many circumstances. As I argue below, one reason to favor it is that it promotes competition of ideas. Another is that majority rule can contribute to political stability just because it institutionalizes the perpetual possibility of upsetting the status quo. Theorists such as Di Palma (1990: 55) and Przeworski (1991: 10–12) note that it is institutionalized uncertainty about the future that gives people who lose in any given round the incentive to remain committed to the process rather than reach for their guns or otherwise become alienated from the political system. This will not happen when there is a single dominant cleavage in the society, as when a majority of the population has identical preference orderings. Such a preference structure will forestall an Arrovian cycle, but at the price of turning loyal opposition (where the democratic system is endorsed though the government of the day is opposed) into disloyal opposition, where those who lose try to overthrow the system itself.

Generalizing this, Nicholas Miller (1983: 735–40) has noted that there is a tension between the notion of stability implicit in the social choice literature since Arrow, where various restrictions on preferences are intended to prevent cycling, and the pluralist idea of stability that turns on the presence of crosscutting cleavages of interest in the population. The periodic turnover of governments is facilitated by just the kind of heterogeneous preferences that create the possibility of cycling. Indeed, students of comparative politics often contend that competitive democracy does not work when heterogeneous preferences are lacking. If the preference-cleavages in the population are not sufficiently crosscutting to produce this result, they propose alternative institutional arrangements, such as Arend Lijphart's "consociational democracy" (1969, 1977), which includes entrenched minority vetoes and forces elites representing different groups to govern by consensus as a cartel, avoiding political competition.[4]

1.1.3 THE LIKELIHOOD OF CYCLES

Closer inspection thus reveals that the possibility of voting cycles is not especially troubling, and it may even be advantageous for the stability of democratic institutions. How likely it is that cycles actually occur is another matter. I have already noted that they are ruled out if an absolute majority has identical preferences. Various other constraints on preferences will also reduce their likelihood or eliminate them (Mueller 1989: 63–66, 81–82). At least one theoretical result suggests that cycles are comparatively unlikely in large populations even when preferences are heterogeneous (Tangian 2000), and an exhaustive empirical study by Gerry Mackie (2003) has revealed almost every alleged cycle identified in the social choice literature to be based on faulty data or otherwise spurious.[5] It may be that democracies turn out to enjoy the best of both worlds. The possibility of cycles gives those who lose in any given election an incentive to remain committed to the system in hopes of prevailing in the future, but the fact that cycles are actually rare means that government policies are not perpetually being reversed (Tullock 1981).[6] In the area of tax policy, for instance, there is undoubtedly a potential coalition to upset every conceivable status quo, as one can see by reflecting on a society of three voting to divide a dollar by majority rule: whatever the distribution, a majority will have an interest in changing it. Yet tax policy remains remarkably stable over time (Witte 1985).

In short, despite the considerable attention to the possibility of voting cycles in the social choice literature, there are few reasons to see them as undermining the attractiveness of majoritarian democracy. Once we abandon the Rousseauist expectation that collective action should be guided by a general will or social welfare function, we can see numerous reasons for thinking that democratic constraints on collective action might nonetheless be desirable. This becomes evident once we remind ourselves that power is exercised willy-nilly in social life; that hamstringing government privileges one set of collective arrangements, however implicitly; that institutions for containing democracy's "irrationality" may be no less susceptible to "irrational" behavior than the legislatures they are intended to limit; that it is unclear how likely cycles are in fact; and that in any case we have not been given good reasons to think that the rational stability prized by democracy's post-Arrovian critics is desirable. On the contrary, there may be good reasons to avoid it.

1.1.4 PRIVILEGING UNANIMITY RULE

If the findings in the public choice literature are less threatening to democracy's legitimacy than is often assumed, what of the more traditional worry about the tyranny of the majority associated with the arguments of Tocqueville and Mill and the countermajoritarian elements that the Framers built into the American Constitution? Tocqueville's forecasts were particularly apocalyptic on this point. "Formerly tyranny used the clumsy weapons of chains and hangmen," he noted, yet "nowadays even despotism, though it seemed to have nothing more to learn, has been perfected by civilization." The possibility of majority tyranny struck him as the greatest threat posed by democracy in America. Quoting Madison's worry in *Federalist* No. 51 that "in a society under the forms of which the stronger faction can readily unite and oppress the weaker, anarchy may truly be said to reign," Tocqueville opined that "if ever freedom is lost in America, that will be due to the omnipotence of the majority driving the minorities to desperation and forcing them to appeal to physical force." The result might be anarchy, as Madison said, "but it will have come as a result of despotism" (Tocqueville [1835] 1969: 255, 260).

An influential theoretical response to this danger put forward by James Buchanan and Gordon Tullock builds on the Framers' impulse to make some rights and liberties more difficult than others to change by majority rule. Deploying the style of reasoning that John

Rawls would later make famous, they asked what decision rules mutually disinterested citizens would choose at a constitutional convention where everyone is uncertain "as to what his own precise role will be in any one of the whole chain of later collective choices that will actually have to be made." Whether selfish or altruistic, each agent is forced by circumstances "to act, from self-interest, *as if* they were choosing the best set of rules for the social group" (Buchanan and Tullock 1962: 78, 96).[7] Thus considered, they argued, there is no reason to prefer majority rule to the possible alternatives. Collective decision making invariably has costs and benefits for any individual, and an optimal decision rule would minimize the sum of "external costs" (the costs to an individual of the legal but harmful actions of third parties) and "decision-making costs" (those of negotiating agreement on collective action). The external costs of collective action diminish as increasingly large majorities are required; in the limiting case of unanimity rule, every individual is absolutely protected since anyone can veto a proposed action. Conversely, decision-making costs typically increase with the proportion required, since the costs of negotiation increase. The choice problem at the constitutional stage is to determine the point at which the combined costs are smallest for different types of collective action, and to agree on a range of decision rules to be applied in different future circumstances (Buchanan and Tullock 1962: 63–77).

At least three kinds of collective action can be distinguished, requiring different decision rules. First is the initial decision rule that must prevail for other decision rules to be decided on. Buchanan and Tullock "assume, without elaboration, that at this ultimate stage … the rule of unanimity holds." Next come "those possible collective or public decisions which modify or restrict the structure of individual human or property rights after these have once been defined and generally accepted by the community." Foreseeing that collective action may "impose very severe costs on him," the individual will tend "to place a high value on the attainment of his consent, and he may be quite willing to undergo substantial decision-making costs in order to insure that he will, in fact, be reasonably protected against confiscation." He will thus require a decision rule approaching unanimity. Last is the class of collective actions characteristically undertaken by governments. For these "the individual will recognize that private organization will impose some interdependence costs on him, perhaps in significant amount, and he will, by hypothesis, have supported a shift of such activities to the public sector." Examples include provision of public education, enforcement of building and fire codes, and maintenance of adequate police forces. For such "general legislation" an individual at the constitutional stage will support less inclusive decision rules, though not necessarily simple majority rule, and indeed within this class different majorities might be agreed on as optimal for different purposes. "The number of categories, and the number of decision-making rules chosen, will depend on the situation which the individual expects to prevail and the 'returns to scale' expected to result from using the same rule over many activities." Requiring high levels of agreement enables people to protect their interests, they say, but this takes time that could be spent on other activities. In effect they come up with a sliding scale. Democracy is best suited to issues of moderate importance on their account. Issues of high importance should be insulated from it, while issues of low importance might even be delegated to administrators (Buchanan and Tullock 1962: 73–77).

This argument is defective in various ways that need not concern us now (see Shapiro 1996: 19–29). The point to note here is that Buchanan and Tullock's initial bias in favor of unanimity rule turns on two dubious assumptions that make democracy look less attractive than it should. First there is the social contract fiction already alluded to: that there could be an initial stage in which only private action prevails in society—without being underwritten by collective institutions. The second defect arises even if we engage in the thought experiment Buchanan and Tullock propose. Unanimity as a decision rule has the unique property, they argue, that if decision-making costs are zero, it is the only rational decision rule for all proposed collective action.[8] But this argument confuses unanimity qua decision rule with unanimity qua social state, that is, a condition in the world where everyone actually wants the same outcome. Douglas Rae has pointed out that from the standpoint of their constitutional convention it makes more sense to assume that we are as likely to be ill disposed toward any future status quo as well disposed toward it, and that in cases where we are ill disposed, a decision rule requiring unanimity will frustrate our preferences. Buchanan and Tullock assume throughout that it is departures from the status quo that need to be justified, but Rae shows that this is not warranted. Externalities over time, or "utility drift" (Rae's term), may change our evaluations of the status quo. We may feel in certain circumstances that failures to act collectively, rather than collective action itself, should shoulder the burden of proof(Rae 1975:1270–94).[9] People may change their minds for other reasons, foreseen or unforeseen, or someone might be opposed to, and not wish to be bound by, a status quo that was the product of the unanimous agreement of a previous generation. Indeed, Rae has shown formally that if we assume we are as likely to be against any proposal as for it, which the condition of uncertainty at the constitutional convention would seem to require, then majority rule or something very close to it is the unique solution to Buchanan and Tul – lock's choice problem (Rae 1969: 40–56, 51).[10]

1.1.5 THE LIKELIHOOD OF MAJORITY TYRANNY

Ultimately it is an empirical question whether majoritarian democracy is more likely than the going alternatives to produce tyranny. Robert Dahl (2002) has recently reminded us that in the century and a half since Tocqueville articulated his fears, the individual rights and political freedoms that he prized have turned out to be substantially better respected in democracies than in nondemocracies. The countries in which there is meaningful freedom of speech and of association, respect for personal and property rights, prohibitions on torture, and guarantees of equality before the law are overwhelmingly the countries that have democratic political systems. Even if we expand the definition of individual rights to include social and economic guarantees, one could not make a credible case that nondemocracies supply these better than do democracies.[11] This subject is, of course, difficult to study empirically, because most of the world's wealthy countries, with the resources for meaningful socioeconomic guarantees, are also democracies, and the failures of the communist systems arguably had at least as much to do with their economies as with their political systems. Yet one would scarcely want the Tocquevillian case to rest on the communist example, where civil and political freedoms were substantially less well respected than in democracies, and the level of social provision was generally low. At a minimum one is bound to conclude that the Tocquevillian case has not been established, and that the converse of it seems more likely to be true, to wit, that democracy is the best known guarantor of individual rights and civil liberties.

On the question whether constitutional courts make a difference among democracies, in the United States there have certainly been eras when the federal judiciary has successfully championed individual rights and civil liberties against the legislative branch of government, that of the Warren Court being the best known.[12] But there have also been eras when it has legitimated racial oppression and the denial of civil liberties (see Smith 1997:165–409). Until recently there has been surprisingly little systematic study of this question beyond the trading of anecdotes. As early as 1956 Dahl had registered skepticism that democracies with constitutional courts could be shown to have a positive effect on the degree to which individual freedoms are respected when compared to democracies without them, a view he developed more fully in his seminal article "Decisionmaking in a Democracy: The Supreme Court as National Policymaker" (Dahl 1997: 279–95). Subsequent scholarship has shown Dahl's skepticism to have been well founded (see Dahl 1956: 105–12, 1989: 188–92, Tushnet 1999, Hirschl 1999). Indeed, there are reasons for thinking that the popularity of independent courts in new democracies may have more in common with the popularity of independent banks than with the protection of individual freedoms. They can operate as devices to signal foreign investors and international economic institutions that the capacity of elected officials to engage in redistributive policies or interfere with property rights will be limited. That is, they may be devices for limiting domestic political opposition to unpopular policies by taking them off the table (Hirschl 2000).

This is not to deny that there may be an appropriate role for second-guessing institutions, such as courts, in majoritarian systems. Ways of thinking about courts that reinforce democracy rather than wall it in are explored in chapter 4. It is to say, however, that the fear that majority rule would become the engine of majority domination has not been borne out historically. Indeed, those on the ideological left who hoped that the "parliamentary road to socialism" would be achieved by the majority appropriation of what they saw as the minority's ill-gotten gains through the ballot box have been sorely disappointed. The reasons for this are taken up in chapter 5.

1.2 DELIBERATIVE CONCEPTIONS OF THE COMMON GOOD

The literature on deliberative democracy is to some extent a reaction to dissatisfaction with the aggregative literature, but not for its inattention to the questions about power and collective action that we have been considering. The aggregative literature concerns itself with how to do the math to solve Rousseau's problem; proponents of deliberative democracy are also in search of the common good. But they hope to get to it by transforming preferences rather than aggregating them. It is not really a Rousseauist project (Rousseau had no faith in deliberation as a useful political device). However, it owes something to his injunction that people should vote not their individual preferences but rather their perceptions of what is good for the society as a whole.[13] The goal is to move us "beyond adversary democracy" (Mansbridge 1980).

People advocate deliberation for different reasons. Some think it inherently worthwhile. More commonly deliberation is valued for instrumental reasons: achieving consensus, discovering the truth, and consciousness – raising are among the usual suspects. Some of the time, at least, deliberation promotes these and related values. But it also has costs. Wasted time, procrastination and indecision, stalling in the face of needed change, and unfair control of agendas are among its frequent casualties. Sometimes by design, sometimes not, deliberation can amount to collective fiddling while Rome burns. If deliberation is not always and everywhere an unmitigated good, how do we determine the conditions under which it is desirable?

Deliberative remedies are put forward in response to various maladies that are perceived as pervading contemporary democracy. Poor quality of decision making, low levels of participation, declining legitimacy of government, and ignorant citizens are among the more frequently mentioned. Advocates of deliberative democracy such as Gutmann and Thompson (1996) and Ackerman and Fishkin (2002) argue for the merits of deliberation by pointing out how little of it there is in contemporary politics dominated by superficial television campaigns and political advertising. The idea is that if we can get away from the soap opera of electoral one-upmanship, more thoughtful and effective political choices will result. Deliberative forums can range from town meetings, to designated deliberation times, to citizen juries and "deliberative polls"—randomly selected groups who become better informed about particular issues and render decisions as to what should be done (Fishkin 1991). On some accounts such entities should inform existing processes; on others they should replace them en route to instituting a more robust participatory politics. The unifying impulse motivating these proposals is that people will modify their perceptions of what society should do in the course of discussing this with others. The point of democratic participation, on this account, is more to manufacture the common good than to discover it. Indeed, deliberative theorists sometimes write as if the activity of searching for the common good is itself the common good (see Shapiro 1996: 109–36). Some deliberative democrats do not go this far, but usually they do assume that if people talk for long enough in the right circumstances, they will agree more often, and this is a good thing.

1.2.1 RECIPROCAL DELIBERATION AS THE COMMON GOOD

One influential account of how deliberation might work in practice has been put forth by Gutmann and Thompson in their much discussed book *Democracy and Disagreement*. There they argue for a view of deliberation that is designed to minimize disagreement when this is possible, and to get people to accommodate themselves to one another's views, maintaining "mutual respect," when it is not. Drawing on the idea of reciprocity, they argue for a view of deliberation in which citizens "aspire to a kind of political reasoning that is mutually justifiable," each making claims that the others will accept. They do not claim that deliberation will vanquish all moral disagreement in politics, but they expect it to reduce disagreement and help people who disagree better to converge on mutually acceptable policies. Even when it does not resolve disagreement, it can "help citizens treat one another with mutual respect as they deal with the disagreements that invariably remain." Gutmann and Thompson claim that the lack of deliberation is not limited to public debate alone. It is also reflected in academic commentary on democracy, which is "surprisingly silent about the need for ongoing discussion of moral disagreement in everyday political life. As a result, we suffer from a deliberative deficit not only in our democratic politics but also in our democratic theory." Moreover, we are "unlikely to lower the deficit in our politics if we do not also reduce it in our theory" (Gutmann and Thompson 1996: 2–12, 52–53, 346).

To know how effective Gutmann and Thompson's deliberative model would be, at either reducing moral disagreement or promoting accommodation of irresolvable differences in American politics, one would have to see it in action in debates among pro-lifers and pro-choicers, parties to the *Mozert v. Hawkins* litigation over school textbooks that parents believe violate their children's free exercise of religion,[14] or protagonists in debates over redistricting, affirmative action, welfare reform, child-support, and the other contentious political issues that Gutmann and Thompson describe. Their claim is that if the various protagonists "seek fair terms of cooperation for their own sake," committing themselves to appeal, in their arguments, "to reasons that are recognizably moral in form and mutually acceptable in content," then such disagreements will be minimized and accommodation will be promoted (Gutmann and Thompson 1996: 53, 57). They report how they believe these and other public policy debates ought to come out when the model is applied, or, in some cases, that it cannot resolve them. This is different, however, from demonstrating that it would actually happen in practice. Gutmann and Thompson do offer qualified praise of some actual deliberative processes, such as the 1990 meetings that were held in Oregon to help set health care priorities for Medicaid recipients (see §1.2.2 below). But they fail to mention any actual deliberative process that does not fall significantly short of their deliberative ideal. Accordingly, the claim that their model would have the beneficial effects claimed for it remains speculative.

Sometimes, perhaps, people might better resolve differences and accommodate themselves to views they reject by more deliberation of the prescribed sort. But what reason is there to suppose that failure to attempt this is the principal reason why the public policy issues they examine are not resolved along the lines Gutmann and Thompson advocate? It is one thing to think that much of what divides people politically is susceptible to rational analysis more often than people realize; quite another to believe that what prevents better resolution of prevailing disagreements is insufficient deliberation of the Gutmann-Thompson sort. They give a plausible account of the nature of some moral disagreements and of possible argumentative strategies for constructive responses to them when protagonists are appropriately inclined, but their account attends too little to the role of power relations and conflicts of interest in politics.

The main reason for Gutmann and Thompson's call for more deliberation is that there seems to be so little of it in the political debate they observe. "In the practice of our democratic politics, communicating by sound bite, competing by character assassination, and resolving political conflicts through self-seeking bargaining too often substitute for deliberation on the merits of controversial issues" (Gutmann and Thompson 1996: 12). But sound-bite politics and media-driven campaigns may well result principally from the powerful American antipathy toward publicly financed elections and the concomitant influence of private money in politics. This would presumably remain in a world of expanded deliberative institutions, given the Supreme Court's 1976 declaration that regulating political expenditures is an unconstitutional interference with free speech.[15] Any credible defense of deliberative democracy in the American context would have to show how deliberative institutions would be any less corrupted than are existing institutions by those with the resources to control agendas and bias decision making, and that it would merit its cost.

Gutmann and Thompson are not alone in treating deliberation as a panacea. Consider, for instance, Bruce Ackerman and James Fishkin's (2002: 129–52) proposal for "deliberation day," to be held a week before national elections. On this proposal all citizens would be paid $150 to show up at their local school or community center to deliberate. According to its proponents this would cost $15 billion a year in public funds—not to mention the indirect costs to the economy. It is hard to see what benefit would result from so vast an expenditure of funds once candidates had been selected, platforms chosen, interest groups deployed, and campaign funds expended. By contrast, $15 billion a year spent to support fledgling third parties or publicly financed elections might attenuate many of the pathologies that lead people to call for more deliberation.[16]

These considerations aside, it is far from clear that deliberation exhibits the felicitous political properties that proponents attribute to it. As Gutmann and Thompson concede at one point, sometimes deliberation can promote disagreement and conflict. The cases they have in mind are moral issues that arouse intense passions, paradigmatically the issues liberals have sought to defuse politically since the seventeenth-century wars of religion. Skeptics of deliberation in these areas proceed from the assumption that there are "moral fanatics as well as moral sages, and in politics the former are likely to be more vocal than the latter." Gutmann and Thompson's response is that although moral argument "can arouse moral fanatics," it can also "combat their claims on their own terms." Deliberation undermines moral extremists, who "must assume that they already know what constitutes the best resolution of a moral conflict without deliberating with their fellow citizens who will be bound by the resolution." In the everyday political forums "the assumption that we know the political truth can rarely if ever be justified before we deliberate with others who have something to say about the issues that affect their lives as well as ours." Accordingly, they conclude with a presumption in deliberation's favor: "By refusing to give deliberation a chance, moral extremists forsake the most defensible ground for an uncompromising position" (Gutmann and Thompson 1996: 44–5).

Alluring as this reasoning might be to many of us, it is difficult to imagine a fundamentalist's being much impressed by it—particularly when she learns that any empirical claims she makes must be consistent with "relatively reliable methods of inquiry." Nor will she be much comforted by Gutmann and Thompson's gloss to the effect that this does not "exclude religious appeals per se" (why not, one wonders?), so long as these do not include taking the Bible literally. The reason for this latter constraint is that "virtually all contemporary fundamentalists subject biblical claims to interpretation, accepting some as literally true and revising the meaning of others. To reject moral claims that rely on implausible premises is therefore not to repudiate religion" (Gutmann and Thompson 1996: 56). If the syllogistic force of this claim was not lost on the fundamentalist in the abstract, surely it would be once it was explained to her that it denies her the right to insist on the literal truth of *any* particular biblical imperative. She will rightly expect to come out on the short end of any deliberative exchange conducted on that terrain. The Gutmann-Thompson model works only for those fundamentalists who also count themselves fallibilist democrats. That, I fear, is an empty class, destined to remain uninhabited.

Gutmann and Thompson are plausibly skeptical of those, like Owen Fiss and Ronald Dworkin, who believe that courts are better suited to achieving principled resolution among contending moral perspectives in the public realm than are other political institutions. Neither a compelling theoretical argument nor any persuasive evidence has ever been adduced in support of this view. Contrary to what they seem to suppose, however, this is scarcely relevant to the standard constitutionalist argument for avoiding, or limiting, public deliberation about intense—particularly religious—differences. This does not turn on any illusion that courts can resolve them in a principled fashion, but rather on the recognition that no one can. The idea is that their explosive potential is so great that it is better, for the welfare of both religious adherents and the democratic polity, if they are kept out of organized politics as much as possible, subjected to what Stephen Holmes (1995: 202–35) describes as "gag rules." Hence the First Amendment's Establishment Clause. That is the serious constitutionalist case against promoting attempts to resolve religious disagreements in the public sphere. Perhaps there is a reply to it from the deliberative democratic perspective, but Gutmann and Thompson do not supply it.

1.2.2 DELIBERATION AND CONFLICTING INTERESTS

Gutmann and Thompson's acknowledgment that deliberation might move politics away from the agreement and accommodation they value skirts the tip of a large iceberg. Beyond the issue of uncompromising religious values, people with opposed interests are not always aware of just how opposed those interests actually are. Deliberation can bring differences to the surface, widening divisions rather than narrowing them.[17] This is what Marxists hoped would result from "consciousness-raising": it would lead workers to discover their interests to be irreconcilably at odds with those of employers, assisting in the transformation of the proletariat from a class-in-itself to a revolutionary class-for-itself. In the event, these hopes proved naive. The general point remains, however, that there is no particular reason to think deliberation will bring people together, even if they hope it will and want it to. A couple with a distant but not collapsing marriage might begin therapy with a mutual commitment to settling some long-standing differences and learning to accommodate one another better on matters that cannot be resolved. Once honest exchange gets underway, however, they might unearth new irreconcilable differences, with the effect that the relationship worsens and perhaps even falls apart in acrimony. Deliberation can reasonably be expected to shed light on human interaction, but this may reveal hidden differences as well as hidden possibilities for convergence. It all depends on what the underlying interests, values, and preferences at stake actually are.

Gutmann and Thompson's inattention to the contending interests at stake is most evidently revealed in their discussion of health care reform in Oregon in the early 1990s. Rationing of health care procedures for the nonelderly poor by the legislature followed a series of "town meetings" in which citizens and various health professionals were asked to rank medical procedures.[18] The object was to find a way of settling disagreements about priorities in health care insurance, given the hard choices that public budget constraints impose. Gutmann and Thompson note that this procedure was flawed because the plan covered only the nonelderly poor. They describe this as a "basic injustice" that "may have adversely influenced the surveys and community meetings, which in any case fell short of the deliberative ideal." Yet they commend the process on the grounds that it "forced officials and citizens to confront a serious problem that they had previously evaded—and to confront it in a cooperative ('first person plural') spirit." They go on to claim that the process helped ameliorate the underlying injustice, because when the legislators "finally saw what treatments on the list would have to be eliminated under the projected budget, they managed to find more resources, and increased the total budget for health care for the poor" (Gutmann and Thompson 1996: 143–44).

Notice that the legislature's decision to appropriate additional funds was unrelated to the substance of the deliberative meetings, which never dealt with what the overall budget should be or how health care resources should be traded off against other demands on the state treasury. It was not a product of reciprocal deliberative exchange whereby citizens with moral disagreements came closer together. It was, rather, a fortunate externality, for the uninsured poor, of the deliberative process—such as it was— in that the publicity it generated helped spotlight their plight in the media and the legislature. If this is the proffered defense of the Oregon process, one would have to compare it to other ways in which the condition of the uninsured poor might have been publicized with similar or better effect—such as publicity campaigns, public protests, or class action lawsuits. This issue, however, does not bear on Gutmann and Thompson's defense of deliberation: that it reduces disagreement and increases mutual accommodation of differences that cannot be resolved.

In fact, as a device for settling disagreements about how hard choices should be made in the rationing of health care resources, the Oregon deliberative process was a notable failure. Gutmann and Thompson acknowledge, as have others, that it is hard to find a relationship between the final rankings of medical procedures and the results of the deliberative process, which eventually became little more than a vehicle for public outrage at attempts to introduce a measure of prudence into Oregon's health care priorities (see Hadorn 1991). Nonetheless, Gutmann and Thompson conclude that the deliberations "evidently helped citizens, legislators, and health care professionals arrive at an improved understanding of their own values—those they shared and those that they did not." But whose values are we really talking about? The "citizens, legislators, and health care professionals" by and large excluded those who would be covered under the Oregon plan: the nonelderly poor. This is not to speak of the injustice which Gutmann and Thompson acknowledge—that in effect this choice was really about "making some poor citizens sacrifice health care that they need so that other poor citizens can receive health care they need even more urgently, while better-off citizens can get whatever treatment they need." Rather, the question is this: why should we attach any legitimacy at all to a deliberative process that involved very few of those whose health care priorities were actually being discussed?[19] Gutmann and Thompson themselves make a similar point in criticizing workfare and welfare reform later in the book. There they suggest the need for participatory processes that "encourage the participation of economically and educationally disadvantaged citizens" (Gutmann and Thompson 1996: 143–4, 303–6). That seems right so far as it goes. But, as I argue in §1.3 below, it needs to be taken further.

Only part of the infirmity in these cases is that those who must live with the results go more or less unrepresented in the decision making; the other part is that most of those making the decisions know that they will never depend on the good whose rationing or provision is under discussion. In countries like Britain and Canada, where the great majority of the population use collectively rationed medical services, their participation in democratic decision making through the political process lends legitimacy to the resulting policies. By contrast in Oregon, upwards of 80 percent of the population is unaffected by the rationing program (see Daniels 1991: 2233–34). The general point here is that the legitimacy of decision-making processes varies with the degree to which they are both inclusive and binding on those who make them. Deliberative processes are not exceptions. Gutmann and Thompson acknowledge this in principle. They define political decisions as collectively binding, adding that "they should therefore be justifiable, as far as possible, to everyone bound by them" (Gutmann and Thompson 1996:13). However, their discussion is not sensitive to the reality that different people are differently bound by collective decisions. When there is great variation in the impact of a decision, then interests diverge in ways that are relevant to the assessment of the decision's legitimacy.

This is most obviously true when there are substantial differences in the capacities of different groups to escape the effects of policies on which they are deciding. Those who can easily avoid them do not have the same kind of interest at stake in a decision as those whose exit costs are prohibitively high. The story of apartheid in American public schools attests eloquently to what happens when this goes unrecognized. Urban public schools are starved of resources by white middle-class voters who opt out either fiscally, to private schools, or physically, to suburban schools (see Hochschild 1984). It should be added that the latter may live in towns that are paragons of deliberative democracy. In 1995, for instance, a statewide Connecticut plan to reduce school segregation was duly deliberated upon at great length in New England town meeting after New England town meeting in which the inner-city residents of Hartford and New Haven had no effective voice at all. As a result, their interests were simply ignored and the plan was easily defeated (see McDermott 1999: 31–53). Gutmann and Thompson place great stress on the importance of adequate elementary and secondary education, like adequate health care, in providing the necessary basic opportunities for living in a democracy. But they seem not to appreciate that as deliberation operates on the ground in what Douglas Rae (1999: 165–92) has described as the "segmented democracies" that Americans increasingly inhabit, it is often an obstacle to providing these goods. When there are great differences in capacity for exit, what is often needed is not widespread deliberation but action to protect the vulnerable.

1.2.3 THE CONTEXT OF DELIBERATION

Another weakness in the deliberative literature concerns its relative inattention to what shapes the terms of deliberation in modern democracies. To the extent that more deliberation would be a healthy thing in the formation of public policy, the principal obstacle often is not the lack of will on the part of people with differing moral convictions to deliberate in ways that can minimize their differences. Rather, the obstacle results from decisions by powerful players who make it their business to shape the terms of public debate through the financial contributions they make available to politicians and political campaigns. Engels once described ballots as "paper stones." In the post–*Buckley v. Valeo* world, when all credible political campaigns require multimillion-dollar war chests to buy the requisite television time to do political battle, public deliberation all too often consists of verbal stones hurled across the airwaves, with victory going to whoever has the most bountiful supply. Granted, this is a long way from what Gutmann and Thompson have in mind when they advocate deliberation, but it is surely curious that a book about the importance of enhancing deliberation in contemporary American politics can ignore the reality it creates.

For instance, in their discussion of the failure of the Clinton administration's attempt at national health care reform, Gutmann and Thompson seek to lay blame on the secret meetings of Hillary Clinton's Task Force on National Health Care Reform, along with other unmentioned factors. Endorsing the claims of critics who, at the time, said that support for the plan would be more difficult to achieve "if the policy makers did not show that they were responding to criticisms and taking into account diverse interests in the process of formulating the plan," they conclude that even when "secrecy improves the quality of a deliberation, it may reduce the chances that a well-reasoned proposal will ever become law" (Gutmann and Thompson 1996: 117). Perhaps the secret meetings contributed something to the failure, along with the Clinton administration's ineptitude in failing to enlist the support of essential Capitol Hill barons like Senators Moynihan and Nunn, their inability to come to grips with the sheer economic scope of the proposal (12 percent of a $3 trillion economy),[20] and the structural deficit inherited from the Reagan and Bush administrations.[21] But how can anyone who lived through the huge amounts of public misinformation that contributed to the steady decline in the bill's popularity, and its eventual abandonment by the administration, not be struck by the importance of the $50 million public relations and lobbying campaign that the medical, insurance, and other corporate establishments waged to kill the legislation?[22]

We need not quarrel with Gutmann and Thompson's contention that secrecy is generally a bad thing in government to ask how much it had to do with the failure of health care reform in 1993 and 1994. Secrecy's importance seems *de minimus* when compared to the way the options were presented in the war of words on television and the activities of political lobbyists. They ensured that important options (notably a Canadian-style single payer system) were never seriously discussed, and that the entire debate came to focus on issues that were irrelevant to the bill's basic goal of achieving universal health care coverage. Arguments about the feasibility of managed competition and the freedom people might or might not have in selecting their own physicians dominated the discussion, as the plight of the 40 million uninsured fell by the wayside. It is difficult to see how any aspect of Gutmann and Thompson's "deliberative deficit" was responsible for this, since the problem had nothing to do with reaching agreement among the contending views or finding an accommodation among those who could not agree. Rather, the problem was that some of what ought to have been the contending views never confronted one another in the public mind. How else is one to explain the fact that a single payer system could not be seriously mooted, even at the start of the public debate, despite a substantial body of academic commentary which suggests that it is by far the most cost-effective way of achieving affordable universal coverage?[23]

For anyone perturbed by the Clinton health care debacle, worrying about how money structured the debate should be high on the list of concerns. Yet Gutmann and Thompson never mention it. Perhaps they would say their book is simply not concerned with this subject, but that is difficult to square with their insistence that their focus is on "the everyday forums of democratic politics," differentiating their deliberative perspective from other academic discussion, which is said to be "insensitive to the contexts of ordinary politics: the pressures of power, the problems of inequality, the demands of diversity, the exigencies of persuasion." As my discussion has indicated, their own account pays surprisingly little attention to these very features of politics. They are heartened by the fact that although "the quality of deliberation and the conditions under which it is conducted are far from ideal in the controversies we consider, the fact that in each case some citizens and some officials make arguments consistent with reciprocity suggests that a deliberative perspective is not utopian" (Gutmann and Thompson 1996: 2–3).

We should not be so easily fortified. Unless it can be shown that these arguments can be made on a sufficient scale and can garner enough institutional force to influence the ways politics is structured by powerful interests, it is difficult to accept the suggestion that deliberation will lead people to converge on the common good through reciprocal recognition of one another's valid claims. The decisive role played by money in politics means that politicians must compete in the first instance for campaign contributions and only secondarily for the hearts and minds of voters. By ignoring this, Gutmann and Thompson attend too little to the ways in which power relations influence what deliberation should be expected to achieve in politics.

Likewise with the Ackerman-Fishkin proposal for "deliberation day." The chances that this could have an impact on actual political options seem negligible. In addition to ignoring the role of campaign expenditures, it ignores candidate selection, conventions, platforms, and interest group activities. Perhaps these difficulties might be mitigated if deliberative mechanisms were injected into the political process much earlier than a week before Election Day and structured to have an impact on the ways in which resource inequalities shape political outcomes. That Ackerman and Fishkin do not even consider such possibilities underscores the extent to which they conflate lack of deliberation with power contexts within which deliberation takes place.

1.3 DELIBERATION IN IDEAL SETTINGS?

Fishkin's proposal for deliberative polls raises comparable worries. They differ from his joint proposal with Ackerman in that they are intended to take place in structured settings in which power inequalities are rendered immaterial: participants are randomly selected and paid for their participation. To be sure, such deliberative polls offer certain advantages, particularly with respect to the trade-off between the costs of deliberation in terms of time and the benefits in terms of sophisticated understanding of complex issues. The idea—which actually goes further than Fishkin suggests—that some political decisions might be devolved to such groups is an innovative one. Perhaps they could develop democratic legitimacy for reasons analogous to those attending the legitimacy of juries. Indeed, one group that organizes such polls, the Jefferson Center in Minneapolis, calls them "citizen juries."[24] Yet proponents of deliberative polls and citizen juries fail to address obvious questions that are pertinent to their democratic legitimacy. Who decides which issues should be presented to these groups for discussion, and, possibly, decision? Who sets the agenda? The "experts" who testify before the randomly selected groups are supposed to be "balanced," but who does the balancing, and who decides what criteria they should use? Participation in deliberative polls and citizen juries might alter people's views, but without satisfactory answers to these questions it is hard to see why we should have much confidence that they have been altered for the better, or that they are owed any particular deference in a democracy.

Other deliberative theories have been developed that abstract from actual politics even more thoroughly than do deliberative polls and citizen juries. Jurgen Habermas's (1979, 1984) "ideal speech situation," for example, appeals to a model of uncoerced speech that is divorced from the power considerations of actual politics, as does Bruce Ackerman's (1980) dialogic model of justice. Proponents of these theories believe that they can establish what political institutions, arrangements, and policies would be agreed upon in ideal deliberative conditions. In this their endeavor is analogous to Rawls's (1971) enterprise of trying to determine what basic structure of political institutions people would chose behind a veil of ignorance designed to factor out self-interest. As my discussion of Buchanan and Tullock in §1.1.4 indicated, you cannot derive something from nothing, and it is scarcely surprising that writers in the Rawlsian tradition reach different results depending on the assumptions about human nature and the causal structure of the social world that are fed into their models (see Shapiro 1986). For present purposes notice that these are solipsistic theories, geared to answering this question: what institutions or arrangements would a rational person choose under specified ideal conditions?[25]

The ideal deliberative theories confront the additional difficulty that if, *per impossible*, this question could be answered unequivocally for one person, then presumably it could be answered for everyone. But what, then, would be left for deliberation? Ideal deliberative theorists are caught on the horns of a dilemma. Either they must concede that their speculations about what would be chosen under ideal deliberative conditions are indeterminate, prompting one to wonder, as with Gutmann and Thompson's speculations, what purpose they can serve in the actual world. Alternatively, they might claim that skepticism about their ability to demonstrate what outcomes authentic deliberation would converge on is misplaced. But in that case deliberation adds no value.

NOTES

1 The most comprehensive and accessible, if somewhat dated, review of this literature is Mueller (1989). See also Shapiro (1996: 16–52) and Przeworski (1999).
2 Riker (1982), Riker and Weigast (1988). On the ways in which multiplying veto points limit the possibilities of governmental action, see Tsebelis (2002).vf
3 Holmes used to taunt his colleagues on the bench by challenging them to name any accepted judicial rule or precedent they liked; he would then show it could be rendered consistent with either outcome in the case at hand. See Menand (2001: 339–347).
4 I leave for §§4.1.3 and 4.1.4 discussion of the empirical difficulties associated with determining whether preferences in a population are mutually reinforcing or crosscutting, and how, if at all, they can be transformed from the former into the latter.
5 See also Green and Shapiro (1994: 98–146).
6 For the argument that institutions reduce the likelihood of cycles, see Shepsle and Weingast (1981).
7 One indicator of the work's influence is that when Buchanan was awarded the Nobel Prize for economics almost a quarter-century after its publication in 1986, the citation singled out "his development of the contractual and constitutional bases for the theory of economic and political decision-making." http://www.nobel.se/economics/laureates/1986/ [9/3/02].

8 This is not strictly true if vote trading is allowed. Under that assumption, and also that there are no decision-making costs, there is no optimal decision rule for the same reason as Coase showed that, in the absence of information costs, wealth effects, external effects, and other blockages to exchange such as free riding, no system of tort liability rules is more efficient than any other. Whatever the system, people will then make exchanges to produce Pareto-optimal results. See Coase (1960: 1–44). Assuming, however, that a pure market in votes does not exist, and Buchanan and Tullock acknowledge that some constraints on it are inevitable, they maintain that unanimity would uniquely be chosen in the absence of decision-making costs (1962: 270–74).

9 See also Barry ([1965] 1990) and Fishkin (1979: 69).

10 When the number of voters is odd, the optimal decision rule is majority rule, n over two, plus one-half; when n is even, the optimal decision rule is either majority rule *(n over two plus one)*, or majority rule minus one (simply n over two).

11 The sociologist Terence Marshall (1965: 78) famously distinguished three types of increasingly comprehensive rights. *Civil* rights include "the rights necessary for individual freedom— liberty of the person, freedom of speech, thought and faith, the right to own property and conclude valid contracts, and the right to justice [the right to assert and defend one's rights]." *Political* rights include "the right to participate in the exercise of political power, as a member of a body invested with political authority or an elector of the members of such a body," and by *social* rights Marshall meant "the whole range from the right to a modicum of economic welfare and security to the right to share in the full social heritage and to live the life of a civilized being according to the standards prevailing in the society." Marshall was more optimistic than the historical record has turned out to warrant in that he conceived of societies as moving from civil to political to social citizenship rights as they modernized.

12 There are terminological issues at stake here on which substantive issues turn. For instance, in the *Lochner* era the Supreme Court struck down much legislation in the name of protecting individual freedoms, but the legislation in question was aimed at increasing social and economic guarantees—promoting civil rights at the expense of social rights, in Marshall's terminology discussed in footnote 12 above. See *Lochner v. New York*, 198 U.S. 45 (1905). For discussion of the *Lochner* era, and for a general discussion of the evolution of American constitutional law through the years of the Warren Court (1953–69), see Tribe (1978).

13 For Rousseau voting was a means of disciplining private interest by getting people to focus on what is best for society as a whole. As he put it, "When a law is proposed in the people's assembly, what is asked of them is not precisely whether they approve the proposition or reject it, but whether it is in conformity with the general will which is theirs; each, by casting his vote, gives an opinion on this question" (Rousseau [1762] 1968: 153).

14 *Mozert v. Hawkins County Board of Education*, 827 F.2d 1058 (6th Cir. 1987).

15 In *Buckley v. Valeo*, 424 U.S. 1 (1976), the Court held, inter alia, that although Congress may regulate financial contributions to political parties or candidates, it cannot otherwise regulate private expenditures on political speech. The Court has since allowed some minor constraints on corporate expenditures in *Austin v. Michigan State Chamber of Commerce*, 110 S. Ct. 1391 (1990), but for all practical purposes the *Buckley* rule makes it impossible to limit privately funded political advertising.

16 Ackerman and Fishkin (2002: 148) insist that "it is a big mistake to view the annualized cost of $15 billion through the lens of standard cost-benefit analysis" on the grounds that its "large" benefits "cannot be reckoned on the same scale as other elements in the cost- benefit equation." Even if we were to concede that the benefits could coherently be declared to be large at the same time as they are said to be incommensurable with their costs, their claim ignores

the point stressed here: that its benefits surely should be weighed against other ways in which such a sum could be spent to enhance American democracy.

17 See Simon (2000) and Sunstein (2002) for discussion of empirical conditions under which deliberation leads to divergence rather than convergence of opinion.

18 The participants were asked to rank categories of treatment by importance and articulate the values that guided their decisions. The state legislature then used the list as a yardstick to appropriate Medicaid funds. The Oregon Plan was intended to expand Medicaid eligibility from 68 percent of those at the federal poverty level to 100 percent, and to finance the increased cost by prudent rationing of procedures. Although Oregon did end up expanding coverage to some 126,000 new members by February 1997, much of this was actually achieved by appropriation of new funds by the legislature rather than from savings generated by the deliberations about rationing priorities. See Daniels (1991) and Montague (1997: 64–66).

19 Daniels (1991: 2234) reports that the meetings were attended predominantly by "college educated, relatively well off, and white" audiences, half of which consisted of health professionals. Of the attendees 9.4 percent were uninsured (whereas 16 percent of the state's population was uninsured at the time), and Medicaid recipients (among other things the only direct representatives of poor children) were underrepresented by half.

20 See Marmor (1994: 2–3, 184).

21 For accounts of the failure, see Hacker (1997) and Skocpol (1997).

22 The $50 million figure is reported by Rinne (1995: 4–5). See also Hamburger and Marmor (1993: 27–32).

23 See the papers collected in Marmor (1994).

24 See the Jefferson Center's Web site at www.jefferson-center.org [9/3/02].

25 On the differences between the Habermas and Rawls, see their exchange: Habermas (1995) and Rawls (1995).

Mark Sagoff — "At the Shrine of Lady Fatima; Why Political Questions Are Not All Economic." In *The Economy of the Earth: Philosophy, Law, and the Environment*. Cambridge: Cambridge University Press.

Reading Commentary

Sagoff asks us to examine the consequences of government officials and business interests thinking of our society as nothing more than a marketplace in which individuals make trades to meet their subjective economic priorities. In this framework, problem-solving efforts in the public arena are aimed primarily at satisfying preferences. The participants intervene to balance costs and benefits in a way that maximizes efficiency and wealth. However, what if the majority of citizens prefer to look at society and its problems in a different way? What if citizens prefer to think in terms of the functioning of institutions and the rule of law? What if they are more concerned about making distinctions between good and evil, right and wrong, innocence and guilt, fairness and injustice, truth and lies? Sagoff highlights the many ways in which citizens in different societies tend to conceive of public decision-making as much more than a search for economic efficiency. If he is right, what do we miss if we do not take this alternative view into account?

For many citizens there are crucial moral, aesthetic and political differences between the risks they take voluntarily (i.e., by smoking) and the risks imposed on them by institutions (i.e., siting a nuclear waste repository in their neighborhood). These differences grow out of relationships with the government and corporate entities that offer little to no chance to shape decisions, both public and private, that significantly affect our lives. Building on Emmanuel Kant, Sagoff argues that citizens are not driven merely by wants but that value judgments (as members of a community) are equally (or even more) important.

This means that a sizable number of citizens evaluate policy recommendations or proposed solutions to environmental problems not solely with reference to what is best for them personally but by what they think is right. This is certainly very different from those who argue that political and economic decisions about environmental protection (or the use of natural resources) should be made primarily in response to consumer preferences. Sagoff argues that environment choices ought to be made on the basis of reason, conceiving of citizens as thinking beings capable of discussing issues on their merits, not just in terms of bundles of preferences in response to their economic wants.

This implies that not all public decisions can be handled via market approaches. Some decisions should be made in response to what can be said in favor of or against them. The protection of wilderness, habitats, water, land

and air quality should not be priced at their marginal value. Instead, their intrinsic value needs to be considered. In this sense, the supposed neutrality of cost-benefit analysis falls short. Without meaningful public deliberation, and without taking account of evolving information, solutions to environmental problems cannot be determined.

CHAPTER 2

AT THE SHRINE OF OUR LADY OF FATIMA; OR, WHY POLITICAL QUESTIONS ARE NOT ALL ECONOMIC

Mark Sagoff

Lewiston, New York, a well-to-do community near Buffalo, is the site of the Lake Ontario Ordinance Works, where the federal government, years ago, disposed of residues from the Manhattan Project. These radioactive wastes are buried but are not forgotten by the residents, who say that when the wind is southerly, radon gas blows through the town. Several parents at a conference I attended there described their terror on learning that cases of leukemia had been found among area children. They feared for their own lives as well. At the other side of the table, officials from New York State and from local corporations replied that these fears were unfounded. People who smoke, they said, take greater risks than people who live near waste disposal sites. One state official spoke about methodologies of rational decision making. This increased the parents' resentment and frustration.

The official told the townspeople that risks they casually accept, for example, by drinking alcohol or by crossing the street, were greater than the risks associated with the buried radioactive residues. He argued that the waste facility brought enough income and employment into the town to compensate for any hazards the residents might face. They remained unimpressed by his estimate of their "willingness to pay" for safety; his risk-benefit analysis left them cold. They did not see what economic theory had to do with the ethical questions they raised. They wanted to talk about manipulation and the distribution of power in our society. They did not care to be lectured about benefits and costs.

If you take the Military Highway (as I did) from Buffalo to Lewiston, you will pass through a formidable wasteland. Landfills stretch in all directions where enormous trucks – tiny in that landscape – incessantly deposit sludge, which great bulldozers, like yellow ants, then push into the ground. These machines are the only signs of life, for in the miasma that hangs in the air, no birds, not even scavengers, are seen. Along colossal power lines that crisscross this dismal land, the dynamos at Niagara push electric power south, where factories have fled, leaving their remains to decay. To drive along this road is to feel the awe and sense of mystery one experiences in the presence of so much energy and so much decadence.

Henry Adams responded in a similar way to the dynamos displayed at the Paris Exposition of 1900. To him the dynamo became a "symbol of infinity"[1] and functioned as the modern counterpart to the Virgin – that is, as the center and focus of power: "Before the end, one began to pray to it; inherited instinct taught the natural expression of man before silent and infinite force"[2]

Adams asks in his essay "The Dynamo and the Virgin" how the products of modern industrial civilization will be compared with those of the religious culture of the Middle Ages. If he could see the landfills and hazardous-waste facilities bordering the power stations and honeymoon hotels of Niagara Falls, he would know the answer. He would understand what happens when efficiency replaces infinity as the central conception of value. The dynamos at Niagara will not produce another Mont-Saint-Michel. "All the steam in the world," Adams writes, "could not, like the Virgin, build Chartres."[3]

At the Shrine of Our Lady of Fatima, on a plateau north of the Military Highway, a larger-than-life sculpture of Mary looks into the chemical air. The original of this shrine stands in central Portugal, where in May 1917 three children said they saw a lady, brighter than the sun, raised on a cloud in an evergreen tree.[4] Five months later, on a wet and cold October day, the lady again appeared, this time before a large crowd. Some in the crowd reported that "the sun appeared and seemed to tremble, rotate violently and fall, dancing over the heads of the throng."[5]

The shrine was empty when I visited it. The cult of Our Lady of Fatima, I imagine, has few devotees. The cult of allocative efficiency, however, has many. Where some people see only environmental devastation, its devotees perceive welfare, utility, and the maximization of wealth. They see the satisfaction of wants. They envision the good life.

As I looked from the shrine over the smudged and ruined terrain, I thought of all the wants and needs that are satisfied in a landscape full of honeymoon cottages, commercial strips, and dumps for hazardous waste. I hoped that Our Lady of Fatima, worker of miracles, might serve, at least for the moment, as the patroness of cost-benefit analysis. I thought of the miracle of perfect markets. The prospect, however, looked only darker in that light.

WHAT WE WANT VERSUS WHAT WE ARE

This book concerns the economic decisions we make about the environment. It also concerns our political decisions about the environment. Some people have suggested that, ideally, these should be the same – that every environmental problem should be understood as an economic one. William Baxter, for example, writes, "All our environmental problems are, in essence, specific instances of a problem of great familiarity: How can we arrange our society so as to make the most effective use of our resources?"[6] He adds:

> To assert that there is a pollution problem or an environmental problem is to assert, at least implicitly, that one or more resources is not being used so as to maximize human satisfactions. In this respect at least environmental problems are economics problems, and better insight can be gained by the application of economic analysis.[7]

On this view, there is really only one problem: the scarcity of resources. Environmental problems exist, then, only if environmental resources could be used more equitably or efficiently so that more people could have more of the things for which they are willing to pay. "To the economist," Arthur Okun writes, "efficiency means getting the most out of a given input." Okun explains: "This concept of efficiency implies that more is better, insofar as the 'more' consists of items people want to buy."[8]

Environmental economists generally define "efficiency" as the "maximum consumption of goods and services given the available amount of resources."[9] On this approach to environmental policy, it is the preferences of the consumer that are important. "The *benefit* of any good or service is simply its value to a consumer."[10] The only values that count or that can be counted, on this view, are those that a market, actual or hypothetical, can price."[11] In principle, the ultimate measure of environmental quality," one text assures us, "is the value people place on these services … or their *willingness to pay*."[12]

Willingness to pay. What is wrong with that? The rub is this. Not all of us think of ourselves primarily as consumers. Many of us regard ourselves as citizens as well. As consumers, we act to acquire what we want for ourselves individually; each of us follows his or her conception of *the good life*. As citizens, however, we may deliberate over and then seek to achieve together a conception of *the good society*.

In a liberal state, we are all free to pursue our personal ideas of the good life, for example, by buying the books we want to read. In a democracy, however, we are also free to pursue our ideal of ourselves as a good society, by trying to convince one another and our political representatives of a particular idea of our national goals and aspirations.

Americans, like citizens of other countries, have national goals and aspirations – a vision of what they stand for as a nation. They believe, for example, that each person must be secure in certain basic rights if he or she is to be able to form preferences that are not merely imposed but are autonomous and express personal values and uncoerced choice. Thus, the freedoms guaranteed by the Constitution are not to be construed as preferences for which they are willing to pay. They are, on the contrary, protections needed to form and express preferences that are truly one's own and that therefore may claim societal recognition and respect.[13]

Americans' conception of their own as a good society, however, includes more than these rights, which, in any event, may be universal and would, perhaps, belong to any good society, not simply theirs. They also recognize in their legislation and, as I shall argue in a later chapter, in their culture – for example, in their literature and art – other national goals and aspirations that are more particular to them. These do not necessarily follow from (although they must be consistent with) an abstract or universal conception of justice or the rights of man. Rather, they may derive from America's history and heritage; they may have to do with the particular role Americans can play in human progress. Americans might object to policies foreign or domestic that benefit them as individuals, if those policies disgrace or depart from national ideals and ethical commitments.

Consumers who have to pay higher prices as a result, for example, nevertheless may favor safety regulations in the workplace, not as a matter of personal self-interest but as a matter of national pride and collective self-respect. I shall argue, likewise, that our environmental goals similarly derive not necessarily from self-interest – not from consumer willingness to pay in markets – but from a common recognition of national purposes and a memory even newcomers adopt of our long historical relationship to a magnificent natural environment.

Our environmental goals – cleaner air and water, the preservation of wilderness and wildlife, and the like – are not to be construed, then, simply as personal wants or preferences; they are not interests to be "priced" by markets or by cost-benefit analysis, but are views or beliefs that may find their way, as public values, into legislation. These goals stem from our character as a people, which is not something we choose, as we might choose a necktie or a cigarette, but something we recognize, something we are.

These goals presuppose the reality of public or shared values we can recognize together, values that are discussed and criticized on their merits and are not to be confused with preferences that are appropriately priced in markets. Our democratic political processes allow us to argue our beliefs on their merits – as distinct from pricing our interests at the margin. Our system of political representation and majority vote may be the best available device for deciding on these values, for "filtering the persuasive from the unpersuasive, the right from wrong, and the good from bad."[14]

WHAT IS COST-BENEFIT ANALYSIS?

In this book, I shall argue that policies for health, safety, and the environment are and ought to be grounded in what Richard Andrews calls the "philosophy of normative constraints." Andrews explains:

> In this conceptual framework, government is not simply a corrective instrument at the margins of economic markets but [a] central arena in which the members of society choose and legitimize … their collective values. The principal purposes of legislative action are to weigh and affirm social values and to define and enforce the rights and duties of members of the society, through representative democracy. The purpose of administrative action is to put into effect these affirmations by the legislature, not to rebalance them by the criteria of economic theory.[15]

Andrews contrasts this conception of government with one "grounded in the language, logic, and values of public investment economics."[16] Those who advocate the primacy of the language, logic, and values of public investment economics generally argue that regulatory decisions, insofar as they are rational, must be based on a conception of markets or on a theory about costs and benefits and, therefore, not necessarily on the goals or purposes set out in legislation. These analysts have mounted a strong a priori, or conceptual, argument to show that efficiency in the allocation of resources is the principal goal of sound regulatory policy.

The argument for the efficiency criterion, as Allen Kneese, for example, states it, begins with a definition of a perfect market, a market, in other words, that meets certain ideal conditions. "All participants in the market" first "are fully informed as to the quantitative and qualitative characteristics of goods and services and the terms of exchange among them."[17]

According to the model, moreover, all valuable assets in the economic system are fully owned, managed, and exchanged in competitive circumstances, which is to say, no individual or firm "can influence any market price significantly by decreasing or increasing the supply of goods and services" it offers.[18] For environmental policy, two corollary conditions are especially important. First: "Individual ownership of all assets plus competition implies that all costs of production and consumption are borne by the producers and consumers directly involved in economic exchanges." Second: "A closely related requirement is that there must be markets for all possible claims."[19]

Markets that meet these conditions will lead, voluntary exchange by voluntary exchange, to a situation, at least in theory, in which all possible gains from voluntary exchange have been exhausted. Kneese points out that theorists who accept a certain value premise have found that "the results of an ideal market are desirable or normative." The premise "is that the personal wants of the individuals in the society should guide the use of resources … and that those personal wants can most efficiently be met through the seeking of maximum profits by all producers."[20]

The argument for allocatory efficiency may proceed in one of two ways, depending on whether one adopts the "ideal market" (Pareto) or the "benefit-cost" (Kaldor–Hicks) definition of efficiency. Those who emphasize the model of the ideal market point out that it provides a number of useful concepts to guide social regulation. This is true because exchange in an ideal market is *not* like what happens in the actual economy. As Kneese notes, the "connection between such a market exchange and the real working economy has always been tenuous at best." The idealized model serves, then, "as a standard against which an actual economy could be judged as a resource-allocation mechanism for meeting consumer preferences."[21]

The government, on this approach, may intervene to make actual markets more perfect, for example, by charging for the use of unowned assets (such as clean air) and by requiring polluters to pay costs associated with the negative side effects, the "spillovers" or "externalities," of the production process. Markets would allocate resources much more efficiently if these "externalized" costs of production were "internalized," which is to say, made subject to the pricing mechanism. As Kneese sees it, "the main source of our environmental problems is the inability of market exchange as it is presently structured to allocate environmental resources efficiently – that is, to price their destructive use appropriately."[22] More generally, as D. W. Pearce writes, "economists have regarded environmental degradation as a particular instance of 'market failure.'"[23]

Many economists, however, see a big problem in the voluntary or "unanimous consent" processes associated with perfect competition in a free market. In a system in which everyone who is affected by a transaction must consent to it, anyone who would lose from and therefore opposes an exchange may veto it, for example, by refusing to sell a property right. This problem, as John Krutilla and Anthony Fisher note, is an old one in welfare economics. They ask, "How is a project or land use policy which results in gains to some individuals and losses to others properly evaluated? In particular, can the gains and losses be algebraically added over all affected individuals to determine the gain (from each of the alternative uses of an area's resources)?"[24]

Many economic theorists observe that we cannot make very much progress if we demand unanimous consent, that is, if we operate entirely within the criterion (associated with Vilfredo Pareto) that allocation A is better than allocation B only if at least one person prefers A to B and *no one* prefers B to A. Fisher and Krutilla observe that any cost-benefit policy prescription, to avoid such a static situation, must adopt the potential-Pareto or Kaldor–Hicks criterion, "according to which the project is efficient, and presumably therefore desirable, if the gains exceed the losses, so that the gainers could compensate the losers and retain a residual gain."[25]

A project or allocation *A* is more efficient than *B* in the Kaldor–Hicks sense, then, if those who prefer *A* would outbid and, therefore, at least theoretically, could compensate those who prefer *B*, and still retain a residual benefit. To apply this criterion, a cost-benefit analyst will turn to markets, real or hypothetical, for data concerning the prices people are willing to pay for commodities and resources. The analyst will then recommend projects and policies that allocate resources efficiently, that is, to those who would pay the highest prices for them and, in that sense, those who benefit from or value them most.

It is worthwhile to note here, although I shall examine this problem more in Chapter 8, that an allocation meeting the Kaldor–Hicks criterion is likely to differ substantially from the outcome that a consensual market would reach. This is true, in part, because those whom a transaction damages are unlikely to consent to it unless compensation not only could be but is actually paid. What is more, people may be unwilling to sell property rights, or they may be willing to sell them only for much more than they would have paid to acquire them in the first place. The outcome of consensual exchange (since willingness to sell might not track with willingness to pay) will depend to a large extent, then, on the way property rights are originally distributed.

In the sort of universal auction or bidding game envisioned by the Kaldor–Hicks test, in contrast, the efficient allocation of resources (as Ronald Coase has shown in a well-known theorem) is invariant with respect to the initial distribution of property rights.[26] Owners of resources in this sort of auction seek to maximize their profit or wealth; they would, therefore, always sell a property right for a bit more than they would have paid for it. I will return, in Chapter 8, to the question whether the cost-benefit criterion, in basing efficiency on willingness to pay rather than willingness to sell, remains in touch with such values as consent and respect for property rights, which are often thought to justify the market as a social institution.

In a formal cost-benefit analysis, the analyst, using market and other data, estimates, on a willing-to-pay basis, the gains and losses associated with all the major effects of a policy, program, or regulation. The analyst must go through a series of steps, which include

> … *identification* of all nontrivial effects, *categorization* of these effects as benefits or costs, *quantitative estimation* of the extent of each benefit or cost associated with an action, translation of these into a *common metric* such as dollars, *discounting* of future costs and benefits into the terms of a given year, and a *summary* of the costs and benefits to see which is greater.[27]

The resulting sums must also be *compared across alternatives*. These tasks are Herculean. An enormous amount of highly skilled work goes into foreseeing the possible consequences of a program (which may differ considerably given other policies and decisions) and estimating the benefits and costs associated with those consequences. A good cost-benefit analysis, which may run into several volumes, can be an impressive document.

In the rest of this chapter, I shall analyze two concepts that are central to the notions of efficiency I have described. First, I shall discuss the concept of an externality as it arises in environmental policy and in economic analysis. Second, I shall investigate the conception of value or valuation on which the cost-benefit approach to social policy is based.

TWO CONCEPTIONS OF EXTERNALITIES

The concept of efficiency, when it is defined in terms of the functioning of a market free of "externalities" and other structural causes of failure, can be used in either a narrow or an expanded sense. When economists speak of "efficiency" in the narrow sense, they use the concept of an externality to refer only to physical side effects, such as pollution, that cause actual damage – the sort defined, for example, by the common law of tort – to person or property. These analysts may then measure the cost of the pollution – to individuals or to society as a whole – in terms of the health damage it causes to people and the economic losses it inflicts upon them.

During the 1950s and 1960s, economists generally worked within this narrow conception of what an externality is, and as a result they showed us how wastefully we had been using many of our environmental resources. They pointed out that publicly owned goods, such as water and air, are overexploited because no one can demand a price for their use; they argued forcefully that sound principles of conservation and management may be founded on the proper functioning of markets. During this period economists spearheaded attempts to conserve natural resources, and we cannot be too grateful for their efforts.

Resource and environmental economists during the 1950s and 1960s generally defined efficiency in the narrow sense, that is to say, in terms of a conception that tied externalities to the physical side effects of market transactions. These economists did not try to estimate on a willing-to-pay basis the "worth" of moral, aesthetic, political, or cultural concerns and convictions. On the contrary, these economists, quite reasonably, associated externalities with the failure *of* markets correctly to price commodities of the sort for which markets are appropriate and which they usually do price. These analysts did not speculate about our failure *to have* markets to cover the goals, values, and concerns that are and ought to be identified and resolved through the political process.

During the 1960s and 1970s, a series of popular statutes set out goals and programs for ending racial discrimination, improving public safety and health, controlling pollution, and enhancing the quality of life. These statutes represented a considered moral judgment about our responsibilities to one another, to future generations, and to the environment. They expressed public disgust with racism, pollution, industrial blight, and other horrors; these laws set out a policy to control and, insofar as possible, eliminate these evils.

Many of the economists who developed the techniques of cost-benefit analysis – E. J. Mishan would be an example – recognized, at least implicitly, the intractability of quantifying benefits associated with the ethical and aesthetic concerns expressed in public law.[28] These economists urged analysts to list these "qualitative" benefits separately to bring them to the attention of public officials. They did not try to "price" ethical and aesthetic values but saw them as the appropriate subject of political deliberation within the legislative process.

During the 1960s and 1970s, however, owing in part to the work of Ronald Coase, which I shall not pause to describe here, economists began to argue that inefficiencies result not so much from spillover damage to unconsenting third parties, for example, damage caused by pollution, as from the inability of these third parties, because of the costs of bargaining, to organize themselves to enter into and thus influence the transactions that affect them. Accordingly, economists began to replace the notion of a *physical spillover* with the notion of a *transaction* or *bargaining cost* as the paradigm of a market failure. In evaluating the overall efficiency of a project or a policy, analysts began to ask not "What is a cause of what?" but "What is a cost of what?"[29] They widened the idea of an externality, then, to include any unpriced benefit or cost, which is to say, anything a person may be willing to pay for but which does not receive a market price that fully reflects willingness to pay.

An efficient policy was understood, at first, to be one that would result from the functioning of markets free of externalities, where externalities were conceived in the old, narrow sense, that is, in terms of physical damage or economic loss to unconsenting third parties. When the notion of an externality expanded to cover *any* unpriced benefit or cost, however, the ideas of an efficient market and an efficient policy widened as well. An efficient market then had to "internalize" or "price" not only the physical damage or property loss a transaction may inflict on unconsenting third parties, but also every belief, argument, or reason those parties might give for or against that transaction, as long as they conceivably were willing to back up those opinions with money.

Theorists who defend cost-benefit analysis as a tool for policymaking face a dilemma. In identifying the costs and benefits of an action, they must construe externalities in either the narrow or the broad sense, that is, narrowly as damage to person or property of the sort recoverable under common law or broadly as any value – economic, moral, aesthetic, or political – that markets leave unpriced. If analysts construe externalities narrowly, they must concede that many policies that appall them and almost all the rest of us for cultural, aesthetic, and ethical reasons might be perfectly efficient when externalities are understood as physical spillovers and "efficiency" is construed, therefore, in the narrow sense.

We might replace our national parks, for example, with tourist traps, we could convert every arcadia into an arcade, without lessening the market value of anyone's property or damaging anyone's lungs. More people might visit the parks, indeed, if they found casinos there. The value of surrounding property would go up. Another Shopper's World need not hurt the ecosystem terribly even if it destroys the habitat of one or more rare or endangered species. Another tract development in place of open land may not injure anyone's health.

As these and other examples suggest, many policies that are efficient in the narrow sense would conflict with and, indeed, reverse many popular environmental laws, such as the Clean Air Act, the Endangered Species Act, and various wilderness acts, the intention of which is plainly ethical and not narrowly economic. We can assume, for example, that no one would suffer any injury or loss remotely recognizable under common law as a result of the commercial development of the habitats of the Colorado squawfish and the Indiana bat, now protected as endangered species. Everyone in Tulare County, California, might benefit financially from the conversion of the little-used wilderness at Mineral King into a Disney resort in the heart of Sequoia National Park.

Many commercially exploitative policies such as these, which surely appear efficient in the narrow sense and which have no nasty spillovers that markets fail to price, are nevertheless so unpopular among the citizenry for cultural and ethical reasons that no one seriously believes Congress will permit them. Laws that protect the natural environment are intended to do just that – not necessarily to balance interests, internalize externalities, maximize benefits, or increase social wealth.

In order to make market and cost-benefit analysis applicable to public values, analysts may appeal to the wider notion of an externality, which includes not simply injury or damage of the sort that might give third parties standing to sue in common law but also any relevant attitude, opinion, argument, or belief that a person might conceivably be willing to back up with money. When analysts expand the notion of an externality in this way to embrace the opinions and beliefs of the citizenry, which are central to environmental legislation, they make a bald attempt not to inform but to replace the political process, an attempt they may not acknowledge. It is for the political process – not for economic analysis – to gather and judge these opinions and beliefs.

Policymakers need to know which beliefs about facts are credible and which arguments about values are sound. The credibility of a belief (e.g., that the earth is round) depends on evidence and expert opinion, not the amount people are willing to bet that it is true. Nor does the soundness of an ethical argument depend on willingness to pay, although economic information, of course, may be relevant. Thus, cost-benefit techniques, when they go beyond the confines of determining efficiency in the narrow sense, do not provide useful information. Rather, they confuse preference with ethical and factual judgment.

Cost-benefit analysis does not, because it cannot, judge opinions and beliefs on their merits but asks instead how much might be paid for them, as if a conflict of views could be settled in the same way as a conflict of interests. Analysts who take this approach, of course, tend to confuse views with interests. They do this by giving political, ethical, and cultural convictions technical names – "bequest values," "existence values," "intangibles," "fragile values" or "soft variables" – as if by this nomenclature they could transform beliefs that have carried the day before legislatures into the data of economic methodology.

I recognize the importance of cost-benefit analysis when it is used narrowly to inform the public and its officials about the actual market costs associated, for example, with reclaiming mined lands or increasing highway safety. To make a wise decision, society must know these costs and recognize the law of diminishing returns. Economic analysis, moreover, may also reveal less expensive ways society may reach its cultural and ethical goals, for example, by replacing "command and control" bureaucratic approaches with market incentives in order to abate pollution.

In Chapter 9 I shall discuss the usefulness of cost-effectiveness analysis and other techniques that differ from cost benefit analysis in the following way: *Cost benefit* analysis fixes our societal goals in advance; it presupposes that any rational policy will seek efficiency (or some mix of efficiency and equity) in the allocation and distribution of resources. *Cost-effectiveness* analysis, in contrast, helps us find the least costly means to achieve societal goals we choose through the political process and approve on moral or cultural grounds.

Although I shall defend some techniques of economic analysis in later chapters, I believe any such technique becomes invidious when used not only to help society achieve its political and ethical objectives – for example, a safer, cleaner environment – but also to determine what those goals are or ought to be. Analysts who do this may be convinced that efficiency in the wide sense is itself a goal worth pursuing, or they may construe efficiency as a meta-value in relation to which all other values may be compared or judged. In Chapters 2, 3, and 4, I shall argue against this expanded conception of efficiency and against the crazy kind of second-guessing of "consumer" preferences that is carried on in its name.

Here it suffices to say that when cost-benefit analysis attempts to do the work of ethical and political judgment, it loses whatever objectivity it might have had and becomes a tool of partisan politics. I shall argue in Chapter 3 that when cost-benefit analysis assigns "shadow" prices to "amenity," "option," "bequest," and other citizen beliefs and values, theoretical "breakthroughs" replace sound judgment and common sense. At that point, economic analysis deteriorates into storytelling and hand-waving likely to convince no one except those partisans who agree with – and possibly have paid for – its results.

EFFICIENCY AND EQUALITY

Policy analysts who favor efficiency as a social goal respect the importance of another social objective, namely, equality, equity, or justice. (Under the broad conception of efficiency, of course, they might construe justice as another "benefit" for which individuals are willing to pay.)[30] I believe that "equality," like "efficiency," may mean two different things. It may refer, first, to the way wealth is distributed: It may mean equality of income or equality of access to resources. It may refer, second, to equality before the law and within the political process. Individuals are treated as equals in this second sense only if their views receive a fair hearing on the basis of the arguments they make for them. No society can grant equality of this kind if it considers only the views of a "vanguard" party or if it dismisses all arguments but those adhering to the "correct" line.

Some commentators argue that policies favored by environmentalists – strict air-quality standards, for example – make consumer products more expensive and thus hurt the poor more than the rich.[31] Environmental protection, then, may sometimes conflict with attempts to enhance social equality in the first sense. There is some evidence, for example, that the average income of Americans who visit the national parks exceeds that of Americans generally.[32] I do not know what to make of these observations. I imagine, however, that no generalizable connection holds between equality in this first sense and programs that attempt to improve the quality or maintain the authenticity of the natural environment.

Equality or justice may also refer to the right of all citizens to argue for their views or opinions *in foro publico* and to have these arguments discussed on their merits without bias or prejudice. This kind of justice may be secured, in part, when all sides to a dispute recognize that finding a solution is a shared rational enterprise in which each has an equal right to participate. This enterprise subscribes to certain virtues such as civility, respect for the opinions of others, and an unwillingness to resort to force. No one can claim a special or privileged access to the "right" decision or to the "correct" methodology. It is "equality" in this sense that is most likely to conflict with efficiency.

Those who favor efficiency as the goal of social policy tend to think of it as a grand value that takes up, incorporates, and balances all other values – because these are simply preferences for which individuals are willing to pay. These partisans, therefore, may regard views opposed to theirs not as arguable opinions deserving a fair hearing but as ideological "wants" that markets fail to price. A policy analyst, if convinced that efficiency or wealth-maximization constitutes the value of values, would not read this book to savor and answer its arguments. He or she would judge the value of the arguments by the sales of the book!

VALUES AS WANTS

Values enter political deliberation and cost-benefit analysis in very different ways. The idea behind deliberation is that the process of negotiation and discussion can be educational. Because people must argue their views on the merits and from a public or intersubjective point of view in order to persuade each other, they may refine or even change their positions to make them plausible representations of the public interest or the general good. Disagreements are likely to turn, therefore, on scientific, technical, and legal considerations, about which both sides may then seek more evidence or better information. In the context of deliberation, in other words, positions are not construed as exogenous variables but are endogenous to the decision-making process. Participants, therefore, may redefine a problem or consider alternatives that permit an unexpected resolution.

Values enter cost-benefit analysis as exogenous variables, that is, as "given" or "arbitrary" preferences for which individuals are willing to pay. An analyst may therefore construe any policy (for example, a decision to allow air pollution in national parks) as a benefit to those who approve and as a cost to those who oppose it. Thus, analysts may construe judgments a person may back up with reasons as if they were preferences of the sort he or she would reveal in a market. The analyst supposes in all such cases that "This is right for the following reasons: ..." and "I believe this because ..." are equivalent to "I want this" and "This is what I prefer."

Value judgments lie beyond criticism if, indeed, they are nothing but expressions of personal preference; they are incorrigible, since every person is in the best position to know what he or she wants. All valuation, according to this approach, happens *in foro interno;* debate *in foro publico,* other than incantations in favor of efficiency or some balance of efficiency and equity, has no point.

This approach denies the educative function of political discussion; from its point of view, the political process is continuous with the kind of trading that goes on in markets. The reasons people give for their views (outside the journals of economic analysis, where *argument* apparently is to be respected) are not to be counted; what counts is how much individuals will pay to satisfy their wants. Those willing to pay the most, for all intents and purposes, have the right view; theirs is the better judgment, the deeper insight, and the more informed opinion.

The assumption that valuation is subjective, that judgments of good and evil are nothing but expressions of desire and aversion, is not unique to the economic theory on which much policy analysis is based. There are some psychotherapists – Carl Rogers is an example – who likewise deny the objectivity or cognitivity of valuation. For Rogers, there is only one criterion of worth: It lies in the "subjective world of the individual. Only he knows it fully."[33] The therapist, according to Rogers, succeeds when the client "perceives himself in such a way that no self-experience can be discriminated as more or less worthy of positive self-regard than any other."[34] The client then "tends to place the basis of standards within himself recognizing that the 'goodness' or 'badness' of any experience or perceptual object is not something inherent in that object, but is a value placed in it by himself."[35]

Rogers points out that "some clients make strenuous efforts to have the therapist exercise the valuing function, so as to provide them with guides for action."[36] The therapist, however, "consistently keeps the locus of evaluation with the client."[37] As long as the therapist refuses to "exercise the valuing function" and as long as he or she practices an "unconditional positive regard"[38] for all the affective states of the client, the therapist remains neutral among the client's values or "sensory and visceral experiences."[39] The therapist accepts all felt preferences as valid and imposes none on the client. The role of the therapist is legitimate, Rogers suggests, because of this neutrality.

Policy analysts sometimes argue that their role in policymaking is legitimate because they are neutral among competing values in the client society. The political economist, according to James Buchanan, "is or should be ethically neutral: the indicated results are influenced by his own value scale only insofar as this reflects his membership in a larger group."[40] The analyst, to maintain his or her value neutrality, might try to derive policy recommendations formally or mathematically from the preferences of all members of society. If theoretical difficulties make such a social welfare function impossible, however, the next best thing, to preserve neutrality, may be to let markets function to transform individual preference orderings into a collective ordering of social states.[41]

The analyst is able, then, to claim that the methods of cost-benefit analysis are neutral among competing preferences. The question remains, however, whether the use of cost-benefit analysis is neutral among competing conceptions of the role of regulation in a good society. The question arises whether reliance on this analytic tool is consistent with the content of current legislation or, indeed, with the idea of democracy and the rule of law.

TWO CONCEPTIONS OF NEUTRALITY

Consider, by way of contrast, what I shall call a Kantian conception of value.[42] The individual, for Kant, is a judge of values, not a mere haver of wants, and the individual judges not merely for himself or herself but as a member of a relevant community or group. The central idea in a Kantian approach to ethics is that some values are more reasonable than others and therefore have a better claim upon the assent of members of the community as such.[43] The world of obligation, like the world of mathematics or the world of empirical fact, is objective – it is public not private – so that the intersubjective virtues and standards of argument and criticism apply.

Kant recognizes that values, like beliefs, are subjective states of mind, but he points out that, like beliefs, they have objective content as well; therefore, values are either correct or mistaken. Thus, Kant discusses valuation in the context not of psychology but of cognition. He believes that a person who makes a value judgment – or a policy recommendation – claims to know what is *right* and not just what is *preferred*. A value judgment is like an empirical or theoretical judgment in that it claims to be *true* not merely to be *felt*.

We have, then, two approaches to social regulation before us. One approach assumes that political and economic decisions about the environment are justified in roughly the same way, that is, in relation to preferences individuals express or would express in their consumer and, possibly, their voting behavior. According to this approach, the policy that may be defended on objective grounds – as the right thing to do – is the policy of maximizing the satisfaction of these preferences; every other policy decision is an application of that one.

The Kantian approach, on the other hand, assumes that policy recommendations in general are to be judged on the basis of reasons rather than wants. This view maintains a notion of the common good as an object posited and understood by reason; this is different from thinking of the public interest as a matter to be measured in terms of subjective wants. The Kantian approach also makes individuals the ultimate sources of policy – but it submits policy to their judgment rather than deriving it from their preferences. This view treats people with respect and concern insofar as it regards them as thinking beings capable of discussing issues on their merits. This is different from regarding people as bundles of preferences capable primarily of revealing their wants.

The Kantian approach assumes that public policies may, in general, be justified or refuted on objective grounds, that is, on the basis of what can be said for or against them, not necessarily on the basis of the intensity of competing desires. The Kantian concedes, nevertheless, that many decisions are either too trivial, too personal, or too knotty to be argued *in foro publico* and thus should be left to some nonpolitical resolution, usually to a market. How many yo-yos should be produced as compared to how many Frisbees? Should pants be cuffed? These questions are so trivial or inconsequential or personal, it is plain markets should handle them. It does not follow from this, however, that we should adopt a market or quasi-market approach to every public question.

A market or quasi-market approach to arithmetic, for example, is plainly inadequate. No matter how much people are willing to pay, three will never be the square root of six. Similarly, segregation is a national curse, and if we are willing to pay for it, that does not make it better but only makes us worse. Similarly, the case for or against abortion rights must stand on the merits; it cannot be priced at the margin.[44] Similarly, the war in Vietnam was a moral debacle, and this can be determined without shadow-pricing the willingness to pay of those who demonstrated against it.[45] Similarly, we do not decide to execute murderers by asking how much bleeding hearts are willing to pay to see a person pardoned and how much hard hearts are willing to pay to see him hanged.

Our failures to make the right decisions in these matters are failures in arithmetic, failures in wisdom, failures in taste, failures in morality – but they are not market failures. There are no relevant markets to have failed.

What separates these questions from those for which markets are appropriate is this: They involve matters of knowledge, wisdom, morality, and taste that admit of better or worse, right or wrong, true or false – and these concepts differ from that of economic optimality. Surely environmental questions – the protection of wilderness, habitats, water, land, and air as well as policy toward environmental safety and health – involve moral and aesthetic principles and not just economic ones. This is consistent, of course, with cost-effective strategies for implementing our environmental goals and with a recognition of the importance of personal freedoms and economic constraints.

The neutrality of the economist, like the neutrality of Rogers's therapist, is legitimate if private preferences or subjective wants are the only values in question. A person should be left free to choose the color of his or her necktie or necklace – but we cannot justify a theory of public policy or private therapy on that basis.

What Rogers's therapist does to the patient the cost-benefit analyst does to society as a whole. The analyst is neutral among our "values" – having first assumed a view of what values are, that is, having assumed a particular theory of the good. This is a theory that fails to treat values as values and therefore fails to treat the persons who have them with respect or concern. It does not treat them even as persons but only as locations at which affective states may be found. And thus we may conclude that the "neutrality" of cost-benefit analysis is no basis for its legitimacy. We recognize this neutrality as an indifference toward value – an indifference so deep, so studied, and so assured that at first one hesitates to call it by its right name.

NOTES

1 Henry Adams, *The Education of Henry Adams,* 2d ed. (Boston: Houghton Mifflin, 1970), p. 380.
2 Ibid., p. 388.
3 For an account, see Joseph A. Pelletier, *The Sun Danced at Fatima* (Worcester, MA: Washington Press, 1951).
4 *New Catholic Encyclopedia* (New York: McGraw-Hill, 1967), p. 856.
5 Richard N. L. Andrews, "Cost-Benefit Analysis and Regulatory Reform," in *Cost-Benefit Analysis and Environmental Regulations: Politics, Ethics, and Methods,* ed. Daniel Swartzman, Richard Liroff, and Kevin Croke (Washington, DC: Conservation Foundation, 1982), pp. 107–135; quotation at p. 112.
6 For discussion, see Jody Freeman, "Collaborative Governance in the Administrative State," *UCLA Law Review* 45 (1997–1998): 1–98.
7 Frank Michelman, "Political Markets and Community Self-determination: Competing Judicial Models of Government Legitimacy," *Indiana Law Journal* 53 (1977–1978): 147–206; quotation at p. 149. Michelman quotes Kenneth Arrow as follows: 'The case for democracy rests on the argument that free discussion and expression of opinion are the most suitable techniques of arriving at the moral imperative implicitly common to all. Voting, from this point of view, is not a device whereby each individual expresses his personal interests, but rather where each individual gives his opinion of the general will." Kenneth Arrow, *Social Choice and Individual Values* (New Haven, CN: Yale University Press, 1963), p. 85.
8 Economists may invoke the truism that only the values of human beings – not those of trees and animals – count in policy formation. These economists therefore correctly inveigh against the odd idea, associated with some speculation in the early 1970s, that objects of nature have interests of their own to be weighed in cost-benefit analysis. This book agrees completely that only human beliefs, commitments, and projects count – that all values are "anthropogenic," that is, they all come from human beings. This book of course strenuously denies that the relevant values are all "anthropocentric," in other words, that they involve only human well-being or welfare. This is because people have all kinds of beliefs, values, and interests that go beyond the effects on them of an outcome or policy. The distinction lies between the logical subject and the logical object of valuation. If the sentence "*S* attaches a value or assigns a value to *p*" is to be relevant to policy, the logical subject *S* must refer to a person. The predicate or object *p*, however, need not refer to the well-being of anyone, including *S*. A person can find any object *p* valuable for intrinsic reasons, that is, reasons that transcend his or her well-being. To be sure, the pain and suffering of animals matters to any decent human being – so they matter in social policy. Human beings, however, assign all the values. In other places, I have argued against the idea that the interests of the environment should matter to environmental policy (though obviously animal welfare is an ethical concern because people have a duty not to be cruel or to cause unnecessary pain to animals). For a diatribe against legal "standing" for trees, and so on, see Mark Sagoff, "On Preserving the Natural Environment," *Yale Law Journal,* 84, no. 2 (December 1974): 205–267.
9 J. G. March, *A Primer in Decision Making* (New York: Free Press, 1994), p. 58.

10 William Baxter, *People or Penguins: The Case for Optimal Pollution* (New York: Columbia University Press, 1974), p. 15. Baxter adds (p. 17), "The question how one organizes society so as to obtain reasonable assurance that resources are deployed effectively, that is, deployed continuously over time so as to yield the maximum aggregate of human satisfactions, is of course the classic and central question to which the science of economics is addressed."

11 Ibid., p. 17.

12 "To the economist, the environment is a scarce resource which contributes to human welfare. The economic problem of the environment is a small part of the overall economic problem: how to manage our activities so as to meet our material needs and wants in the face of scarcity." A. Myrick Freeman, "The Ethical Basis of the Economic View of the Environment," *The Environmental Ethics and Policy Book* (Belmont, CA: Wadsworth, 1994), p. 307.

13 Arthur M. Okun, *Equality and Efficiency: The Big Tradeoff* (Washington, DC: Brookings Institution, 1975), p. 2. One thinks of Jim Henson's magnificent puppet Miss Piggy and her immortal epithet, "More is More." On the other hand, Madonna's "material girl" presents a more complex and provocative approach to the "more is better" theme.

14 Joseph Seneca and Michael Taussig, *Environmental Economics*, 2d ed. (Englewood Cliffs, NJ: Prentice-Hall, 1979), p. 6. The passage reads: *"Efficiency* is defined as maximum consumption of goods and services given the available amount of resources or, what is logically equivalent, the use of a minimum amount of resources to produce or make available for consumption a given amount of goods and services. *Equity* refers to a just distribution of total goods and services among all consumer units." Note that these economists see the resource base as "given" – allocation is a zero-sum game. In fact, resources are a function of technology. The point of economic analysis is to help increase resources, not just to divvy them up.

15 Ibid. There is a contradiction here. To assure the maximum production and consumption of goods and services given resources, society would have to let trades take place in response to competitive market prices. To "value" goods in terms of "the most the individual is willing to pay," in contrast substitutes centralized planning by economists who measure this "most." There is no reference to markets that depend on competitive prices.

16 James R. Kahn, *The Economic Approach to Environmental and Natural Resources* (Fort Worth, TX: Dryden Press, 1998).

17 A. Myrick Freeman, Robert H. Haveman, and Allen V. Kneese, *The Economics of Environmental Policy* (New York: Wiley, 1973), p. 23.

18 R. J. Hammond, *Benefit-Cost Analysis and Water Pollution Control* (Stanford, CA: Stanford University Press, 1960).

19 This is Kneese's second condition; the first is that markets are competitive, for example, non-monopolistic. Allen V. Kneese, "Environmental Policy," in *The United States in the 1980s*, ed. Peter Duignan and Alvin Rabushka (Stanford, CA: Hoover Institution Press, 1980), pp. 253–283; quotation at p. 256. Kneese has stated essentially the same argument in Allen Kneese and Blair Bower, *Environmental Quality and Residuals Management* (Baltimore: Johns Hopkins University Press, 1979), pp. 4–5.

20 Kneese, "Environmental Policy," p. 256.

21 Ibid.

22 Ibid., p. 259.

23 David William Pearce, *Environmental Economics* (London: Longmans, 1976), p. 1.

24 Eban S. Goodstein, *Economics and the Environment*, 4th ed. (New York: Wiley, 2005); Charles D. Kolstad, *Environmental Economics* (New York: Wiley, 2000).

25 H. P. Green, "Cost-Benefit Assessment and the Law," *George Washington Law Review* 45, no. 5 (August 1977): 904–905; see also E. J. Mishan, *Cost-Benefit Analysis* (New York: Praeger, 1976), pp. 160–166.

26 For an introduction to and discussion of this approach to the Coase theorem, see Duncan Kennedy, "Cost-Benefit Analysis of Entitlement Problems: A Critique," *Stanford Law Review* 33 (1981): 387–445.

27 Carl Rogers, "A Theory of Therapy, Personality, and Interpersonal Relationships, as Developed in the Client Centered Framework," in *Psychology: A Study of a Science,* ed. S. Koch (New York: McGraw-Hill, 1959), vol. 3, p. 210.

28 Ibid., p. 208.

29 Carl Rogers, *Client Centered Therapy* (Boston: Houghton Mifflin, 1965), p. 150.

30 Ibid.

31 Ibid.

32 Rogers, "A Theory of Therapy," p. 208.

33 Ibid., pp. 523–524-

34 James Buchanan, "Positive Economics, Welfare Economics, and Political Economy," *Journal of Law and Economics* 2, no. 127 (1959): 124–138.

35 A. Randall, "What Mainstream Economists Have to Say about the Value of Biodiversity," in *Biodiversity,* ed. E. O. Wilson and Frances M. Peter (Washington, DC: National Academy Press, 1988), p. 217. The assimilation of what is good with what the individual wants captures the change in emphasis in America from the moral and social to the psychological and subjective. See Philip Reiff, *The Triumph of the Therapeutic* (New York: Harper & Row, 1966). One way to understand the ethos of cost-benefit analysis is to recognize it as a station on the road America has followed from the moral self-castigation of the *Education of Henry Adams* to the self-absorbed neuroses depicted in the movies of Woody Allen.

36 For a discussion of social orderings and preference relations, see A. K. Sen, *Collective Choice and Social Welfare* (San Francisco: Holden-Day, 1970), and K. J. Arrow, *Social Choice and Individual Values,* 2d ed. (New York: Wiley, 1983), chap. 2.

37 Immanuel Kant, *Foundations of the Metaphysics of Morals,* ed. R. Wolff, trans. L. Beck (Indianapolis: Bobbs-Merrill, 1969). I follow the interpretation of Kantian ethics of W. Sellars, *Science and Metaphysics* (New York: Humanities Press, 1968), chap. 7, and W. Sellars, "On Reasoning about Values," *American Philosophical Quarterly* 17 (1979): 81–101.

38 See Alasdair MacIntyre, *After Virtue* (Notre Dame, IN: University of Notre Dame Press, 1981).

39 For the suggestion that property rights to have an abortion be traded in markets as a solution to the political controversy, see Hugh H. Macauley and Bruce Yandle, *Environmental Use and the Market* (Lexington, MA: Lexington Books, 1977). These authors write (pp. 120–121): "There is an optimal number of abortions, just as there is an optimal level of pollution, or purity… Those who oppose abortion could eliminate it entirely, if their intensity of feeling were so strong as to lead to payments that were greater at the margin than the price anyone would pay to have an abortion."

40 For this suggestion, see Charles J. Cicchetti, A. Myrick Freeman III, Robert H. Haveman, and Jack L. Knetsch, "On the Economics of Mass Demonstrations: A Case Study of the November 1969 March on Washington," *American Economic Review* 61 (1971): 179–195. The authors use the Clawson-Knetsch-Hotelling travel-cost method to measure political opposition to the Vietnam War. Had they the data, they would also factor in the cost of postage on letters to Congress.

41 William Simon, "Homo Psychologious: Notes on a New Legal Formalism," *Stanford Law Review* 32 (1980): 495.

42 Ibid.

43 Rieff, *The Triumph of the Therapeutic*, p. 52.

44 MacIntyre, *After Virtue*, p. 22. The idea here is that some theories of political economy take ruler-subject relations seriously, and some do not, except insofar as these relations may be revealed in a market. For this distinction, see Edward Nell, 'The Revival of Political Economy," *Social Research* 39 (1972): 32–53, and John Gurley, "The State of Political Economics," *American Economic Review* 61 (1971): 53–63. Gurley writes (pp. 54–55): "Political economics ... studies economic problems by systematically taking into account, in a historical context, the pervasiveness of ruler-subject relations in society... It is these pervasive relations of power and authority that lead to conflict, disharmony and disruptive change." Conventional welfare economics of the sort I criticize here seeks to understand and arbitrate conflict without understanding it in this context.

45 I lift this idea from Gunnar Myrdal, *The Political Element in the Development of Economic Theory*, trans. Paul Streeter (London: Routledge and Kegan Paul, 1953), esp. p. 54. For discussion, see Hannah Arendt, *The Human Condition* (Chicago: University of Chicago Press, 1958), sect. 6.

Lawrence Susskind and Connie Ozawa — "Mediated Negotiation in the Public Sector: The Planner as Mediator." *Journal of Planning Education and Research.*

Reading Commentary

Public resource allocation decisions are frequently made by legislative and administrative agencies without much input from the groups likely to be affected by these decisions. As a consequence, discontented parties invariably emerge, seeking redress through the courts (in costly and time-consuming litigation) or through direct political action, both of which have widespread negative consequences. This cycle of confrontation undermines trust in government. It also gives greater weight to procedural and legal rules than to considerations of fairness, efficiency or creative problem-solving. A quick perusal of recent media coverage confirms that this continues to be a serious problem. The question, then, is how we might do better. Who can break these deadlocks, and how?

Susskind and Ozawa (1984) suggest an alternative path, documenting the ways in which urban and regional planners (and the mediation efforts they are particularly suited to lead) can help guide us away from adversarial and adjudicatory processes. They illustrate with in-depth case studies of economic development, energy transitions and water management in which mediation can make a contribution. They are especially optimistic about using mediation to reduce the adverse impacts of negotiated agreements on underrepresented groups; to enhance awareness that joint gains have yet to be maximized; and to reduce the long-term spillover effects of negotiated settlements or undesirable precedents.

Different stakeholders tend to approach problems with different time horizons and perspectives. This is evident in disagreements between business and environmental interests over the allocation of fixed resources, the ranking of public policy priorities and the setting and enforcement of health and safety standards. In broad terms, business interests and pro-development groups tend to work within a shorter time horizon, are less risk averse, believe short-term gains outweigh long-term impacts and are convinced that subsequent repairs can recreate what existed before or provide something better. Environmentalists tend to view issues in the opposite way, taking a longer time horizon, worrying more about risk, believing long-term gains outweigh short-term impacts and believing that what is destroyed now cannot be recreated later. Both groups tend to include hard-liners who are unwilling to settle as well as many pragmatists who are willing to explore common ground. The pragmatists tend to be frustrated whenever government agencies are unresponsive to the core interests and primary needs of the contending parties.

Mediation efforts can help break these deadlocks by assisting in a number of ways, including (1) identifying all the parties with a stake in the outcome of a problem-solving effort; (2) ensuring that relevant interest groups are appropriately represented; (3) providing a safe space for the parties to explore fundamentally different assumptions and values; (4) facilitating agreements on the boundaries and time horizon of the impacts that need to be discussed; (5) allowing stakeholders to jointly undertake the weighing, scaling and judging of costs and benefits; (6) empowering stakeholders to develop a number of feasible solutions; (7) accounting for fair compensation and mitigation actions; (8) guaranteeing the legality and financial feasibility of agreements; and (9) holding parties accountable to their commitments.

Planners as mediators can foster better understanding between contending parties, encourage constructive communication instead of tactical misrepresentation and frame policies in ways that are less threatening to conflicting stakeholders. They can do this by catalyzing the search for mutual gains while emphasizing joint fact-finding and information sharing. This approach makes it more likely that policy proposals will gain wider acceptance.

MEDIATED NEGOTIATION IN THE PUBLIC SECTOR: THE PLANNER AS MEDIATOR

Lawrence Susskind and Connie Ozawa

INTRODUCTION

While most public resource allocation decisions are made by legislative and administrative agencies, individuals and groups affected by such decisions continue to demand a greater voice than the electoral process allows. While referendums, initiatives, formal participation in administrative hearings, and lobbying offer opportunities for more direct involvement, they fall short of providing the "stakeholding interests" with the control they seek. Referenda oversimplify the range of alternatives, failing to reflect the diversity of positions held by the electorate. Public hearings rarely guarantee full or fair consideration of minority viewpoints. Almost all of the traditional participatory supplements fail to reconcile the conflicting claims of contending parties.

In many instances, discontented parties can seek redress through the courts. Unfortunately, judicial decision making is deficient in several important respects. Litigation is costly and time consuming. It often precludes access by the poorest groups and is obviously inappropriate in situations that demand immediate attention. The adversarial nature of litigation undermines relationships, pitting parties against one another and inevitably erodes whatever goodwill existed. Court rulings are limited by procedural and substantive constraints. Indeed, the essence of many allocation conflicts may be left unexamined because of the constraints imposed by adjudicatory rules and procedures. The legal process tends to give greater consideration to conformance with procedural ground rules and legal precedent than to fairness and efficiency. Recently, experiments with mediated negotiation—face-to-face negotiations involving teams representing key stakeholding interests and an impartial mediator—indicate that traditional resource allocation decision-making processes may be supplemented in a manner likely to yield informed and durable agreements. In Connecticut, mediated negotiation was used to reach consensus on the allocation of federal block grant funds to public and private social service providers (Watts 1983). Federal agencies, including the Environmental Protection Agency, have experimented with mediation in the rule-making process (Baldwin 1983). Mediation was used to resolve a crisis in the funding of the state unemployment compensation fund in Wisconsin (Bellman 1983), to settle a number of water policy disputes in the western United States (Kennedy and Lansford 1983; Folk-Williams 1982), and to resolve dozens of land use and facility-siting disputes across the country (Susskind et al. 1983; Talbot 1983). These and other instances of mediated negotiation indicate that the parties with recognized stakes in the outcome of disputes can be involved in decision making without infringing on the authority and responsibility of elected officials.

The skills that successful mediators use appear strikingly similar to many of the process-management skills that planners have been taught for decades. The art of persuasion and the creative accomodation of competing interests, coupled with the technical skills of design—the identification of options, the generation of alternative plans or policies, and the assessment of various alternatives—as well as a concern for the implementation of agreements or recommendations are common to the mediator and the planner. Our analysis of recent cases of mediated negotiation in the public sector underscores these similarities and suggests that it might be instructive for planners to think of themselves as playing mediating roles.

In this paper, we want to demonstrate the advantages of thinking about the planner's role in this way. We begin by describing three instances in which mediated negotiation was used to resolve resource allocation disputes in the public sector: a Negotiated Investment Strategy (NIS) experiment in Columbus, Ohio; the conversion of an oil-fired power plant to coal in Massachusetts; and a water development "battle" in the Denver area. These examples illustrate the procedures involved in mediated negotiation, particularly the roles played by mediators.

We then review the changing conceptualization of the planner's role over the past three decades, identifying the weaknesses of prevailing models of practice, and indicating why the model of *planner as mediator* is more satisfying especially given the now generally shared presumption that planners have a major role to play in the process of implementation. The third section of our paper presents a brief overview of the four realms in which mediated negotiation has been used with some success: (1) labor relations (i.e., collective bargaining), (2) international relations, (3) community dispute resolution, and (4) environmental dispute resolution. We highlight variations in the roles played by mediators in each of these four realms to underscore those aspects of mediation practice that might be most helpful to planners. We conclude with some observations about the problems of educating planners to be effective mediators and by responding to three challenges likely to be directed at our call for planners to think of themselves as mediators: (1) What authority and credibility will planners have if they claim to be able to mediate?, (2) What are the employment opportunities for mediator-planners?, and (3) Can the model of the planner as a mediator be reconciled with the planner's traditional commitment to social change?

SOME ILLUSTRATIVE CASES

THE COLUMBUS NEGOTIATED INVESTMENT STRATEGY (NIS)

The Columbus NIS experiment was one of several efforts by the Charles F. Kettering Foundation to develop a more effective approach to coordinating intergovernmental policies. The experiment, begun in 1979, focused on complex development issues ranging from site-specific land use matters to metropolitan-wide social service policies.

Three teams representing the city, state, and federal governments participated in the negotiations. The Mayor of Columbus (Tom Moody), Governor James Rhodes, and Federal Regional Commission Chair-person Douglas Kelm (assisted by White House Intergovernmental Affairs Advisor Jack Watson) appointed the leaders of the three teams, who, in turn, selected additional team members. The only guidelines for the selection of the team members were that the teams be kept to a workable size and that individuals with appropriate expertise and authority be included.

The mediators, Lawrence Susskind and Frank Keefe, were recommended by the Kettering Foundation and the Ford Foundation and approved by team leaders Tom Moody, Lt. Governor George Voinovich, and Federal Regional Council representative Fran Ryan. Susskind, a professor of Urban Studies and Planning at the Massachusetts Institute of Technology, had been active in public participation experiments for several years. Keefe, a city planner by training, had been Director of the Office of State Planning (a cabinet-level position) in Massachusetts.

The negotiating teams met prior to the first formal negotiating session to jointly establish ground rules for the NIS process. They made a number of decisions concerning team membership, media involvement, recordkeeping, inter-team communication, the relationship of the NIS effort to other ongoing intergovernmental negotiations, and the process for ratifying agreements. They also delineated the roles which the mediators would play.

The agenda for the negotiations was formulated through a multi-step process. First, the teams met separately to prepare written statements summarizing their primary concerns about growth and development in the Columbus area and identifying obstacles to effective intergovernmental planning and policy coordination. These statements formed the basis of a joint meeting held in October 1979. Based on this discussion, the mediators drafted a list of presumably shared concerns. The mediators' draft was circulated to the three teams, commented upon, and revised. The final version of the agenda included eight topics: transportation, human services, fair housing, community development, local water quality management, minority business development, employment and job training, and the needs of special populations. Each team then prepared position papers on each agenda item. Prior to the first negotiation session, the mediators combined these position papers into a "consolidated briefing paper" summarizing the issues at stake, the perspectives of each team on each issue, and specific items to be negotiated at scheduled bargaining sessions.

Face-to-face negotiations continued over three, two-day sessions held in mid – December, 1979, late January and early March, 1980. Each session was followed by a meeting of various tripartite committees (i.e., comprised of one appointee designated by each team leader). These committees hammered out detailed language elaborating points of agreement reached at the full negotiating sessions and suggested ways of handling points of disagreement. A rush of tripartite committee meetings followed the March negotiating session and final reports from each of the eight tripartite committees were sent to the mediators by mid-April.

The mediators consolidated these reports and distributed to all team members a "draft agreement." After a final revision, an agreement was signed at a final NIS session held on April 30, 1980. According to the mediators, the participants "express(ed) both pride and relief" (Susskind and Keefe 1980, p. 12). The Columbus agreement not only itemized projects (potentially totalling more than $500 million in public and private investments) and policy changes relevant to each of the eight items on the agenda, but also outlined procedures for implementing the terms of the agreement, assigned responsibility for monitoring specific items to particular team members, and included provisions for resolving subsequent conflicting interpretations of the document.

By 1982, sixty percent of the more than eighty points outlined in the ninety-one page final agreement had been implemented (Bersch 1982). Some items had become impossible to complete when the Reagan Administration dismantled relevant federal programs. All in all, the mediated agreement appears to have withstood both the test of time and the transition in federal leadership.

Mediators Susskind and Keefe played central roles in the negotiations. The structure of the NIS process depended largely on the ability of the mediators to identify shared concerns and common interests among the negotiating teams from the agenda setting stage through the ratification of a final written agreement. They also met separately with each team and each tripartite committee to unsnarl intra and inter-team differences that threatened to block agreement. Perhaps, most importantly, the ability of the mediators to generate a "single text" from the various proposals submitted by the teams and tripartite committees served to bridge differences among divergent interests.

The team leaders played important behind-the-scenes "mediating" roles. They facilitated intra-team negotiations necessary to the development of bargaining positions (N.B., This was especially important for the federal team which included personnel from seventeen different agencies) and took responsibility for making sure that constituent groups or agency heads were kept informed as the process unfolded.

When the final document was signed, the key steps for implementation had been spelled out and all appropriate decisionmakers were "on-board."

BRAYTON POINT COAL CONVERSION

Authorized by the Energy Supply and Environmental Coordination Act (ESECA) of 1974, the Department of Energy (DOE) notified the New England Power Company (NEPCO) that it would be required to burn coal instead of oil at three units of its Brayton Point electric generating facility (Smith 1983). Anticipating a reduction in efficiency levels, NEPCO contested the DOE's estimates of the cost of conversion and the steps necessary to meet state air quality standards, and announced that it would challenge the DOE order in court. A legal contest would have brought in the Environmental Protection Agency which would have argued against coal conversion of Brayton Point because of the adverse air quality impacts. Prospects for a smooth and timely conversion appeared poor. The ESECA program divided regulatory responsibility among a number of state and federal agencies and made no provision for settling conflicts among them.

In April 1977, the Center for Energy Policy, a non-profit organization in Boston interested in promoting creative processes for developing energy policies in the northeastern U.S., persuaded the principal parties to enlist the assistance of a mediator. They arranged a meeting attended by officials of NEPCO, DOE, EPA, and the Massachusetts Department of Environmental Quality (DEQE). Although the DOE agreed to participate, it also continued to pursue its objectives through formal regulatory channels.

The negotiations proceeded in three phases. First, the parties agreed on an agenda which set the order of issues to be considered and named the parties to discuss each issue. Second, several months of technical and quantitative analyses were undertaken to establish a shared factual basis for the conversion decision. Finally, bilateral discussions between NEPCO and DEQE produced new standards for particulate and sulfur emissions for the Brayton Point plant.

Eleven months after the negotiations formally began, an agreement was reached. Since the key parties had agreed to include only the regulatory agencies and NEPCO in the negotiations, public hearings were held to allow excluded groups to express their concerns. The final agreement specified a phased conversion plan for Brayton Point, set maximum levels for the sulfur content of coal used at the facility, and prescribed particulate emission standards.

The negotiated agreement appeared to satisfy all the participating parties, especially the DOE which recognized the distinct advantage of achieving the conversion without litigation. An attempt to include the concerns and opinions of other interested parties was made through collateral procedures (i.e., public hearings) required under formal regulatory rules. Nonetheless, the negotiated agreement could be criticized for apparently neglecting to consider the interests of populations in other parts of the northeast susceptible to increased sulfate levels and acid rain. Some observers have suggested that the federal officials involved in the negotiations were "standing in" for the interests of the broader public. This was not obvious, however, from the way they handled themselves in the Brayton Point negotiations.

The mediator in the Brayton Point case was David O'Connor, a staff member at the Center for Energy Policy. He acted in several capacities during the course of the negotiations; taking, overall, a rather active part in the process. Foremost, O'Connor functioned as a manager of the negotiating sessions. He guided the participants through the process of establishing ground rules (for setting agenda, introducing issues, making proposals, handling the press, documenting discussions, and formulating agreements), scheduled and convened the meetings, and moderated discussions to ensure that all parties had an adequate opportunity to present their positions. O'Connor also served as an information resource. He attempted to explain and review technical and legal matters to ensure that the parties shared a common understanding of the issues and their implications. O'Connor regularly provided encouragement to the group, identifying points of agreement and stressing the progress of the negotiations as they unfolded. This psychological support was crucial to sustaining the momentum of the negotiations. O'Connor also acted as a confidant, meeting with each of the parties in private sessions and helping them articulate their interests and experiment with new positions and proposals without jeopardy. Finally, through these private meetings O'Connor was able to gain a more comprehensive understanding of the interests of the parties and the basic conflicts underlying their dispute. He used this privileged vantage point to develop what were often pivotal suggestions.

FOOTHILLS WATER TREATMENT PROJECT

The Foothills Water Treatment Project was designed initially to treat 125 million gallons of water a day, with an ultimate capacity of 500 million gallons per day. A proposal to construct this facility and a dam and reservoir on the South Platte River twenty-five miles southwest of Denver, Colorado, sparked a fierce dispute among federal and local agencies and environmental groups (Burgess 1983). The groups debated the merits of the proposal, projected impacts on air quality and ground patterns, and sought to estimate future raw water needs for the Denver area.

The parties were embroiled in the dispute when Congressional Representative Pat Schroeder suggested, in May 1977, the use of mediation to help settle the conflict. Her overtures were repeatedly rebuffed by one of the key project proponents, the Denver Water Board. Prospects for mediation improved when Congressman Tim Wirth, a well-known environmental advocate who publicly supported the proposed facility, volunteered to mediate.

Wirth began preparations for the negotiations by consulting privately with officials from the principal agencies. Formal negotiations began with a series of meetings arranged by Wirth, involving the Army Corps of Engineers, the Environmental Protection Agency (EPA), and the Denver Water Board. Eventually the Denver Water Board and the EPA agreed to continue discussions following a thorough study of the proposed project and possible alternatives by the Corps.

The Corps' report went beyond a simple examination of the proposal and alternatives. It ultimately addressed all the major issues in the controversy and provided the basis for a tentative settlement among the three agencies. This proposal was then circulated to a coalition of environmentalists. Although a few objections were raised by members of this group, a final agreement was reached with only minor alterations.

The final settlement was viewed by the parties involved as containing victories and concessions for all. On the whole, the parties expressed satisfaction with the outcome, believing that the agreement provided greater gains than those likely to be achieved through court action or a federally imposed decision.

Nonetheless, the settlement can be criticized on several grounds. First, as a consequence of the failure to ensure the participation of all interested parties in the negotiations, a discontented faction within the environmental coalition contested the settlement in court. Implementation of the negotiated settlement, therefore, was not ensured. Also, the reduced capacity of the water facility that was finally agreed upon may cause severe water shortages in the Denver area in the future. Ratepayers and homeowners may find that they have to pay a premium to expand the facility at a later time. This suggests that, perhaps, other interest groups (i.e., homeowners) with a stake in the outcome were not adequately represented in the negotiations. Finally, at least one decision-maker, the judge presiding over the case, before the court, strongly objected to the informal negotiations. He was angered by what he perceived to be a preemption of his powers and responsibilities. (The negotiated settlement was ratified, albeit indirectly, through various formal regulatory processes.)

Throughout the negotiations, which were never referred to as "mediation," Wirth performed a number of functions. He initiated the dialogue among the key parties which ultimately led to the settlement, orchestrated meetings and manipulated media coverage to create and maintain momentum, and occasionally pressured the parties to accept compromise. Wirth's reported behavior at the evening meetings which produced the tentative settlement among the Army Corps, EPA, and the Denver Water Board was quite assertive. He insisted that the parties continue negotiating long into the night until an agreement was reached.

The acceptance of Congressman Wirth as a mediator merits special attention because of his public stand on the issues in dispute and his official status. Despite his pro-environmental image, Wirth had written to the Bureau of Land Management in 1977 supporting one version of the Foothills Project. While local organizations and actors believed that he represented their interests, officially, Wirth was accountable only to his congressional constituents.

In any case, his status as a Congressman allowed him to bring both subtle and direct pressure to bear on the negotiating parties, especially the federal agencies. Wirth's biases were, perhaps, not easy to decipher, but he could hardly be considered a "neutral" mediator.

A NEW CONCEPTION OF THE PLANNER'S ROLE

Before moving to our proposal for a new conception of the planner's role, we will first review some of the changing conceptions of the role of the planner in the United States over the past three decades. Substantial shifts have occurred as a result of changing socioeconomic and political conditions, changes in the make-up of the planning profession, as well as adaptations resulting from careful reflection on past experiences (Susskind 1974).

At the core of the earliest (and most endurable) conception of the planner's role is the notion of technical expertise. The technician-planner was conceived as an apolitical designer of plans and alternative futures who worked within a framework of goals, objectives, and resource limitations established by elected decisionmakers. This conception of the planner's role assumes that analytic skills (the planner's "expertise") can be used to discover the "best" solution to a particular problem. Over time the presumed responsibilities of the planner-technician have broadened to include an examination of the consequences of alternative plans (impacts), but such evaluations still are presumed to be objective statements developed through the planner's use of specialized analytic tools.

Benveniste has helped to deepen our conception of the planner-technician's role (in *The Politics of Expertise*) by describing the political dimensions of even the most selfconsciously objective expert's practice. He points out that the technical expertise of the planner is the source of his or her power, and that the planner's overt role remains one of advisor to decision-makers. But astute planners, aware of the power of the relationship between themselves and the decision-maker, understand that their own values infuse the formulation and analysis of plans and policy recommendations.

The role of the technician-planner presupposes a centralized and stable political environment and assumes the planner's loyalty to the individual or organization seeking his or her advice. Neither condition is always true. As a consequence, the technician's role has been the cause of much frustration experienced by many practicing planners. Planners propose courses of action, but the acceptance and implementation of their recommendations are almost always beyond the scope of their influence.

A radically different vision of the planner's role was put forward by Paul Davidoff in his seminal article on advocacy planning (1965). Davidoff argued that planners and their designs are not value-neutral and that their implicit biases could be confronted if planners would affirm their values openly. He wrote: "Appropriate planning action cannot be prescribed from a position of value neutrality, for prescriptions are based on desired objectives" (Davidoff 1965, p. 331).

The advocacy model asserts that the development of plans and policies ought not be the exclusive responsibility of government agencies. On the contrary, special interest groups, especially disadvantaged segments of society, should be represented by trained planners committed to integrating the needs and desires of their clients into proposed alternatives to official policy. Proponents of the advocacy model believe that there is no unitary public interest and that the role of the planner is to give voice to the many "publics" whose welfare is at stake in any public resource allocation decision. Planning would then be transformed into a competitive marketplace of alternatives. Under the terms of the advocacy model, the planner is responsible for pointing out the biases inherent in the plans offered by opposing groups, informing the public about the concerns and views of his or her client group, and helping the client clarify and transform ideas into technically feasible alternatives.

Usually, the advocate planner is seen as challenging the system from outside, opposing the plans put forward by government agencies as well as proposing alternatives. But in such a model, the influence of the advocate planner ends as soon as a particular plan has been proposed. Advocate planners have not claimed a role in the ultimate implementation of plans, presuming that implementation is the responsibility of government agencies with appropriate jurisdiction (and that any failure to implement them can be challenged politically or through litigation).

Advocacy planning has been challenged by Peattie and others who, while concurring with the basic premises of the advocacy model and its advantages, criticize it for its practical limitations. Which segment of the community should the planner represent? How can the planner effectively identify his or her client's interest when even small subneighborhoods are rarely ethnically or economically homogeneous? Peattie offers this evaluation of the advocacy approach:

> "In effect, advocacy planning for the local community miniaturizes, but does not eliminate, the problems of conflicting interest which inhere in the planning activities of citywide agencies" (Peattie 1968, p. 84).

The advocacy model has also been criticized for downplaying the importance of scientific analysis. Klosterman recently noted that the denial of a factual basis for planning (rooted in the scientific methods of analysis that many planners employ) leaves decision making entirely in the hands of contending political actors. "This gives planners no basis for criticizing the outcomes of political processes which selectively promote the interests of the few over those of the many" (Klosterman 1983, p. 217). Decisions become merely a contest among power blocks to determine whose *preferences* should prevail.

In addition to the technician-planner and the advocate planner, other conceptualizations of the planner's role have been suggested. In 1969, Rabinovitz described two additional roles for planners operating in "competitive" or "fragmented" political environments. The mobilizer-planner, she suggested, is a technician who develops plans and policy recommendations and then works within the system to generate support for them. Rabinovitz' broker/ negotiator acts as a liaison between contending power blocks in a community, attempting to engineer palatable compromises. Finally, planners can serve as communicators, facilitating interaction among contending groups and within agencies (Forester 1980), or as educators (Alexander 1979).

Most of these conceptions share the presumption that the planner's responsibilities end once a plan, policy, or program has been accepted by the decision-makers or key power blocks. The planner presumably abandons the process at this point and returns to earlier steps in the process. Although planners expect to be summoned at some later point to evaluate the results of programs or projects (with an eye toward formulating improvements), they rarely play a continuing role in implementation. It is unfortunate since most public policies and projects flounder after a direction has been set-partly because it is difficult to orchestrate all the necessary actors and partly because dissatisfied segments of the community create obstacles to implementation.

In our view, implementation failures are a consequence of the planning profession's hesitancy to stress the important role that planners can play during implementation, especially in building and maintaining a durable consensus and in resolving disagreements. These tasks are central to our view of the planner as mediator. The mediator-planner performs some of the same roles as Rabinovitz' broker/negotia – tor. However, the mediator-planner does more than bring contending parties together around a common plan. The mediator-planner commits to a continuing role, seeking to ensure that the interests of all parties affected by a policy or a plan are taken into account from beginning to end.

Moreover, mediation is a method of involving disputing parties in developing mutually satisfactory responses to their differences. The mediator-planner encourages contending stakeholders to explore their differences—seeking zones of overlapping interest or possible items to trade, striving to maximize "joint gains" (Raiffa 1982). As Raiffa points out, most disputes involve multiple issues. Because disputing parties seldom value all issues equally, trade-offs are possible. By emphasizing the possibility of "joint gains" and mutually satisfying arrangements, the mediator-planner enhances the chances of implementing a particular course of action.

Of the three mediation efforts described above, only one (the Columbus case) involved trained planners. Yet the situations faced by the mediators in all these cases were similar to those that planners confront all the time. In our view, planners would do well to equip themselves as mediators and to become more familiar with the techniques of consensus-building and dispute resolution.

FINDING AN APPROPRIATE MODEL

Approaches to mediation are as varied as the personalities of the mediators involved. Labor mediators tend to be highly dispassionate about the negotiation process. They assume that the parties have an equal understanding of the rules of collective bargaining and equal experience in piecing together a mutually satisfactory package of concessions under the pressure of a deadline. International mediators, in contrast, tend to be more flamboyant. Henry Kissinger, controversial as he is, illustrates the considerable activism typical of international mediators. The process of mediating international disputes is usually ad hoc—there are few rules.

We have compiled a list of the functions that most mediators perform some, if not all, of the time. At a minimum, mediators are catalysts for action. Intervention by an outsider (at the request of the contending parties) creates a certain amount of pressure for interaction. Mediators are also facilitators. They usually schedule meetings, moderate discussions, and encourage participants to continue talking.

In addition, as illustrated in the three cases described earlier, mediators often assume very active roles. In the Columbus NIS case, the mediators controlled much of the communication among the parties, coordinating the exchange of concessions and, at times, served to mask the bargaining strengths and weaknesses of one side from the other. From the outset, Congressman Wirth was aggressive in shaping a role for himself in the Foothills negotiation. He persuaded the parties to come to the bargaining table and, later, encouraged them (sometimes using rather heavy-handed tactics) to make concessions. Wirth's manipulation of the media also helped to sustain momentum. O'Connor functioned as an educator, attempting to ensure that the participants understood the technical and legal aspects of the Brayton Point controversy. He held private sessions, allowing the participants to try out new proposals without jeopardizing their bargaining stand vis-a-vis the other parties. As a proposer of alternatives, he integrated his knowledge of the issues and the interests of the parties to devise new courses of action.

The roles that mediators play in bringing parties to an agreement vary, in part, according to the arenas in which mediation is used. We will analyze mediation in four realms: labor relations, international relations, community dispute resolution, and environmental dispute resolution. Since the role of any mediator is shaped by both the context in which negotiations take place as well as by the particular issues at stake, our examinations of mediator roles highlight both. Table 1 offers a comparison. Our goal is to provide a basis for characterizing strategies for mediator-planners.

LABOR RELATIONS

Mediation in collective bargaining is highly institutionalized. Federal legislation sets a legal framework. Labor unions are well-established and have adopted organizational structures that make it easy to represent the interests of labor in dealing with management. A nationwide mediator referral system has been established. The Federal Mediation and Conciliation Service (FMCS), an independent federal governmental agency, and the American Arbitration Association (AAA), a private, non-profit organization, provide mediation and arbitration services in both public and private labor disputes.

Mediation in collective bargaining situations takes place in a well-structured environment The issues are generally clear (until recently, wages and benefits, working conditions and union recognition were the primary agenda items), the stakeholding parties are easily identifiable, representatives are readily selected, and the relationships among the parties are ongoing and long-term. Power relationships are well-defined by past experience and the interests of disputing parties (in reaching agreement) are usually symmetrical—that is, the costs of nonsettlement increase with time for both labor and management. Negotiations regularly make reference to various conventions such as the inflation rate, the consumer price index and other economic indicators by which both sides can assess the reasonableness of demands and offers Further, the outcome of negotiations in labor disputes is fairly predictable, ending with either the signing of a contract or a strike.

The highly structured context of labor mediation has important implications for the role of the mediator. Traditionally, the mediator in labor disputes is preoccupied with concerns about the mediation *process*, ensuring that legal deadlines are met and that procedures for filing grievances conform to established standards. Labor mediators cherish their public image of neutrality; believing, probably correctly, that neutrality is the key to their success Labor mediators generally leave responsibility for the quality of agreements completely up to the negotiating parties.

INTERNATIONAL RELATIONS

In international disputes, the context of mediation tends to be rather loosely defined. Although disputing parties are usually identifiable, the issues may be highly complex, involving spillover effects on countries not directly involved in the negotiations. International mediation, especially regarding trade agreements, often involves large delegations. Delegation members are usually quite conversant with the dynamics of mediated negotiation. International mediation also frequently occurs at times of crisis during which a further deterioration of relationships can lead to disasterous results affecting the entire world. In the United States, international mediation is the terrain of the State Department and, in most cases, the appointment of a mediator is an overtly political decision.

International mediators generally tend to play a much more dynamic role than their labor counterparts, asserting more control over the proceedings and exhibiting more concern about the terms of settlement (i.e., the quality of agreements). In international disputes, the neutrality of the mediator is less of an issue. Zartman and Berman point out that "nothing requires the third party ... to be subtle and indirect, except for the general requirements of effectiveness" (Zartman and Berman 1982, p. 78). It is acceptable, according to Zartman and Berman, for the mediator to assume an active posture during negotiations, pointing out benefits that will flow from a particular solution or new possibilities for resolving the problem at hand, or even offering inducements for a negotiated settlement.

Discussing the intervention by Henry Kissinger in the Middle East, Dean Pruitt asserts that a number of mediation strategies were employed that appear to exceed the conventional roles of "catalyst" or "facilitator" (Rubin 1981). Kissinger directly controlled all communications between the disputing parties, actively persuaded parties to make concessions, allowed the parties to direct their anger toward him rather than against their opponents, coordinated the exchange of concessions, made his own proposals for settlement, and created and maintained the momentum for talks. Moreover, Kissinger was conspicuously linked to the United States whose interests in settling the Israelm-Arab conflict were only thinly disguised (Rubin 1981, p. 274).

COMMUNITY DISPUTE RESOLUTION

The use of mediation in community disputes has gained popularity over the past two decades. The National Center for Dispute Settlement of the American Arbitration Association and a network of Neighborhood Justice Centers, sponsored by the U.S. Department of Justice, as well as a number of smaller private, non-profit organizations, offer alternatives to the courts for dealing with disputes among family members and between landlords and tenants, administrators and students, merchants and consumers, clients and agencies, and racial groups.

Proponents contend that mediation is preferable to court settlements from both the perspective of the parties involved and the community at large (Alper 1981). They point out

395

Table 1 Models of Mediation

Elements	Labor Disputes	International Disputes	Community Disputes	Environmental Disputes
Source of Conflict	Contract expires	A "crisis" arises involving two or more nations.	A civil case is referred by the court; neighbors or family members seek action on a complaint.	A regulatory permit is requested or an alleged violation of existing rules or a "dangerous situation" is detected.
Process is Triggered by	A deadlock in negotiations: the NLRB steps in.	The leadership of one country announcing a position or stating a concern.	One or more parties seeks assistance from a community dispute resolution center.	One or more parties involved in a dispute or an "observer" suggests mediation.
Identification of Stakeholders	Institutionalized through: (1) Unions or employee reps, and (2) Management.	Self-identification (usually requiring some "proof" of stake) and identification by other involved nations.	Party who initiates mediation process identifies other responsible parties. These parties in turn may identify still others.	Parties already involved identify additional parties; public notice and active search for additional stakeholders may be necessary.
Representation	Union reps formally authorized by union membership. Management reps appointed by management.	Appointment of a negotiating delegation.	Parties usually represent themselves. Sometimes represented by legal counsel.	Ad hoc groups may be formed and asked to select negotiating representatives. If many groups are involved, it may be necessary to devise a system of "pyramiding" representation.

Familiarity of Parties with Negotiation	High for both parties.	High for all parties.	Usually low for all parties.	Variable. Some parties may be relatively experienced. Most will have low familiarity with negotiating techniques.
Number of Stakeholding Parties	Usually bilateral	Usually bilateral, could be multilateral.	Mostly bilateral.	Usually multilateral.
Mediator Selection	Appointed by NLRB, AAA, or FMCS with the approval of the parties.	Suggested by one of distributing parties; agreed upon by all.	Appointed by community dispute resolution center.	Suggested by one or more disputing parties or by an "outsider," subject to approval by all parties.
Agenda Setting	Generally circumscribed by federal legislation.	Open to negotiation.	Set by complaint subject to revision by other parties.	Open negotiation.
Approval of Draft Agreement Required by:	Union membership.	Official governmental bodies (i.e., Congress, Parliament, etc.).	Usually none.	Public officials or public agencies whose support is necessary for implementation.

that voluntary mediation has advantages in terms of access, timeliness of resolution, and efficiency (measured in terms of the proportion of cases settled) (McGillis and Mullen 1977). Mediation demonstrates respect for the disputants by allowing them direct participation in the resolution of their differences, unencumbered by legal rules of evidence and technical burdens of proof. Assigning guilt or blame is less important than resolving the underlying conflicts which gave rise to the dispute. The community suffers less demoralization if disputes are amicably resolved and residents feel a greater sense of control (because problems can be solved *within* the community).

The issues in community disputes tend to be unambiguous. Disputants are well-defined (although weaker groups—such as tenants in areas lacking active tenants' organization — may require organizational and educational assistance) and, if parties do not participate directly, representatives who can reasonably ensure the implementation of agreements can be readily selected. In the context of community disputes, mediators fulfill the roles of "listener, suggester, and the formulator of the final agreement to which both sides have contributed" (Alper et al. 1981, p. 131). The relative immaturity of the community dispute resolution field translates into flexible rules for mediation, varying with the needs of the particular conflicts and personalities involved. Mediators bring to the process as much substantive information as they can to help the parties achieve a resolution of their underlying differences. Nonpartisanship is clearly a critical factor in determining the effectiveness of a mediator, although familiarity with the community and even the parties involved can help enhance the mediator's credibility.

ENVIRONMENTAL DISPUTE RESOLUTION

Mediation was introduced into the environmental field in 1973, when two mediators attempted to settle a lengthy controversy involving construction of a dam on the Snoqualmie River in Washington state (Dembart and Kwartler 1980). A number of private, non-profit organizations have since sprung up across the United States with an interest in developing the theory and expanding the practice of environmental mediation. Among these are the New England Environmental Mediation Center in Boston, Massachusetts; the Institute for Environmental Negotiation at the University of Virginia, Charlottesville, Virginia; ACCORD, (formerly the Rocky Mountain Center on the Environment), the Center for Environmental Problem Solving in Boulder, Colorado; and the Institute for Environmental Mediation in Seattle, Washington. Although the growth of environmental mediation has been rapid, it has not occurred without considerable debate concerning its theory and practice (Cormick and Patton 1977; McCrory 1980; Stulberg 1980; and Susskind 1980).

Environmental disputes are characterized by substantial complexity, often heavy reliance on technical data and analysis, diffuse and unrepresentable interests (such as the interests of future generations), and substantial "externalities." Power relationships among interested parties tend to vary considerably, especially in terms of access to information, ability to manipulate the media and public opinion, and availability of resources to garner public support. The outcomes of environmental disputes can also have substantial implications for parties not represented in the negotiations, especially since such disputes involve what are, for all practical purposes, "irreversible" effects. The implementation of agreements often presents formidable obstacles when the cooperation of elected or appointed officials not involved in the negotiations is necessary. Finally, for the most part, environmental mediation is ad hoc, unlinked to formal processes of decision making. Thus, mediated settlments require the stamp of approval of formal decision makers. Environmental mediators operate under no standard code of ethics and without the benefit of a higher authority that can order the participation of key parties.

Consequently, environmental mediators serve a broad and variable range of functions. Like labor mediators, they help by scheduling, chairing, and recessing meetings; arranging joint and separate sessions; setting the location of meetings; proposing a sequence for the discussion of agenda items; keeping records; and imposing deadlines (Susskind and Weinstein 1980, p. 348). They facilitate communication between meetings, identifying points of agreement among disputants through confidential-talks, advising participants about positions they might take, and helping to prompt concessions and compromise.

More controversial aspects of the environmental mediator's role include helping to dentify appropriate stakeholding interests in a dispute, ensuring adequate representation throughout a negotiation, helping to generate alternatives and options for resolving differences, helping to establish agreement on technical points (i.e., by proposing and arranging joint fact-finding), and concern about the outcome of the negotiation process. Some have argued that the preponderance of unrepresentable and unorganized stakeholders in environmental disputes place special demands on the mediator (Susskind 1981). The environmental mediator's concern with the quality of negotiated agreements also stems from his or her obligation to assist the parties in developing a *stable* and implementable agreement (Susskind and Ozawa 1983). Although it has been suggested that a credo be developed to better define the environmental mediator's responsibilities (Susskind 1981; Center for Environmental Problem Solving 1982), the mediator's role in environmental disputes is still a subject of experimentation and debate.

Among these four models, the labor mediation model appears perhaps the least appropriate to the planner's situation. The context of mediation in collective bargaining is more highly structured and institutionalized than the context in which most planning occurs. The international model might be considered inspirational. Not unlike the milieu of international relations, planning usually takes place among many actors brought together on an ad hoc basis. In most cases, however, international mediators probably have less direct influence in shaping the settlement and generating a consensus than do planners.

The planner's role in developing alterna tives, identifying the implications of particular actions, and building an informed consensus appears closest to the role of the mediator in community and environmental disputes. Community and environmental mediation stress the quality of the negotiated agreements and their implementability. Because of the relative unfamiliarity of the parties with the negotiation process and, especially in environmental disputes, when technical analyses are essential, mediators in community and environmental disputes often serve as important sources of information. The role of the mediator in these arenas thus more nearly parallels the role of the planner.

Moreover, just as the mediator must remain available to help untangle problems that arise during implementation so, too, the planner ought to remain involved. Indeed, the professional planner's claim to a unique competence is a specialized capacity to overcome the obstacles to implementing a community-wide consensus.

CONCLUDING OBSERVATIONS

We anticipate several objections to our conception of the mediator-planner. In closing, we will attempt to respond to some of the more obvious concerns and put forth a few ideas about training mediator-planners.

First, it might be argued that planners do not have the credibility (or the authority) to mediate public resource allocation disputes since they are typically government employees and, as such, their impartiality is likely to be suspect in the eyes of various stakeholders. This, of course, would apply as well to consultants paid by government.

Although conventional models of mediation stress mediator neutrality, this concept may be less of a problem in practice than in theory. In the Foothills case, Congressman Wirth was by no means neutral since his position favoring the construction of the facility was well-known. Yet, his credibility was unchallenged by the participants. He established his nonpartisanship by the way he handled himself throughout the negotiations. If Wirth were perceived as partial, it can be argued, discontented parties would have withdrawn.

With regard to authority, it is often argued that the mediation process itself confers authority on the mediator. Cormick describes the process in this way: "Mediation is a voluntary process. The mediator has no authority to impose a settlement. The mediated dispute is settled when the parties themselves reach what they consider to be a workable solution" (Susskind and Weinstein 1980, p. 314). The willingness of key parties to participate and their mutual approval of the mediator is the mediator's primary source of credibility. Conversely, the mediator's authority is constrained by the ability of participants to enter and abandon negotiation at will, and usually by an explicit definition of his or her role predetermined by the negotiators. Further, experience has shown that the authority to mediate public disputes need not derive from a legislative or electoral mandate. The mediators in the Columbus NIS and the Brayton Point cases were appointed—their only credentials were their interest in assisting the parties and the willingness of the parties to accept their aid. Planners should not worry about their credibility or authority as mediators; rather, they should focus on how best to present themselves, handle confidences, and apply the skills of a mediator.

A second criticism of the mediator-planner that might be raised is that the demand for mediation in public sector disputes is not sufficient to support a full-time career for professional planners interested in mediation. Although the use of mediated negotiation in the public sector is growing steadily, the instances in which parties seek mediation are likely to remain few until the merits of mediated negotiation are more widely appreciated. Meanwhile, however, planners can help to publicize the mediation option until demand creates full-time employment opportunities. Moreover, while state and local planning agencies hit by cutbacks are not in the position to hire mediation specialists, planners with mediation skills may be more attractive in times of austerity. Finally, planners need not think of themselves solely as mediators. They can specialize in a variety of techniques while reserving their mediation skills for appropriate moments.

Finally, skeptics may view the mediator-planner concept as failing to incorporate a critical element of traditional planning: a concern for the redistribution of power and wealth. While mediation *per se* may do little to correct existing injustices, we believe that mediation can be an important tool for social reform. In some instances, groups out of power or at the bottom of the income scale have achieved more through mediated negotiation than by direct action or litigation.

Mediated negotiation can improve the position of disadvantaged groups in certain important ways. In contrast to litigation and other adversarial methods of dispute resolution, mediated negotiation emphasizes joint fact-finding and information sharing. Since a lack of information is often a major weakness of less powerful groups, especially in technologically sophisticated disputes, increased access to information may represent a significant gain. The extent to which information is actually shared may depend on the ability of the mediator to encourage cooperation. Nonetheless, mediated negotiation at least offers an opportunity for information sharing not available in adversarial proceedings. Mediated negotiation is also less costly (most of the time) than sustained political action or litigation. Groups that lack the financial resources to engage in other forms of dispute resolution can suggest mediated negotiation, and perhaps thereby gain a greater chance to influence public resource allocation decisions.

We also believe that consensual approaches to dispute resolution leave less advantaged groups in a better position to influence policy-making in the future. Face-to-face negotiation is less belligerent and more conducive to constructive communication (through which contending groups may learn to understand, and not simply attempt to discredit, the positions and interests of opponents). Armed with an improved understanding of the interests and concerns of various groups, tactical alliances can be formed and future propositions can be framed in ways less threatening to parties with potentially conflicting interests—thereby improving the likelihood of such proposals gaining acceptance.

We do not presume that all disputes arising over public resource allocation can (or should) be mediated. Dispute resolution can only occur if contending parties are willing to explore possible joint gains or to make concessions. A dispute over the construction of a nuclear power plant, for example, may not be negotiable if the utility company cannot guarantee absolute safety and if opponents are unwilling to accept any degree of risk. Siting plants elsewhere but within the radius of industry's "preferred locations" (in terms of markets for electric power) will probably not appease protestors whose homes would still lie within potentially higher risk zones. Nor would compensation be satisfactory to those whose ideological commitments are nonnegotiable. Further research is necessary to determine exactly which disputes can (and ought to) be resolved through mediated negotiation. In any case, the parties involved retain veto power over the mediation process, since participation is voluntary.

If planners are to add the skills of mediation and consensus building to their repertoire, how should their professional training be augmented? Presently, few people are educated to function as effective negotiators, much less as effective mediators. Indeed, not all persons are temperamentally suited to serve as mediators.

With this in mind, we urge that planning students be made more aware of the procedures and opportunities for employing mediation. Students interested in mediation techniques ought to have access to skill training. Law schools have already begun to add such courses to their curricula. Planning departments should do likewise. The training of planners as mediators can also be accomplished through mid-career training. Short courses are currently available in many parts of the country.

Given the current political and economic climate in the United States, planning ought to be on the lookout for new roles for their graduates. They also ought to concentrate on new roles that build on the skills and commitments of existing members of the profession. The mediator-planner embodies some of the functions of traditional planners, emphasizing the planner's role in ensuring successful implementation of plans and policies. Most importantly, the concept of the planner as mediator provides a compelling metaphor-one likely to attract the very best minds to planning at a time when the field has lost much of its luster.

REFERENCES

Alexander, E.R. 1979. Planning Roles and Contexts. In *Introduction to Urban Planning*, eds. A. Catanese and J. Snyder. New York: McGraw Hill.

Alper, B.S. and Nichols, L.T. 1981 *Beyond the Courtroom*. Cambridge, MA: Lexington Books.

Bacow, L. and Wheeler, M. 1983. *Environmental Dispute Resolution*. New York: Plenum Publishers.

Baldwin, N.J. 1983. Negotiated Rulemaking: A Case Study of Administrative Reform. Unpublished M.A. thesis. Department of Urban Studies and Planning, Massachusetts Institute of Technology.

Bellman, H. 1983. Talk given at Harvard Law School, Program on Negotiation.

Benveniste, G. 1972. *The Politics of Expertise*. Berkeley, CA: The Glendessary Press.

Bersch, D. 1982. The Negotiated Investment Strategy (NISI: An Evaluation of the Columbus, Ohio Experiment. Ohio State University, Department of City and Regional Planning.

Burgess, H. Forthcoming. The Foothills Water Project: A Case Study of Environmental Mediation. In *Resolving Environmental Regulatory Disputes*, eds., L. Susskind, L. Bacow and M. Wheeler. Cambridge, MA: Schenkman Publishing Co.

Catanese, A.J. and Farmer, W.P., eds. ——— 1979. *Personality, Politics, and Planning*. Beverly Hills: Sage Publications.

Center for Environmental Problem Solving (ROMCOE). 1982. *Workshop Summary*. Environmental Conflict Management Practitioners' Workshop, Florissant, Colorado, October 27–29, 1982.

Charles F. Kettering Foundation. 1981. *Mediation and New Federalism*. Proceedings on the Negotiated Investment Strategy. Washington, D.C. July 8, 1981.

Cormick, G.W. and Patton, L. 1977. Environmental Mediation: Potentials and Limitations. *Environmental Comment*. 1316.

Davidoff, P. 1965. Advocacy and Pluralism in Planning. *Journal of the American Institute of Planners*. 31:331–338.

Dembart, L. and Kwartler, R. 1980. The Snoqualmie River Conflict: Bringing Mediation into Environmental Disputes. In *Roundtable Justice: Case Studies in Conflict Resolution*, ed., R.B. Goldmann. Boulder, CO: Westview Press.

Folk-Williams, J.A. 1982. *Water in the West*. Sante Fe, NM: Western Network.

Forester, J. Critical Theory and Planning Practice. *Journal of the American Planning Association*. 46:275–286.

Goldman, R. 1980. *Roundtable Justice: Case Studies in Conflict Resolution*. Boulder, CO: Westview Press.

Kennedy, W.J.D. and Lansford, H. 1983. The Metropolitan Water Round Table Resource Allocation Through Conflict Management. *Environmental Impact Assessment Review*. 4.

Klosterman, R.E. 1983. Fact and Value in Planning. *Journal of the American Planning Association*. 49:216–225

McCrary, J.P. 1981 Environmental Mediation — Another Piece for the Puzzle. *Vermont Law Review*. 6:49–84.

M cG i I Ms, D. and Mullen, J. 1977. *Neighborhood Justice Centers: An Analysis of Potential Models.* Washington, D C.: U.S. Department of Justice, National Institute of Law Enforcement and Criminal Justice

Mernitz, S. 1980. *Mediation of Environmental Disputes.* New York: Praeger Publishers.

Peattie, L.R. 1968. Reflections on Advocacy Planning. *Journal of the American Institute of Planners.* 34:80–88.

Rabinovitz, F. F. 1969. *City Politics and Planning.* New York: Atherton Press.

Raiffa, H. 1982. *The Art and Science of Negotiation.* Cambridge, MA: Harvard University Press.

Rubin, J.Z., ed. 1981. *Dynamics of Third Party Intervention: Kissinger in the Middle East.* New York: Praeger Press.

Smith, D. Forthcoming. Brayton Point Coal Conversion. In *Resolving Environmental Regulatory Disputes,* eds., L. Susskind, et al. Cambridge, MA: Schenkman Publishing Co.

Stulberg, J. 1981. The Theory and Practice of Mediation: A Reply to Professor Susskind. *Vermont Law Review.* 6:85–117.

Susskind, L.E. 1974. The Future of the Planning Profession. In *Planning in America: Learning from Turbulence,* ed., D.R. Godschalk. Washington, D.C.: American Institute of Planners.

———. 1981. Environmental Mediation and the Accountability Problem. *Vermont Law Review.* 6:1–47.

Susskind, L.E.; Bacow, L.; and Wheeler, M. 1983. *Mediation in Environmental Regulatory Disputes.* Cambridge, MA: Schenkman Publishers.

Susskind, L.E. and Keefe, F. 1980. A Negotiated Investment Strategy for Columbus, Ohio, Volumes I and II.

Susskind, L.E. and Ozawa, C. 1983. Mediated Negotiation in the Public Sector: Mediator Accountability and the Public Interest Problem. *American Behavioral Scientist.* 27:255–279.

Susskind, L.E. and Weinstein, A. 1980. Toward a Theory of Environmental Dispute Resolution. *Boston College Environmental Law Review.* 9:311–357.

Talbot, A.R. 1983. *Settling Things: Six Case Studies in Environmental Mediation.* Washington, D.C.: The Conservation Foundation.

Watts, S. 1983. Description and Assessment of the Connecticut Negotiation Investment Strategy Experiment. Working Paper, Program on Negotiation at Harvard Law School.

Zartman, I. W. and Berman, M.R. 1982. *The Practical Negotiator.* New Haven: Yale University Press.

Eugene A. Rosa, Ortwin Renn and Aaron McCright — "Risk Governance: A Synthesis." In *Risk Society Revisited: Social Theory and Governance*. Philadelphia: Temple University Press.

Reading Commentary

Many environmental decisions need to take risk into account. How do you plan for a serious flood? Or a nuclear disaster? Rosa, Renn and McCright lay out a strategy for managing risky decisions that they call risk governance. They move away from the dominant approach to managing risks in terms of the probability of a disaster occurring times its effect. They think this is too simplistic. There are more factors at play than just cause and effect. In particular, they are concerned about addressing uncertainty, complexity and ambiguity. Is it possible to truly understand the impact climate change will have on a city in 50 years? To address the shortcomings of the conventional approach to risk estimation and risk management, they urge involving many different stakeholders—states, experts, NGOs and the public—to be certain that as many perspectives as possible are taken into account.

This article outlines the various stages in risk governance: preassessment, interdisciplinary risk estimation, characterization and evaluation, management and participation. Public engagement is central to every step. The authors believe that the public should help identify risks, frame the issues, make clear how they perceive different risks, determine the acceptability of risks and design a management strategy.

The authors favor collective decision-making as the best way to address societal risks. Since society will bear the burden, the public should be involved in deciding how risks should be managed. Finally, none of the decisions they are talking about are straightforward. There are no right answers. Trade-offs can be better understood through involving the public. While this makes the process more complicated, they think it will lead to better outcomes. Do you agree?

Chapter 9
RISK GOVERNANCE A SYNTHESIS

Eugene A. Rosa, Ortwin Renn and Aaron McCright

We are more inclined to engage in a risk if we believe that the odds of a truly catastrophic outcome are negligible.

—Robert Meyer, "Why We Still Fail to Learn from Disasters" (2010)

BEYOND GOVERNMENT: THE NEED FOR COMPREHENSIVE GOVERNANCE

Risk governance, as we defined it in the Introduction, is a broad rubric referring to a complex of coordinating, steering, and regulatory processes conducted for collective decision making involving uncertainty. Risk sets this collection of processes in motion whenever the risk affects multiples of people, collectivities, or institutions. Governance comprises both the institutional structure (formal and informal) and the policy process that guide and restrain collective activities of a group, society, or international community. Its aim is to regulate, reduce, or control risk problems.

This chapter addresses each stage of the risk governance process: pre-assessment, interdisciplinary risk estimation, characterization and evaluation, management, and communication/participation. Furthermore, we explicate the design of risk communication and participation to address the challenges raised by the three risk characteristics of complexity, uncertainty, and ambiguity. Finally, this chapter concludes with basic lessons for risk governance. Before beginning those tasks, we identify a fundamental change in the actors now participating in governance.

In recent decades, the handling of collectively relevant risk problems has shifted. The shift is from traditional state-centric approaches, with hierarchically organized governmental agencies as the dominant locus of power, to multilevel governance systems, where the political authority for handling risk problems is distributed among separately constituted public bodies (cf. Lidskog 2008; Lidskog et al. 2011; Rosenau 1992; Wolf 2002). These bodies are characterized by overlapping jurisdictions that do not match the traditional hierarchical order of state-centric systems (cf. Hooghe and Marks 2003; Skelcher 2005). They consist of multi-actor alliances that include not only traditional actors such as the executive, legislative, and judicial branches of government, but also socially relevant actors from civil society. Prominent among those actors are industry, science, and nongovernmental organizations. The result of the governance shift is an increasingly multilayered and diversified sociopolitical landscape. It is a landscape populated by a multitude of actors whose perceptions and evaluations draw on a diversity of knowledge and evidence claims, value commitments, and political interests. Their goal, of course, is to influence processes of risk analysis, decision making, and risk management (Irwin 2008; Jasanoff 2004).

405

Institutional diversity can offer considerable advantages when complex, uncertain, and ambiguous risk problems need to be addressed. First, risk problems that vary in scope can be managed across different institutions or at different levels. Second, an inherent degree of overlap and redundancy makes nonhierarchical adaptive and integrative risk governance systems more resilient and therefore less vulnerable. Third, the larger number of actors facilitates experimentation and learning (Klinke and Renn 2012; Renn 2008c: 177ff.; Renn, Klinke, and van Asselt 2011). There are also disadvantages. They include the possible com-modification of risk, the fragmentation of the risk governance process, more costly collec-tive risk decision making, the potential loss of democratic accountability, and paralysis by analysis. The paralysis shows itself in the inability to make decisions due to unresolved cogni-tive and normative conflicts and lack of accountability vis-à-vis multiple responsibilities and duties (Charnley 2000; Garrelts and Lange 2011; Lyall and Tait 2004).

Thus, understanding the dynamics, structures, and functionality of risk governance processes requires a general and comprehensive understanding of procedural mechanisms and structural configurations. The standard model of risk analysis consisting of three components—risk assessment (including identification), management, and communication—is too narrowly focused on private or public regulatory bodies. It fails to consider the full range of governance actors engaged in processes for governing risk. Furthermore, it ignores stakeholder and public involvement as a core feature in the stage of communication and deliberation, another key element of governance.

Despite new attempts to develop new models and frameworks of risk governance, there is still a need to link these conceptual frameworks to actual case studies. Such a link allows us to test past experiences and explore their usefulness for designing more informed and robust risk management strategies (Renn and Walker 2008). As Timo Assmuth (2011:167), a senior researcher at the Finnish Environment Institute, concludes, "With complex risk and risk-benefit issues such as those of Baltic Sea fish [that are threatened not only by overfishing, but also by toxic contaminants), a narrow and rigid assessment and management approach based on illusory certainty and on a sectorized and top-down governance and deliberation style needs to be complemented by a broader, more flexible and evolutionary approach."

FROM GOVERNMENT TO GOVERNANCE

The term *government* typically refers to a civil body defined as a sovereign state, whether it is of an international body, of a nation, of a jurisdiction within a nation, or of a local domain. The governing structure of the modern nation-state is the most obvious example of gov-ernment. While "governing" is the province of governments, the purview of governance is, after John Dewey, the public sphere.[1] The public sphere is much broader than govern-ment, consisting as it does of a set of processes conducted by a wide variety of social actors. Governance choices in modern societies are generally conceptualized as a mutual interplay among governmental institutions, economic forces, and civil-society interests (mediated, e.g., by NGOs). Generally, governance embodies a nonhierarchically organized structure in which there is no superordinate authority. It consists of government and nongovernment actors bringing about policies that are binding to all participants (cf. Lidskog 2008; Rosenau 1992; Wolf 2002). In this perspective, nongovernmental actors play an increasingly relevant role due to their decisive advantages in flexibility and in gaining information or resources compared with governmental agencies (Kern and Bulkeley 2009).

As noted in the Introduction, the concept of governance came into fashion again in the 1980s in circles concerned with international development. It was soon adopted in other domains. During the past decade, the term has experienced tremendous popularity in the fields of, among others, international relations, various policy sciences (including subfields referred to as European studies and comparative political science), environmental studies, and risk research. The idea of governance thus has been dusted off and reintroduced to enlarge the perspective on policy, politics, and polity by acknowledging the key role of other actors in managing and organizing societal and political solutions. The revitalization of the term *governance* and its growth in use is best understood as a response to new challenges, such as globalization, increased international cooperation (e.g., the European Union), technological changes (e.g., international communication), the increased engagement of citizens, the rise of NGOs, and the changing role of the private sector. Augmenting these societal challenges are the increased complexity of policy issues and the resulting difficulty in implementing decisions with confidence and legitimacy (Pierre and Peters 2000; Walls et al. 2005).

Many classical theories of regulation presume a hierarchical orientation in which government is the central actor wielding the medium of power. In economic policy theories, the central actor is the market, whose medium is money. Both theoretical orientations are focused on a dominant actor that exercises power and control. The governance perspective views things very differently. It generates and implements collectively binding policy solutions in a complex context of multi-actor networks and processes. Power is distributed among the variety of actors in the multi-actor networks. Added to government and markets are new civic actors, such as expert groups and NGOs, as well as ad hoc coalitions of citizens from unclear ranks of supporters, posing the difficulty of determining whom they represent. Governance also includes the role of non-elected actors, such as civil servants, scientific and policy experts, think tanks, and a broad range of committees active in various ways in policy processes. The governance perspective thus draws attention to the diversity of actors, the diversity of their roles, the diversity in the logics of action, the manifold relationships among them, and all kinds of dynamic networks emerging from these relationships. Scholars who subscribe to the governance perspective examine social networks and the roles of the various actors in these dynamics as a way to understand policy development and political decision making.

Some authors differentiate between horizontal and vertical governance Benz and Eberlein 1999; Lyall and Tait 2004). The horizontal level includes all actors within a defined geographical, functional, or political administrative segment—for example, a governmental agency, stakeholder groups, industry, and a science association at the national level. The vertical level describes the links between these horizontal levels (such as the institutional relationships between the local, regional, state, and international levels). When various levels are involved, which is often the case, the notion of multi-level governance is advanced. In such a context, "government" is no longer a single entity (Rauschmayer et al. 2009).

We should note that governance includes both a descriptive component and a normative component. In a descriptive use of the term, it refers to the complex web of manifold interactions between heterogeneous actors in a particular policy domain. It also refers to the resultant decision or policy. Governance is then an observation and description of the approach to characterizing the scale and scope of problem solving. Here is a descriptive definition of governance: "structures and processes for collective decision-making involving governmental and non-governmental actors" (Nye and Donahue 2000: 12).

Normatively, governance refers to the model or framework of how we ought to organize or manage collective decisions. The highly influential white paper of the European Commission on Governance (2001) propagates such a normative perspective on "good governance." In that white paper, a direct response to the BSE (mad cow disease) crisis, governance is presented as a prescriptive model, where transparency, stakeholder participation, accountability, and policy coherence are adopted as key principles.

With that general background, we can now focus on risk governance more specifically. It involves the translation of the substance and core principles of governance to the context of risk and problem solving (International Risk Governance Council 2005; Klinke and Renn 2012; Renn 2008c; Renn and Walker 2008; Renn, Klinke, and van Asselt 2011). It refers to a body of scholarly ideas over how to deal with challenging public risks. These ideas have been informed by forty years of interdisciplinary research drawing from sociological, psychological, and political science research on risk, from science and technology studies and from research by policy scientists and legal scholars.[2] This considerable body of knowledge provides a convincing, theoretically demanding, and empirically sound basis for critiquing the standard risk analysis model that consists of two dimensions: the probability of occurrence and the consequences if a risk is realized. Many risks cannot be calculated on the basis of quantitative probabilities and effects alone, as we demonstrated in Chapters 7–8. Risks too often are multidimensional. Too often they have qualitative features that are not easily amenable to quantification. Too often they exist in contexts that are ambiguous or intractably complex. In short, they have all the characteristics of systemic risks.

This means that regulatory models that build on the standard model not only are inadequate but also constitute an obstacle to dealing with risk responsibly. These limitations in standard risk analysis are the reason for the recent adoption of a governance framework to understand and manage risk. Risk governance does accommodate the qualitative features of risk. But it also comprehends the various ways in which many actors—individuals and institutions, public and private—deal with risks surrounded by deep uncertainty, complexity, or ambiguity.[3] It includes formal institutions and regimes and informal arrangements. It refers to the totality of actors, rules, conventions, processes, and mechanisms concerned with how relevant risk information is collected, analyzed, and communicated and how regulatory decisions are taken (International Risk Governance Council 2005, 2007; van Asselt 2007). However, risk governance is more than just descriptive shorthand for a complex, interacting network in which collectively binding decisions are taken to deal with a particular set of societal issues. The aim is that risk governance—similar to governance in general—provides a conceptual as well as a normative basis for how to deal responsibly with uncertain, complex, or ambiguous risks in particular (van Asselt and Renn 2011).

FROM SIMPLE TO SYSTEMIC RISKS

At the core of risk governance is the recognition that risk comes in many flavors. In 1995, the Dutch Health Council phrased it this way: "not all risks are equal" (quoted in van Asselt and

Renn 2011: 440). The modern concept of risk can be traced to the seventeenth century in the work of such mathematical luminaries as Blaise Pascal, Pierre de Fermat, and Christiaan Huygens, and to the emergence of the word *probability* (Hacking 1975). But with the exception of the insurance industry, the application of risk to decisions about investments, technology, and the environment is a child of the twentieth century (Knight 1921; Starr 1969). The predominant formulation adopted for application has been to treat risk in terms of probability and effects, of dose and response, or of agent and consequences. This dominant framing of risk underlies what has been called the technocratic, decisionistic, and economic models of risk assessment and management (cf. Löfstedt 1997; Millstone et al. 2004; Renn 2008c: 10). This framing of risk, like all models of reality, abbreviates the many facets of risk into a simple two-dimensional cause and consequence or dose and response model. This is not much of a problem for simple risks, such as annual auto fatalities, where considerable actuarial data is available to estimate probabilities. The cause for the risk is generally well known; the potential negative consequences are fairly obvious; and the uncertainty is low, so there is little ambiguity about the interpretation of the risk. Simple risks are ostensible and recurrent and therefore grounded in ontological reality. They follow the process of long-term frequency distributions, meaning they are less affected by current or future disturbances. As a consequence, a whole toolbox of statistics is available, and their application is meaningful. Examples include a wide variety of accidents, such as with automobiles and airplanes, and regularly recurring natural events, such as seasonal flooding.

But many risks are not simple and cannot be calculated as a function of probability and effects. This alternative view of risk, shared by an increasing group of risk scholars, explicitly challenges the idea of risk inherited from scholars such as Frank Knight (1921) in which risk is restricted to numerically defined probability distributions (Aven and Renn 2009a). As we pointed out earlier, many risks have qualitative features that are not easily captured in the standard two-dimensional risk model. For example, the aesthetic and functional loss associated with the extinction of polar bears would be difficult to fit into that model.

Risk governance highlights the importance of recognizing the key roles that uncertainty,[4] complexity, and ambiguity play in many societal risks. Ironically, it is a consistent practice in most cases to assess and manage risks with those characteristics as if they were simple and amenable to the standard risk model. The conventional assessment and management routines do not do justice to the confounding characteristics of such risks. This practice leads to a whole range of difficulties, among them unjustified amplification or irresponsible attenuation of the risk, sustained controversy, deadlocks, legitimacy problems, unintelligible decision making, trade conflicts, and border conflicts. The main message from Chapters 7–8 is that we urgently need to develop better conceptual and operational approaches to understand and characterize, let alone manage, non-simple risks.

LESSONS FOR RISK GOVERNANCE

Risk governance is a paradigm, an orienting perspective of highly contextualized practices for understanding and managing risks. Hence, it is not a model in the strict sense of the word. However, it does incorporate the results of model applications where appropriate. The aims of risk governance are more fundamental. Risk governance aims for a paradigm shift from the conventional practices in the field. It also seeks to orient risk professionals to a broader concept of risk (van Asselt and Renn 2011). It is a dynamic, adaptive learning, and decision-making process of continuous and gradual learning and adjustments that permits prudent handling of complexity, scientific uncertainty, or sociopolitical ambiguity. Adaptive and integrative capacity in risk governance processes encompasses a broad array of structural and procedural means and mechanisms by which politics and society can handle collectively relevant risk problems. In practical terms, adaptive and integrative capacity is the ability to design and incorporate the necessary steps in a risk governance process that allow risk managers to reduce, mitigate, or control the occurrence of harmful outcomes resulting from collectively relevant risk problems in an effective, efficient, and fair manner (Brooks and Adger 2005). The adaptive and integrative quality of the process requires the capacity to learn from previous and similar risk handling experiences to inform and deal with current risk problems. It also has the capacity for adaptively applying these lessons to cope with future potential risk problems and surprises.

A significant institutional development for the promotion of risk governance was the founding of the International Risk Governance Council (IRGC) in 2002. The IRGC (2005: 5) is an independent organization whose "work includes developing concepts of risk governance, anticipating major risk issues, and providing risk governance policy recommendations for key decision makers." In 2005, it suggested a process framework of risk governance (International Risk Governance Council 2005; Renn 2008c; Renn and Walker 2008). The framework structures the risk governance process into four phases: (1) pre-assessment; (2) appraisal; (3) characterization and evaluation; and (4) risk management. Communication and stakeholder involvement were conceptualized as constant companions to all four phases of the risk governance cycle. The framework suggests a phase-by-phase process beginning with pre-assessment and ending with risk management. Each phase is further subdivided into functional components that need to be included to complete each step. Furthermore, there is a strict separation between knowledge acquisition and decision making and between physical and non-physical impacts (a distinction between risk assessment and so-called concern assessment—that is, a scientific investigation of people's perceptions and concerns associated with the respective risk).

Andreas Klinke and Ortwin Renn (2012) have proposed some alterations to the IRGC risk governance model, because it appears too rigid to be applied to systemic risks. They recognized a need for a comprehensive risk governance model with additional adaptive and integrative capacity. The modified framework suggested by Klinke and Renn (2012) consists of the following interrelated activities: pre-estimation, interdisciplinary risk estimation, risk characterization, risk evaluation, and risk management. This requires the ability and capacity of risk governance institutions to use resources effectively. Appropriate resources include institutional and financial means, as well as social capital (e.g., strong institutional mechanisms and configurations, transparent decision making, allocation of decision making authority, formal and informal networks that promote collective risk handling, and education), technical resources (e.g., databases, computer software and hardware, research facilities) and human resources (e.g., skills, knowledge, expertise, epistemic communities). Hence, the adequate involvement of experts, stakeholders, and the public in the risk governance process is a crucial dimension to produce and convey adaptive and integrative capacity in risk governance institutions (cf. Pelling et al. 2008; Stirling 2008). Since the social acceptance of any response of risk governance to risk problems associated with complexity, uncertainty, or ambiguity is critical, risk handling and response strategies need to be flexible, and the risk governance approaches need to be iterative and inclusionary.

PRE-ESTIMATION

Risk, while an ontologically real phenomenon, is understood only via mental constructions resulting from how people perceive uncertain phenomena. Those perceptions, interpretations, and responses are shaped by social, political, economic, and cultural contexts (cf. International Risk Governance Council 2005; Luhmann 1993; Organization for Economic Cooperation and Development 2003). At the same time, those mental constructions are informed by experience and knowledge about events and developments in the past that were connected with real consequences. That the understanding of risk is a social construct with real consequences is contingent on the presumption that human agency can prevent harm. Our understanding of risk as a construct has major implications for how risk is considered. While risks have an ontological status, understanding them is always a matter of selection and interpretation. What counts as a risk to someone may be destiny explained by religion for a second party or even an opportunity for a third party. Although societies over time have gained experience and collective knowledge of the potential effects of events and activities, one neither can anticipate all potential scenarios nor be worried about all of the many potential consequences of a proposed activity or an expected event. At the same time, it is impossible to include all possible options for intervention. Therefore, societies always have been and always will be *selective* in what they choose to consider and what they choose to ignore.

Pre-estimation, therefore, involves *screening* to winnow from a large array of actions and problems that are risk candidates. Here it is important to explore what political and societal actors (e.g., governments, companies, epistemic communities, and NGOs) as well as citizens identify as risks. Equally important is to discover what types of problems they identify and how they frame them in terms of risk and in terms of uncertainty, complexity, and ambiguity. This step is referred to as *framing*, how political and societal actors rely on schemes of selection and interpretation to understand and respond to those phenomena that are relevant risk topics (Kahneman and Tversky 2000; Nelson et al. 1997; Reese 2007). According to Robert Entman (1993:52), "To frame is to *select some aspects of a perceived reality and make them more salient in a communication text, in such a way as to promote a particular problem definition, casual interpretation, moral evaluation, and/or treatment recommendation* for the item described." Perceptions and interpretations of risk depend on the frames of reference (Daft and Weick 1984).

Framing implies that pre-estimation requires a multi-actor and multi-objective governance structure. Governmental authorities (national, supranational, and international agencies), risk producers, opportunity takers (e.g., industry), those affected by risks and benefits (e.g., consumer organizations, local communities, and environmental groups), and interested parties (e.g., the media or experts) are all engaged. They will often debate about the appropriate frame to conceptualize the problem. What counts as risk may vary greatly among these actor groups.

INTERDISCIPLINARY RISK ESTIMATION

For political and societal actors to arrive at reasonable decisions about risks in the public interest, it is not enough to consider only the results of risk assessments, scientific or otherwise. To understand the concerns of affected people and various stakeholders, information about their risk perceptions and their concerns about the direct consequences if the risk is realized is essential and should be taken into account by risk managers.

Interdisciplinary risk estimation consists of a systematic assessment not only of the risks to human health and the environment and but also of related concerns, as well as social and economic implications (cf. International Risk Governance Council 2005; Renn and Walker 2008). The interdisciplinary risk estimation process should be informed by scientific analyses. Yet in contrast to traditional risk regulation models, the scientific process includes the biophysical sciences as well as the social and economic sciences. The interdisciplinary risk estimation comprises two activities. The first is *risk assessment*, which produces the best estimate of the likelihood and the physical harm that a risk source may induce. The second is *concern assessment*, or identifying and analyzing the issues that individuals or society as a whole link to a certain risk. For this purpose, the research repertoire of the social sciences, such as survey methods, focus groups, econometric analysis, macroeconomic modeling, and structured hearings with stakeholders may be used.

In 2003, the Wissenschaftliche Beirat der Bundesregierung Globale Umweltveränderungen (German Advisory Council on Global Environmental Change; WGBU) offered a set of criteria to characterize risks that go beyond the classic components of probability and extent of damage. The WBGU isolated and validated eight measurable risk criteria through a rigorous process of interactive surveying. Experts from the biophysical sciences and the social sciences were asked to characterize risks based on the dimensions that they would use for substantiating a judgment on tolerance to risk. Their input was subjected, through discussion sessions, to a comparative analysis. To identify the eight definitive criteria, the WBGU distilled the experts' observations down to those that appeared most influential in the characterization of different types of risk. In addition, alongside the expert surveys the WBGU performed a meta-analysis of the major insights gleaned from existing studies of risk perception and evaluated the risk management approaches adopted by countries including the United Kingdom, Denmark, the Netherlands, and Switzerland. The WBGU's long exercise of deliberation and investigation pinpointed the following eight physical criteria for the evaluation of systemic risks:

1. *Extent of damage*, or the adverse effects arising from a risk—measured in natural units such as deaths, injuries, or production losses.
2. *Probability of occurrence*, an estimate of the relative frequency of a discrete or continuous loss function that could arise from the manifestation of a risk.
3. *Incertitude*, an overall indicator of the degree of remaining uncertainties inherent in a given risk estimate.
4. *Ubiquity*, which defines the geographic spread of potential damages and considers the potential for damage to span generations.
5. *Persistence*, which defines the duration of potential damages, also considering potential impact across the generations.

6. *Reversibility*, or the possibility of restoring the situation, after the event, to the conditions that existed before the damage occurred (e.g., reforestation, the cleaning of water).
7. *Delay effect*, which characterizes the possible extended latency between the initial event and the actual impact of the damage it caused. The latency itself may be of a physical, chemical, or biological nature.
8. *Potential for mobilization*, understood as violations of individual, social, or cultural interests and values that generate social conflicts and psychological reactions among individuals or groups of people who feel that the consequences of the risk have been inflicted on them personally. Feelings of violation may also result from perceived inequities in the distribution of costs and benefits.

Interestingly, the final criterion—mobilization—was the only one aimed at describing public response (whether it is acceptance or outrage) that found favor among all the experts consulted by the WBGU. Subsequently, a research team at the Center of Technology Assessment in Baden-Württemberg concluded that a broader understanding of this important criterion for the evaluation of systemic risks is needed. The team unfolded the compacted *mobilization index* and divided it into four major, identifiable elements that drive the potential for mobilization (Klinke and Renn 2002):

1. The *inequity and injustice* associated with the distribution of risks and benefits over time, space, and social status.
2. The psychological *stress and discomfort* associated with the risk or the risk source, as measured by psychometric scales.
3. The *potential for social conflict and mobilization*, which translates into the degree of political or public pressure on agencies with responsibility for risk regulation.
4. The likely *spill-over effects* that may occur when highly symbolic losses have repercussions in seemingly unconnected areas, such as financial markets or the loss of credibility in management institutions.[5]

The four social criteria can be used by the risk analyst to assess the extra effect that the manifestation of a risk—or even its mere presence—may have on psychological or social responses among subject groups. They can be used to gauge how much such social side effects are likely to stretch beyond the expected impact ascertained through consideration of the risk in the context of the other seven physical criteria.

A similar decomposition has been proposed by the United Kingdom's Environment Agency (1998). For example, in the report *A Strategic Approach to the Consideration of "Environmental Harm"* (Pollard et al. 2000), Environment Agency researchers propose two main criteria for the assessment of mobilization potential, each of which is subdivided into three further criteria: anxiety, divided into dread, unfamiliarity, and notoriety; and *discontent*, divided into unfairness, imposition, and distrust.

In a recent draft document, the U.K. Treasury Department (Her Majesty's Treasury 2004) recommended a risk classification that includes hazard characteristics, the traditional risk assessment variables such as probability and extent of harm, indicators on public perception, and the assessment of social concerns. The document offers a tool for evaluating public concerns against six factors that are centered on the hazards leading to a risk, the risk's effects, and its management:

1. Perception of familiarity and experience with the hazard.
2. Understanding the nature of the hazard and its potential effects.
3. Repercussions of the risk's effects on intergenerational, intragenerational, or social equity.
4. Perception of fear and dread in relation to a risk's effect.
5. Perception of personal or institutional control over the management of a risk.
6. Degree of trust in risk management organizations.

The methodology for including social criteria in the formal risk evaluation process definitely needs further refinement, since it is an area of risk analysis that remains in its infancy. Fortunately, risk governance bodies and management agencies around the world are preparing different approaches to an extended evaluation process. Their work is vital to the ultimate success of risk governance in the modern world, because gaining a better understanding of the social and psychological criteria that affect human reactions to risk is a crucial requirement for the characterization and evaluation of systemic risk.

RISK EVALUATION

A heavily disputed task in the risk governance process concerns the procedure for evaluating the societal acceptability or tolerability of a risk. In classical approaches, risks are ranked and prioritized based on a combination of probability (how likely is it that the risk will be realized?) and impact (what are the consequences if the risk does occur?) (Klinke and Renn 2002; Klinke and Renn 2012; Renn 2008c: 149ff.). However, as described above, in situations of uncertainty, complexity, and ambiguity, risks cannot be treated only in terms of likelihood (probability) and (quantifiable) impacts. This standard two-dimensional model ignores many important features of risk. Values and issues such as reversibility, persistence, ubiquity, equity, catastrophic potential, controllability, and voluntariness need to be integrated into risk evaluation. Furthermore, risk-related decision making is neither about risks alone nor usually about a single risk. Evaluation requires risk-benefit evaluations and risk-risk tradeoffs. So by definition, risk evaluation is multidimensional. To evaluate risks, the first step is to characterize the risks on all of the dimensions that matter to the affected populations. Once the risks are characterized in a multidimensional profile, their acceptability can be assessed.

Furthermore, there are competing, legitimate viewpoints over evaluations, over whether there are or could be adverse effects, and if there are, whether these risks are tolerable or even acceptable. As noted above, drawing the lines between "acceptable," "tolerable," and "intolerable" risks is one of the most controversial and challenging tasks in the risk governance process. The U.K. Health and Safety Executive developed a procedure for chemical risks based on risk-risk comparisons (Löfstedt 1997). Some Swiss cantons such as Basel-Landschaft experimented with roundtables consisting of industry, administrators, county officials, environmentalists, and neighborhood groups. As a means for reaching consensus, two demarcation lines were drawn between the area of tolerable and acceptable risk and between acceptable and intolerable risks. Irrespective of the selected means to support this task, the judgment on acceptability or tolerability is contingent on making use of a variety of different knowledge sources. In other words, it requires taking the interdisciplinary risk estimation seriously.

Risk evaluations, an epistemological issue, generally rely on causal and principal presuppositions as well as worldviews (cf. Goldstein and Keohane 1993). Causal beliefs refer to the scientific evidence from risk assessment on whether, how, and to what extent the risk might cause harm. This dimension emphasizes cause-effect relations and provides guidance for which strategy is most appropriate to meet the goal of risk avoidance, risk reduction, or adaptation. But risks typically embed normative issues, too. Looming below all risks is the question of how safe is safe enough, implying a normative or moral judgment about acceptability of risk and the tolerable burden that risk producers can impose on others. The results of the assessment can provide hints about what kind of mental images are present and which moral judgments guide people's perceptions and choices. Of particular importance is the perception of just or unjust distribution of risks and benefits. How these moral judgments are made and justified depends to a large degree on social position, cultural values, and worldviews. The judgments also depend on shared ontological (a belief in the state of the world) and ethical (a belief of how the world should be) convictions. This collection of forces influences thinking and evaluation strategies. The selection of strategies for risk management therefore is understandable only within the context of broader worldviews. Hence, society can never derive acceptability or tolerability from the assessment evidence alone. Facts do not speak for themselves. Furthermore, the evidence is essential not only to reflect the degrees of belief about the state of the world regarding a particular risk, but also to know whether a value might be violated or not (or to what degree).

In sum, risk evaluation involves the deliberative effort to characterize risks in terms of acceptability and tolerability in a context of uncertainty, ambiguity, and complexity. Such contexts often imply that neither the risks nor the benefits can be clearly identified. Multiple dimensions and multiple values are always in play and must be considered. Finally, risk evaluations may shift over time. But the harsh reality is this: notwithstanding uncertainty, complexity, and ambiguity, decisions must be made. It may well be possible at a certain point in time to agree whether risks are acceptable, tolerable, or intolerable. When the tolerability or acceptability of risks is heavily contested, that, too, is highly relevant input in the decision-making process.

RISK MANAGEMENT

Risk management starts with a review of the output generated in the previous phases of interdisciplinary risk estimation, characterization, and risk evaluation. If the risk is acceptable, no further management is needed. Tolerable risks are those for which the benefits are judged to be worth the risk but risk reduction measures are necessary. If risks are classified as tolerable, risk management needs to design and implement actions that either render these risks acceptable or sustain that tolerability in the long run by introducing risk reduction strategies, mitigation strategies, or strategies aimed at increasing societal resilience at the appropriate level. If the risk is considered intolerable, notwithstanding the benefits, risk management should be focused on banning or phasing out the activity creating the risk. If that is not possible, management should be devoted to eliminating or mitigating the risk in other ways or to increasing societal resilience. If the risk is contested, risk management can aim at finding ways to create consensus. If that is impossible or highly unlikely, the goal would be to design actions that increase tolerability among the parties most concerned or to stimulate alternative course of action.

A variety of ways exist to design the process of identifying risk management options in contexts of uncertainty, complexity, and ambiguity (Klinke and Renn 2002; Renn 2008c: 173ff.). In those situations, routine risk management within risk assessment agencies and regulatory institutions is inappropriate for this category, since the risk problems are not sufficiently known or are contested. Klinke and Renn (2002) have developed a decision tree for the various combinations of these contextual factors. In a case where complexity is dominant and uncertainty and ambiguity are low, the challenge is to invite experts to deliberate with risk managers to understand complexity. Flood risk management may be a good example of this. Although the occurrence of certain types of floods follows a random pattern, one can address vulnerability and design emergency management actions well in advance. The major challenge is to determine the limit to which one is willing to invest in resilience. And once the complexity is well understood, it is a question of political will to implement the desired level of protection. Heiko Garrelts and Hellmuth Lange (2011: 207–208), for example, emphasize the need for state decisiveness in such cases: "for all the indispensability of participatory approaches—for reasons of integrating citizen's expertise, for reasons of the additional need for legitimacy in face of existing future uncertainty—it is the state that remains the institutional guarantor for ensuring that problems can be addressed from diverging perspectives. The ability of state agencies to intervene with sanctions and directives addresses the question of ultimate responsibility, which is all too often overlooked by participation oriented approaches."

In cases of high complexity, low uncertainty, and low ambiguity, Garrelts and Lange (2011) suggest a reversal from governance to government. We do not argue that this conclusion holds for all risk management situations. It is conceivable that the strategy is inapplicable to situations, such as flooding, where the effects of climate change complicate the matter or where societal actors resist particular flood risk management options, such as higher dikes or the dismantling of settlements in flood plains, for aesthetic or cultural reasons (cf. Wisner et al. 2004).

The second node in the Klinke and Renn (2002) decision tree concerns risk problems that are characterized by high uncertainty but low ambiguity. They argue that expanded knowledge acquisition may help to reduce uncertainty. If uncertainty cannot be reduced (or can be reduced only in the long run) by additional knowledge, however, they argue for what they call "precaution-based risk management." Precaution-based risk management explores a variety of options: containment, diversification, monitoring, and substitution. The focal point here is to find an adequate and fair balance between over-cautiousness and insufficient caution. This argues for a reflective process involving stakeholders to ponder concerns, economic budgeting, and social evaluations.

For risk problems that are highly ambiguous (regardless of whether they are low or high on uncertainty and complexity), the Klinke and Renn (2002) decision tree recommends "discourse-based management." Discourse management requires a participatory process involving stakeholders, especially the affected public. The aim of such a process is to produce a collective understanding among all stakeholders and the affected public about how to interpret the situation and how to design procedures to collectively justify binding decisions on acceptability and tolerability that are considered legitimate. In such situations, the task of risk managers is to create a condition where those who believe that the risk is worth taking and those who believe otherwise are willing to respect each other's views and to construct and create strategies acceptable to the various stakeholders and interests. But deliberation is not a guarantee for a smooth risk management process. Rolf Lidskog, Ylva Uggla, and Linda Soneryd (2011) argue that complexity and ambiguity are grounds for continuous conflict that is difficult, if not impossible, to resolve. The reduction of complexity simultaneously implies reducing the number of actors as relevant or legitimate participants. The resolution of ambiguity requires broad representation of all actors involved in the case, so it is difficult to find the perfect or, perhaps, even optimal path between functionality and inclusiveness. In any case, our response to this inherent conflict is to invest in structuring an effective and efficient process of inclusion (whom to include) and closure (what counts as evidence and the adopted decision-making rules) (Aven and Renn 2010: 181ff.; Renn 2008c: 284ff.).

In sum, neither the characterization of the systemic risk at hand (uncertain, complex, or ambiguous) nor the contingent evaluation of the risk (acceptable, tolerable, intolerable, disputed) results in a simple typology for risk management. Nevertheless, decisions must be made, for even non-decisions about risk are decisions—often with far-reaching consequences. Characterizations and evaluations of systemic risks do provide some guidance for risk management. They provide guidance about how to design a process that appears to be sensible, that prioritizes risks, and that delineates reasonable options for different contexts. It is clear that the traditional risk management style, resting on a two-dimensional assessment model, is not just inadequate to address systemic risks, but also may exacerbate the governance challenge by fueling societal controversies over risk.

RISK COMMUNICATION AND PARTICIPATION

Effective communication across all relevant interests is one of the key challenges in risk governance. It is not a distinct stage (in contrast to how it is often treated in the risk literature) but central to the entire governance process. Positively framed, communication is at the core of any successful risk governance activity. Negatively framed, a lack of communication destroys risk governance. Early on, risk communication was predicated on the view that disagreements between experts and citizens over risks were due to the lack of accurate knowledge by citizens. The solution was sought in the so-called deficit model of communication and engagement. The solution to the gap in knowledge between experts and laypeople was education and persuasion of the deficient public (Fischhoff 1995). Implied in this solution was the belief that an educated public would perceive and evaluate risks the same way as experts. However, this deficit model has been subjected to considerable criticism. For one thing, increased knowledge often elevated citizens' concerns about risk, creating an even greater diversion between citizens and experts. For another, as Nick Pidgeon and his colleagues (2005:467) aptly phrased it, "One of the most consistent messages to have arisen from social science research into risk over the past 30 years is that risk communication …

needs to accommodate far more that an simple one-way transfer of information. ... [T]he mere provision of 'expert' information is unlikely to address public and stakeholder concerns or resolve any underlying societal issues." Third, research on risk controversies has demonstrated that in general, the public does not always misunderstand science. Furthermore, experts and governments may also misunderstand public perceptions (Horlick-Jones 1998; Irwin and Wynne 1996).

The important point to emphasize is that risk communication and trust are delicately interconnected processes. A large volume of literature demonstrates the connection between trust in the institutions managing risks and citizens' perceptions of the seriousness of risks (Earle and Cvetkovich 1996; Luhmann 1980; Poortinga and Pidgeon 2003; Whitfield et al. 2009). Communication breakdowns can easily damage trust. However, communication strategies that misjudge the context of communication, in terms of the level of and reasons for distrust, may boomerang, resulting in increased distrust (Löfstedt 2005).

Communication strategies proliferate. We define communication as meaningful interactions in which knowledge, experiences, interpretations, concerns, and perspectives are exchanged (cf. Löfstedt 2003). Communication in the context of risk governance refers to exchanges among policymakers, experts, stakeholders, and affected publics. The aim of communication is to provide a better basis for responsible governing of uncertain, complex, or ambiguous risks. Its aim also is to enhance trust and social support. Depending on the nature of the risks and the context of governing choices, communication will serve various purposes. It might serve to share information about the risks and possible ways to handle them. It might support building and sustaining trust among various actors where particular arrangements or risk management measures become acceptable. It might result in actually engaging people in risk-related decisions, through which they gain ownership of the problem.

However, communication in the context of risk governance is not simple. It is not just a matter of having accurate assessments of risks. It is not just a matter of bringing people together. It is not just a matter of effective communication. It requires all of these features and more. Also required is a set of procedures for facilitating the discourse among various actors from different backgrounds so they can interact meaningfully in the face of uncertainty, complexity, or ambiguity.

Meaningful communication features multiple actors. The highly Influential U.S. National Research Council report (Stern and Fineberg 1996) is an important milestone in the recognition of the need for risk decision making as an inclusive multi-actor process. It also was a germinal precursor to the idea of risk governance with an emphasis on the coordination of risk knowledge and expertise with citizens' and other stakeholders' priorities. Scholars subscribing to the idea of risk governance share its procedures and normative position that the involvement of interested and affected parties in collective decision making about risk is needed (see, e.g., De Marchi 2003; Irwin 2008; Jasanoff 2004; Rosa and Clark 1999; Stirling 2007).

One key challenge to risk governance is that of inclusion: which stakeholders and publics should be included in governance deliberations? The inclusion challenge has deep implications. Contrary to the conventional paradigm in which risk topics are usually identified by experts, with the analytic-deliberative process underpinning risk governance, public values and social concerns are key agents for identifying and prioritizing risk topics. Inclusion means more than simply including relevant actors. That is the outmoded practice of "public hearings," where relevant actors are accorded a fairly passive role. Inclusion means that actors play a key role in framing (or pre-assessing) the risk (International Risk Governance Council 2005; Renn and Schweizer 2009; see also Roca, Gamboa, and Tàbara 2008). Inclusion should be open to input from civil society and adaptive at the same time (Stirling 2004). Crucial questions in this respect are, Who is included? What is included? What is the scope and mandate of the process? (see also Renn and Schweizer 2009),

Inclusion can take many different forms: roundtables, open forums, negotiated rulemaking exercises, mediation, or mixed advisory committees consisting of scientists and stakeholders (Renn 2008c:332ff.; Rowe and Frewer 2000; Stoll-Kleemann and Welp 2006). Due to a lack of agreement on method, social learning promoted by structured and moderated deliberations is required to find out what level and type of inclusion is appropriate in the particular context and for the type of risk involved. The methods that are available have contrasting strengths and weaknesses (Pidgeon et al. 2005).

A focus on inclusion is defended on several grounds (cf. Roca, Gamboa, and Tàbara 2008). First, we argue that in view of uncertainty, complexity, or ambiguity, there is a need to explore various sources of information and to incorporate various perspectives. It is important to know what the various actors label as risk problems and which most concern them. Here inclusion is interpreted to be a means to an end—a procedure for integrating all relevant knowledge and for including of all relevant concerns. Second, from a democratic perspective, actors affected by the risks or the ways in which the risks are governed have a legitimate right to participate in deciding about those risks. Here inclusion is interpreted not just as a means but also as an end in itself. At the same time, inclusion is a means to agree on principles and rules that should be respected in the processes and structures of collective decision making. Third, the more actors who are involved in the weighing of the heterogeneous pros and cons of risks, the more socially robust the outcome will be. When uncertainty, complexity, or ambiguity reigns, there is no simple decision rule. In that view, inclusion also is a way to organize checks and balances between various interest and value groups in a plural society. Hence, inclusion is intended to support the coproduction of risk knowledge, the coordination of risk evaluation, and the design of risk management.

Social learning is also required here. And it is not simply a matter of degree, where more inclusion equals more learning and therefore better risk governance. The degree and type of inclusion may vary depending on the phase of governance process and the risk context. In each phase and context, careful thought is needed about the kind and degree of inclusion necessary. Hence, differentiation is not an exception, It is the rule.

The task of inclusion is to organize productive and meaningful communication among a range of actors who have divergent interests but complementary roles. The cumulative empirical analyses suggest that providing a platform for the inclusion of a variety of stakeholders—to deliberate over their concerns and exchange arguments—can help to deescalate conflicts and legitimize final decisions. Nevertheless, however careful the establishment of the platform and the decision rule about inclusion, there will always be some disenfranchised or disappointed actors in society (Beierle and Cayford 2002; National Research Council 2008).

CONCLUSION

In this chapter, we have explored the genesis and analytical scope of risk governance. We have argued for a broader, paradigmatic turn from government to governance. We argued that in the context of risk, the idea governance is conceptualized as both a descriptive and normative activity: as a description of how decisions are made and as a normative model for improving structures and processes of risk decisions. Risk governance draws attention to the fact that many risks are not simple; they cannot all be calculated as a function of probability and effect or consequence. Many risks embed complex tradeoffs of costs and benefits. Systemic risks are characterized by their complexity, uncertainty, and ambiguity. Risk governance underscores the need to ensure that societal choices and decisions adequately address these complicating features. However, conventional risk characterization typically treats, assesses, and manages such risks as if they were simple. This practice has led to many failures to deal adequately with risks such as genetic engineering, nuclear energy, the global financial crisis, global warming, and cyber-terrorism, demonstrating an urgent need to develop alternative concepts and approaches to deal with uncertain, complex, or ambiguous risks. We have argued that a risk governance framework is the promising alternative.

We have also outlined the idea of adaptive and integrative risk governance, organizing governance into five phases: pre-estimation, interdisciplinary risk estimation, risk characterization, risk evaluation, and risk management. Each phase entails communication processes and public involvement. We also outlined the challenges that confront each phase.

Our analysis demonstrates that risk governance is not simply an opportunistic buzzword, but a disciplined argument for a paradigm shift. Paradigms and reforms shift not just in the abstract but also in practice. Such fundamental transitions are not easy. Yet we hope that, by combining the insights of risk theory with an argument for governance, this volume helps to stimulate and facilitate that shift. We also hope it will nudge a change in risk practice.

To lean on an old saw, the proof of any pudding is in the eating. In this chapter, we have laid out the recipe for a new pudding: a system of governance that both reflects underlying changes in how societies are attempting to manage risks and one that supersedes the government approach of the past. The eating can only come when that recipe is put to use. In Chapter 10, we describe an analytic-deliberative procedure for accomplishing just that.

NOTES

1 The possible range of governance often has been provocatively termed *governance by government, governance with government*, and *governance without government* (Rosenau 1992. 1995), which emphasizes the decreasing role of the nation-state.

2 Sec the review of literature in van Asselt and Renn 2011.

3 Ambiguity is a condition in which the presence of a risk is unclear, its scope is ill-defined, and it is not clear whether the risk is serious enough to characterize and manage. Ambiguity also refers to the plurality of legitimate viewpoints for evaluating decision outcomes and justifying judgments about their tolerability and acceptability. So ambiguity refers to a context of vagueness and the existence of multiple values and perspectives.

420

4 Some like-minded authors prefer to reconceptualize risk in a way that renders the addition of "uncertain" superfluous. For example. Terje Aven and Ortwin Renn (2009a: 2) suggest that risk be redefined as a reference to "uncertainty about and severity of the consequences (or out-comes) of an activity with respect to something that humans value" (see also Rosa 2003, 2010). We agree with such definitions and use them, as well.
5 These spillover effects have been the main target of the social amplification of risk frame-work. This theory was developed by a research team at Clark University in the late 1980s (e.g., Kasperson et al. 1988) and has been shown to be effective across a range of applications (Rosa 2003).

Garrett Hardin — "The Tragedy of the Commons." *Science Journal.*

Reading Commentary

In this seminal article, Hardin argues that we have exceeded the earth's carrying capacity due to overpopulation. While this may seem reminiscent of Thomas Malthus's focus on the danger of unchecked population growth, Hardin (somewhat grudgingly) recognizes the futility of trying to impose population controls. He therefore shifts his focus to natural resource management.

Hardin draws from an older essay, published in 1833 by economist William Forster Lloyd, describing the negative effects of unregulated grazing on the commons (i.e., public land). Hardin uses Lloyd's illustration of the "tragedy of the commons" to demonstrate how decisions made by rational individuals in pursuit of their self-interest (e.g., herders allowing their cattle to graze on public lands without restraint) can lead to outcomes that are not in the collective best interest:

> *[t]herein is the tragedy. Each man is locked into a system that compels him to increase his herd without limit—in a world that is limited [...] Freedom in a commons brings ruin to all.*

The "commons" has come to represent any shared or unregulated resource. Hardin is not arguing against shared resources. Rather, he believes we cannot afford to leave them unregulated. How does Hardin suggest we regulate shared resources? Not through privatization or appeals to the conscience, but through "mutual coercion," or regulation that we all agree is in our collective best interest.

Hardin's article is not without controversy. Despite his emphasis on the need for regulation, he concludes that the tragedy of the commons can only be ended through population control. He argues in this article that welfare policies and global principles developed to ensure just outcomes (like the United Nations Universal Declaration of Human Rights) are against our collective interest because they will inadvertently lead to population growth.[1] Interestingly, he recognizes the inherent injustice of limiting some people's access to resources, but asserts that "injustice is preferable to total ruin."

We have included this provocative essay not only as a challenge to you as you develop your personal theory of environmental problem-solving but also because Hardin's essay has come to represent a call for strong (state and/or market) regulation of common-pool resources. Although never explicitly mentioned, his argument for regulation is very much in line with the Hobbesian argument for a *Leviathan*—in both cases, people agree (by entering into a social contract according to Thomas Hobbes, or through mutual coercion according to Hardin) to strong regulation by a central authority.

THE TRAGEDY OF THE COMMONS

Garrett Hardin

The population problem has no technical solution; it requires a fundamental extension in morality.

Garrett Hardin

At the end of a thoughtful article on the future of nuclear war, Wiesner and York (1) concluded that: "Both sides in the arms race are ... confronted by the dilemma of steadily increasing military power and steadily decreasing national security. *It is our considered professional judgment that this dilemma has no technical solution.* If the great powers continue to look for solutions in the area of science and technology only, the result will be to worsen the situation."

I would like to focus your attention not on the subject of the article (national security in a nuclear world) but on the kind of conclusion they reached, namely that there is no technical solution to the problem. An implicit and almost universal assumption of discussions published in professional and semipopular scientific journals is that the problem under discussion has a technical solution. A technical solution may be defined as one that requires a change only in the techniques of the natural sciences, demanding little or nothing in the way of change in human values or ideas of morality.

In our day (though not in earlier times) technical solutions are always welcome. Because of previous failures in prophecy, it takes courage to assert that a desired technical solution is not possible. Wiesner and York exhibited this courage; publishing in a science journal, they insisted that the solution to the problem was not to be found in the natural sciences. They cautiously qualified their statement with the phrase, "It is our considered professional judgment. ..." Whether they were right or not is not the concern of the present article. Rather, the concern here is with the important concept of a class of human problems which can be called "no technical solution problems," and, more specifically, with the identification and discussion of one of these.

It is easy to show that the class is not a null class. Recall the game of tick-tack-toe. Consider the problem, "How can I win. the game of tick-tack-toe?" It is well known that I cannot, if I assume (in keeping with the conventions of game theory) that my opponent understands the game perfectly. Put another way, there is no "technical solution" to the problem. I can win only by giving a radical meaning to the word "win." I can hit my opponent over the head; or I can drug him; or I can falsify the records. Every way in which I "win" involves, in some sense, an abandonment of the game, as we intuitively understand it. (I can also, of course, openly abandon the game—refuse to play it. This is what most adults do.)

The class of "No technical solution problems" has members. My thesis is that the "population problem," as conventionally conceived, is a member of this class. How it is conventionally conceived needs some comment. It is fair to say that most people who anguish over the population problem are trying to find a way to avoid the evils of overpopulation without relinquishing any of the privileges they now enjoy. They think that farming the seas or developing new strains of wheat will solve the problem—technologically. I try to show here that the solution they seek cannot be found. The population problem cannot be solved in a technical way, any more than can the problem of winning the game of tick-tack-toe.

WHAT SHALL WE MAXIMIZE?

Population, as Malthus said, naturally tends to grow "geometrically," or, as we would now say, exponentially. In a finite world this means that the per capita share of the world's goods must steadily decrease. Is ours a finite world?

A fair defense can be put forward for the view that the world is infinite; or that we do not know that it is not. But, in terms of the practical problems that we must face in the next few generations with the foreseeable technology, it is clear that we will greatly increase human misery if we do not, during the immediate future, assume that the world available to the terrestrial human population is finite. "Space" is no escape (2).

A finite world can support only a finite population; therefore, population growth must eventually equal zero. (The case of perpetual wide fluctuations above and below zero is a trivial variant that need not be discussed.) When this condition is met, what will be the situation of mankind? Specifically, can Bentham's goal of "the greatest good for the greatest number" be realized?

No—for two reasons, each sufficient by itself. The first is a theoretical one. It is not mathematically possible to maximize for two (or more) variables at the same time. This was clearly stated by von Neumann and Morgenstern (5), but the principle is implicit in the theory of partial differential equations, dating back at least to D'Alembert (1717–1783).

The second reason springs directly from biological facts. To live, any organism must have a source of energy (for example, food). This energy is utilized for two purposes: mere maintenance and work. For man, maintenance of life requires about 1600 kilocalories a day ("maintenance calories"). Anything that he does over and above merely staying alive will be defined as work, and is supported by "work calories" which he takes in. Work calories are used not only for what we call work in common speech; they are also required for all forms of enjoyment, from swimming and automobile racing to playing music and writing poetry. If our goal is to maximize population it is obvious what we must do: We must make the work calories per person approach as close to zero as possible. No gourmet meals, no vacations, no sports, no music, no literature, no art ... I think that everyone will grant, without argument or proof, that maximizing population does not maximize goods. Bentham's goal is impossible.

In reaching this conclusion I have made the usual assumption that it is the acquisition of energy that is the problem. The appearance of atomic energy has led some to question this assumption. However, given an infinite source of energy, population growth still produces an inescapable problem. The problem of the acquisition of energy is replaced by the problem of its dissipation, as J. H. Fremlin has so wittily shown (4). The arithmetic signs in the analysis are, as it were, reversed; but Bentham's goal is still unobtainable.

The optimum population is, then, less than the maximum. The difficulty of defining the optimum is enormous; so far as I know, no one has seriously tackled this problem. Reaching an acceptable and stable solution will surely require more than one generation of hard analytical work—and much persuasion.

We want the maximum good per person; but what is good? To one person it is wilderness, to another it is ski lodges for thousands. To one it is estuaries to nourish ducks for hunters to shoot; to another it is factory land. Comparing one good with another is, we usually say, impossible because goods are incommensurable. Incommensurables cannot be compared.

Theoretically this may be true; but in real life incommensurables *are* commensurable. Only a criterion of judgment and a system of weighting are needed. In nature the criterion is survival. Is it better for a species to be small and hideable, or large and powerful? Natural selection commensurates the incommensurables. The compromise achieved depends on a natural weighting of the values of the variables.

Man must imitate this process. There is no doubt that in fact he already does, but unconsciously. It is when the hidden decisions are made explicit that the arguments begin. The problem for the years ahead is to work out an acceptable theory of weighting. Synergistic effects, nonlinear variation, and difficulties in discounting the future make the intellectual problem difficult, but not (in principle) insoluble.

Has any cultural group solved this practical problem at the present time, even on an intuitive level? One simple fact proves that none has: there is no prosperous population in the world today that has, and has had for some time, a growth rate of zero. Any people that has intuitively identified its optimum point will soon reach it, after which its growth rate becomes and remains zero.

Of course, a positive growth rate might be taken as evidence that a population is below its optimum. However, by any reasonable standards, the most rapidly growing populations on earth today are (in general) the most miserable. This association (which need not be invariable) casts doubt on the optimistic assumption that the positive growth rate of a population is evidence that it has yet to reach its optimum.

We can make little progress in working toward optimum poulation size until we explicitly exorcize the spirit of Adam Smith in the field of practical demography. In economic affairs, *The Wealth of Nations* (1776) popularized the "invisible hand," the idea that an individual who "intends only his own gain," is, as it were, "led by an invisible hand to promote … the public interest" (5). Adam Smith did not assert that this was invariably true, and perhaps neither did any of his followers. But he contributed to a dominant tendency of thought that has ever since interfered with positive action based on rational analysis, namely, the tendency to assume that decisions reached individually will, in fact, be the best decisions for an entire society. If this assumption is correct it justifies the continuance of our present policy of laissez-faire in reproduction. If it is correct we can assume that men will control their individual fecundity so as to produce the optimum population. If the assumption is not correct, we need to reexamine our individual freedoms to see which ones are defensible.

TRAGEDY OF FREEDOM IN A COMMONS

The rebuttal to the invisible hand in population control is to be found in a scenario first sketched in a little-known pamphlet (6) in 1833 by a mathematical amateur named William Forster Lloyd (1794–1852). We may well call it "the tragedy of the commons," using the word "tragedy" as the philosopher Whitehead used it (7): "The essence of dramatic tragedy is not unhappiness. It resides in the solemnity of the remorseless working of things." He then goes on to say, "This inevitableness of destiny can only be illustrated in terms of human life by incidents which in fact in volve unhappiness. For it is only by them that the futility of escape can be made evident in the drama."

The tragedy of the commons develops in this way. Picture a pasture open to all. It is to be expected that each herdsman will try to keep as many cattle as possible on the commons. Such an arrangement may work reasonably satisfactorily for centuries because tribal wars, poaching, and disease keep the numbers of both man and beast well below the carrying capacity of the land. Finally, however, comes the day of reckoning, that is, the day when the long-desired goal of social stability becomes a reality. At this point, the inherent logic of the commons remorselessly generates tragedy.

As a rational being, each herdsman seeks to maximize his gain. Explicitly or implicitly, more or less consciously, he asks, "What is the utility *to me* of adding one more animal to my herd?" This utility has one negative and one positive component.

1) The positive component is a function of the increment of one animal. Since the herdsman receives all the proceeds from the sale of the additional animal, the positive utility is nearly +1.

2) The negative component is a function of the additional overgrazing created by one more animal. Since, however, the effects of overgrazing are shared by all the herdsmen, the negative utility for any particular decisionmaking herdsman is only a fraction of − 1.

Adding together the component partial utilities, the rational herdsman concludes that the only sensible course for him to pursue is to add another animal to his herd. And another; and another. … But this is the conclusion reached by each and every rational herdsman sharing a commons. Therein is the tragedy. Each man is locked into a system that compels him to increase his herd without limit—in a world that is limited. Ruin is the destination toward which all men rush, each pursuing his own best interest in a society that believes in the freedom of the commons. Freedom in a commons brings ruin to all.

Some would say that this is a platitude. Would that it were! In a sense, it was learned thousands of years ago, but natural selection favors the forces of psychological denial (8). The individual benefits as an individual from his ability to deny the truth even though society as a whole, of which he is a part, suffers.

Education can counteract the natural tendency to do the wrong thing, but the inexorable succession of generations requires that the basis for this knowledge be constantly refreshed.

A simple incident that occurred a few years ago in Leominster, Massachusetts, shows how perishable the knowledge is. During the Christmas shopping season the parking meters downtown were covered with plastic bags that bore tags reading: "Do not open until after Christmas. Free parking courtesy of the mayor and city council." In other words, facing the prospect of an increased demand for already scarce space, the city fathers reinstituted the system of the commons. (Cynically, we suspect that they gained more votes than they lost by this retrogressive act.)

In an approximate way, the logic of the commons has been understood for a long time, perhaps since the discovery of agriculture or the invention of private property in real estate.

But it is understood mostly only in special cases which are not sufficiently generalized. Even at this late date, cattlemen leasing national land on the western ranges demonstrate no more than an ambivalent understanding, in constantly pressuring federal authorities to increase the head count to the point where over – grazing produces erosion and weed – dominance. Likewise, the oceans of the world continue to suffer from the survival of the philosophy of the commons. Maritime nations still respond automatically to the shibboleth of the "freedom of the seas." Professing to believe in the "inexhaustible resources of the oceans," they bring species after species of fish and whales closer to extinction (9).

The National Parks present another instance of the working out of the tragedy of the commons. At present, they are open to all, without limit. The parks themselves are limited in extent—there is only one Yosemite Valley—whereas population seems to grow without limit. The values that visitors seek in the parks are steadily eroded. Plainly, we must soon cease to treat the parks as commons or they will be of no value to anyone.

What shall we do? We have several options. We might sell them off as private property. We might keep them as public property, but allocate the right to enter them. The allocation might be on the basis of wealth, by the use of an auction system. It might be on the basis of merit, as defined by some agreed upon standards. It might be by lottery. Or it might be on a first-come, first – served basis, administered to long queues. These, I think, are all the reasonable possibilities. They are all objectionable. But we must choose—or acquiesce in the destruction of the commons that we call our National Parks.

POLLUTION

In a reverse way, the tragedy of the commons reappears in problems of pollution. Here it is not a question of taking something out of the commons, but of putting something in—sewage, or chemical, radioactive, and heat wastes into water; noxious and dangerous fumes into the air; and distracting and unpleasant advertising signs into the line of sight. The calculations of utility are much the same as before. The rational man finds that his share of the cost of the wastes he discharges into the commons is less than the cost of purifying his wastes before releasing them. Since this is true for everyone, we are locked into a system of "fouling our own nest," so long as we behave only as independent, rational, free-enterprisers.

The tragedy of the commons as a food basket is averted by private property, or something formally like it. But the air and waters surrounding us cannot readily be fenced, and so the tragedy of the commons as a cesspool must be prevented by different means, by coercive laws or taxing devices that make it cheaper for the polluter to treat his pollutants than to discharge them untreated. We have not progressed as far with the solution of this problem as we have with the first. Indeed, our particular concept of private property, which deters us from exhausting the positive resources of the earth, favors pollution. The owner of a factory on the bank of a stream—whose property extends to the middle of the stream—often has difficulty seeing why it is not his natural right to muddy the waters flowing past his door. The law, always behind the times, requires elaborate stitching and fitting to adapt it to this newly perceived aspect of the commons.

The pollution problem is a consequence of population. It did not much matter how a lonely American frontiersman disposed of his waste. "Flowing water purifies itself every 10 miles," my grandfather used to say, and the myth was near enough to the truth when he was a boy, for there were not too many people. But as population became denser, the natural chemical and biological recycling processes became overloaded, calling for a redefinition of property rights.

HOW TO LEGISLATE TEMPERANCE?

Analysis of the pollution problem as a function of population density uncovers a not generally recognized principle of morality, namely: *the morality of an act is a function of the state of the system at the time it is performed* (10). Using the commons as a cesspool does not harm the general public under frontier conditions, because there is no public; the same behavior in a metropolis is unbearable. A hundred and fifty years ago a plainsman could kill an American bison, cut out only the tongue for his dinner, and discard the rest of the animal. He was not in any important sense being wasteful. Today, with only a few thousand bison left, we would be appalled at such behavior.

In passing, it is worth noting that the morality of an act cannot be determined from a photograph. One does not know whether a man killing an elephant or setting fire to the grassland is harming others until one knows the total system in which his act appears. "One picture is worth a thousand words," said an ancient Chinese; but it may take 10,000 words to validate it. It is as tempting to ecologists as it is to reformers in general to try to persuade others by way of the photographic shortcut. But the essense of an argument cannot be photographed: it must be presented rationally—in words.

That morality is system-sensitive escaped the attention of most codifiers of ethics in the past. "Thou shalt not ..." is the form of traditional ethical directives which make no allowance for particular circumstances. The laws of our society follow the pattern of ancient ethics, and therefore are poorly suited to governing a complex, crowded, changeable world. Our epicyclic solution is to augment statutory law with administrative law. Since it is practically impossible to spell out all the conditions under which it is safe to burn trash in the back yard or to run an automobile without smog-control, by law we delegate the details to bureaus. The result is administrative law, which is rightly feared for an ancient reason—*Quis custodiet ipsos custodes?*—"Who shall watch the watchers themselves?" John Adams said that we must have "a government of laws and not men." Bureau administrators, trying to evaluate the morality of acts in the total system, are singularly liable to corruption, producing a government by men, not laws.

Prohibition is easy to legislate (though not necessarily to enforce); but how do we legislate temperance? Experience indicates that it can be accomplished best through the mediation of administrative law. We limit possibilities unnecessarily if we suppose that the sentiment of *Quis custodiet* denies us the use of administrative law. We should rather retain the phrase as a perpetual reminder of fearful dangers we cannot avoid. The great challenge facing us now is to invent the corrective feedbacks that are needed to keep custodians honest. We must find ways to legitimate the needed authority of both the custodians and the corrective feedbacks.

FREEDOM TO BREED IS INTOLERABLE

The tragedy of the commons is involved in population problems in another way. In a world governed solely by the principle of "dog eat dog"—if indeed there ever was such a world—how many children a family had would not be a matter of public concern. Parents who bred too exuberantly would leave fewer descendants, not more, because they would be unable to care adequately for their children. David Lack and others have found that such a negative feedback demonstrably controls the fecundity of birds (11). But men are not birds, and have not acted like them for millenniums, at least.

If each human family were dependent only on its own resources; *if* the children of improvident parents starved to death; it, thus, overbreeding brought its own "punishment" to the germ line—*then* there would be no public interest in controlling the breeding of families. But our society is deeply committed to the welfare state (12), and hence is confronted with another aspect of the tragedy of the commons.

In a welfare state, how shall we deal with the family, the religion, the race, or the class (or indeed any distinguishable and cohesive group) that adopts overbreeding as a policy to secure its own aggrandizement (13)? To couple the concept of freedom to breed with the belief that everyone born has an equal right to the commons is to lock the world into a tragic course of action.

Unfortunately this is just the course of action that is being pursued by the United Nations. In late 1967, some 30 nations agreed to the following (14):

> The Universal Declaration of Human Rights describes the family as the natural and fundamental unit of society. It follows that any choice and decision with regard to the size of the family must irrevocably rest with the family itself, and cannot be made by anyone else.

It is painful to have to deny categorically the validity of this right; denying it, one feels as uncomfortable as a resident of Salem, Massachusetts, who denied the reality of witches in the 17th century. At the present time, in liberal quarters, something like a taboo acts to inhibit criticism of the United Nations. There is a feeling that the United Nations is "our last and best hope," that we shouldn't find fault with it; we shouldn't play into the hands of the archconservatives. However, let us not forget what Robert Louis Stevenson said: "The truth that is suppressed by friends is the readiest weapon of the enemy." If we love the truth we must openly deny the validity of the Universal Declaration of Human Rights, even though it is promoted by the United Nations. We should also join with Kingsley Davis (15) in attempting to get Planned Parenthood-World Population to see the error of its ways in embracing the same tragic ideal.

CONSCIENCE IS SELF-ELIMINATING

It is a mistake to think that we can control the breeding of mankind in the long run by an appeal to conscience. Charles Galton Darwin made this point when he spoke on the centennial of the publication of his grandfather's great book. The argument is straightforward and Darwinian.

People vary. Confronted with appeals to limit breeding, some people will undoubtedly respond to the plea more than others. Those who have more children will produce a larger fraction of the next generation than those with more susceptible consciences. The difference will be accentuated, generation by generation.

In C. G. Darwin's words: "It may well be that it would take hundreds of generations for the progenitive instinct to develop in this way, but if it should do so, nature would have taken her revenge, and the variety *Homo contracipiens* would become extinct and would be replaced by the variety *Homo progenitivus*" (16).

The argument assumes that conscience or the desire for children (no matter which) is hereditary—but hereditary only in the most general formal sense. The result will be the same whether the attitude is transmitted through germ cells, or exosomatically, to use A. J. Lotka's term. (If one denies the latter possibility as well as the former, then what's the point of education?) The argument has here been stated in the context of the population problem, but it applies equally well to any instance in which society appeals to an individual exploiting a commons to restrain himself for the general good—by means of his conscience. To make such an appeal is to set up a selective system that works toward the elimination of conscience from the race.

PATHOGENIC EFFECTS OF CONSCIENCE

The long-term disadvantage of an appeal to conscience should be enough to condemn it; but has serious short-term disadvantages as well. If we ask a man who is exploiting a commons to desist "in the name of conscience," what are we saying to him? What does he hear?—not only at the moment but also in the wee small hours of the night when, half asleep, he remembers not merely the words we used but also the nonverbal communication cues we gave him unawares? Sooner or later, consciously or subconsciously, he senses that he has received two communications, and that they are contradictory: (i) (intended communication) "If you don't do as we ask, we will openly condemn you for not acting like a responsible citizen"; (ii) (the unintended communication) "If you *do* behave as we ask, we will secretly condemn you for a simpleton who can be shamed into standing aside while the rest of us exploit the commons."

Everyman then is caught in what Bateson has called a "double bind." Bateson and his co-workers have made a plausible case for viewing the double bind as an important causative factor in the genesis of schizophrenia (*17*). The double bind may not always be so damaging, but it always endangers the mental health of anyone to whom it is applied. "A bad conscience," said Nietzsche, "is a kind of illness."

To conjure up a conscience in others is tempting to anyone who wishes to extend his control beyond the legal limits. Leaders at the highest level succumb to this temptation. Has any President during the past generation failed to call on labor unions to moderate voluntarily their demands for higher wages, or to steel companies to honor voluntary guidelines on prices? I can recall none. The rhetoric used on such occasions is designed to produce feelings of guilt in noncooperators.

For centuries it was assumed without proof that guilt was a valuable, perhaps even an indispensable, ingredient of the civilized life. Now, in this post-Freudian world, we doubt it.

Paul Goodman speaks from the modern point of view when he says: "No good has ever come from feeling guilty, neither intelligence, policy, nor compassion. The guilty do not pay attention to the object but only to themselves, and not even to their own interests, which might make sense, but to their anxieties" (*18*).

One does not have to be a professional psychiatrist to see the consequences of anxiety. We in the Western world are just emerging from a dreadful two-centuries-long Dark Ages of Eros that was sustained partly by prohibition laws, but perhaps more effectively by the anxiety-generating mechanisms of education. Alex Comfort has told the story well in *The Anxiety Makers* (19); it is not a pretty one.

Since proof is difficult, we may even concede that the results of anxiety may sometimes, from certain points of view, be desirable. The larger question we should ask is whether, as a matter of policy, we should ever encourage the use of a technique the tendency (if not the intention) of which is psychologically pathogenic. We hear much talk these days of responsible parenthood; the coupled words are incorporated into the titles of some organizations devoted to birth control. Some people have proposed massive propaganda campaigns to instill responsibility into the nation's (or the world's) breeders. But what is the meaning of the word responsibility in this context? Is it not merely a synonym for the word conscience? When we use the word responsibility in the absence of substantial sanctions are we not trying to browbeat a free man in a commons into acting against his own interest? Responsibility is a verbal counterfeit for a substantial *quid pro quo*. It is an attempt to get something for nothing.

If the word responsibility is to be used at all, I suggest that it be in the sense Charles Frankel uses it (*20*). "Responsibility," says this philosopher, "is the product of definite social arrangements." Notice that Frankel calls for social arrangements—not propaganda.

MUTUAL COERCION

MUTUALLY AGREED UPON

The social arrangements that produce responsibility are arrangements that create coercion, of some sort. Consider bank-robbing. The man who takes money from a bank acts as if the bank were a commons. How do we prevent such action? Certainly not by trying to control his behavior solely by a verbal appeal to his sense of responsibility. Rather than rely on propaganda we follow Frankel's lead and insist that a bank is not a commons; we seek the definite social arrangements that will keep it from becoming a commons. That we thereby infringe on the freedom of would-be robbers we neither deny nor regret.

The morality of bank-robbing is particularly easy to understand because we accept complete prohibition of this activity. We are willing to say "Thou shalt not rob banks," without providing for exceptions. But temperance also can be created by coercion. Taxing is a good coercive device. To keep downtown shoppers temperate in their use of parking space we introduce parking meters for short periods, and traffic fines for longer ones. We need not actually forbid a citizen to park as long as he wants to; we need merely make it increasingly expensive for him to do so. Not prohibition, but carefully biased options are what we offer him. A Madison Avenue man might call this persuasion; I prefer the greater candor of the word coercion.

Coercion is a dirty word to most liberals now, but it need not forever be so. As with the four-letter words, its dirtiness can be cleansed away by exposure to the light, by saying it over and over without apology or embarrassment. To many, the word coercion implies arbitrary decisions of distant and irresponsible bureaucrats; but this is not a necessary part of its meaning. The only kind of coercion I recommend is mutual coercion, mutually agreed upon by the majority of the people affected.

To say that we mutually agree to coercion is not to say that we are required to enjoy it, or even to pretend we enjoy it. Who enjoys taxes? We all grumble about them. But we accept compulsory taxes because we recognize that voluntary taxes would favor the conscienceless. We institute and (grum – blingly) support taxes and other coercive devices to escape the horror of the commons.

An alternative to the commons need not be perfectly just to be preferable. With real estate and other material goods, the alternative we have chosen is the institution of private property coupled with legal inheritance. Is this system perfectly just? As a genetically trained biologist I deny that it is. It seems to me that, if there are to be differences in individual inheritance, legal possession should be perfectly correlated with biological inheritance—that those who are biologically more fit to be the custodians of property and power should legally inherit more. But genetic recombination continually makes a mockery of the doctrine of "like father, like son" implicit in our laws of legal inheritance. An idiot can inherit millions, and a trust fund can keep his estate intact. We must admit that our legal system of private property plus inheritance is unjust—but we put up with it because we are not convinced, at the moment, that anyone has invented a better system. The alternative of the commons is too horrifying to contemplate. Injustice is preferable to total ruin.

It is one of the peculiarities of the warfare between reform and the status quo that it is thoughtlessly governed by a double standard. Whenever a reform measure is proposed it is often defeated when its opponents triumphantly discover a flaw in it. As Kingsley Davis has pointed out (27), worshippers of the status quo sometimes imply that no reform is possible without unanimous agreement, an implication contrary to historical fact. As nearly as I can make out, automatic rejection of proposed reforms is based on one of two unconscious assumptions: (i) that the status quo is perfect; or (ii) that the choice we face is between reform and no action; if the proposed reform is imperfect, we presumably should take no action at all, while we wait for a perfect proposal.

But we can never do nothing. That which we have done for thousands of years is also action. It also produces evils. Once we are aware that the status quo is action, we can then compare its discoverable advantages and disadvantages with the predicted advantages and disadvantages of the proposed reform, discounting as best we can for our lack of experience. On the basis of such a comparison, we can make a rational decision which will not involve the unworkable assumption that only perfect systems are tolerable.

RECOGNITION OF NECESSITY

Perhaps the simplest summary of this analysis of man's population problems is this: the commons, if justifiable at all, is justifiable only under conditions of low-population density. As the human population has increased, the commons has had to be abandoned in one aspect after another.

First we abandoned the commons in food gathering, enclosing farm land and restricting pastures and hunting and fishing areas. These restrictions are still not complete throughout the world.

Somewhat later we saw that the commons as a place for waste disposal would also have to be abandoned. Restrictions on the disposal of domestic sewage are widely accepted in the Western world; we are still struggling to close the commons to pollution by automobiles, factories, insecticide sprayers, fertilizing operations, and atomic energy installations.

In a still more embryonic state is our recognition of the evils of the commons in matters of pleasure. There is almost no restriction on the propagation of sound waves in the public medium. The shopping public is assaulted with mindless music, without its consent. Our government is paying out billions of dollars to create supersonic transport which will disturb 50,000 people for every one person who is whisked from coast to coast 3 hours faster. Advertisers muddy the airwaves of radio and television and pollute the view of travelers. We are a long way from outlawing the commons in matters of pleasure. Is this because our Puritan inheritance makes us view pleasure as something of a sin, and pain (that is, the pollution of advertising) as the sign of virtue?

Every new enclosure of the commons involves the infringement of somebody's personal liberty. Infringements made in the distant past are accepted because no contemporary complains of a loss. It is the newly proposed infringements that we vigorously oppose; cries of "rights" and "freedom" fill the air. But what does "freedom" mean? When men mutually agreed to pass laws against robbing, mankind became more free, not less so. Individuals locked into the logic of the commons are free only to bring on universal ruin; once they see the necessity of mutual coercion, they become free to pursue other goals. I believe it was Hegel who said, "Freedom is the recognition of necessity."

The most important aspect of necessity that we must now recognize, is the necessity of abandoning the commons in breeding. No technical solution can rescue us from the misery of overpopulation. Freedom to breed will bring ruin to all. At the moment, to avoid hard decisions many of us are tempted to propagandize for conscience and responsible parenthood. The temptation must be resisted, because an appeal to independently acting consciences selects for the disappearance of all conscience in the long run, and an-increase in anxiety in the short.

The only way we can preserve and nurture other and more precious freedoms is by relinquishing the freedom to breed, and that very soon. "Freedom is the recognition of necessity"—and it is the role of education to reveal to all the necessity of abandoning the freedom to breed. Only so, can we put an end to this aspect of the tragedy of the commons.

NOTE

1 The author is professor of biology. University of California, Santa Barbara. This article is based on a presidential address presented before the meeting of the Pacific Division of the American Association for the Advancement of Science at Utah State University, Logan, 25 June 1968.

433

REFERENCES

1. J. B. Wiesner and H. F. York, *Sci. Amer.* 211 (No. 4), 27 (1964).
2. G. Hardin, *J. Hered.* 50, 68 (1959); S. von Hoemor, *Science* 137, 18 (1962).
3. J. von Neumann and O. Morgenstem, *Theory of Games and Economic Behavior* (Princeton Univ. Press, Princeton, N.J., 1947), p. 11.
4. J. H. Fremlin, *New Sci.*, No. 415 (1964), p. 285.
5. A. Smith, *The Wealth of Nations* (Modem Library, New York, 1937), p. 423.
6. W. F. Lloyd, *Two Lectures on the Checks to Population* (Oxford Univ. Press, Oxford, England, 1833), reprinted (in part) in *Population, Evolution, and Birth Control*, G. Hardin, Ed. (Freeman, San Francisco, 1964), p. 37.
7. A. N. Whitehead, *Science and the Modern World* (Mentor, New York, 1948), p. 17.
8. G. Hardin, Ed. *Population, Evolution, and Birth Control* (Freeman, San Francisco, 1964), p. 56.
9. S. McVay, *Sci. Amer.* 216 (No. 8), 13 (1966).
10. J. Fletcher, *Situation Ethics* (Westminster, Philadelphia, 1966).
11. D. Lack, *The Natural Regulation of Animal Numbers* (Clarendon Press, Oxford, 1954).
12. H. Girvetz, *From Wealth to Welfare* (Stanford Univ. Press, Stanford, Calif., 1950).
13. G. Hardin, *Perspec. Biol. Med.* 6, 366 (1963).
14. U. Thant, *Int. Planned Parenthood News*, No. 168 (February 1968), p. 3.
15. K. Davis, *Science* 158, 730 (1967).
16. S. Tax, Ed., *Evolution after Darwin* (Univ. of Chicago Press, Chicago, 1960), vol. 2, p. 469.
17. G. Bateson, D. D. Jackson, J. Haley, J. Weak– land, *Behav. Sci.* 1, 251 (1956).
18. P. Goodman, *New York Rev. Books* 10(8), 22 (23 May 1968).
19. A. Comfort, *The Anxiety Makers* (Nelson, London, 1967).
20. C. Frankel, *The Case for Modem Man* (Harper, New York, 1955), p. 203.
21. J. D. Roslansky, *Genetics and the Future of Man* (Appleton-Century-Crofts, New York, 1966), p. 177.

Elinor Ostrom — "Reflections on the Commons." In *Governing the Commons.* **Cambridge: Cambridge University Press.**

Reading Commentary

This excerpt is the first chapter of *Governing the Commons*, another seminal theoretical piece in the field of environmental policy-making. Ostrom starts by briefly describing three "models"—Hardin's "tragedy of the commons" (which you just read), the prisoner's dilemma (taken from game theory), and Mancur Olson Jr.'s (1965) *Logic of Collective Action.* All three are often used to justify strong government regulation of natural resource utilization. Ostrom challenges the key assumption behind all three models, namely, that rational individuals lack the agency and foresight to make decisions in their collective best interest. She points to community-regulated fishery practices in Alanya, Turkey, as an example of successful self-governance.

Ostrom's work is often mistakenly seen as a call for decentralized community-based approaches to all resource management problems because it criticizes both privatization (i.e., reliance on market forces) and strong government regulation. However, she acknowledges that, in many cases, privatization and/or state regulation may be more effective than community-based or voluntary approaches. Her attention to community-based strategies of resource management is simply out of a recognition that they are undertheorized and, therefore, often dismissed by policy analysts. In other words, Ostrom suggests that we need to strike a balance between reliance on state-, market – and community-based approaches to natural resource management. She advocates institutional arrangements that fit the unique constraints of each resource management situation. Much of Ostrom's later work—her Institutional Analysis and Development framework (2005), for example, and the variables she outlines in analyzing social-ecological systems (2011, earlier excerpt included)—emphasizes the importance of institutional design in relation to context.

Why did we end, rather than start, with this excerpt? We believe that you cannot do environmental problem-solving unless you can figure out how to get people to cooperate in managing shared resources. Ostrom suggests that communities may be best equipped to design the institutions needed to govern their resources because they have local knowledge, can quickly adapt to changing circumstances and can avoid the principal-agent problem that often makes enforcement so difficult. We similarly believe that democratic engagement in environmental policy-making can lead to "fairer, more stable, wiser and more efficient" (Susskind and Cruikshank 2006) outcomes. As such, we have tried to introduce you to works that have influenced our own theory of environmental problem-solving.

CHAPTER 1

REFLECTIONS ON THE COMMONS

Elinor Ostrom

Hardly a week goes by without a major news story about the threatened destruction of a valuable natural resource. In June of 1989, for example, a *New York Times* article focused on the problem of overfishing in the Georges Bank about 150 miles off the New England coast. Catches of cod, flounder, and haddock are now only a quarter of what they were during the 1960s. Everyone knows that the basic problem is overfishing; however, those concerned cannot agree how to solve the problem. Congressional representatives recommend new national legislation, even though the legislation already on the books has been enforced only erratically. Representatives of the fishers argue that the fishing grounds would not be in such bad shape if the federal government had refrained from its sporadic attempts to regulate the fishery in the past. The issue in this case – and many others – is how best to limit the use of natural resources so as to ensure their long-term economic viability. Advocates of central regulation, of privatization, and of regulation by those involved have pressed their policy prescriptions in a variety of different arenas.

Similar situations occur on diverse scales ranging from small neighborhoods to the entire planet. The issues of how best to govern natural resources used by many individuals in common are no more settled in academia than in the world of politics. Some scholarly articles about the "tragedy of the commons" recommend that "the state" control most natural resources to prevent their destruction; others recommend that privatizing those resources will resolve the problem. What one can observe in the world, however, is that neither the state nor the market is uniformly successful in enabling individuals to sustain long-term, productive use of natural resource systems. Further, communities of individuals have relied on institutions resembling neither the state nor the market to govern some resource systems with reasonable degrees of success over long periods of time.

We do not yet have the necessary intellectual tools or models to understand the array of problems that are associated with governing and managing natural resource systems and the reasons why some institutions seem to work in some settings and not others. This book is an effort to (1) critique the foundations of policy analysis as applied to many natural resources, (2) present empirical examples of successful and unsuccessful efforts to govern and manage such resources, and (3) begin the effort to develop better intellectual tools to understand the capabilities and limitations of self-governing institutions for regulating many types of resources. To do this, I first describe the three models most frequently used to provide a foundation for recommending state or market solutions. I then pose theoretical and empirical alternatives to these models to begin to illustrate the diversity of solutions that go beyond states and markets. Using an institutional mode of analysis, I then attempt to explain how communities of individuals fashion different ways of governing the commons.

THREE INFLUENTIAL MODELS

THE TRAGEDY OF THE COMMONS

Since Garrett Hardin's challenging article in *Science* (1968), the expression "the tragedy of the commons" has come to symbolize the degradation of the environment to be expected whenever many individuals use a scarce resource in common. To illustrate the logical structure of his model, Hardin asks the reader to envision a pasture "open to all." He then examines the structure of this situation from the perspective of a rational herder. Each herder receives a direct benefit from his own animals and suffers delayed costs from the deterioration of the commons when his and others' cattle overgraze. Each herder is motivated to add more and more animals because he receives the direct benefit of his own animals and bears only a share of the costs resulting from overgrazing. Hardin concludes:

> Therein is the tragedy. Each man is locked into a system that compels him to increase his herd without limit – in a world that is limited. Ruin is the destination toward which all men rush, each pursuing his own best interest in a society that believes in the freedom of the commons. (Hardin 1968, p. 1,244)

Hardin was not the first to notice the tragedy of the commons. Aristotle long ago observed that "what is common to the greatest number has the least care bestowed upon it. Everyone thinks chiefly of his own, hardly at all of the common interest" *(Politics,* Book II, ch. 3). Hobbes's parable of man in a state of nature is a prototype of the tragedy of the commons: Men seek their own good and end up fighting one another. In 1833, William Forster Lloyd (1977) sketched a theory of the commons that predicted improvident use for property owned in common. More than a decade before Hardin's article, H. Scott Gordon (1954) clearly expounded similar logic in another classic: "The Economic Theory of a Common-Property Research: The Fishery." Gordon described the same dynamic as Hardin:

> There appears then, to be some truth in the conservative dictum that everybody's property is nobody's property. Wealth that is free for all is valued by no one because he who is foolhardy enough to wait for its proper time of use will only find that it has been taken by another. ... The fish in the sea are valueless to the fisherman, because there is no assurance that they will be there for him tomorrow if they are left behind today. (Gordon 1954, p. 124)

John H. Dales (1968, p. 62) noted at the same time the perplexing problems related to resources "owned in common because there is no alternative!" Standard analyses in modern resource economics conclude that where a number of users have access to a common-pool resource, the total of resource units withdrawn from the resource will be greater than the optimal economic level of withdrawal (Clark 1976, 1980; Dasgupta and Heal 1979).

If the only "commons" of importance were a few grazing areas or fisheries, the tragedy of the commons would be of little general interest. That is not the case. Hardin himself used the grazing commons as a metaphor for the general problem of overpopulation. The "tragedy of the commons" has been used to describe such diverse problems as the Sahelian famine of the 1970s (Picardi and Seifert 1977), firewood crises throughout the Third World (Norman 1984; Thomson 1977), the problem of acid rain (R. Wilson 1985), the organization of the Mormon Church (Bullock and Baden 1977), the inability of the U.S. Congress

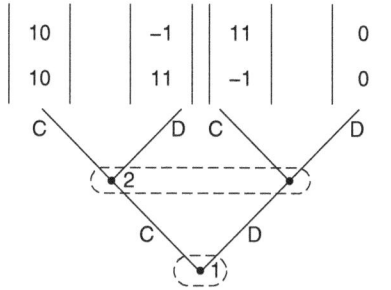

Figure 1.1 Game 1: The Hardin herder game.

to limit its capacity to overspend (Shepsle and Weingast 1984), urban crime (Neher 1978), public-sector/private-sector relationships in modern economies (Scharpf 1985, 1987, 1988), the problems of international cooperation (Snidal 1985), and communal conflict in Cyprus (Lumsden 1973). Much of the world is dependent on resources that are subject to the possibility of a tragedy of the commons.

THE PRISONER'S DILEMMA GAME

Hardin's model has often been formalized as a prisoner's dilemma (PD) game (Dawes 1973, 1975).[1] Suppose we think of the players in a game as being herders using a common grazing meadow. For this meadow, there is an upper limit to the number of animals that can graze on the meadow for a season and be well fed at the end of the season. We call that number L. For a two-person game, the "cooperate" strategy can be thought of as grazing $L/2$ animals for each herder. The "defect" strategy is for each herder to graze as many animals as he thinks he can sell at a profit (given his private costs), assuming that this number is greater than $L/2$. If both herders limit their grazing to $L/2$, they will obtain 10 units of profit, whereas if they both choose the defect strategy they will obtain zero profit. If one of them limits his number of animals to $L/2$, while the other grazes as many as he wants, the "defector" obtains 11 units of profit, and the "sucker" obtains −1. If each chooses independently without the capacity to engage in a binding contract, each chooses his dominant strategy, which is to defect. When they both defect, they obtain zero profit. Call this the Hardin herder game, or Game 1. It has the structure of a prisoner's dilemma game.[2]

   The prisoner's dilemma game is conceptualized as a noncooperative game in which all players possess complete information. In noncooperative games, communication among the players is forbidden or impossible or simply irrelevant as long as it is not explicitly modeled as part of the game. If communication is possible, verbal agreements among players are presumed to be nonbinding unless the possibility of binding agreements is explicitly incorporated in the game structure (Harsanyi and Selten 1988, p. 3). "Complete information" implies that all players know the full structure of the game tree and the payoffs attached to outcomes. Players either know or do not know the current moves of other players depending on whether or not they are observable.

In a prisoner's dilemma game, each player has a dominant strategy in the sense that the player is always better off choosing this strategy – to defect – no matter what the other player chooses. When both players choose their dominant strategy, given these assumptions, they produce an equilibrium that is the third-best result for both. Neither has an incentive to change that is independent of the strategy choice of the other. The equilibrium resulting from each player selecting his or her "best" individual strategy is, however, not a Pareto-optimal outcome. A Pareto-optimal outcome occurs when there is no other outcome strictly preferred by at least one player that is at least as good for the others. In the two-person prisoner's dilemma game, both players prefer the (cooperate, cooperate) outcome to the (defect, defect) outcome. Thus, the equilibrium outcome is Pareto-inferior.

The prisoner's dilemma game fascinates scholars. The paradox that individually rational strategies lead to collectively irrational outcomes seems to challenge a fundamental faith that rational human beings can achieve rational results. In the introduction to a recently published book, *Paradoxes of Rationality and Cooperation*, Richmond Campbell explains the "deep attraction" of the dilemma:

> Quite simply, these paradoxes cast in doubt our understanding of rationality and, in the case of the Prisoner's Dilemma suggest that it is impossible for rational creatures to cooperate. Thus, they bear directly on fundamental issues in ethics and political philosophy and threaten the foundations of the social sciences. It is the scope of these consequences that explains why these paradoxes have drawn so much attention and why they command a central place in philosophical discussion. (Campbell 1985, p. 3)

The deep attraction of the dilemma is further illustrated by the number of articles written about it. At one count, 15 years ago, more than 2,000 papers had been devoted to the prisoner's dilemma game (Grofman and Pool 1975).

THE LOGIC OF COLLECTIVE ACTION

A closely related view of the difficulty of getting individuals to pursue their joint welfare, as contrasted to individual welfare, was developed by Mancur Olson (1965) in *The Logic of Collective Action*. Olson specifically set out to challenge the grand optimism expressed in group theory: that individuals with common interests would voluntarily act so as to try to further those interests (Bentley 1949; Truman 1958). On the first page of his book, Olson summarized that accepted view:

> The idea that groups tend to act in support of their group interests is supposed to follow logically from this widely accepted premise of rational, self-interested behavior. In other words, if the members of some group have a common interest or object, and if they would all be better off if that objective were achieved, it has been thought to follow logically that the individuals in that group would, if they were rational and self-interested, act to achieve that objective. (Olson 1965, p. 1)

Olson challenged the presumption that the possibility of a benefit for a group would be sufficient to generate collective action to achieve that benefit. In the most frequently quoted passage of his book, Olson argued that unless the number of individuals is quite small, or unless there is coercion or some other special device to make individuals act in their common interest, *rational, self-interested individuals will not act to achieve their common or group interests.* (Olson 1965, p. 2; emphasis in original)

Olson's argument rests largely on the premise that one who cannot be excluded from obtaining the benefits of a collective good once the good is produced has little incentive to contribute voluntarily to the provision of that good. His book is less pessimistic than it is asserted to be by many who cite this famous passage. Olson considers it an open question whether intermediate-size groups will or will not voluntarily provide collective benefits. His definition of an intermediate-size group depends not on the number of actors involved but on how noticeable each person's actions are.

The tragedy of the commons, the prisoner's dilemma, and the logic of collective action are closely related concepts in the models that have defined the accepted way of viewing many problems that individuals face when attempting to achieve collective benefits. At the heart of each of these models is the free-rider problem. Whenever one person cannot be excluded from the benefits that others provide, each person is motivated not to contribute to the joint effort, but to free-ride on the efforts of others. If all participants choose to free-ride, the collective benefit will not be produced. The temptation to free-ride, however, may dominate the decision process, and thus all will end up where no one wanted to be. Alternatively, some may provide while others free-ride, leading to less than the optimal level of provision of the collective benefit. These models are thus extremely useful for explaining how perfectly rational individuals can produce, under some circumstances, outcomes that are not "rational" when viewed from the perspective of all those involved.

What makes these models so interesting and so powerful is that they capture important aspects of many different problems that occur in diverse settings in all parts of the world. What makes these models so dangerous – when they are used metaphorically as the foundation for policy – is that the constraints that are assumed to be fixed for the purpose of analysis are taken on faith as being fixed in empirical settings, unless external authorities change them.[3] The prisoners in the famous dilemma cannot change the constraints imposed on them by the district attorney; they are in jail. Not all users of natural resources are similarly incapable of changing their constraints. As long as individuals are viewed as prisoners, policy prescriptions will address this metaphor. I would rather address the question of how to enhance the capabilities of those involved to change the constraining rules of the game to lead to outcomes other than remorseless tragedies.

THE METAPHORICAL USE OF MODELS

These three models and their many variants are diverse representations of a broader and still-evolving theory of collective action. Much more work will be needed to develop the theory of collective action into a reliable and useful foundation for policy analysis. Considerable progress has been made during the past three decades by theorists and empirically oriented social scientists. The sweeping conclusions of the first variants of this theory have given way to a more qualified body of knowledge involving many more variables and explicit base conditions.

As an evolving, rather than completed, theory, it provokes disagreement regarding the importance or insignificance of some variables and how best to specify key relationships.[4] The results from more recent work, particularly work focusing on the dynamic aspects of relevant empirical settings, have begun to generate more optimistic predictions than did earlier models; see, in particular, the work of Axelrod (1981, 1984) and Kreps and Wilson (1982). This is one of the most exciting areas in the social sciences, for although considerable cumulation has already occurred, some deep questions remain unanswered. Some of these puzzles are key to understanding how individuals jointly using a common-pool resource might be able to achieve an effective form of governing and managing their own commons. These puzzles are examined in Chapter 2.

Much that has been written about common-pool resources, however, has uncritically accepted the earlier models and the presumption of a remorseless tragedy (Nebel 1987). Scholars have gone so far as to recommend that "Hardin's 'Tragedy of the Commons' should be required reading for all students … and, if I had my way, for all human beings."[5] Policy prescriptions have relied to a large extent on one of the three original models, but those attempting to use these models as the basis for policy prescription frequently have achieved little more than a metaphorical use of the models.

When models are used as metaphors, an author usually points to the similarity between one or two variables in a natural setting and one or two variables in a model. If calling attention to similarities is all that is intended by the metaphor, it serves the usual purpose of rapidly conveying information in graphic form. These three models have frequently been used metaphorically, however, for another purpose. The similarity between the many individuals jointly using a resource in a natural setting and the many individuals jointly producing a suboptimal result in the model has been used to convey a sense that further similarities are present. By referring to natural settings as "tragedies of the commons," "collective-action problems," "prisoner's dilemmas," "open-access resources," or even "common-property resources," the observer frequently wishes to invoke an image of helpless individuals caught in an inexorable process of destroying their own resources. An article in the December 10, 1988, issue of *The Economist* goes so far as to assert that fisheries can be managed successfully only if it is recognized that "left to their own devices, fisherman will overexploit stocks," and "to avoid disaster, managers must have effective hegemony over them."

Public officials sometimes do no more than evoke grim images by briefly alluding to the popularized versions of the models, presuming, as self-evident, that the same processes occur in all natural settings. The Canadian minister of fisheries and oceans, for example, captured the color of the models in a 1980 speech:

> If you let loose that kind of economic self-interest in fisheries, with everybody fishing as he wants, taking from a resource that belongs to no individual, you end up destroying your neighbour and yourself. In free fisheries, good times create bad times, attracting more and more boats to chase fewer and fewer fish, producing less and less money to divide among more and more people.

(Romeo LeBlanc, speaking at the 50th anniversary meeting of the United Maritime Fishermen, March 19, 1980; quoted by Matthews and Phyne 1988)

The implication, of course, was that Canadian fisheries universally met that description – an empirically incorrect inference.[6] But many observers have come to assume that most resources are like those specified in the three models. As such, it has been assumed that the individuals have been caught in a grim trap. The resulting policy recommendations have had an equally grim character.

CURRENT POLICY PRESCRIPTIONS

LEVIATHAN AS THE "ONLY" WAY

Ophuls (1973, p. 228) argued, for example, that "because of the tragedy of the commons, environmental problems cannot be solved through cooperation ... and the rationale for government with major coercive powers is overwhelming." Ophuls concluded that "even if we avoid the tragedy of the commons, it will *only* be by recourse to the tragic necessity of Leviathan" (1973, p. 229; emphasis added).[7] Garrett Hardin argued a decade after his earlier article that we are enveloped in a "cloud of ignorance" about "the true nature of the fundamental political systems and the effect of each on the preservation of the environment" (1978, p. 310). The "cloud of ignorance" did not, however, prevent him from presuming that the only alternatives to the commons dilemma were what he called "a private enterprise system," on the one hand, or "socialism," on the other (1978, p. 314). With the assurance of one convinced that "the alternative of the commons is too horrifying to contemplate" (1968, p. 1,247), Hardin indicated that change would have to be instituted with "whatever force may be required to make the change stick" (1978, p. 314). In other words, "if ruin is to be avoided in a crowded world, people must be responsive to a coercive force outside their individual psyches, a 'Leviathan,' to use Hobbes's term" (Hardin 1978, p. 314).

The presumption that an external Leviathan is necessary to avoid tragedies of the commons leads to recommendations that central governments control most natural resource systems. Heilbroner (1974) opined that "iron governments," perhaps military governments, would be necessary to achieve control over ecological problems. In a less draconian view, Ehrenfeld (1972, p. 322) suggested that if "private interests cannot be expected to protect the public domain then external regulation by public agencies, governments, or international authorities is needed." In an analysis of the problems involved in water resource management in developing countries, Carruthers and Stoner (1981, p. 29) argued that without public control, "overgrazing and soil erosion of communal pastures, or less fish at higher average cost," would result. They concluded that "common property resources *require* public control if economic efficiency is to result from their development" (1981, p. 29; emphasis added).[8] The policy advice to centralize the control and regulation of natural resources, such as grazing lands, forests, and fisheries, has been followed extensively, particularly in Third World countries.

One way to illustrate these proponents' image of centralized control is to modify the Hardin herder game using the assumptions that underlie this policy advice. The proponents of centralized control want an external government agency to decide the specific herding strategy that the central authority considers best for the situation: The central authority will decide who can use the meadow, when they can use it, and how many animals can be grazed. Let us assume that the central authority decides to impose a penalty of 2 profit units on anyone who is considered by that authority to be using a defect strategy. Assuming that the central agency knows the sustainable yield of the meadow *(L)* and can unfailingly discover and penalize any herder using the defect strategy, the newly restructured game imposed by the central authority is represented in Game 2. Now, the solution to Game 2 is (cooperate, cooperate). Both players receive 10 profit units each, rather than the zero units they would have received in Game 1. If an external authority accurately determines the capacity of a common-pool resource, unambiguously assigns this capacity, monitors actions, and unfailingly sanctions noncompliance, then a centralized agency can transform the Hardin herder game to generate an optimally efficient equilibrium for the herders. Little consideration is given to the cost of creating and maintaining such an agency. This is seen as exogenous to the problem and is not included as a parameter of Game 2.[9]

The optimal equilibrium achieved by following the advice to centralize control, however, is based on assumptions concerning the accuracy of information, monitoring capabilities, sanctioning reliability, and zero costs of administration. Without valid and reliable information, a central agency could make several errors, including setting the carrying capacity or the fine too high or too low, sanctioning herders who cooperate, or not sanctioning defectors. The implications of all forms of incomplete information are interesting. However, as an example, I shall focus entirely on the implications arising from a central agency's incomplete information about the herders' strategies. The implicit assumption of Game 2 is that the central agency monitors all actions of the herders costlessly and imposes sanctions correctly.

In Game 3, we assume that the central agency has complete information about the carrying capacity of the meadow, but incomplete information about the particular actions of the herders. The central agency consequently makes errors in imposing punishments. Let us assume that the central agency punishes defections (the correct response) with probability y and fails to punish defections with probability $1 - y$ (the erroneous response). Let us also assume that the central agency punishes cooperative actions (the erroneous response) with probability x and does not punish cooperative actions (the correct response) with probability $1 - x$. The payoff parameters are illustrated in Figure 1.3.

A central agency with complete information would make no errors in its punishment level; in that case, $x = 0$ and $y = 1$. Game 2 would then be a special case of Game 3 in which

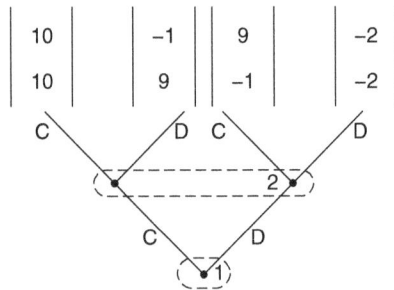

Figure 1.2 Game 2: The central-authority game with complete information.

$x = 0$ and $y = 1$. However, if the central agency does not have complete information about the actions of the herders, it imposes both types of sanctions correctly with a probability of 0.7 $(x = 0.3, y = 0.7)$. An example of the specific payoffs for this game is shown as Game 4 in Figure 1.4. Given this payoff structure, the herders again face a prisoner's dilemma game. They will defect (overgraze) rather than cooperate (graze within the carrying capacity). In Game 4, as in the original Game 1, the equilibrium outcomes for the herders were $(0, 0)$. In a game in which a central agency sanctions correctly with a probability of 0.7, the equilibrium outcomes are $(-1.6, -1.6)$. The equilibrium of the regulated game has a lower value than that of the unregulated game. Given the carrying capacity and profit possibilities of Game 1, the central agency must have sufficient information so that it can correctly impose sanctions with a probability greater than 0.75 to avoid pushing the herders to the (D, D) equilibrium.[10]

PRIVATIZATION AS THE "ONLY" WAY

Other policy analysts, influenced by the same models, have used equally strong terms in calling for the imposition of private property rights whenever resources are owned in common (Demsetz 1967; O. Johnson 1972). "Both the economic analysis of common property resources and Hardin's treatment of the tragedy of the commons" led Robert J. Smith (1981, p. 467) to suggest that "the *only* way to avoid the tragedy of the commons in natural resources and wildlife is to end the common-property system by creating a system of private property rights" (emphasis added); see also the work of Sinn (1984). Smith stressed that it is "by treating a resource as a common property that we become locked in its inexorable destruction" (1981, p. 465). Welch advocated the creation of full private rights to a commons when he asserted that "the establishment of full property rights is necessary to avoid the inefficiency of overgrazing" (1983, p. 171). He asserted that privatization of the commons was the optimal solution for all common-pool problems. His major concern was how to impose private ownership when those currently using a commons were unwilling to change to a set of private rights to the commons.

Those recommending the imposition of privatization on the herders would divide the meadow in half and assign half of the meadow to one herder and the other half to the second herder. Now each herder will be playing a *game against nature* in a smaller terrain, rather than a game against another player in a larger terrain. The herders now will need to invest in fences and their maintenance, as well as in monitoring and sanctioning activities to enforce their division of the grazing area (B. Field 1984, 1985b). It is presumed that each herder will now choose $X/2$ animals to graze as a result of his own profit incentive.[11] This assumes that the meadow is perfectly homogeneous over time in its distribution of available fodder. If rainfall occurs erratically, one part of the grazing area may be lush with growth one year, whereas another part of the area may be unable to support $X/2$ animals. The rain may fall somewhere else the next year. In any given year, one of the herders may make no profit, and the other may enjoy a considerable return. If the location of lush growth changes dramatically from year to year, dividing the commons may impoverish both herders and lead to overgrazing in those parts where forage is temporarily inadequate. Of course, it will be possible for the herder who has extra fodder in one year to sell it to the other herder. Alternatively, it will be possible for the herders to set up an insurance scheme to share the risk of an uncertain environment. However, the setup costs for a new market or a new insurance scheme would

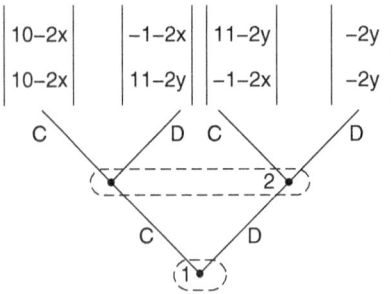

Figure 1.3 Game 3: The central-authority game with incomplete information.

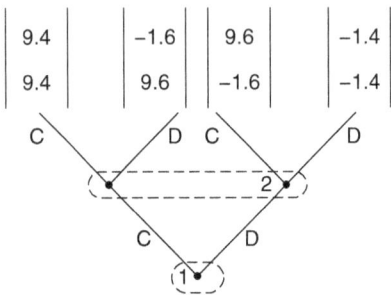

Figure 1.4 Game 4: An example of the central-authority game with incomplete information.

be substantial and will not be needed so long as the herders share fodder and risk by jointly sharing a larger grazing area.

It is difficult to know exactly what analysts mean when they refer to the necessity of developing private rights to some common-pool resources (CPRs). It is clear that when they refer to land, they mean to divide the land into separate parcels and assign individual rights to hold, use, and transfer these parcels as individual owners desire (subject to the general regulations of a jurisdiction regarding the use and transfer of land). In regard to nonstationary resources, such as water and fisheries, it is unclear what the establishment of private rights means. As Colin Clark has pointed out, the "'tragedy of the commons' has proved particularly difficult to counteract in the case of marine fishery resources where the establishment of individual property rights is virtually out of the question" (1980, p. 117). In regard to a fugitive resource, a diversity of rights may be established giving individuals rights to use particular types of equipment, to use the resource system at a particular time and place, or to withdraw a particular quantity of resource units (if they can be found). But even when particular rights are unitized, quantified, and salable, the resource *system* is still likely to be owned in common rather than individually.[12] Again, referring to fisheries, Clark has argued that "common ownership is the fundamental fact affecting almost every regime of fishery management" (1980, p. 117).

THE "ONLY" WAY?

Analysts who find an empirical situation with a structure presumed to be a commons dilemma often call for the imposition of a solution by an external actor: The "only way" to solve a commons dilemma is by doing X. Underlying such a claim is the belief that X is necessary and sufficient to solve the commons dilemma. But the content of X could hardly be more variable. One set of advocates presumes that a central authority must assume continuing responsibility to make unitary decisions for a particular resource. The other presumes that a central authority should parcel out ownership rights to the resource and then allow individuals to pursue their own self-interests within a set of well-defined property rights. Both centralization advocates and privatization advocates accept as a central tenet that institutional change must come from outside and be imposed on the individuals affected. Despite sharing a faith in the necessity and efficacy of "the state" to change institutions so as to increase efficiency, the institutional changes they recommend could hardly be further apart.

If one recommendation is correct, the other cannot be. Contradictory positions cannot both be right. I do not argue for either of these positions. Rather, I argue that both are too sweeping in their claims. Instead of there being a single solution to a single problem, I argue that many solutions exist to cope with many different problems. Instead of, presuming that optimal institutional solutions can be designed easily and imposed at low cost by external authorities, I argue that "getting the institutions right" is a difficult, time-consuming, conflict-invoking process. It is a process that requires reliable information about time and place variables as well as a broad repertoire of culturally acceptable rules. New institutional arrangements do not work in the field as they do in abstract models unless the models are well specified and empirically valid and the participants in a field setting understand how to make the new rules work.

Instead of presuming that the individuals sharing a commons are inevitably caught in a trap from which they cannot escape, I argue that the capacity of individuals to extricate themselves from various types of dilemma situations *varies* from situation to situation. The cases to be discussed in this book illustrate both successful and unsuccessful efforts to escape tragic outcomes. Instead of basing policy on the presumption that the individuals involved are helpless, I wish to learn more from the experience of individuals in field settings. Why have some efforts to solve commons problems failed, while others have succeeded? What can we learn from experience that will help stimulate the development and use of a better theory of collective action – one that will identify the key variables that can enhance or detract from the capabilities of individuals to solve problems?

Institutions are rarely either private or public – "the market" or "the state." Many successful CPR institutions are rich mixtures of "private-like" and "public-like" institutions defying classification in a sterile dichotomy. By "successful," I mean institutions that enable individuals to achieve productive outcomes in situations where temptations to free-ride and shirk are ever present. A competitive market – the epitome of private institutions – is itself a public good. Once a competitive market is provided, individuals can enter and exit freely whether or not they contribute to the cost of providing and maintaining the market. No market can exist for long without underlying public institutions to support it. In field settings, public and private institutions frequently are intermeshed and depend on one another, rather than existing in isolated worlds.

AN ALTERNATIVE SOLUTION

To open up the discussion of institutional options for solving commons dilemmas, I want now to present a fifth game in which, the herders themselves can make a binding contract to commit themselves to a cooperative strategy that they themselves will work out. To represent this arrangement within a noncooperative framework, additional moves must be overtly included in the game structure. A binding contract is interpreted within noncooperative game theory as one that is unfailingly enforced by an external actor – just as we interpreted the penalty posited earlier as being unfailingly enforced by the central authority.

A simple way to represent this is to add one parameter to the payoffs and a strategy to both herders' strategy sets.[13] The parameter is the cost of enforcing an agreement and will be denoted by e. The herders in Game 5 must now negotiate prior to placing animals on the meadow. During negotiations, they discuss various strategies for sharing the carrying capacity of the meadow and the costs of enforcing their agreement. Contracts are not enforceable, however, unless agreed to unanimously by the herders. Any proposal made by one herder that did not involve an equal sharing of the carrying capacity and of enforcement costs would be vetoed by the other herder in their negotiations. Consequently, the only feasible agreement – and the equilibrium of the resulting game – is for both herders to share equally the sustainable yield levels of the meadow and the costs of enforcing their agreement so long as each herder's share of the cost of enforcement is less than 10.[14]

Further, in Game 5, players can *always* guarantee that the worst they will do is the (defect, defect) outcome of Game 1. They are not dependent on the accuracy of the information obtained by a distant government official regarding their strategies. If one player suggests a contract based on incomplete or biased information, the other player can indicate an unwillingness to agree. They determine their own contract and ask the enforcer to enforce only that on which they have agreed. If the enforcer should decide to charge too much for its services [any number equal to or greater than $Pi(C, C) - Pi(D, D)$, $i = 1, 2$], neither player would agree to such a contract.

The "solution" of a commons-dilemma game through instrumentalities similar to Game 5 is not presented as the "only way" to solve a commons dilemma. It is merely one way. But this way has been almost totally ignored in both the policy-analysis literature and the formal-theory literature. Contemplating such an option raises numerous questions. First, might it be possible for the herders to hire a private agent to take on the role of enforcer? This is not as farfetched as it might seem at first. Many long-term business exchanges have the structure of a prisoner's dilemma.[15] Businesses are hesitant to accept promises of future performance rather than enforceable contracts, especially when beginning new business relationships. To reduce enforcement costs, however, a frequent practice is to use a private arbitrator rather than a civil court as the mechanism to achieve enforcement.[16] In N-person settings, all professional athletic leagues face problems similar to those illustrated here. During the play of a professional game, the temptation to cheat and break the rules is ever present. Further, accidents do happen, and rules get broken, even by players who were intending to follow the rules. Athletic leagues typically employ private monitors to enforce their rules.[17]

As soon as we allow the possibility of a private party to take on the role of an external enforcer, the nature of the "solution" offered by Game 5 to the commons dilemma begins to generate a rich set of alternative applications. A self-financed contract-enforcement game

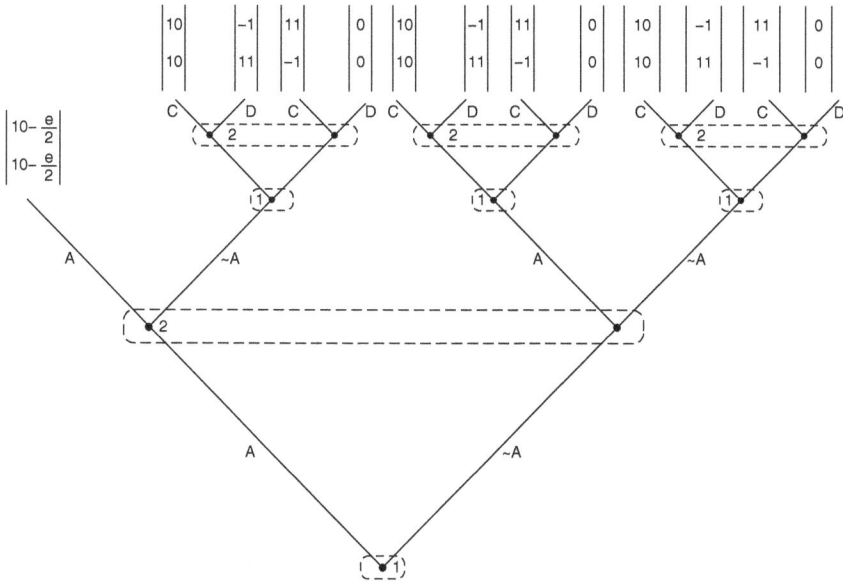

Figure 1.5 Game 5: Self-financed contract-enforcement game.

allows the participants in the situation to exercise greater control over decisions about who will be allowed to graze and what limits will be placed on the number of animals, as compared with either Game 2 or Game 3. If the parties use a private arbitrator, they do not let the arbitrator impose an agreement on them. The arbitrator simply helps the parties find methods to resolve disputes that arise within the set of working rules to which the parties themselves have agreed. Arbitrators, courts, and other arrangements for enforcement and dispute resolution make it possible for individuals to initiate long-term arrangements that they could not otherwise undertake.[18] Further, as soon as one thinks about a "solution" like Game 5, it is a small step to thinking about the possibility of several arbitrators offering enforcement services at varying charges during the negotiation stage. The payoff-dominant equilibrium is to agree on that arbitrator who will enforce the contract at the lowest e.

The key difference between Game 5 and Games 2 and 3 is that the participants themselves design their own contracts in Game 5 in light of the information they have at hand. The herders, who use the same meadow year after year, have detailed and relatively accurate information about carrying capacity. They observe the behavior of other herders and have an incentive to report contractual infractions. Arbitrators may not need to hire monitors to observe the activities of the contracting parties. The self-interest of those who negotiated the contract will lead them to monitor each other and to report observed infractions so that the contract is enforced. A regulatory agency, on the other hand, always needs to hire its own monitors. The regulatory agency then faces the principal–agent problem of how to ensure that its monitors do their own job.

The proponents of the central-authority "solution" presume that such agencies have accurate information and are able to change incentives to produce something like Game 2. It is difficult for a central authority to have sufficient time-and-place information to estimate accurately both the carrying capacity of a CPR and the appropriate fines to induce cooperative behavior. I believe that situations like that in Game 3, in which incomplete information leads to sanctioning errors, occur more frequently than has been presumed in the policy literature. The need for external monitors and enforcers is particularly acute when what is being enforced is a decision by an external agent who may impose excess costs on participants.

A further problem for consideration is that games in which enforcers have been arranged for by mutual agreement may be mistaken by analysts and public officials for games in which there have been *no* agreements about how to cooperate and enforce agreements. In other words, some examples of a "Game 5" may be mistaken for a "Game 1."[19] These situations may be construed to be "informal," carrying a presumption that they are not lawful. This goes to fundamental presumptions about the nature of governments as external authorities governing over societies.

As will be seen in the later discussion of empirical cases, users of CPRs have developed a wide diversity in their own agreements, which are enforced by many mechanisms. Some of the enforcement mechanisms are external governmental agencies. Some enforcement mechanisms involve members of the users' community who have been employed as monitors and enforcers. Some enforcement mechanisms involve the users themselves as their own monitors. When the enforcement mechanism is not an external governmental agency, some analysts presume that there is no enforcement. That is why Game 5 is mistaken for Game 1.

A self-financed contract-enforcement game is no panacea. Such institutional arrangements have many weaknesses in many settings. The herders can overestimate or underestimate the carrying capacity of the meadow. Their own monitoring system may break down. The external enforcer may not be able to enforce ex post, after promising to do so ex ante. A myriad of problems can occur in natural settings, as is also the case with the idealized central-regulation or private-property institutions.

The structure of the institutional arrangements that one finds in natural settings is, of course, far more complicated than the structure of any of the extremely simple games presented here for discussion. What I attempt to do with these simple games is to generate different ways of thinking about the mechanisms that individuals may use to extricate themselves from commons dilemmas – ways different from what one finds in much of the policy literature. To challenge this mind-set, one needs only simple mechanisms that illustrate alternatives to those that normally are presented as the dominant solutions.

AN EMPIRICAL ALTERNATIVE

Game 5 illustrated a theoretical alternative to centralization or privatization as ways to solve CPR problems. Let us now briefly consider a solution devised by participants in a field setting – Alanya, Turkey – that cannot be characterized as either central regulation or privatization. The inshore fishery at Alanya, as described by Fikret Berkes (1986b), is a relatively small operation. Many of the approximately 100 local fishers operate in two – or three-person boats using various types of nets. Half of the fishers belong to a local producers' cooperative. According to Berkes, the early 1970s were the "dark ages" for Alanya. The economic viability of the fishery was threatened by two factors: First, unrestrained use of the fishery had led to hostility and, at times, violent conflict among the users. Second, competition among fishers for the better fishing spots had increased production costs, as well as the level of uncertainty regarding the harvest potential of any particular boat.

Early in the 1970s, members of the local cooperative began experimenting with an ingenious system for allotting fishing sites to local fishers. After more than a decade of trial-and-error efforts, the rules used by the Alanya inshore fishers are as follows:

- Each September, a list of eligible fishers is prepared, consisting of all licensed fishers in Alanya, regardless of co-op membership.
- Within the area normally used by Alanya fishers, all usable fishing locations are named and listed. These sites are spaced so that the nets set in one site will not block the fish that should be available at the adjacent sites.
- These named fishing locations and their assignments are in effect from September to May.
- In September, the eligible fishers draw lots and are assigned to the named fishing locations.
- From September to January, each day each fisher moves east to the next location. After January, the fishers move west. This gives the fishers equal opportunities at the stocks that migrate from east to west between September and January and reverse their migration through the area from January to May (Berkes 1986b, pp. 73–4).

The system has the effect of spacing the fishers far enough apart on the fishing grounds that the production capabilities at each site are optimized. All fishing boats also have equal chances to fish at the best spots. Resources are not wasted searching for or fighting over a site.[20] No signs of overcapitalization are apparent.

The list of fishing locations is endorsed by each fisher and deposited with the mayor and local gendarme once a year at the time of the lottery. The process of monitoring and enforcing the system is, however, accomplished by the fishers themselves as a by-product of the incentive created by the rotation system. On a day when a given fisher is assigned one of the more productive spots, that fisher will exercise that option with certainty (leaving aside last-minute breakdowns in equipment). All other fishers can expect that the assigned fisher will be at the spot bright and early. Consequently, an effort to cheat on the system by traveling to a good spot on a day when one is assigned to a poor spot has little chance of remaining undetected. Cheating on the system will be observed by the very fishers who have rights to be in the best spots and will be willing to defend their rights using physical means if necessary. Their rights will be supported by everyone else in the system. The others will want to ensure that their own rights will not be usurped on the days when they are assigned good sites. The few infractions that have occurred have been handled easily by the fishers at the local coffeehouse (Berkes 1986b, p. 74).

Although this is not a private-property system, rights to use fishing sites and duties to respect these rights are well defined. And though it is not a centralized system, national legislation that has given such cooperatives jurisdiction over "local arrangements" has been used by cooperative officials to legitimize their role in helping to devise a workable set of rules. That local officials accept the signed agreement each year also enhances legitimacy. The actual monitoring and enforcing of the rules, however, are left to the fishers.

Central-government officials could not have crafted such a set of rules without assigning a full-time staff to work (actually fish) in the area for an extended period. Fishing sites of varying economic value are commonly associated with inshore fisheries (Christy 1982; Forman

1967), but they are almost impossible to map without extensive on-site experience. Mapping this set of fishing sites, such that one boat's fishing activities would not reduce the migration of fish to other locations, would have been a daunting challenge had it not been for the extensive time-and-place information provided by the fishers and their willingness to experiment for a decade with various maps and systems. Alanya provides an example of a self-governed common-property arrangement in which the rules have been devised and modified by the participants themselves and also are monitored and enforced by them.

The case of the Alanya inshore fishery is only one empirical example of the many institutional arrangements that have been devised, modified monitored, and sustained by the users of renewable CPRs to constrain individual behavior that would, if unconstrained, reduce joint returns to the community of users. In addition to the case studies discussed in Chapters 3, 4, and 5, productive CPR institutional arrangements have been well documented for many farmer-managed irrigation systems, communal forests, inshore fisheries, and grazing and hunting territories.[21]

Game 5 and empirical cases of successfully governed CPRs provide theoretical and empirical alternatives to the assertion that those involved cannot extricate themselves from the problems faced when multiple individuals use a given resource. The key to my argument is that some individuals have broken out of the trap inherent in the commons dilemma, whereas others continue remorsefully trapped into destroying their own resources.[22] This leads me to ask what differences exist between those who have broken the shackles of a commons dilemma and those who have not. The differences may have to do with factors *internal* to a given group. The participants may simply have no capacity to communicate with one another, no way to develop trust, and no sense that they must share a common future. Alternatively, powerful individuals who stand to gain from the current situation, while others lose, may block efforts by the less powerful to change the rules of the game. Such groups may need some form of external assistance to break out of the perverse logic of their situation.

The differences between those who have and those who have not extricated themselves from commons dilemmas may also have to do with factors *outside* the domain of those affected. Some participants do not have the autonomy to change their own institutional structures and are prevented from making constructive changes by external authorities who are indifferent to the perversities of the commons dilemma, or may even stand to gain from it. Also, there is the possibility that external changes may sweep rapidly over a group, giving them insufficient time to adjust their internal structures to avoid the suboptimal outcomes. Some groups suffer from perverse incentive systems that are themselves the results of policies pursued by central authorities. Many potential answers spring to mind regarding the question why some individuals do not achieve collective benefits for themselves, whereas others do. However, as long as analysts presume that individuals cannot change such situations themselves, they do not ask what internal or external variables can enhance or impede the efforts of communities of individuals to deal creatively and constructively with perverse problems such as the tragedy of the commons.

POLICY PRESCRIPTIONS AS METAPHORS
Policy analysts who would recommend a single prescription for commons problems have paid little attention to how diverse institutional arrangements operate in practice. The centrists

presume that unified authorities will operate in the field as they have been designed to do in the textbooks – determining the best policies to be adopted for a resource based on valid scientific theories and adequate information. Implementation of these policies without error is assumed. Monitoring and sanctioning activities are viewed as routine and nonproblematic.

Those advocating the private-property approach presume that the most efficient use patterns for CPRs will actually result from dividing the rights to access and control such resources. Systematic empirical studies have shown that private organization of firms dealing in goods such as electricity, transport, and medical services tends to be more efficient than governmental organization of such firms; for a review of this literature, see De Alessi (1980). Whether private or public forms are more efficient in industries in which certain potential beneficiaries cannot be excluded is, however, a different question. We are concerned with the types of institutions that will be most efficient for governing and managing diverse CPRs for which at least some potential beneficiaries cannot be excluded. Privatizing the ownership of CPRs need not have the same positive results as privatizing the ownership of an airline. Further, privatizing may not mean "dividing up" at all. Privatization can also mean assigning the exclusive right to harvest from a resource system to a single individual or firm.

Many policy prescriptions are themselves no more than metaphors. Both the centralizers and the privatizers frequently advocate oversimplified, idealized institutions – paradoxically, almost "institution-free" institutions. An assertion that central regulation is necessary tells us nothing about the way a central agency should be constituted, what authority it should have, how the limits on its authority should be maintained, how it will obtain information, or how its agents should be selected, motivated to do their work, and have their performances monitored and rewarded or sanctioned. An assertion that the imposition of private property rights is necessary tells us nothing about how that bundle of rights is to be defined, how the various attributes of the goods involved will be measured, who will pay for the costs of excluding nonowners from access, how conflicts over rights will be adjudicated, or how the residual interests of the right-holders in the resource system itself will be organized.

An important lesson that one learns by carefully studying the growing number of systematic studies by scholars associated with "the new institutionalism" is that these "institutional details" are important.[23] Whether or not any equilibria are possible and whether or not an equilibrium would be an improvement for the individuals involved (or for others who are in turn affected by these individuals) will depend on the particular structures of the institutions. In the most general sense, all institutional arrangements can be thought of as games in extensive form. As such, the particular options available, the sequencing of those options, the information provided, and the relative rewards and punishments assigned to different sequences of moves can all change the pattern of outcomes achieved. Further, the particular structure of the physical environment involved also will have a major impact on the structure of the game and its results. Thus, a set of rules used in one physical environment may have vastly different consequences if used in a different physical environment.

POLICIES BASED ON METAPHORS CAN BE HARMFUL

Relying on metaphors as the foundation for policy advice can lead to results substantially different from those presumed to be likely. Nationalizing the ownership of forests in Third

World countries, for example, has been advocated on the grounds that local villagers cannot manage forests so as to sustain their productivity and their value in reducing soil erosion. In countries where small villages had owned and regulated their local communal forests for generations, nationalization meant expropriation. In such localities, villagers had earlier exercised considerable restraint over the rate and manner of harvesting forest products. In some of these countries, national agencies issued elaborate regulations concerning the use of forests, but were unable to employ sufficient numbers of foresters to enforce those regulations. The foresters who were employed were paid such low salaries that accepting bribes became a common means of supplementing their income. The consequence was that nationalization created *open-access resources* where limited-access *common-property resources* had previously existed. The disastrous effects of nationalizing formerly communal forests have been well documented for Thailand (Feeny 1988a), Niger (Thomson 1977; Thomson, Feeny, and Oakerson 1986), Nepal (Arnold and Campbell 1986; Messerschmidt 1986), and India (Gadgil and Iyer 1989). Similar problems occurred in regard to inshore fisheries when national agencies presumed that they had exclusive jurisdiction over all coastal waters (Cordell and McKean 1986; W. Cruz 1986; Dasgupta 1982; Panayoutou 1982; Pinkerton 1989a).

A CHALLENGE

An important challenge facing policy scientists is to develop theories of human organization based on realistic assessment of human capabilities and limitations in dealing with a variety of situations that initially share some or all aspects of a tragedy of the commons. Empirically validated theories of human organization will be essential ingredients of a policy science that can inform decisions about the likely consequences of a multitude of ways of organizing human activities. Theoretical inquiry involves a search for regularities. It involves abstraction from the complexity of a field setting, followed by the positing of theoretical variables that underlie observed complexities. Specific models of a theory involve further abstraction and simplification for the purpose of still finer analysis of the logical relationships among variables in a closed system. As a theorist, and at times a modeler, I see these efforts at the core of a policy science.

One can, however, get trapped in one's own intellectual web. When years have been spent in the development of a theory with considerable power and elegance, analysts obviously will want to apply this tool to as many situations as possible. The power of a theory is exactly proportional to the diversity of situations it can explain. All theories, however, have limits. Models of a theory are limited still further because many parameters must be fixed in a model, rather than allowed to vary. Confusing a model – such as that of a perfectly competitive market – with the theory of which it is one representation can limit applicability still further.

Scientific knowledge is as much an understanding of the diversity of situations for which a theory or its models are relevant as an understanding of its limits. The conviction that all physical structures could be described in terms of a set of perfect forms – circles, squares, and triangles – limited the development of astronomy until Johannes

Kepler broke the bonds of classical thought and discovered that the orbit of Mars was elliptical – a finding that Kepler himself initially considered to be no more than a pile of dung (Koestler 1959). Godwin and Shepard (1979) pointed out a decade ago that policy scientists were doing the equivalent of "Forcing Squares, Triangles and Ellipses into a Circular Paradigm" by using the commons-dilemma model without serious attention to whether or not the variables in the empirical world conformed to the theoretical model. Many theoretical and empirical findings have been reported since Godwin and Shepard's article that should have made policy scientists even more skeptical about relying on a limited set of models to analyze the diversity of situations broadly referred to as CPR problems. Unfortunately, many analysts – in academia, special-interest groups, governments, and the press – still presume that common-pool problems are all dilemmas in which the participants themselves cannot avoid producing suboptimal results, and in some cases disastrous results.

What is missing from the policy analyst's tool kit – and from the set of accepted, well-developed theories of human organization – is an adequately specified theory of collective action whereby a group of principals can organize themselves voluntarily to retain the residuals of their own efforts. Examples of self-organized enterprises abound. Most law firms are obvious examples: A group of lawyers will pool their assets to purchase a library and pay for joint secretarial and research assistance. They will develop their own internal governance mechanisms and formulas for allocating costs and benefits to the partners. Most cooperatives are also examples. The cases of self-organized and self-governed CPRs that we consider in Chapter 3 are also examples. But until a theoretical explanation – based on human choice – for self-organized and self-governed enterprises is fully developed and accepted, major policy decisions will continue to be undertaken with a presumption that individuals cannot organize themselves and always need to be organized by external authorities.

Further, all organizational arrangements are subject to stress, weakness, and failure. Without an adequate theory of self-organized collective action, one cannot predict or explain when individuals will be unable to solve a common problem through self-organization alone, nor can one begin to ascertain which of many intervention strategies might be effective in helping to solve particular problems. As discussed earlier, there is a considerable difference between the presumption that a regulatory agency should be established and the presumption that a reliable court system is needed to monitor and enforce self-negotiated contracts. If the theories being used in a policy science do not include the possibility of self-organized collective action, then the importance of a court system that can be used by self-organizing groups to monitor and enforce contracts will not be recognized.[24]

I hope this inquiry will contribute to the development of an empirically supported theory of self-organizing and self-governing forms of collective action. What I attempt to do in this volume is to combine the strategy used by many scholars associated with the "new institutionalism" with the strategy used by biologists for conducting empirical work related to the development of a better theoretical understanding of the biological world.

As an institutionalist studying empirical phenomena, I presume that individuals try to solve problems as effectively as they can. That assumption imposes a discipline on me. Instead of presuming that some individuals are incompetent, evil, or irrational, and others are omniscient,

I presume that individuals have very similar limited capabilities to reason and figure out the structure of complex environments. It is my responsibility as a scientist to ascertain what problems individuals are trying to solve and what factors help or hinder them in these efforts. When the problems that I observe involve lack of predictability, information, and trust, as well as high levels of complexity and transactional difficulties, then my efforts to explain must take these problems overtly into account rather than assuming them away. In developing an explanation for observed behavior, I draw on a rich literature written by other scholars interested in institutions and their effects on individual incentives and behaviors in field settings.

Biologists also face the problem of studying complex processes that are poorly understood. Their scientific strategy frequently has involved identifying for empirical observation the simplest possible organism in which a process occurs in a clarified, or even exaggerated, form. The organism is not chosen because it is representative of all organisms. Rather, the organism is chosen because particular processes can be studied more effectively using this organism than using another.

My "organism" is a type of human situation. I call this situation a CPR situation and define exactly what I mean by this and other key terms in Chapter 2. In this volume, I do not include all potential CPR situations within the frame of reference. I focus entirely on small-scale CPRs, where the CPR is itself located within one country and the number of individuals affected varies from 50 to 15,000 persons who are heavily dependent on the CPR for economic returns. These CPRs are primarily inshore fisheries, smaller grazing areas, groundwater basins, irrigation systems, and communal forests. Because these are relatively small-scale situations, serious study is more likely to penetrate the surface complexity to identify underlying similarities and processes. Because the individuals involved gain a major part of their economic return from the CPRs, they are strongly motivated to try to solve common problems to enhance their own productivity over time. The effort to self-organize in these situations may be somewhat exaggerated, but that is exactly why I want to study this process in these settings. Further, when self-organization fails, I know that it is not because the collective benefits that could have been obtained were unimportant to the participants.

There are limits on the types of CPRs studied here: (1) renewable rather than nonrenewable resources, (2) situations where substantial scarcity exists, rather than abundance, and (3) situations in which the users can substantially harm one another, but not situations in which participants can produce major external harm for others. Thus, all asymmetrical pollution problems are excluded, as is any situation in which a group can form a cartel and control a sufficient part of the market to affect market price.

In the empirical studies, I present a synopsis of important CPR cases that have aided my understanding of the processes of self-organization and self-governance. These cases are in no sense a "random" sample of cases. Rather, these are cases that provide clear information about the processes involved in (1) governing long-enduring CPRs, (2) transforming existing institutional arrangements, and (3) failing to overcome continued CPR problems. These cases can thus be viewed as a collection of the most salient raw materials with which I have worked in my effort to understand how individuals organize and govern themselves to obtain collective benefits in situations where the temptations to free-ride and to break commitments are substantial.

From an examination and analysis of these cases, I attempt to develop a series of reasoned conjectures about how it is possible that some individuals organize themselves to govern and manage CPRs and others do not. I try to identify the underlying design principles of the institutions used by those who have successfully managed their own CPRs over extended periods of time and why these may affect the incentives for participants to continue investing time and effort in the governance and management of their own CPRs. I compare the institutions used in successful and unsuccessful cases, and I try to identify the internal and external factors that can impede or enhance the capabilities of individuals to use and govern CPRs.

I hope these conjectures contribute to the development of an empirically valid theory of self-organization and self-governance for at least one well-defined universe of problematical situations. That universe contains a substantial proportion of renewable resources heavily utilized by human beings in different parts of the world. It is estimated, for example, that 90% of the world's fishermen and over half of the fish consumed each year are captured in the small-scale, inshore fisheries included within the frame of this study (Panayoutou 1982, p. 49). Further, my choice of the CPR environment for intensive study was based on a presumption that I could learn about the processes of self-organization and self-governance of relevance to a somewhat broader set of environments.

Given the similarity between many CPR problems and the problems of providing small-scale collective goods, the findings from this volume should contribute to an understanding of the factors that can enhance or detract from the capabilities of individuals to organize collective action related to providing local public goods. All efforts to organize collective action, whether by an external ruler, an entrepreneur, or a set of principals who wish to gain collective benefits, must address a common set of problems. These have to do with coping with free-riding, solving commitment problems, arranging for the supply of new institutions, and monitoring individual compliance with sets of rules. A study that focuses on how individuals avoid free-riding, achieve high levels of commitment, arrange for new institutions, and monitor conformity to a set of rules in CPR environments should contribute to an understanding of how individuals address these crucial problems in some other settings as well.

Let me now give a brief sketch of how this book is organized. In Chapter 2, I define what I mean by a CPR situation and individual choice in a CPR situation. Then I examine a series of crucial questions that any theory of collective action must answer. To conclude the chapter, I examine two assumptions that have framed prior work and discuss the alternatives that frame my analysis. The empirical part of this volume is contained in Chapters 3, 4, and 5, where I examine specific cases of long-enduring CPR institutions and resources, the origin and development of CPR institutions, and CPR failures and fragilities. At the end of each empirical chapter, I consider what can be learned from the cases in that chapter that will contribute toward the development of a better theory of self-organization related to CPR environments. In Chapter 6, I pull together the theoretical reflections contained at the ends of Chapters 3, 4, and 5 and address the implications of these conjectures for the design of self-organizing and self-governing institutions.

NOTES

1 Attributed to Merrill M. Flood and Melvin Dresher and formalized by Albert W. Tucker (R. Campbell 1985, p. 3), the game is described (Luce and Raiffa 1957, p. 95) as follows: "Two suspects are taken into custody and separated. The district attorney is certain that they are guilty of a specific crime, but he does not have adequate evidence to convict them at a trial. He points out to each prisoner that each has two alternatives: to confess to the crime the police are sure they have done, or not to confess. If they both do not confess, then the district attorney states he will book them on some very minor trumped-up charge such as petty larceny and illegal possession of a weapon, and they will both receive minor punishment; if they both confess they will be prosecuted, but he will recommend less than the most severe sentence; but if one confesses and the other does not, then the confessor will receive lenient treatment for turning state's evidence whereas the latter will get 'the book' slapped at him. In terms of years in a penitentiary, the strategic problem might be reduced" to the following:

	Prisoner 2	
Prisoner 1	**Not confess**	**Confess**
Not confess	1 year each	10 years for prisoner 1 3 months for prisoner 2
Confess	3 months for prisoner 1 10 years for prisoner 2	8 years each

R. Kenneth Godwin and W. Bruce Shepard (1979), Richard Kimber (1981), Michael Taylor (1987), and others have shown that commons dilemmas are not always prisoner's dilemma (PD) games. Dawes (1973, 1975) was one of the first scholars to show the similarity of structure.

2 Hardin's model easily translates into the prisoner's dilemma structure. Many problems related to the use of common-pool resources (CPRs) do *not* easily translate. Simple games such as "chicken" and "assurance" games are better representations of some situations (M. Taylor 1987). More complex games involving several moves and lacking dominant strategies for the players are better able to capture many of the problems involved in managing CPRs.

3 Hardin recommends "mutual coercion, mutually agreed upon" as a solution to the problem, but what "mutual agreement" means is ambiguous given his emphasis on the role of central regulators; see Orr and Hill (1979) for a critique.

4 A howling debate raged for some time, for example, regarding whether the number of participants involved was positively, negatively, or not at all related to the quantity of the good provided (Buchanan 1968; Chamberlin 1974; Frohlich and Oppenheimer 1970; McGuire 1974). Russell Hardin (1982) resolved the controversy to a large extent by pointing out that the effect of the number of contributors was largely dependent on the type of collective benefits being provided – whether or not each unit of the good was subtractable. Thus, the initial debate did not lead to clarification until implicit assumptions about the type of good involved had been made explicit.

5 J. A. Moore (1985, p. 483), reporting on the education project for the American Society of Zoologists.

6 See, for example, Berkes (1987), Berkes and Kislalioglu (1989), Berkes and Pocock (1981), A. Davis (1984), K. Martin (1979), Matthews and Phyne (1988). For strong critiques of Canadian policy, see Pinkerton (1989a,b) and Matthews (1988).

7   Michael Taylor (1987) analyzes the structure of Hobbes's theory to show that Hobbes proposed the creation of a Leviathan in order to avoid the equilibrium of situations structured like prisoner's dilemmas. See also Sugden (1986).

8   Stillman (1975, p. 13) points out that those who see "a strong central government or a strong ruler" as a solution implicitly assume that "the ruler will be a wise and ecologically aware altruist," even though these same theorists presume that the users of CPRs will be myopic, self-interested, and ecologically unaware hedonists.

9   The form of regulation used in Game 2 would be referred to in the resource economics literature as a "pure quota scheme." Alternative regulatory instruments that are frequently proposed are a "pure licensing scheme" and a "pure tax scheme." As Dasgupta and Heal (1979) point out, however, it is "the" government in each of these schemes that takes control of the resource and sets up the regulatory scheme. "The idea, in each case, is for the government to take charge of the common property resource and to introduce regulations aimed at the attainment of allocative efficiency" (Dasgupta and Heal 1979, p. 66). All of the models of these various schemes assume that the costs of sustaining these systems are nil (as in Game 2). Dasgupta and Heal repeatedly stress that these costs are *not* nil in field settings and may affect whether or not any of them actually will solve a commons problem or the relative efficiency of one scheme versus another. But Dasgupta and Heal's careful warnings about the importance of the relative costs of various constitutional arrangements are rarely heeded in the policy literature.

10 More accurately, the sum of the two types of errors must be less than 0.50, given the fixed parameters of this game, for the restructured game to have a (C, C) equilibrium. I am grateful to Franz Weissing, who suggested this particular analysis for illustrating the problem of incomplete information on the part of a central agency.The last two decades of work in social-choice theory also have revealed other problems that may be involved in any system where a collective choice about policy must be reached through mechanisms of collective choice. Even if complete information is available about the resources, problems associated with cycling and/or agenda control can also occur (McKelvey 1976, 1979; Riker 1980; Shepsle 1979a).

11 This overlooks the fact that in a dynamic setting the decision whether to manage the meadow at a sustainable level or to "mine" it rapidly will depend delicately on the discount rate used by the private owner. If the discount rate is high, the private owner will "overuse" a commons just as much as will a series of unorganized co-owners. See Clark (1977) for a clear statement of how overexploitation can occur under private property.

12 And it should be pointed out that the private-rights system is itself a *public* institution and is dependent on public instrumentalities for its very existence (Binger and Hoffman 1989).

13 My thanks again to Franz Weissing, who suggested this symmetric version of the contract-enforcement game. I had originally modeled Game 5 giving one herder the right to offer a contract, and the second herder only the right to agree or not agree to it.

14 See the interesting paper by Okada and Kleimt (1990), in which they model a three-player contract-enforcement game using the rule that any two (or three) persons who agree can set up their own contract to be enforced by an external agent. They conclude that three persons will not make use of a costless enforcement process, whereas two may. The article helps to illustrate how very subtle changes in conditions make important differences in results.

15 Williamson (1983) argues, however, that the numbers of actual unresolved PD situations in long-term business relationships have been exaggerated because economists have overlooked the contracts that businesses negotiate to change the structure of incentives related to long-term contracts.

16 Much of the literature in the new institutional economics tradition has stressed the importance of private orderings in the governance of long-term private contracts (Galanter 1981; Williamson 1979, 1985).

17 When considerable competition exists among arbitrators for the job of monitoring and enforcing, one can assume that arbiters are strongly motivated to make fair decisions. If there is no competition, then one faces the same problem in presuming fair decisions as one does in relation to a public bureau with monopoly status.

18 Simply iterating the PD game is not a guaranteed way out of the dilemma. The famous "folk theorem" that cooperation is a possible perfect equilibrium outcome is sometimes misrepresented as asserting that cooperation is the only equilibrium in repeated games. In addition to the "all cooperate at every iteration" equilibrium, many other equilibria are also possible. Simple repetition without enforceable agreements does not produce a clear result (Güth, Leininger, and Stephan 1990).

19 Private orderings frequently are mistaken for *no* order, given the absence of an official formal legislative or court decision. See Galanter (1981) for a review of the extensive literature on private orderings.

20 The formal game-theoretical structures and outcomes of this and three other sets of rules for allocating fishing sites are analyzed by Gardner and E. Ostrom (1990).

21 See, for example, the cases contained in National Research Council (1986), McCay and Acheson (1987), Fortmann and Bruce (1988), Berkes (1989), Pinkerton (1989a), Ruddle and Akimichi (1984), Coward (1980), and Uphoff (1986c). In addition to these collections, see citations in F. Martin (1989) for the extensive literature contained in books, monographs, articles, and research reports. There are also common-property institutions that break down when challenged by very rapid population growth or changes in the market value of the products harvested from the CPR. As discussed in Chapter 5, however, fragility of common-property systems is much more likely when these systems are not recognized by the formal political regimes of which they are a part.

22 That the "remorseless logic" was built into Hardin's assumptions, rather than being an empirical result, was pointed out by Stillman (1975, p. 14): "But the search for a solution cannot be found within the parameters of the problem. Rather, the resolution can only be found by changing one or more of the parameters of the problem, by cutting the Gordian knot rather than untying it."

23 See Shepsle (1979a, 1989a), Shepsle and Weingast (1987), Williamson (1979, 1985), North and Weingast (1989), and North (1981).

24 One can search the development literature long and hard, for example, without finding much discussion of the importance of court systems in helping individuals to organize themselves for development. The first time that I mentioned to a group of AID officials the importance of having an effective court system as an intervention strategy to achieve development, there was stunned silence in the room. One official noted that in two decades of development work she had never heard of such a recommendation being made.

Scenario Assignment:
Public Participation Strategies

You are a consultant hired by a municipal planning agency. Your firm was selected, after a competitive bidding process, to produce the new master plan for a small city of about 60,000 that has been in decline over the past few decades. The city is racially divided, with an elderly Caucasian population and a more recent Latin American immigrant population. The elected city council includes mostly conservative white businessmen and a very outspoken (and Republican) young African American real estate developer. The city is located along a fairly large river that has been polluted by a badly designed regional wastewater treatment system, a now-defunct industrial plant—which is an unremediated Superfund site—and the runoff from a range of large-scale agricultural activities upstream. Unless and until the river is cleaned up, downtown redevelopment and new investment along the river's edge are unlikely.

The city planning department has been told to produce a new master plan (the old one is almost 15 years out of date). The city council wants a plan that will ensure long-term economic growth and short-term pollution reduction. The only way to encourage new economic investment is to convince people that the greening of the city and a new sustainable pattern of investment will attract additional residents with money to spend.

1. What kind of public participation strategy will your firm recommend as part of the new master planning effort?
2. In general, there is no tradition of public participation in the city. Assume you have a budget of $50,000 to support whatever public involvement activities you choose. What techniques will best help educate and involve all the segments and factions in the city?
3. How will you justify your public engagement strategy to a skeptical city council?

Scenario Assignment:
Regional Consensus Building

You are the executive director of a metropolitan planning agency in the Southwest. Your board includes most of the chief elected officials of the cities and towns in the region along with the heads of numerous stakeholder groups. Growth has mushroomed over the past two decades, although the recent economic slowdown has brought new development to a halt. There is some hope that housing development will pick up, especially on the rural fringe. Open space has been eaten up at a frightening rate. Of even greater concern is that water supplies are clearly insufficient to sustain another round of development given the demands for water from the industrial, residential, agricultural and conservation "sectors." Battles over water allocations and investment in new water supplies are likely to tear communities apart.

Your board has attached top priority to formulating a regional water strategy that balances the demands and interests of all the competing groups. You have identified a senior staff member to facilitate this effort. With the board's approval you have appointed a 12-member blue ribbon advisory group to formulate a regional water strategy. You are pretty sure you have the money you need to staff a 12-to–18-month effort. You assume that the advisory group will tap appropriate (volunteer) academic and industry experts from universities and businesses in the region. You are worried, though, that some of the most extreme environmental and industry groups (which were purposefully left off the advisory group) will try to sabotage the effort. You are also worried that the state government will try to undermine what you are trying to do. The state does not think very highly of regional planning, preferring to do everything on a state-wide basis. Finally, the major newspaper in the region has already ridiculed the idea of a regional water strategy, arguing that senior water rights are held by those who have always held them (by law) and that nothing can be done to change that. They also point out that your agency is powerless to do anything about the economic forces at work.

1. What strategy will you urge the advisory group and your staff director to follow? What can they do to generate an informed regional consensus that will have the political backing needed to make a difference?
2. If you want to advocate some kind of "blue ribbon advisory committee," how would you suggest it be structured? If not, what other public engagement/public education technique would you use?

Scenario Assignment:
Environmental Dispute Resolution

You work for the Chemical Manufacturers' Association (CMA) of America. This is a trade association dedicated to educating the public about the important contributions that your member companies make to the economy and to the quality of everyday life. The CMA has decided to expand its environmental staff to demonstrate that the CMA is concerned about sustainability and environmental quality. You took them at their word when you decided to join the staff several years ago. Computer chip manufacturing plants are facing increased local opposition to expansion. Even though you are convinced that these plants do what is necessary to protect abutters from the dangers associated with water pollution or air pollution caused by leaks from the plant, you have been unable to generate anything that is taken seriously by your critics and opponents. It is true that there could be a fire at a plant and that the impacts of such an accident could be serious. But the same is true of almost any manufacturing facility. CMA members have adopted a Good Neighbor Pledge indicating their promise to minimize the risks of pollution and to accept full responsibility for any adverse impacts their facilities might cause.

The latest battle in the Southwest is between America's largest chip-making company and a group of environmentalists who have decided to take a stand against the proposed expansion of a plant that has been in place for more than two decades. The opponents claim to have evidence that shows an abnormally high rate of certain cancers in the area around the plant. While they might be correct, there is nothing to tie the incidence of cancer to what is going on at the plant. The leaders of the CMA have asked you to formulate a plan for meeting with and working out differences with the critics of this site. You know that all eyes nationally will be on what happens. So, you need to think in terms of a pilot process that could be repeated at many other sites.

1. How do you propose to tackle this conflict in a way that might lead to a workable agreement to proceed with expansion of the plant?
2. What principles would you use to guide the design of a dispute resolution process given that the conflict involves groups with radically different values and perhaps a range of hidden agendas?

End of Unit IV Written Assignment:
Public Interest and Group Decision-Making

There is an ongoing debate between political philosophers and dispute resolution professionals regarding the most appropriate means of conceptualizing the public interest (with regard to the use of natural resources or patterns of urban development). Some political philosophers believe "deliberative polling," which provides a snapshot of what the "average citizen" prefers, should be sufficient for elected officials to determine what actions to take in the public interest. Dispute resolution professionals argue the public interest can best be understood as the product of a consensus-building dialogue among contending interests (not individuals) and that public officials armed with polling data can never know or produce on their own the public interest.

1. In light of what you read and heard in Unit IV, what is your view of this debate?
2. What should be most important, in your view, in assessing the relevant contributions that various public participation tools and techniques can make to environmental planning?
3. What is your reaction to the notion that a neutral facilitator can add value in important ways to environmental planning efforts?

First Example Response to Assignment:
Public Interest and the Consensus Building Approach

Public process tools, consensus building included, have enormous potential to resolve disputes, bring people together and fundamentally alter the dynamics of a situation. As a community mediator/facilitator, I have worked with clients who, 30 minutes prior to mediation, brawled in front of the court and afterward walked to lunch together, laughing. As an environmental mediator, I have managed consensus-building processes that resolved decade-old disputes, bringing stakeholders together around novel solutions. While I firmly believe that consensus-building processes offer the most value as a "public participation tool," many other tools, including deliberative polling, offer a way to explore the "public interest" and provide other benefits to public processes. The nature of the tool used also heavily depends on the specifics of the process goals as well as the available resources, money, time and skill. In almost all public participation processes, neutral facilitators add enormous value, with their benefits outweighing the additional financial costs.

Political philosophers claim that deliberative polling—a process through which "random" citizens engage in discussion (i.e., deliberation) with (competing or neutral) experts and material to develop informed and "reflective" opinions—sufficiently gauges the public interest. Alternative dispute resolution professionals argue that consensus building among competing interests best uncovers this public interest.

Deliberative polling entails several fundamental flaws that diminish its capacity to identify the public interest. Deliberative polling, like aggregative polling, assumes the existence of a "median voter" (Shapiro 2005). In this model, voters simply need more information to assume the role of this median voter and make the "right" choice, with lack of information the chief villain of effective, democratic decision-making. Moreover, by searching for the interests of this median voter, deliberative polling assumes equal weight to the interests of all citizens across issues, rather than emphasizing the interests of those likely most affected by the decision under deliberation.

Shapiro raises three additional challenges to deliberative polling, questioning who sets the agenda of the deliberation, who ensures the balance of the presented materials and experts, and whether the final outcome actually improves the status quo process outcome.

Agenda setting represents one of the consequential, if not the most consequential, phases of a public process. By setting the bounds of discussion, the agenda setter has already demarcated acceptable from unacceptable interests and solutions. An "expert" may set the options for discussion within a deliberative poll that reflect their own view of what is right rather than the public interest. Furthermore, while these experts should supposedly be neutral or evenly balanced, nonobjective opinions subtly infiltrate most expert judgments (Susskind and Dunlap 1981). Finally, the outcome of such a deliberative polling effort may not significantly improve the ultimate outcome, especially if the process managers bound the range of solutions developed by the "public interest" to preset, conventional choices.

The consensus-building approach rectifies many of the problems facing deliberative polling. The consensus-building approach does not presume the existence of median voter and opens the process to competing interests (Drazkiewicz, Challies and Newig 2015; Innes and Booher 1999). Consensus building, at its best, recognizes the power of competing self-interests to realize better outcomes for all (i.e., improving the public interest) through dialogue, creativity and trade-offs. Furthermore, a robust consensus-building approach offers solutions for Shapiro's criticisms of deliberative polling. Self-organization within a consensus-building approach allows stakeholders to set their own agendas and frame the issue to align with their own interests (Innes and Booher 1999). In addition, while power dynamics within a consensus-building approach must be managed, there is no perception of stakeholder neutrality; it is in fact the intermingling and resulting reactions among competing interests that catalyze creative, mutual gain solutions. Last, a consensus-building approach allows the stakeholders to form joint consensus proposals that ensure everyone can live with the final outcome—a promise deliberative polling cannot deliver.

However, despite the benefits of consensus building over deliberative polling as a means to determine public interest, many factors determine how to assess the relevant contributions of public participation tools to environmental planning. I apply two primary assessment lenses to these tools: resource requirements and results.

While not a maxim, there is often a trade-off between the resources required for a public participation tool and the quality of its results. In the case of deliberative polling versus a consensus-building approach, consensus building offers an improved realization of public interest, but a comprehensive consensus-building approach usually takes substantial time, money and expertise (Susskind and Cruikshank 2006). Deliberative polling, in contrast, provides a much quicker and cheaper way to conceptualize the public interest if resources are limited. The relative contribution of other public participation tools should also be partially assessed in this light. In an ideal situation, officials would endow public processes with the resources for the most effective participation tools for the given situation. However, almost all such processes face considerable resources constraints. In these cases, tools requiring fewer resources contribute more than do more effective, resource-intensive alternatives. The EPA's public participation guidelines divide tools into those that inform, generate/obtain input and build consensus/seek agreement. Tools within the first two categories still offer value in many public processes while tending to require less resources and can also be used to supplement an existing consensus-building approach.

In terms of results, Susskind's four evaluation measures of consensus-building success can be extended to assess the contributions of another participation tool: (1) Is the process and outcome fair in the eyes of stakeholders? (2) Is it efficient in terms of process and outcome (representing an improvement over parties' best alternatives to a negotiated agreement)? (3) Does it produce a stable outcome/agreement? (4) Does it produce a wise outcome? While not all tools will achieve these four outcomes, or are intended to, these metrics still provide an ideal against which to compare. Innes and Booher (1999) also raise the concept of "emancipatory knowledge," that is, knowledge freed from constraints and convention. The best participation tools will enable liminal spaces and creative arenas that allow stakeholders to explore new modes of thought, to develop truly novel solutions and approaches. Again, these criteria provide insight for considering the contributions of all public participation tools. The success of any tool should be considered on what plans, reports and regulations it produces but also based on the political capital it builds, the networks it creates and more.

One of the many ways professional neutrals can add value to environmental planning efforts is through the creation of a process that encourages the development of such "second-order" outcomes, including social capital, political capital, intellectual capital and innovation (Mandarano 2008). The establishment of a good process by a skilled neutral allows for these good outcomes and good outputs (ibid.). Professional neutrals also assist the environmental process by management of stakeholders outside of the at-the-table process through back-table conferencing and negotiation (Carson 2008; Susskind and Ozawa 1984). Much, if not the majority, of neutrals' work in long-term processes occurs between the formal stakeholder meetings. In this function, neutrals assuage concerns, mollify frustrations, solicit feedback about the last meeting to shape the agenda for the next meeting and encourage parties to forge forward together.

Neutrals also play a key role before and after the official process—before, through the creation of a situation/stakeholder assessment that shapes the stakeholders at the table and the framework of the process, and after, by ensuring smooth implementation of the stakeholders' proposal. In addition,

Forester and Stitzel (1989) raise the issue of the "negotiator's dilemma," in which parties do not reveal or prioritize their most important interests for fear of weakening their bargaining power. This can result in suboptimal agreements for all parties. Neutrals can assist parties in identifying their interests and priorities and trading across them, thereby creating value.

Acquiring professional neutrals does make a process more expensive, but this cost should be lower than the counterfactual case of failure—usually a legal challenge to the proposed product from upset parties, development of a suboptimal solution that leaves value on the table or the inability to create any product at all. Conveners must carefully consider whether they can accept and afford these risks if they decide not to hire a neutral to manage the process.

I argue that the consensus-building approach outperforms deliberative polling in determining the public interest through the collision of competing interests rather than assuming the existence of the median voter. However, as the contribution of public participation tools should be considered both on their results and required resources, deliberative polling and other tools offer value in the face of resources constraints. Neutrals, despite their additional cost, can play an invaluable role in furthering the success of environmental planning processes. As a mediator, I have seen the consensus-building approach result in near-miracles. People may or may not be altruistic at heart, but the consensus-building approach has the power to channel the worst in all of us into the best for all of us.

References

Carson, Lyn. 2008. "The IAP2 Spectrum: Larry Susskind in Conversation with IAP2 Members." *International Journal of Public Participation* 2, no. 2: 67–84.

Drazkiewicz, Anna, Edward Challies and Jens Newig. 2015. "Public Participation and Local Environmental Planning: Testing Factors Influencing Decision Quality and Implementation in Four Case Studies from Germany." *Land Use Policy* 46: 211–22.

Forester, J., and David Stitzel. 1989. "Beyond Neutrality." *Negotiation Journal* 5: 251–64.

Innes, Judith, and David Booher. 1999. "Consensus Building and Complex Adaptive Systems—A Framework for Evaluating Collaborative Planning." *APA Journal* 65, no. 4: 412–23.

Shapiro, I. 2005. *The State of Democracy Theory*. Princeton: Princeton University Press.

Susskind, L., and L. Dunlap. 1981. "The Importance of Nonobjective Judgments in Environmental Impact Assessments." *Environmental Impact Assessment Review* 2, no. 4: 335–66.

Susskind, Lawrence, and Connie Ozawa. 1984. "Mediated Negotiation in the Public Sector: The Planner as Mediator." *Journal of Planning Education and Research* 4, no. 1: 5–15.

Susskind, Lawrence, and Jeffrey Cruikshank. 2006. *Breaking Robert's Rules*. New York: Oxford University Press.

Second Example Response to Assignment:
Democracy and Environmental Decision-Making

As it becomes clearer and clearer that today's democracies are failing to meet the environmental needs of present and future citizens, political philosophers and professionals are considering alternative ways of conceptualizing the public interest in the hopes of enhancing policy-making. On the one hand, some believe that a "snapshot" of the average citizen, achieved through a deliberative polling process, will enhance officials' ability to act in the public interest. On the other hand, others argue that only through a dialogic process aimed at consensus between relevant stakeholders can the public interest in any one issue be met. This essay argues that deliberative polling is only a weak improvement on the status quo, and that consensus-based dialogues offer more hope for environmental policy makers. However, inherent to consensus building is the risk of undermining pluralism by silencing adversarial voices, and I therefore also lay out a role for a facilitator that I see as central to ensuring that agonistic politics thrive.

Despite the intention to fill knowledge gaps among participants, deliberative polling is problematic in that it assumes that respondents have finite, known interests (that are in turn identifiable by pollsters), and that they are able to choose between and rank or quantify these interests. Further, it assumes that the public interest can be found by aggregating responses to exogenously identified interests. The issues discussed, the framing of the agenda and the choice of "experts" who educate the respondents are all subjective, and as Shapiro (2006) argues, it is hard to see why this method would be owed any deference in a democracy. It is almost inevitable that those developing a poll will use what Fung (2007) calls "stylised facts" that in some way deviate from the facts of the world, thus rendering their democratic conception false and "out" of "pragmatic equilibrium."

Consensus building seeks to move beyond mere interest aggregation toward collective learning through the collaborative development of creative policy bundles that meet the interests of all stakeholders involved. This process tends to result in more legitimate, lasting and mutually beneficial outcomes. Consensus building holds at its heart the notion that the public interest is not exogenous but rather endogenous to the dialogue between contending interests. The method also posits the possibility for each group to move beyond the public interest as they may presently interpret it. An ideal against which to gauge the success of a consensus-building process is Jürgen Habermas's concept of "emancipatory knowledge," which Innes and Booher (1999) define as knowledge that transcends the "self-fulfilling rationalities" that societies and institutions develop. Moving beyond these blinders is crucial in environmental policy in particular, and a consensus-building approach where stakeholders are equally informed, listened to and represented is much more likely to produce emancipatory knowledge. At a table where various representatives have a range of "knowledges" about an environmental problem, from the panoramic to the local, and various types of expertise, from the scientific to the traditional, a poll would merely reflect the divisions among participants. In contrast, a consensus approach builds, from individual interests, a public interest that may well be different from that envisioned by each stakeholder group at the commencement of the process.

Just as it might reveal hidden possibilities for convergence, a consensus-building process may reveal hidden differences. This outcome is equal in significance as a matter of public interest, and must be accepted as a potential result. If an attempt at consensus produces disagreement and conflict, this cannot be seen as inherently bad, particularly if the alternative is a hollow agreement that would only foster distrust. In the same vein, consensus building has the power to silence, and thus to reinforce, unequal power contexts if not properly facilitated. The power of consensus building is its ability to transform preferences, but this could easily happen to the exclusion of some, especially if the powerful define the process.

In order for a consensus approach to be conducive to emancipatory knowledge and agonistic politics, a facilitator is absolutely crucial. Facilitators must be equipped with more than just a bundle of process skills. In the choosing of a facilitator, the priority must be on context-based professional knowledge and an understanding of the range of cultural and institutional dynamics at work. Then, on top of this, conveners should look for an individual who is persuasive and creative—these skills are essential in accommodating competing interests in the design of alternatives. In environmental disputes, the stakeholders are often unorganized or unrepresentable (particularly in the case of nonhumans). It is the facilitator's responsibility to ensure that these interests are represented to the best of his or her ability and to structure the dialogue so that the force of arguments can outweigh the quantity of resources in terms of intellectual, social and financial capital.

The assumption that the facilitator be neutral is worth reconsidering, however. I would argue that key aspects of the facilitator's role would be compromised by true neutrality, and that to promise it would be unethical. The facilitator must be able to identify and alleviate knowledge and power imbalances around the negotiating table. One could argue that this constitutes a deviation from neutrality, traditionally understood, and I would respond that addressing power disparities is ultimately more important than a neutrality that merely accepts the status quo. Political order is the expression of power relations. The practice of democratic politics, therefore, must constitute forms of power that are compatible with democratic values. Even if the facilitator makes an offer of assistance in this regard to all parties, which he or she should do, the impetus for this offer is not a neutral one, in that it targets the incapacities of particular parties with the aim, even the hope, of altering the balance of power. This is not to say that the facilitator should be partisan, and what I have described is not partisanship—rather, it is an ethics of facilitation. It does not privilege environmental groups over business interests, or wetlands over homeowners. Rather, mediators must privilege equity and the preservation of adversarial politics, and in doing so they create a space between pure partisanship and the ideal of neutrality within which stakeholders have a voice, have the opportunity to use it and have the opportunity to solve problems creatively and in collaboration with others.

When it comes to assessing the contribution that public participation techniques make to environmental policy, the output (consensus, tally, etc.) is not the most important factor to consider. Rather, public participation processes must be judged by the extent to which they provide an arena where differences can be confronted. Consensus building is almost paradoxical in that, by postulating a public sphere where power can be suspended and consensus thereby reached, it potentially disavows the agonism that is central to democracy. If conflicts cannot take an agonistic form (a struggle between adversaries), they are more likely to become antagonistic (a struggle between enemies) and pose a danger to democratic societies (Mouffe 2013). So, although a consensus process is far better than polling when it comes to acting in the public interest, it also carries within its own rationality the threat of inhibiting the critical role that passion and emotion play in making democracy possible. In light of this, the first ethical priority of all involved must be to ensure that all "democratic forms of individuality and subjectivity" (Mouffe 2000, 95) remain available to all stakeholders. As Day, Gunton and Williams (2003) show in the case of British Columbia, agonistic politics—legal action, civil disobedience, boycotting—shifted the provincial balance of power and ultimately brought the resource extraction firms to the table in the first place. Consensus building must reinforce the capacity of all actors to engage in agonistic politics—it cannot be seen as a means to silence or to tame (for lack of a better word) a radical voice, and if this is the result, I would deem it an unsuccessful process.

A consensus is just a temporary stabilization of power, with all the implications of inequality that entails. In a democratic community, that power can and should shift, and this is why retaining the agonistic component of politics is also essential. In fact, agonism and consensus building are, to me, mutually reinforcing. Groups only embark on consensus building when they have reached some sort of political impasse. A political impasse only arises when stakeholders have the power to speak up when they disagree with how they are spoken for under the present conception of the public interest. Stakeholders can only speak up if we legitimize the confrontation of difference. Consensus building, at its strongest, legitimizes difference in the pursuit of mutually beneficial outcomes, thus laying out a path to deeper democracy.

Note

1 This argument is further developed in Hardin's article "Lifeboat Ethics: The Case Against Helping the Poor" (1974).

References

Day, J. C., T. I. Gunton and P. W. Williams. 2003. "Evaluating Collaborative Planning: The British Columbia Experience." *Environments* 31, no. 3: 1–11.
Fung, Archon. 2007. "Democratic Theory and Political Science: A Pragmatic Method of Constructive Engagement." *American Political Science Review* 101, no. 3:
Innes, Judith, and David Booher. 1999. "Consensus-building and Complex Adaptive Systems—A Framework for Evaluating Collaborative Planning." *Journal of the American Planning Association* 65, no. 4: 412–23.
Mouffe, Chantal. 2000. *The Democratic Paradox*. London: Verso.
Mouffe, Chantal. 2013. *Agonistics: Thinking the World Politically*. London: Verso.
Shapiro, Ian. 2006. *The State of Democratic Theory*. Princeton: Princeton University Press.

FINAL EXAM

Questions

1. The city has made a commitment to incorporate potential climate risks into its review of all proposed projects in the areas of the city prone to flooding. Along these lines, it has received a grant of $250,000 from a local foundation to underwrite the work of a "blue ribbon" science advisory committee to advise the city over the next 12 months on how to ensure that climate risks are managed appropriately in low-lying areas. A number of activist groups in the city are not included on the blue ribbon committee, even though they have been working for several years to promote more ambitious climate adaptation policies. They feel they should be given a leadership role in formulating the city's Flood Risk Management and Climate Adaptation Plan. In response, the mayor has announced that he will hold four "town hall meetings"—one every two months—in different locations to ensure that all groups have a chance to present their ideas on "flood risk management." He has also promised to hire a professional polling firm to undertake a scientific poll of public opinion on flood risk and climate adaptation. He will use the results of the poll and the town hall meetings to inform the work of his science advisory committee. Drawing upon what we have covered in this book, write a 250-to-500-word editorial or op-ed that will be published in the local newspaper either supporting or challenging the city's public engagement strategy on this issue.

2. Habitat conservation is important, especially in ecologically significant areas. Five years ago, the state government initiated a river restoration strategy and allocated substantial funding and professional staff to support its work. The focus of this initiative has been almost entirely on the larger rivers that run through the state and support all kinds of ecological, commercial and residential activities in the surrounding watershed. Your environmental management firm has been asked by the state legislature to organize an assessment of the river restoration program. Even before you collect any scientific evidence, you need to think about the structure of your evaluation. What kinds of things should you be trying to measure? Why? What approaches to data gathering and analysis will you use? What problems or difficulties do you anticipate? How will you try to handle them?

3. It would be foolhardy to make major investments in public infrastructure without trying to forecast and calibrate their likely impacts on humans and the environment. In this book, we have looked at a range of analytical tools that can be used to assess new infrastructure investments before or after they are made. Imagine you have been asked by the mayor of Boston to analyze the likely impacts of possible private investments in burying almost all power lines in the city. Which analytic method(s) would probably be most appropriate? Why? What nonobjective judgments would need to be made as part of any such analysis? How could you insulate such judgments from political and other challenges?

4. The new president's administration will be appointing a great many senior staff to multiple federal agencies. These individuals will have opportunities to initiate a range of administrative changes in policies and programs. On what basis can and should the new administration justify administrative (not legislative) shifts in national environmental and energy policy? That is, how should it decide and defend its environmental priorities? Should not environmental policy just reflect the "best science available" at a given moment? If something more than science ought to provide the basis for environmental policy-making by administrative agencies, what additional types of evidence and arguments should come into play? Why?

5. Significant differences in institutions, politics, culture and values among countries limit the transferability of environmental policies and programs from one nation to another. Does this mean that environmental policy is absolutely context-dependent? If so, why? If not, what can nations learn from one another's environmental management approaches and how transferable are environmental policies? Please select a specific example of environmental management or sustainability policy in a developing country to illustrate your answers.

6. Many states in America complain that the Environmental Protection Agency has preempted a wide range of policy-making and regulatory responsibilities. They want to leave these to state government. Fracking is one example of an environmental regulatory issue that remains almost entirely under the control of the states. There are very few federal regulations that apply. As might be expected, tates have adopted different regulatory regimes. New York has forbidden any fracking, while Texas relies heavily on private companies to self-regulate. Are you in favor of or opposed to such decentralized environmental regulation? Why? Draw on the discussions we had about policy-making, the philosophical underpinning of environmental policy and various uses of environmental assessment techniques.

7. Environmental planning is different from other kinds of planning in some respects yet similar in many ways. Select an environmental planning situation and describe what you see as the professional challenges associated with this situation that are unique to *environmental* planning. Describe what you see as the key features of this situation that are similar to any other (i.e., not specifically environmental) planning issue. Are the ethical considerations involved in environmental planning—and in the situation you have chosen to discuss—different from those involved in planning issues in general? Please explain.

8. Brownfield cleanup efforts are often contentious. One source of disagreement is how clean restored areas need to be. Some observers argue that since we have the necessary technology, we should restore contaminated areas to their "pristine" state to protect public health and re-establish lost ecosystem services. Others, taking a more pragmatic view, believe we need only reduce the risks to human health to "safe" levels and select appropriate uses (like parking lots) for remediated areas. Where do you come down on this question, especially when concerns about environmental (in)justice are involved? That is, if the decision about how clean is clean enough appears to vary depending on the income of the stakeholders involved, would that change your answer to the question?

9. In many developing countries, encouraging economic growth is paramount. Efforts to implement strict environmental regulations in these countries are often seen as a threat to economic growth. What is your view on this question? Select a particular country in which to discuss this issue. What would be the specific advantages and disadvantages of holding back on implementing aggressive environmental regulations in this country? Do you feel the same way about *all* environmental regulations or only certain kinds? (Which ones?) Is this issue of equal concern at both the national and the local level?

10. Scenario planning is one way of dealing with the substantial levels of uncertainty surrounding efforts to promote sustainable development. Is scenario planning a sufficient remedy given the many difficulties raised by uncertainty? What else can be done to deal with uncertainty in promoting sustainable development? Select a specific environmental or sustainable development policy and highlight the ways in which uncertainties of various kinds make it difficult to pursue this policy. Explain how scenario planning would help in this context. Offer at least one other idea for handling the uncertainty involved.

Sample Responses to Select Exam Questions

1. **The new President's administration will be appointing a great many senior staff to multiple federal agencies. These individuals will have opportunities to initiate a range of administrative changes in policies and programs. On what basis can and should the new administration justify administrative (not legislative) shifts in national environmental and energy policy? That is, how should it decide and defend its environmental priorities? Should not environmental policy just reflect the "best science available" at a given moment? If something more than science ought to provide the basis for environmental policy-making by administrative agencies, what additional types of evidence and arguments should come into play? Why?**

As the new administration takes shape and prepares to lead the many executive branch agencies responsible for the administrative running of this country, they will be planning and steering significant shifts in national energy and environment policy. Although I personally disagree with many of the declared stances the administration is taking and the policies it is advocating, they have the legal right to change the environmental priorities of this country within the boundaries of administrative discretion. As they do so, they will need to justify their administrative shifts in policy through the interpretation of existing legislative statutes and within the current structure of the Administrative Procedures Act. For certain policies and stances, they will be required to conduct prescribed rule-making processes. For others, the process is much more fluid, as they will be approving and rejecting permits, writing and setting standards, and providing or rescinding funding to various educational institutions, companies, groups, subagency sections and research institutions.

Many of their priorities will be based upon nonobjective values that are easily challenged if the courts and legislature are able and willing to do so. Yet, as the administration moves forward, they are bound in their discretion by existing legal statutes, precedents and legislation including a recent Supreme Court case requiring the EPA to address greenhouse gas emissions in the short term. Also, the sheer bureaucracy and slow-moving nature of the government and judiciary means that many priorities will be slow to take off and be completed. The administrative discretion given to the executive branch helps it be flexible and proactive in making choices without needing to seek judicial or legislative approval on every single decision. Discretion is important in providing a way to examine and apply current political realities to the real world, but it can also be an extremely dangerous power in the hands of partisan and unscrupulous managers.

As the new administration plans its shifts in policies and legislation, in order to be most efficient, it needs to ground them in unchanging, standardized values and stances that they can consistently operate off of and use as support for their policies. Such values and stances can come from a variety of sources. One potential source is the Republican Platform 2016, which would give them the support of the Republican Party as a whole as its members collectively agreed upon it. By using a source based upon the consensus of the party, the new administration can insulate itself from party infighting and present a united front to the country and the Democratic minority in Congress. This would also protect them, potentially, in the standard of review for executive branch determinations of fact against a ruling that their actions are arbitrary and capricious.

I believe that environmental policy should reflect the best science available as well as a clear consideration of the economic reality that this country faces. Yet, environmental science and economic analyses are not fool-proof—they can sometimes be unclear, overly technical and also often disagreed with. I do not believe that explicit religious value judgments belong in the governing of federal policy, as there is a clear definition of church and state in this country; nevertheless, individuals can base their values and morals upon their religious background and moral upbringing. Therefore, I believe that it is virtually impossible for nonobjective values to be excluded completely from environmental policy and regulation. Environmental policy-making is therefore never completely scientific in determination. As science deniers gain a foothold in environment and energy policy, they will be balancing their moral judgments against a growing scientific body. As they do so, it is very likely that they will be cherry-picking scientific and economic findings to present and justify their stances. Guided by nonobjective values and the inherent discretion provided to the executive branch in rulemaking and permitting, there is little to stop the new administration from making widespread changes across the government.

2. **Habitat conservation is important, especially in ecologically significant areas. Five years ago, the state government initiated a river restoration strategy and allocated substantial funding and professional staff to support its work. Its focus has been almost entirely on the larger rivers that run through the state and support all kinds of ecological, commercial and residential activities in the surrounding watershed. Your environmental management firm has been asked by the state legislature to organize an assessment of the river restoration program. Even before you collect any scientific evidence, you need to think about the structure of your evaluation. What kinds of things should you be trying to measure? Why? What approaches to**

data gathering and analysis will you use? What problems or difficulties do you anticipate? How will you try to handle them?

In order to adequately assess the river restoration program, it is important to construct a robust, public engagement–focused evaluation of both the process by which the river restoration program has been pursuing its activities and the outcomes of the program. According to Howlett and Perl, "both policies and programs can succeed or fail either in substantive terms—that is, delivering or failing to deliver the goods—or in procedural terms—as being legitimate or illegitimate, fair or unfair, just or unjust" (2009, 182).

The process of the river restoration program can be judged based on four criteria: fairness in the eyes of the various parties, efficiency in that it maximizes benefits given time and financial constraints and leaves no joint gains unclaimed, stability over a long period of time and wisdom based on available knowledge (Susskind and Cruikshank 1987). A strong public engagement process can help satisfy these four criteria, as people are more likely to be both informed about the program and engaged actively in trying to seek good outcomes.

Outcomes can also be judged by the four criteria listed above, with a few additions. The program can also be measured for its environmental, economic, social and health impacts. These criteria are more conducive to numerical measurement than the others. One can estimate the number of acres of river restored or the change in a local community's GDP, for example. These types of criteria are important in order to see if the program is accomplishing its goals, but they should also be balanced by the less tangible criteria to ensure the durability of the program in the long term. If people do not think the program is fair, for example, this could incite such problems as political backlash, which the state government surely does not want.

In order to gather data and conduct analysis based on the above criteria, it is important to have a strong evaluative process, which can also be judged on how fair, efficient, stable and wise it is as well. Any analysis includes subjective judgments, which is why it should stand up to public critique. According to Howlett and Perl, judgments about success and failure are malleable. Failure itself is a judgment (2009, 183).

Because of the subjective nature of evaluation, it is best to include representatives from as many different stakeholder groups as possible in evaluating the process and results. This can take the form of an advisory committee overseeing the evaluation. It could also take the form of a consensus conference where a panel of people representing the various stakeholder groups picks experts and engages in dialogue about the program. This group of people can help define the criteria that matter to them in judging the river restoration program. Because evaluations are key for identifying ways to change policies and programs in the future, involving the community in the evaluation process can help them shape their own future.

There are several problems that I can anticipate with my evaluation. As stated above, I try to overcome the problem of subjective judgments by involving as many different interests as possible in the process. This helps get the community bought into the process and also diffuses blame if something goes wrong. There are problems that come along with engaging large numbers of people in an evaluation, however. A great amount of time, effort and resources are required to maintain an evaluation committee or run a consensus conference. This will likely take some tenacity to pull off, justifying to the state that its money should be spent engaging in a public evaluation process over a cheaper and less time-intensive one-person assessment. Based on the potential benefits of getting the community involved in shaping their own future, however, I think I can convince them.

3. **Many states in the United States complain that the Environmental Protection Agency (EPA) has preempted a wide range of environmental policy-making and regulatory responsibilities. They want to leave these to state government. Fracking is one example of an environmental regulatory issue that remains almost entirely under the control of the states. There are very few federal regulations that apply. As might be expected, the states have adopted very different regulatory regimes. New York has forbidden any fracking, while Texas relies heavily on private companies to self-regulate. Are you in favor of or opposed to such decentralized environmental regulation? Why? Draw, to the extent possible, on the discussions we had during the semester about public policy-making, the philosophical underpinning of environmental policy and the uses of various environmental assessment techniques.**

Many US states argue that they deserve greater control over the management of their environmental resources, accusing the EPA of usurping their policy-making and regulatory authority. While this policy leads to discontent at the state level, overall centralized environmental authority in the United States serves to prevent a policy race to the bottom, protect the most vulnerable, avoid ethical lapses and allow joint collaboration on commons management problems. However, states should be granted certain leeway in environmental management to accommodate preferences, facilitate local management and tap into local knowledge.

Unified, federal authority over environmental regulations serves to forestall several potential environmentally disastrous outcomes. US policy-making is a highly political, subjective process (Howlett and Perl 2009). To realize economic growth, many states are likely to slash environmental regulations to compete for businesses, which would serve to catalyze similar policy deregulation in other states. At the end of this race to the bottom, no state has the economic advantage and is worse off (Cohen 2006). Furthermore, it is possible that through environmental deregulation processes, states are likely to pass the brunt of environmental costs to the most vulnerable populations, while those already advantaged, and politically powerful, realize the majority of gains. Federal protections for minority and other vulnerable groups play a powerful role in forestalling such outcomes. States pursue many such harmful environmental agendas using "holistic" cost-benefit analyses (CBAs) to realize utilitarian agendas (Cohen 2006). However, such CBAs often ignore the nonuse and nonconsumptive benefits of the environmental system, leading to short-term gains but much greater long – (and near-) term costs. Federal oversight restricts such deleterious decisions. As US policy bases its analyses on a utilitarian framework, state-level environmental regulation at both the state and federal level tend to further these utilitarian environmental ideas rather than those of competing philosophies, such as deep ecology, that place much greater value (and necessity) on environmental integrity. While federal authority does not further ideals of a competing philosophy, it does check the overambitiousness of short-term utilitarianism.

Finally, when making the majority of their environmental policy decisions, states act, reasonably, at the state level. However, many, if not most, environmental systems transcend state boundaries, and as such, actions in one state impact residents in others. The EPA's authority helps prevent states from enacting policies that will harm others. Furthermore, operation at the state level, without oversight, is likely to lead to "tragedy of the commons" problems as all states seek to maximize their use of the environment (Hardin 1968). EPA authority also allows the United States to better operate as a single environmental entity to work against many global "tragedy of the commons" issues, such as global warming.

However, there is justification for leaving bounded environmental discretion to state actors. Ostrom (1990) argues that local actors can collaborate to manage shared resources in the absence of "Leviathan" (government) and notes that government oversight of local commons can actually inhibit effective local management. Allowing certain environmental authority at the state level catalyzes innovate ideas, ripe for policy pinching. For example, stakeholders in Puget Sound successfully piloted scenario planning as a resource management approach and were able to broadly share their lessons (UW Urban Ecology Research Lab. 2008). Finally, many of the above arguments, in support of federal oversight, stem from the normally stricter standards set by the EPA as well as its enforcement role. If the EPA were to (hypothetically) be run by an anti-environmentalist who worked to dismantle environmental protection standards, federalism and state authority would allow states the power to preserve their natural resources at a level above that required (and desired) by an environmentally unfriendly EPA.

4. **It would be foolhardy to make major investments in public infrastructure without trying to forecast and calibrate their likely impacts on humans and the environment. In this book, we have looked at a range of analytical tools that can be used to assess new infrastructure investments before or after they are made. Imagine you have been asked by the mayor of Boston to analyze the likely impacts of possible private investments in burying almost all power lines in the city. Which analytic method(s) would probably be most appropriate? Why? What nonobjective judgments would need to be made as part of any such analysis? How could you insulate such judgments from political and other challenges?**

Given that all these methods involve many nonobjective judgments, I would recommend to the mayor that this assessment involve representatives from all major stakeholder groups (including the mayor's office) to collaborate on the assessment. Collectively, we would conduct these analyses together, along with any necessary outside experts, then ultimately the results of the analyses would be one form of input into the overall assessment, and not isolated, objective, predictive models. Furthermore, there is no complete insulation from political or other challenges, and so the best we can do is to get contributions on the analysis inputs and outputs from a representative group of stakeholders and make the assumptions as transparent as possible during the analysis process and presentation. Getting stakeholder buy-in (including from the mayor's office) for both the assessment design and outcome is one way to safeguard against undue influence of one party over any of the nonobjective judgments made along the way (Susskind and Dunlap 1981).

Here is a summary of the following methods I would propose we use to conduct this joint fact-finding effort to evaluate the potential impacts of investments in moving power lines underground. Each has its own usefulness in this context and its own nonobjective judgments that we must be aware of when conducting our analysis.

Method	Nonobjective judgments involved
EIA: A way to compare environmental impacts of alternatives (i.e., underground power lines vs. no underground; underground some places vs. no underground other places, etc.) Prevents environmentally damaging plans from being proposed, so it can be a useful tool for making sure we are evaluating more realistic plans (Momtaz and Kabir) Good for the "how" (i.e., evaluating the impacts of various ways of moving the power underground so it minimizes environmental impacts) (Momtaz and Kabir)	**EIA:** Judgments are made in every part of the process (Susskind and Dunlap). For example: Which research to use about gains or losses to efficiency (since there is conflicting research) How far back to look in terms of damage caused by current system How do you rank significance? How do you discount things? How broad a scope do you extend to include environmental impacts? For example, do you look at the carbon emissions from traffic caused by the installation? Which part of the city do you examine? Boston is an old city, and the streets and infrastructure vary significantly.
Cost-Benefit Analysis: (Ashford and Caldartt; Sagoff) Clearly lays out economic, environmental costs and health/well-being costs and benefits Good for comparing magnitudes of relative gains and losses from each plan Helpful for determining who the winners and losers might be and can help with strategizing on how the losses to the losers can be offset with the gains to the gainers (i.e., insurance)	Just looks at gains and losses from an efficiency perspective (as revealed through WTP), which is not an objective evaluation or even comprehensive of someone's values (Sagoff) Many assumptions in terms of what is included in the equation, how it is weighted and how the nondollar amounts are converted to dollar amounts (Ashford and Caldartt) Discounting involves lots of assumptions about the future and future use (Ashford and Caldartt)
Risk Assessment (RA) and Risk Management (RM): Good way to start a conversation on what kinds of risks the city is already taking on and what kinds of risks they might be willing to take on (either during the construction process or after) When RA and RM are considered to be inherently coordinated/linked, then RA is a good way to plan RM and then evaluate whether the RM is working or not RA with a keen focus on risk perception helps consider what kinds of options there might need to be for people affected by the implementation or outcome to avoid pushback and could help design RM (Callison)	Making a lot of assumptions about the inputs and their magnitudes: there is just a lot we do not know about the hazards and the associated risks, specifically the potential impacts and the probability of the harm (Kunreuther et al. 1996) Which studies to include? Lots of conflicting information about levels of hazard and probability of hazard etc. and reliable data can be hard to get (European Environmental Agency 1998) What geographic or temporal scope to include? Decisions about the risk to whom? What time to include? RA depends on the RM, but RM relies on people, and people and context change over time, which changes RA. So, RA is not fixed [...]

CONCLUSIONS

Some environmental problem-solving books appear to suggest that the only thing you need in order to solve an environmental problem is the right tool or method. By now you realize we do not accept that premise. Environmental Impact Assessments (EIA), for example, can be used to help decision makers, including stakeholders, think about the advantages and disadvantages of alternative designs, technologies or locations (and they should!), but a final decision about whether to proceed with a project and whether the unavoidable adverse impacts should be accepted, is a political choice, not a technical one. And, in almost every case, there are sure to be disagreements among groups who favor or oppose various versions of the project or policy being considered. They can use cost-benefit analysis to justify their positions, but, as we discussed in Unit III, many of the decisions related to the design of the tool are subjective and the results of a CBA will not be definitive.

Philosophical or, more appropriately, ethical, considerations will always come into play. And, while elected and appointed officials, or the courts if it comes to that, will have the final say in a democracy, there is almost always room for informal consultation with a wide range of stakeholder groups prior to the government making a final decision. Our view is that environmental problem-solving requires finding a balance between science and politics and debating the ethical choices involved. This is best accomplished through open public deliberation. The credibility of the judgments that emerge also depends on transparency and accountability. That is, the officials and experts involved need to make explicit the information and analyses they are relying on. The stakeholders and the general public need to take advantage of the opportunities offered to challenge the technical basis for government (or industry) decision-making.

Both process and outcome are important. Since there are no "correct" answers, the decisions that officials make (on behalf of the public) must be justified. All possible justifications are rooted in philosophical or ethical assumptions about the role and responsibility of citizens to each other, our responsibilities for the maintenance of a functioning ecosystem, and our responsibilities to future generations. Each person offering a "solution" needs to explain why they have taken the position they have, and what interests they have at stake. In the process, they will reveal their personal theory of practice. Whatever analytic tools they use, and whatever local knowledge they rely on, they must be accountable for the "nonobjective" judgments embedded in any technical argument they make.

All environmental decision-making takes place in a context of substantial uncertainty. No one knows enough about the complex interactions among the pieces and parts of our socioecological system to predict with certainty the effects of the actions they advocate. That does not mean that anyone should ignore what science can tell them. It does mean, though, that it would probably be smart to declare support for a particular "solution" on a contingent basis. The implication that follows is that every new policy, program or project should be accompanied by a serious monitoring commitment, along with a promise to support appropriate adjustments as needed. As we have said elsewhere in this book, it often makes sense to view all environmental problem-solving efforts as experiments. While we should be clear about what we think is going to happen, we should be open to counterintuitive or unanticipated results. Since substantial uncertainty exists (and cannot be eliminated), we should be modest in our claims and open to new information that can guide us to a better version of what we had in mind. This adds up, in our minds, to an approach to environmental problem-solving best thought of as collaborative adaptive management.

We started this book with environmental policy-making because existing policies set the stage for all subsequent environmental problem-solving efforts. Understanding why and how specific policies came into effect is the first step in environmental problem-solving. When you have a clear sense of how policy is made, you will then be in a good position to lobby for changes in policy or governmental practice. We next moved to environmental ethics, as illustrated by a series of philosophical debates. These debates between economic growth and sustainable development, utilitarianism and deep ecology, formal science and indigenous knowledge, and the authority of elected officials versus broad-gauged public involvement are likely to continue indefinitely. Where you stand on them now, though, and the ethical reasoning that leads you to the stands that you take, will underlie any environmental problem-solving you do. These debates are also at the heart of your personal theory of environmental practice (that we hope you will continue to try to clarify).

We also reviewed the most common methods of environmental analysis, pointing out how each works and underscoring the inherent subjective assumptions that affect the design and use of these tools. We favor scenario planning because it embraces the uncertainty that cannot be eliminated in environmental problem-solving. We ended with a discussion of collective decision-making because no single individual, regardless of his or her political position or status, should have the power to impose a solution to any environmental problem. There is too much at stake. Given that political and economic interests are almost always in conflict, we advocate a consensus-building approach to environmental problem-solving that attempts to resolve conflicting interests. Even though we know that most people think the chances of reaching consensus are slim, we underscore why and how it is actually quite realistic, and most importantly, significantly useful for our communities, especially if problem-solving is facilitated by trained mediators.

You now have a base to build on, but you are not finished! You are well-positioned to continue to learn from your own experience. Every time you see or read about an environmental controversy, look beyond the headlines and formal position statements that dominate the media reports. We hope that reading this book has given you more perspective on the complexity of environmental problem-solving and some ideas on a potential approach based on your own theory of practice.

BIOGRAPHIES

Lawrence Susskind has been a member of the MIT faculty for almost 50 years. He founded MIT's Environmental Policy and Planning Group and teaches a range of undergraduate and graduate subjects including Introduction to Environmental Policy and Planning, Water Diplomacy, International Environmental Negotiation, Participatory Action Research and Negotiation and Dispute Resolution in the Public Sector. Susskind is the author of twenty books in eight languages including *Environmental Diplomacy* (with Saleem Ali, Oxford University Press, 2015), *Better Environmental Policy Studies: How to Conduct More Effective Analyses* (with Ravi Jain and Andrew Martyniuk, Island Press, 2013), *Negotiating Environmental Agreements: How to Avoid Escalating Confrontation Needs Costs and Unnecessary Litigation* (with Paul Levy, Island Press, 2013), *Water Diplomacy* (with Shafiqul Islam, Resources for the Future Press, 2015), and *Managing Climate Risks in Coastal Communities* (with Danya Rumore, Carri Hulet and Patrick Field, Anthem Press, 2015). The founder and first editor of the premier peer-reviewed professional journal dealing with environmental and social impact assessment, he won the 2007 Global Environmental Award presented by the International Association for Impact Assessment. As a practitioner, Professor Susskind has helped to mediate more than 50 complex environmental disputes on five continents. He has been involved in efforts to resolve completing natural resource management claims in the Arctic, Mexico, Chile, Malaysia and several regions of the United States and Canada.

Bruno Verdini, a recipient of MIT's D'Arbeloff Award for Excellence in Education, designed and teaches MIT's top-ranked courses "The Art and Science of Negotiation" & "Leadership in Negotiation: Advanced Applications" for which 600+ STEM students from 20+ MIT departments apply every year. With the inspiring mentorship of Professor Susskind, Dr. Verdini received MIT's first ever and only Ph.D. in Negotiation, Communication, Diplomacy, and Leadership. His work, which explores how to improve transboundary natural resource management between developed and developing countries, won Harvard Law School's Raiffa Award for the best research of the year in negotiation, competitive decision-making, mediation and dispute resolution, the only time in the history of the award that it has been won by an MIT member and by someone born and raised in Latin America. A former government official, he has conducted executive and MBA trainings in communication, leadership, and conflict resolution for professionals from over 80 different countries, and has been involved with the teams negotiating financial, technical, and scientific cooperation agreements between stakeholders across all continents, including partnerships with the International Energy Agency (IEA), International Atomic Nuclear Agency (IANA), International Renewable Energy Agency (IRENA), International Energy Forum (IEF), United Nations Industrial Development Organization (UNIDO), World Bank and World Economic Forum.

Jessica Gordon is an international environmental policy expert. Her work focuses on climate change, natural resource management and food security for organizations such as Oxfam and the United Nations Development Program. She was awarded fellowships from the Fulbright Program, Social Science Research Council and the East-West Center. She holds a PhD in Environmental Policy and Planning at Massachusetts Institute of Technology (MIT), a Master's in Environmental Science from the Yale School of the Environment and a BA in International Development from Brown University.

Yasmin Zaerpoor holds a PhD in Environmental Policy and Planning at the Massachusetts Institute of Technology (MIT). Her doctoral research focuses on transboundary water negotiation and, more broadly, collaborative approaches to natural resource management. Before starting her PhD, she conducted research on urban policy in the South Asia region as a consultant at the World Bank. She holds an MS in urban planning from Columbia University and a BS in Animal Physiology and Neuroscience from University of California, San Diego.

REFERENCES

Beierle, Thomas C. 1998. "Public Participation in Environmental Decisions: An Evaluation Framework Using Social Goals." Discussion Paper 99–06, Resources for the Future.

Cohen, Steven. 2014. *Understanding Environmental Policy*. 2nd ed. New York: Columbia University Press.

Corburn, Jason. 2005. "Local Knowledge in Environmental Health Policy." In *Street Science*. Cambridge, MA: MIT Press, 25–45.

Costanza, Robert, Rudolf de Groot, Paul Sutton, Sander van der Ploeg, Sharolyn J. Anderson, Ida Kubiszewski, Stephen Farber and R. Kerry Turner. 2014. "Changes in the Global Value of Ecosystem Services." *Global Environmental Change* 26: 152–8.

DesJardins, Joseph R. 2013. *Environmental Ethics: An Introduction to Environmental Philosophy*. 5th ed. Belmont: Wadsworth.

Hardin, Garrett. 1968. "The Tragedy of the Commons." *Science* 162, no. 3859: 1243–8.

Howlett, Ramesh, and Anthony Perl. 2009. *Studying Public Policy*. 3rd ed. Oxford: Oxford University Press.

Kunreuther, Howard, and Paul Slovic. 1996. "Challenges in Risk Assessment and Risk Management." *Annals of the American Academy of Political and Social Science* 545: 8–13.

Ludwig, Donald. 2000. "Limitations of Economic Valuation of Ecosystems." *Ecosystems* 3: 31–5.

Ostrom, Elinor. 1990. "Reflections on the Commons." In *Governing the Commons*. Cambridge: Cambridge University Press, 1–28.

———. 2012. "The Future of the Commons: Beyond Market Failure and Government Regulations." In *The Future of Commons: Beyond Market Failure and Government Regulations*, edited by Elinor Olstrom. London: Institute of Economic Affairs, 68–83.

Pearce, David, Giles Atkinson and Susana Mourato. 2006. *Cost-Benefit Analysis and the Environment: Recent Developments*. Paris: OECD.

Rosa, Eugene A., Ortwin Renn and Aaron McCright. 2014. "Risk Governance: A Synthesis." In *The Risk Society Revisited: Social Theory and Governance*, edited by Aaron McCright and Ortwin Renn. Philadelphia: Temple University Press, 150–69.

Sagoff, Mark. 2008. "At the Shrine of Lady Fatima or Why All Political Questions Are Not Economic." In *The Economy of the Earth: Philosophy, Law, and the Environment*. Cambridge: Cambridge University Press, 24–45.

Shapiro, Ian. 2009. "Aggregation, Deliberation, and the Common Good." In *The State of Democratic Theory*. Princeton: Princeton University Press.

Sterman, J. D. 1991. "A Skeptic's Guide to Computer Models." In *Managing a Nation: The Microcomputer Software Catalog*, edited by Gerald O. Barney, W. Brian Kreutzer, Martha J. Garrett. Boulder, CO: Westview Press, 209–29.

Susskind, Lawrence. 2009. "The Environment and Environmentalism." In *Local Planning: Contemporary Principles and Practice*, edited by Gary Hack, Eugenie L. Birch, Paul H. Sedway and Mitchell J. Silver. Washington, DC: ICMA Press, 74–80.

Susskind, Lawrence, and Connie Ozawa. 1984. "Mediated Negotiation in the Public Sector: The Planner as Mediator." *Journal of Planning Education and Research* 4, no. 5: 5–15.

Susskind, Lawrence, and Jeffrey Cruikshank. 2006. *Breaking Robert's Rules*. New York: Oxford University Press.

Susskind, Lawrence, Ravi K. Jain and Andrew O. Martyniuk. 2011. *Better Environmental Policy Studies*. Washington, DC: Island Press.

UW Urban Ecology Research Lab. 2008. Puget Sound Future Scenarios. University of Washington. http:// urbaneco.washington.edu/wp/wpcontent/uploads/2012/09/scenarios_report.pdf.

Van Buuren, Arwin, and Sibout Nooteboom. 2009. "Evaluating Strategic Environmental Assessment in The Netherlands: Content, Process, and Procedure as Indissoluble Criteria for Effectiveness." *Impact Assessment and Project Appraisal* 27, no. 2: 145–54.

INDEX

www.ingramcontent.com/pod-product-compliance
Lightning Source LLC
Chambersburg PA
CBHW030633270326
41929CB00007B/53